CLARENDON STUDIES IN THE HISTORY OF ART

General Editor: Dennis Farr

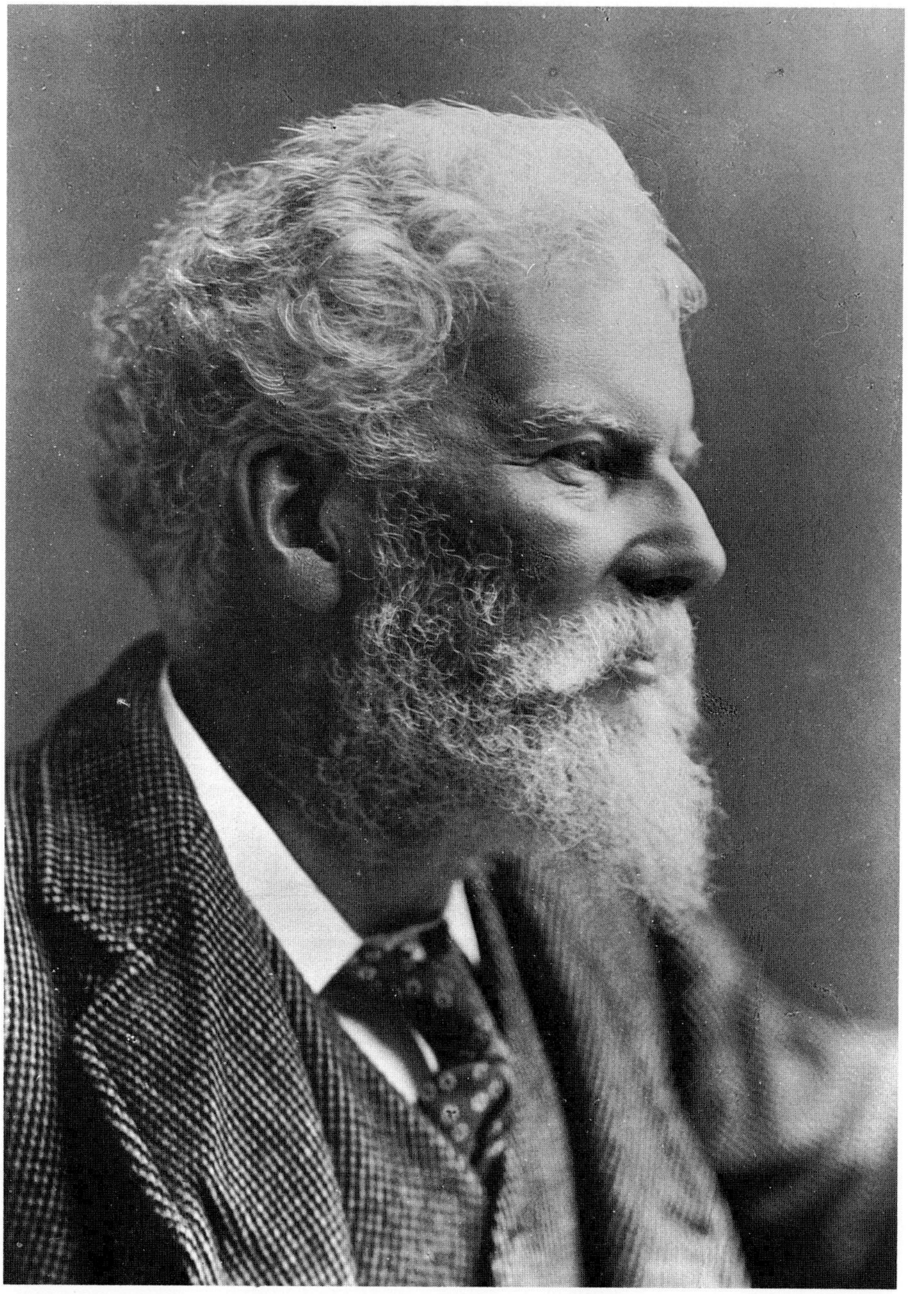

1. Alfred Waterhouse, portrait photograph, *c.*1890

Alfred Waterhouse
1830–1905

BIOGRAPHY OF A PRACTICE

COLIN CUNNINGHAM
and
PRUDENCE WATERHOUSE

CLARENDON PRESS · OXFORD
1992

Oxford University Press, Walton Street, Oxford OX2 6DP
Oxford New York Toronto
Delhi Bombay Calcutta Madras Karachi
Petaling Jaya Singapore Hong Kong Tokyo
Nairobi Dar es Salaam Cape Town
Melbourne Auckland
and associated companies in
Berlin Ibadan

Oxford is a trade mark of Oxford University Press

Published in the United States
by Oxford University Press, New York

British Library Cataloguing in Publication Data
Data available

Library of Congress Cataloging in Publication Data
Cunningham, Colin.
Alfred Waterhouse, 1830–1905—biography of a practice / Colin Cunningham and Prudence Waterhouse.
(Clarendon studies in the history of art)
Includes bibliographical references and index.
1. Waterhouse, Alfred, 1830–1905. 2. Artists—England—Biography. 3. Architecture, Modern—19th century—England. I. Waterhouse, Prudence. II. Title. III. Series.
NA997.W36C86 1992
720'.92—dc20
[B] 91-44581
ISBN 0-19-817511-6

Typeset by Pentacor PLC, *High Wycombe, Bucks.*

Printed in Great Britain by
Butler & Tanner Ltd
Frome and London

for Ray, Alisoun,
and Bessie

ACKNOWLEDGEMENTS

ALFRED WATERHOUSE'S works are numerous and widely scattered, and in preparing this biography we have had help and encouragement from a wide range of people, from members of his family, from owners and occupiers of his buildings, from companies and corporations (and their successors) with which he did business, from fellow scholars, librarians, and archivists, and, especially, from our own families who have lived patiently with this project for years. Without their assistance much of what we have brought together would have remained unrecorded, and we owe them all a debt of gratitude, even though space does not permit us to mention them all by name. We owe a very special debt to David Waterhouse, who not only welcomed us into his own office and made us free of the original office archive, but has also provided generous help with the illustrations for this volume. We are also grateful to the British Academy for a generous grant that has made possible the inclusion of colour reproductions. We would also like to thank the following: A. Allan, Messrs Allen & Appleyard, Knutsford, Finch Allibone, Jill Allibone, Sr. Aloysius and the staff of Allerton Priory School, J. H. G. and Joan Archer, H. Godwin Arnold, Captain J. Asbury, Jean M. Ayton and the staff of Manchester Central Library archives department, R. E. R. Banks (Natural History Museum), Barclay's Bank Property Division, Darlington, J. E. Barlow and the staff of the National Children's Home, Evenley, Miss E. Bassom, D. L. Baxter and the staff of the Cambridge Union Society, B. A. Bentley (Commercial Union Properties), Professor T. J. Benton, R. E. Berry, B. Blair, Professor B. Blomberg, E. J. Bonner and the staff of Lady Barn House School, Ken Brand, John Brandon-Jones, Professor C. N. L. Brooke, Dr P. Bullock, librarian of Balliol College, Oxford, B. J. Burns (National Westminster Bank premises division), Matthew Butler, R. M. Button, Mrs J. Campbell and the staff of Darlington Library, the headmaster and staff of Carmel College, Darlington, Mrs J. Carpenter, T. H. Carter, K. E. Cassir, Dr N. Chalmers and the staff of the Natural History Museum, Mrs V. Chapman, Bridget Cherry, J. Christie-Miller, Mr and Mrs R. O. Collins, Gordon Collis (Prudential Assurance Company), H. M. Colvin, I. K. S. Cooke, Nicholas Cooper, A. J. Coulson, Mrs Marjorie Cox, R. Crewdson, Mrs R. Crook, Professor J. Mordaunt Crook, Miss C. H. Cruft, Mr Curran and the staff of Grinkle Park Hotel, Revd Br. Cyril, CFX, and the staff of the Xaverian College, Michael Darby, Mrs Davies and the staff of Bingley Library, M. J. Davies, W. Davies, Christopher Dean, E. M. Denton, Mrs S. Denyer, R. Dingwall, Revd David Dixon, D. Dossett, Ted and Beryl Earle, Lady Laura Easthaugh, Mr and Mrs E. A. Eden, R. H. Essex (Natural History Museum), E. S. P. Evans and the staff of Liverpool City Planning Office, Stuart Evans, Mrs D. Evered, Mrs N. Evetts and the staff of the Picton Library, Liverpool, R. F. Farrow, Revd Br. Felix, OH, and staff of St Cuthbert's Hospital, Pilmore, R. S. Fitzgerald, Mrs J. M. D. Forster and the staff of Leeds University archives department, Adrian Forty, Norman Foster, Mrs V. Fox of Leighton Buzzard Library, Norman Franklin and the late Jill Franklin, Mrs W. Gaskell and the staff of Girton College Library, W. R. Gerrity (Refuge Assurance Company), J. Gilmour, Ian Gordon

(Prudential Assurance Company), A. Greek (Prince Park Housing Co-operative), K. W. Griffiths and the staff of the Frank Wise School, Banbury, C. A. Griggs, Revd Tony Griggs, Mrs C. Hall, archivist of Caius College, Cambridge, Maya Hambly, Dr and Mrs C. Hardman, E. J. P. Hardman and the staff of Clayesmore School, Mr and Mrs W. Harnett, Mr and Mrs V. Harney, John Harris, Hathernware Ceramics Ltd., N. Hellawell, Edward Hibbert, J. Hilton of the Atkinson Library, Southport, The Most Revd A. E. Hornyak, OSBM, Peter Howell, the late Edward Hubbard, Barbara Humphrey, W. S. Hutton of Pembroke College, Cambridge, D. Jackson (W. H. Robinson & Co.), Paul A. Jarvis, Simon Jervis, Mrs J. W. Johnson, Mr and Mrs K. M. Johnston, Miss A. M. Jones and the Staff of Rye St Anthony School, K. Jones (Liverpool Social Services), Lord Kaberry of Adel, Frank Kelsall, H. Kinrade and the staff of Mossley Park Hospital, Jill Lever and the staff of the RIBA Drawings Collection, Dr P. Linehan, Colin Maber, Sally MacDonald, Jane MacGregor, Professor R. MacLeod, J. E. Makin, P. Makin, Sally Maltby, Elfrida Manning, the librarian of Mansfield College, Oxford, Revd G. G. Marshall, Colonel and Mrs R. A. St. G. Martin, Angela Mason and archives staff of the RIBA Library, the late Mrs I. M. Mathison, R. J. McKelvey and the staff of the Northern College, P. McNiven, W. McVicker, J. T. Merriden, Robin Middleton, Dr G. F. Mills, Mrs Ethne Morris, Mrs N. Morris and the staff of Park Hospital, Liverpool, C. Murfin (Neighbourhood Housing Services, Liverpool), Dr S. Muthesius, Anne Neary, archivist of Gonville and Caius College, Cambridge, L. H. Nixon (Prudential Assurance Company), M. J. O'Dell, D. O'Sullivan, the librarian and the committee of management of the Oxford Union Society, A. E. B. Owen, Miss C.-A. Parker, F. E. Payne, Mrs M. Pemberton-Piggott, Mr Philips, headmaster of St Andrew's School, Pangbourne, C. J. Pickford and the staff of Bedford County Record Office, Portsmouth City Art Galleries, D. J. Prescott, T. Price, the Chairman and Directors of the Prudential Assurance Corporation, their staff, and the managers of their District Offices, the Principal of Prior Pursglove Sixth Form College, Anthony Quiney, Pat Read, the headmaster and staff of Reading Grammar School, the staff of Reading University Archives Department, David Rhodes, Miss C. B. Richard, Margaret Richardson, Anne Riches, the Rt. Hon. R. Rhodes-James, MP, E. H. Roberts and the staff of Crookhey Hall School, the Viscount Rochdale, Katy Rogers, the Royal Institution of Chartered Surveyors and Sir Idris Pearce (Past President), the headmistress and staff of the Royal Victoria School for the Blind, Newcastle, Andrew Saint, E. J. Sawtell and the staff of St Margaret's School, Bushey, J. Sharples, Mr and Mrs C. E. W. Sheepshanks, Teresa Sladen, W. S. Slomka, Mrs M. Smith and the staff of Hurworth Grange community centre, Mrs M. Smith and the staff of Middlesbrough Library, Tricia Smoothy, G. Snell and the Committee of Management of the National Liberal Club, Michael Snodin, P. H. Southcott and the staff of the National Westminster Bank Staff College, Gavin Stamp, Julia Standing, D. A. Stockfis, Dr M. Stratton, Staff of HM Prison Strangeways, Dr D. Sturdy, the late Miss B. Sullivan, Andrew Tate, Dr Jeremy Taylor, Lady Joanna Temple, Revd Bill Thomas, William B. Thompson, Robert Thorne, Peter Traynor (Prudential Assurance Company), Julian Treuherz, T. Turner and the staff of Manchester Town Hall, the officer in charge and the staff of the Turner Memorial Home, D. H. Tyrell and the staff of Cleveland County archives department, the district engineer and staff of University College Hospital, D. C. W. Verey, Clive Wainwright, Chris Wakeling, David Walker, A. B. Ware and the staff of Brathay Hall, the late Ursula M. Waterhouse, Mr and Mrs Watson, G. Watson-Broughton, Alexandra, Lady Wedgwood, Revd R. A. C. West, the Duke of Westminster, Adam White of Leeds City Art Gallery, D. E. Wickham and the staff of the Clothworkers' Company archives department,

Dr and Mrs R. D. Willcock, J. Williams and the staff of Chetham's Hospital School, Mr and Mrs Peter Willis, R. V. J. Windsor and the staff of All Nations Christian College, Revd J. Wolstencroft, Doreen Yarwood, Andrew Younger.

ILLUSTRATION ACKNOWLEDGEMENTS

WE are grateful to the following for kind permission to reproduce drawings and photographs:

Bedford County Record Office: 114, 116

Besses o' th' Barn United Reformed Church, by kind permission of the Minister and Elders: 32

British Architectural Library: 115

Building News: 45

CAS Ltd., Manchester: Col. Pl. 6

Colin Cunningham: Col. Pl. 4; 19, 20, 25, 53, 59, 60, 64, 65, 66, 68, 76, 77, 90, 99, 100, 101, 102, 103, 104, 105, 106, 107, 108, 109, 110, 111, 112, 117, 120, 124, 141, 145, 146, 151, 152, 155, 156, 157, 164, 165, 166, 167, 168, 170, 171, 176, 177, 188, 189, 190, 201, 203, 204, 205

Country Life: 186

Field, Fisher, Waterhouse, by kind permission: 46

Gonville and Caius College, Cambridge, by kind permission of the Master and Fellows: 69, 70, 71, 162

Guildhall Library, London: 39

The late Edward Hubbard, by kind permission: 91

Hulton Picture Library: 37

Lloyds Bank, Cambridge, by kind permission: 158

Manchester Central Reference Library: 31, 38, 49, 52, 54

Manchester City Engineer's Department: 47, 55

Manchester City Museum and Art Gallery: 22

Manchester Crown Court: 33

Manchester University School of Architecture: 29, 30, 48, 50, 51, 149

Motor Cycle Traders' Benevolent Fund: 118

National Liberal Club, by kind permission: 133, 136

Private Collections: Col. Pls. 2, 8, 9; 1, 2, 3, 4, 5, 6, 7, 8, 9, 10, 11, 12, 13, 14, 15, 16, 17, 18, 26, 27, 35, 36, 40, 41, 42, 43, 44, 73, 84, 98, 122, 138, 142, 144, 148, 159, 160, 174, 192, 193, 194, 195, 196, 197, 198, 199, 202

Prudential Corporation: Col. Pls. 5, 7

Royal Academy: Col. Pl. 1

Royal Commission on the Historical Monuments of England: 23, 63, 72, 74, 75, 78, 82, 87, 89, 93, 94, 113, 125, 126, 127, 128, 129, 130, 131, 132, 134, 135, 137, 169, 183

RIBA Drawings Collection: Col. Pl. 3; 28, 34, 56, 57, 58, 61, 62, 67, 79, 80, 81, 83, 85, 86, 92, 95, 96, 97, 119, 121, 123, 139, 140, 147, 150, 153, 154, 161, 163, 172, 173, 175, 178, 179, 180, 181, 182, 184, 185, 187, 191, 200

St. Paul's School, by kind permission of the Governors: 143

Walter Scott, Bradford: 88

Victoria & Albert Museum: 21, 24

Plates 32 and 33 photography by Mack of Manchester. Col. Pl. 6 photography by CAS Ltd., Manchester. All photography from the Private Collections, RIBA Drawings Collection, and Pls. 45, 46, 115, and 133 by Geremy Butler.

CONTENTS

PART IV THEMES IN WATERHOUSE'S CAREER

EPILOGUE

APPENDICES

LIST OF COLOUR PLATES

between pages 204 and 205

LIST OF PLATES

All drawings are by Alfred Waterhouse unless otherwise stated. Most of the watercolours and sketches are precisely dated, and the date is recorded for each. Where drawings and perspectives of buildings are dated this is recorded with the word 'dated'; but, for the many undated originals, an approximate year only is given. Accurate dates for photographs are given when known; otherwise the beginning date of each commission is given in brackets.

LIST OF FIGURES

All these plans have been redrawn to a consistent format by Prudence Waterhouse, working where possible from original drawings (those in public collections are noted below). For clarity of reproduction there has been some simplification in the major public buildings. The smaller house plans are in some cases based on sketches made on site or during demolition; where possible this information has been supplemented by reference to other sources.

ABBREVIATIONS

AW Journal	'The Story of the Founding of Yattendon Court', MS notes by Alfred Waterhouse in private collection
DC	RIBA Drawings Collection, The British Architectural Library
MCL	Manchester Central Reference Library
MW *Journal*	*Extracts from the Journal of Mary Waterhouse 1825–1880*, edited by her son Edwin Waterhouse (1907). Printed for private circulation only
OLB	Office Letter Books: bound volumes of copied letters from the office. There are three volumes surviving: vol. I, 18 Sept. 1856–14 Oct. 1862. Private collection; vol. II, 16 Nov. 1864–9 May 1865. RIBA; vol. III, 23 Nov. 1866–21 Mar. 1867. RIBA Library
RIBA	Royal Institute of British Architects
RIBA Library	British Architectural Library of the RIBA
SK	Sketchbook. The sketchbooks numbered on the covers 1 to 11 cover the years 1855–91
Thompson letters	Letters from AW to Joseph Thompson held in the Manchester Central Reference Library, Archives Department

Introduction

ANY architect who was a Royal Academician as well as President of the Royal Institute of British Architects for three years, who designed mansions, clubhouses, colleges, churches, and town halls and who left a substantial archive of drawings and documents ought to have provided a ready quarry for historians. Yet appreciation of Waterhouse's work has been slow in coming. The aggressiveness of his harsh red terracotta and the relative coarseness of his ornament still sometimes militate against enjoyment. However, if his buildings are not all to our taste, it must be recognized that they were very much to the taste of his nineteenth-century patrons. His career reached a peak in the 1870s and 1880s and though his success was built on the gothic style he remained the leading British architect throughout the uneasy period of stylistic vacuum that accompanied the end of high Victorian gothic. And his was no narrow success based on the personal favour of a few. A great talent for planning, a consistent theory of design, delight in the picturesque, and an exemplary efficiency in practice are all features of his work; and the extent of his practice suggests that a study of his career may lead to some reassessment of the later Victorian architectural scene. A fuller understanding of his buildings should allow us to appreciate their confident vigour and solidity and to learn more of what Victorian architectural practice was about.

Waterhouse's working life covered the second half of the nineteenth century and spanned the building worlds of London and the provinces. It grew through the declining years of the gothic revival and outlived the generation of stylistic confusion that followed. Indeed Waterhouse, whose early works can be related to Loudon, who made his name by gothic, was still in practice when the Arts and Crafts movement was in full flood. In retirement at Yattendon he could have read of the completion of the Glasgow School of Art. Though he was fully of an age before the Modern Movement, at least one of its founding fathers, Berlage, acknowledged him in a claim that the lessons of the Amsterdam Stock Exchange were all in his work. Yet in the critical heritage he has been bypassed, dismissed, or forgotten.

Lethaby's analysis in his book on Philip Webb was, as usual, accurate. 'The current of the time set away from his methods towards the softer effects of stylism'; and it is certainly method as much as style that gives buildings such as Manchester Town Hall their quality. Almost, Waterhouse can be seen as an anti-stylist, an intelligent eclectic whose business was building; and the enormous success of this approach tells us much about later Victorian attitudes to architecture. For some, the evolution of the stylistic climate may look more fascinating, but from that Waterhouse seems to stand aloof. He was, however, unwilling rather than unable to be absorbed in what he would have seen as secondary considerations. By the time he began to practise in 1854 (or at least by about 1860) detailed knowledge of the available styles was well established. An array of scholarly works could prevent the stylistic solecisms to which a previous generation had been liable. The iron rule of copyism was beginning: and it is to Waterhouse's credit rather than otherwise that, in an age obsessed by the published precedent, he could keep himself free from its tyranny. He would have been the first to acknowledge the influence of Pugin and Ruskin; but would never have countenanced their control of his designs. His own imaginative insight was not dominated by precedent but incorporated the lessons of an upbringing in the Protestant work ethic. The result was that he was particularly well equipped to design buildings to suit the new needs of the age.

Waterhouse's theory of design followed the Vitruvian triad, commodity, firmness, and delight, in strict order. He had a clear belief in the primacy of the functional efficiency of the plan. Structural soundness was important but secondary; and the form of the exterior followed from the first two. His façades reveal the organization of the plan, and the structural systems are openly expressed. Decoration was an essential adjunct but genuinely subject to the constraints of material or cost. He often repeated simple elements, a means of building decently but economically. In short, Waterhouse was no radical crusading architect; rather he set out to provide, and market, a reliable service. His buildings were not manifestos of a stylistic principle, but working edifices designed for particular patrons. His whole method went far beyond the formal considerations that are the first, and sometimes the only, ones appreciated by those who do no more than look at buildings from the outside. He was, in the true sense, a complete architect; and function and finance as well as form have to be taken into account in assessing his buildings.

Conventionally he is seen as a goth; but his version of gothic was a thoroughly modern one, and at the height of his career, in the 1870s and 1880s, he evidently preferred a sort of Romanesque revival. Yet he was equally happy with the French Renaissance, competent in the classical language, and did not disdain to work in the more nearly anonymous modes demanded by small budgets or in alterations to existing structures. However, he was not simply a stylistic cavalier; and his relationship to the currents of style is more complex and more interesting. The plan of a building was to a great extent dictated by the function; but the outward form would be the product of his own imagination married to the preferences of his clients. He can be accused of giving in too readily to their wishes; but the logic of his approach is inescapable. They were the clients, and he their obedient servant. Those who wanted classical piles in the 1860s would probably not turn to him – he was not an out and out reactionary – nor would those with an advanced taste look to him for a novel quirk of fashion. He was stylistically conservative, adept at producing works that were thoroughly up to date but not aesthetically risky. He was a master of the Victorian preference for comfort and solid worth, in every sense a 'safe' man. Whether his clients were private individuals spending tens of thousands on a mansion or corporations spending hundreds of thousands on a town hall, this was a very attractive feature. It was probably the key to his success as a professional architect and is important in reaching a balanced assessment of high and late Victorian architecture.

To produce designs that are simultaneously modern while satisfying a patron's sense of history and permanence is no mean achievement. It was a task that Waterhouse fulfilled admirably, and he is an excellent architectural mirror to his age. In the best sense of compromise, his works and life echo the tensions and contradictions of the world of the Victorian middle classes. In no sense a stylistic extremist, he was never able to abandon style altogether. Of provincial dissenting stock, he rose to join the metropolitan establishment as a member of the Church of England and squire of Yattendon. Politically of Liberal leanings, he retained many friends and connections among the Liverpool Tories. A devoted husband and paterfamilias, he epitomized the successful Victorian businessman. Unlike the entrepreneurs of the first industrial generation he was no self-made man, but benefited from his father's very comfortable income. Yet he worked hard to increase his patrimony and died richer even than Scott.

For most of the time he was too busy building to theorize, and has left little record of his thinking. He would have preferred to be a painter, and his love of the picturesque, like his early membership of the National Trust and campaigns against litter, are also important. They reflect the Victorian urban obsession with the countryside. Yet in spite of that he will be remembered for his contribution to the urban scene: for the Prudential buildings, Manchester Town Hall, the National Liberal Club, and the Natural History Museum. In the 1870s and early 1880s he was indisputably the leading architect of the kingdom. Now, it is unfortunate that so many of his lesser buildings have already vanished as the process of industrialization and urbanization continues to sear our city centres and engulf what were once salubrious suburbs. None the less, enough remains, and remains usable and comfortable, to establish his claim as a highly successful practitioner, a professional, architect to the first age of Capital.

This book is the story of his practice much more than of the man, for he was intensely private about his home life, leaving little record of his relations with his brothers and sisters and as little of his enjoyment as paterfamilias of life with his beloved Bessie and their five children. However, a biography of a practice needs to do more than survey the key monuments. The great extent of Waterhouse's practice (several schemes occupied him for over thirty years) makes it impossible to write a simple chronological tale or to offer a tidy typological study. In all but a tiny minority of cases his buildings were not conceived and designed complete and then built. More often designing and building were organically related with new ideas and modifications being gradually worked out. Since he commonly had several dozen commissions on hand at any one time, and tended to produce only two or three drawings for each at a time, a precise chronology would be almost impossible to unscramble. (A complete chronological list is, however, provided by the catalogue of commissions in Appendix 1.) A neat typological survey seemed equally unsatisfactory in view of the extraordinary diversity of his work, from High Anglican churches to dog kennels. Besides, neither approach would readily accommodate his free attitude to styles.

We have chosen instead to present his career in three stages. A broad division of early, middle, and late corresponds adequately with his stylistic development, from the uncertain eclecticism of his early years, through his own modern gothic maturity, to a bolder, if more mechanical, eclecticism in his later years. Chronologically this approach has a certain logic in that the first stage equates roughly with his years in Manchester. From 1864, when he moved to London, until the opening of the Manchester Town Hall in 1878 is the period of his maturity, and the high-water mark of his career. The remaining two decades of his career were a period when the quantity of work in his office could only have been handled by a man who could see at once how a building might be designed, and who could draw on a full range of precedents to clothe it appropriately. Even so we have needed to treat chronology as freely as Waterhouse treated styles in order to bring together important related commissions. In the final part we have highlighted key themes in a series of chapters that provide a concentrated analysis of what it is that makes Waterhouse important, and we show how those themes developed across the whole of his *œuvre*. Inevitably some major themes are cited more than once (though with concentration on different aspects), and some attractive minor schemes get little detailed attention.

Waterhouse's importance as an architect may be seen from his buildings alone. His importance to the profession is equally a part of the man; and he was one of relatively few nineteenth-century architects who managed to combine a high level of designing skill with real business efficiency and acumen. It has been too easy to dismiss his buildings as merely efficient without pausing to consider how, without his diplomatic handling of clients and legendary efficiency, such masterpieces as the Natural History Museum could ever have been built at all. With this in mind it has also seemed important throughout to consider the architect's relations with clients and the problems of developing each design from the original concept to the finished building. However, we have always aimed to give primacy to the buildings since they are the record by which he deserves, and would have liked, to be judged.

Part One

THE EARLY YEARS

1

Quaker Roots

ALFRED WATERHOUSE was born in Liverpool on 19 August 1830. It was in many ways an auspicious time for a future architect. The mill towns, already equipped with steam power, were producing more and more goods on bigger and better machines. Eighteen-thirty saw the opening of the Liverpool and Manchester railway, which really began to open up the north-west to trade and transport. The great seaport of Liverpool was linked to manufacturing Manchester in what was to develop as a key nexus of inland trade by rail and canal. Young Alfred was, therefore, born into a city that was beginning a phase of enormous economic growth. He was also born into a family that was already established at the centre of the cotton trade, at that time Britain's greatest. His grandfather was one of the leading cotton brokers in the town. Nicholas Waterhouse, 1766–1823, had started as an apprentice in Bolton with a manufacturer of fustian, and had then worked in a Manchester cotton warehouse before founding his own business in Liverpool.[1] Like many another Quaker of his time, through shrewdness and honest dealing, he became a very rich man, and married into another commercial dynasty, the Rogers family of Manchester. His wife Ann bore him fourteen children, most of whom became, or married, substantial citizens in Liverpool or Manchester; so there was a large body of uncles, aunts, and cousins for the young Alfred to grow up among.

Alfred's father, Alfred senior, 1798–1873, was the third son of Nicholas Waterhouse and had followed him into the cotton broking business. As a firm Quaker he married, inside the Society of Friends, Mary Bevan of Tottenham, 1805–80,[2] and a little over a year later she bore him a son, Alfred junior, the first of a family of five brothers and three sisters.[3] Soon after Alfred's birth they moved from their first home at Stone Hill, Liverpool, to a house designed by Cunningham and Holme, Oakfield at Aigburth (Pl. 2). In 1843 Alfred senior ceased to be a partner in the family business. The official explanation was that he had long thought that when a 'competence was secured' it was better to give way for others to succeed. He considered that a life of comparative leisure was more conducive to the 'welfare of the soul, than one spent in more than amassing enough'.[4] Mary gave as a reason for his unhappiness in the business that, after West Indian planters changed from cotton to sugar production, he wanted nothing to do with the sale of rum, which was being shipped to Liverpool for the firm to dispose of.

[1] 'The Root of All Evil?' Neville H. Newhouse, *Friends Quarterly* (Oct. 1981).

[2] From the time she was 19 years old until almost her death Mary kept a journal which is the main source for the early history of the family (MW *Journal*).

[3] The family consisted of Alfred (1830–1905), Ellen (1832–76) m. Wilson Crewdson, Maria (1834–1905), Katharine (1836–98) m. George T. Redmayne, Theodore (1838–91), William (1839–64), Edwin (1841–1917), Sylvanus Bevan (1844).

[4] Quaker *Annual Monitor* (1875), in an assessment of his life.

Mary viewed the prospect of retirement in 1843 with apparent equanimity, finding comfort in her husband's ability to 'lay aside the desire for greater acquisition, and to settle down contentedly on what, with the blessing, will doubtless be sufficient'. During the first year of his retirement she noted that there was 'little more than enough to meet expenses'. However, this was only in comparison with her relations, for there is no evidence at all of deprivation, and they evidently kept at least three servants.[5] Oakfield, though not large, seems to have been a comfortable villa in the Tudor style, and Edwin later recalled grounds that were 'laid out with much taste'.[6] It was in an area of Liverpool where high walls hid the spacious and leafy gardens of the well-to-do middle classes, several of whom were related, and more of whom will have been members of the Society of Friends; and it provided a secure home for the whole of Alfred's childhood.

The family moved from Aigburth in 1849, perhaps partly because Alfred had already left home, perhaps partly for reasons of health. Mary's brother, William, had already had to move away on account of his health;[7] and the death of their youngest child, Sylvanus, in 1844, coupled with the fact that young Theodore was a sickly child, may have spurred them to leave the smoky north. They had been searching for a new house since 1847, spending that summer and autumn in Dorking, Sandgate, and Reigate. They returned to the search the next year, staying with relations among the Friends at Tottenham. By this time, however, Alfred senior seems to have settled on Bristol, with its strong colony of Quakers, as a likely spot; and, when a tenant was found for Oakfield, they moved, first into Redland Hall, a house with 'four large sitting-rooms, seven good bedrooms, and two attics, with a very large entrance hall'.[8] Within a year they had moved on to Sneyd Park (Pl. 3), on which they took a lease early in 1850. It was an old house in need of some repair, and was not really what Mary considered suitable. Still she was glad to be settling down at last, and there was consolation in being among Friends, with two meeting houses within easy distance. To judge from a sketch by William Waterhouse, Sneyd Park was large and stood in wooded grounds. Edwin described it as a 'lovely situation with gardens and fields stretching down southwards to the river Avon and the wooded slopes beyond'.[9] The continuous upheavals of their younger days seem to have drawn the family very close, and they remained to a great extent dependent on each other throughout their lives. At Sneyd Park, however, they settled down for a while with a governess and various tutors for the younger children.

Alfred meanwhile was busy with his pupillage in Manchester, though the growing railway network now gave him ready access to this comfortable family home. His independence had begun several years before the move from Aigburth, when he was sent away to the small but fashionable Quaker school, Grove House, at Tottenham, where the fees were £100 a year. The building was a large house in generous grounds where the chief game was hockey. Lessons included the classics, a little history, more French and German, the French being particularly well taught. Weekly recitation sessions, mostly in prose, alternated with mechanical drawing and map making. Of course there was no music taught in a Quaker school of this era but exuberant outbursts of song from Alfred were not unknown. Thomas Hodgkin describes him as 'a regular singing bird'. His greater accomplishment, which drew admiration from master and pupils, was drawing.[10]

5 MW *Journal*, 81.

6 Ibid. A sketch of the house by Waterhouse is reproduced (ibid., facing p. 69); and an elevation survives among the Waterhouse papers.

7 Ibid. 77.

8 *Theodore Waterhouse 1838–1891. Notes on his Life and Extracts from his Letters and Papers* (privately published, 1894), 5 (Journal, 27 Apr. 1849).

9 MW *Journal*, 86.

10 This description is drawn from an unpublished memoir by Thomas Hodgkin, Waterhouse's future brother-in-law.

His time at Grove House was important because it brought him into contact with a wider circle of young Quakers. Many influential families and later clients were among the pupils at that time. There were Waterhouse cousins from Liverpool; Wilson Crewdson, who married his sister Ellen; and the Hodgkin brothers, Thomas and John Eliot, who lived nearby in Bruce Grove, and whose sister was to become Alfred's wife. Until the first half of the nineteenth century, Quakers still held very strongly together. As late as 1857, the yearly meeting discussed whether the plainness of speech, behaviour, and apparel which their principles demanded necessarily involved peculiarity.[11] They had stood out against all worldly titles, all judicial oaths, and against all forms of war. They were also against all music, private or public, all theatre, and all 'vain sports' such as hunting and shooting. One result of these principles was that almost all the professions were closed to them. Their difficulty about oaths prevented them becoming Justices of the Peace, although there were many among their number who were eminently suitable; and the result of this exclusion, as with other dissenters, led them into banking, commerce, industry, and railway building all of which were thriving areas of the early nineteenth-century economy. Many soon amassed huge fortunes, which allowed them to acquire lands and forward their philanthropic projects in education and health. But they were also drawn into a tight network of mutual support within the Society of Friends.[12]

Yet, however close the Quaker family ties, through their religion and way of life, their increasing fortune began to have its effect on the younger generation. They found it difficult to reconcile the strict beliefs and customs of their parents with the strong desire for conformity in the new level of society to which their parents' wealth had brought them. Probably the first Quaker to accept a title was Edward Fry who took a knighthood in 1877 when he became a judge of the High Court,[13] and others were to follow, including Joseph Whitwell Pease, one of Waterhouse's most important clients.

By the mid-century, there was little to distinguish many Quakers from the bulk of the well-to-do middle classes except a preference for simplicity in dress. There seems, however, to have been no objection to the best quality in either clothing or surroundings. This brought the new generation much closer to the established professional classes, and it is no surprise that all Alfred's unmarried brothers and sisters were baptized into the Church of England, at Penge in 1864. It is a clear indication of their established status, though they still regarded themselves as members of the Society of Friends. Their parents seem to have stuck more closely to the old traditions, and Mary even reverted to the peculiar Quaker dress in the 1870s after her husband's death. However, she did compromise on some rules, and when her daughters finally persuaded her to let them have a piano she came to enjoy picking out hymn tunes, especially Moody and Sankey's. Alfred himself was not baptized with his brothers—that came later—though he does not appear to have been a particularly staunch supporter of the local meetings.[14] However, the closely knit communities of Friends in and

[11] A very clear account of their situation in the 18th cent. and up until this time is given in *The Eliot Papers* (compiled from family papers by Eliot Howard and privately published 1895). 'Some of the principal points of "Testimony" were:—Against the changing fashions of the world in the matter of dress; theoretically, they never varied their costumes, but practically, they were always about half a century behind the world—against all respect of persons, as shown in the plural form of address, "you" instead of "thou",—and against the recognition of heathen gods in the names of the days of the week and of the months, hence their conversation could not fail to be peculiar.'

[12] The strength of this connection is shown by the way in which Alfred senior was able to retire from the cotton business on the understanding that his eldest son, then only 13, should be apprenticed to it when he finished school, and, on coming of age, should become a partner if he proved sufficiently businesslike. When Waterhouse turned to architecture and Theodore to the law, the place was kept for the third son, William.

[13] Edward Fry was son of Lewis Fry of Goldney, Bristol, an early client. His wife Mariabella and Alfred's wife Elizabeth were sisters.

[14] Alfred was baptized in 1877 with his wife and eldest child Paul. However, both his family and Maria retained their names as members of the Reading Monthly Meeting up to 1904; Alfred died the following year.

around centres of commerce and industry produced a complicated network of ties of kinship and belief.

There were links, too, to the various nonconformist congregations, all of which could provide an immediate source of business for any young man about to embark on a career. Waterhouse's progress was designed to build on just this foundation. He had wanted to be an artist but his parents thought this too frivolous and insecure a path and so nudged him into architecture.[15] He therefore went to Manchester in 1848 to study architecture and after three months trial was apprenticed for five years in the office of Richard Lane, himself a Quaker and one of the leading architects of the city. In Manchester Waterhouse had contact with his wider family through his uncle, Henry Waterhouse, who lived at 23 Ardwick Green. In fact he designed an extension to the dining-room of his uncle's Georgian terraced house, possibly the first commission he ever undertook.[16] We know from his mother's journal that he went with his cousins to a prayer meeting at their Independent Chapel, perhaps an intimation of a desire to explore other beliefs but also probably the basis of contact with his future friend and client Joseph Thompson.

Manchester was obviously congenial, and at the end of his pupillage we find Waterhouse setting up in practice there. He had an address in the Polygon, which in the early 1850s was still a relatively fashionable area. Meanwhile his parents were on the move again. Sneyd Park had become too expensive and in the spring of 1855 they moved into a smaller house in Clifton. From here their daughter Ellen was married to Wilson S. Crewdson of Manchester, Alfred's school friend, which would have given him another family member close by. In August their parents were in Dover, from whence Alfred and his brother Theodore set out for the Paris *Exposition*. On their return they found by enquiry 'at Capper, Son and Moore's shop that our present home is 29 York Terrace, Regent's Park'.[17] This was the family home while Alfred's brothers were being educated: Theodore went to the office of his solicitor uncle, William Bevan, in Old Jewry and then to University College, graduating in 1857, and Edwin to University College School.[18] Finally in 1858, after some searching, their parents bought Whiteknights, in Whiteknights Park (now the campus of Reading University). The move was a hardship for the unmarried sisters Maria and Katherine in leaving brothers and friends in London, but the former were certainly frequent visitors. There was ample room for them it seems; Mary would have preferred a 'far snugger place'. Even after two years there she still felt insecure and feared that her husband might tire 'even of this place',[19] if he was not found work to do. She was to end her days there in 1880, her husband having died in 1873.

[15] This is confirmed in a letter to Prudence Waterhouse from Maurice Waterhouse, a grandson of Waterhouse.

[16] The drawings survive in a private collection.

[17] AW SK 10 Sept. 1855–19 Feb. 1858.

[18] *Theodore Waterhouse*. Thomas Hodgkin in his unpublished memoir mentions that University College was then the only place of higher education for Friends.

[19] MW *Journal*, 157.

2

Becoming an Architect: Apprenticeship and Early Travels

DURING the fifteen years or so up to the end of Waterhouse's pupillage Manchester was 'a mighty reservoir receiving the industrial products of the district for which it is the centre and distributing them by innumerable mills to every nook and corner of the habitable globe'.[1] The city amassed new buildings, and new types of buildings, in a wide range of styles. There was Francis Goodwin's Grecian Ionic town hall; Charles Barry had built the Greek revival Royal Manchester Institution in 1835, and two Gothic Commissioners' churches, St Matthew's, Campfield, and All Saints', Stand. C. R. Cockerell had built the Classical Bank of England, and J. E. Gregan an Italianate bank in St Ann Street for Sir Benjamin Heywood. In 1853 Edward Walters was building the Free Trade Hall in the Italian Renaissance style, though he had also tried Gothic for his Cavendish Street Chapel and Romanesque for the Knott Mill Chapel. Pugin built the Gothic church of St Wilfrid, in 1842; and Scott began his work in the area with Holy Trinity, Stretford (1840) and St Mark, Worsley (1845). Edmund Sharpe built two churches of particular interest, since they were the first to be faced entirely in terracotta: St Stephen at Lever Bridge, Bolton (1842), and, in Manchester itself, Holy Trinity, Platt Lane (1845); and Thomas Rickman, a Quaker and friend of the Waterhouse family, had produced his first iron church in Liverpool as long ago as 1815. Civil engineers were already making iron a familiar material for architects; and in Manchester there was a proliferation of iron-framed warehouses for storage of the huge quantities of goods on which the city's prosperity was based. The need to present a face of prosperity to the world gave architects the opportunity for stylistic adventures on their façades.

This was the booming commercial city to which Waterhouse came in 1848 to train as an architect.[2] His employer Richard Lane was a classicist at heart, although his Deaf and Dumb School of 1835 wore Elizabethan dress. His most notable buildings were Salford Town Hall (Doric), 1825, the Friends' Meeting House (Ionic), 1828, and the new fronts of the Infirmary and Lunatic Asylum in Piccadilly of 1835.[3] His office, on the corner of Chapel Walks and Cross Street, was evidently a

[1] Alfred Brothers, FRAS, *Manchester As It Was* (Manchester, 1878).

[2] Barry had been advised to try Manchester as a likely place to launch a career by Sir John Soane as long ago as the 1820s (Marcus Whiffen, 'The Architecture of Sir Charles Barry in Manchester and Neighbourhood' in J. H. G. Archer (ed.), *Art and Architecture in Victorian Manchester* (Manchester, 1985).

[3] Richard Lane (fl. 1815–58) was by 1848 well established in the town having earlier in his career worked with Francis Goodwin (1784–1835). Other buildings including the first Corn Exchange and the hall for the Natural History Society were all in forms of the Ionic order. He also built in forms of Gothic, but St Simon's, Salford, and the Pendleton Congregational Chapel did not enhance his reputation.

thriving one with access to some of the best commissions in town. Here Waterhouse learnt to produce drawings in the fine crisp lines and pale tints that mark the style of the Regency (Pls. 17, 18). He was to move in his own work towards a firmer style with richer colouring, but without losing any of his master's clarity of line. The training included theory as well as practice and Waterhouse was set to copy long extracts from H. W. Inwood's book *Of the Resources of Design in the Architecture of Greece etc.*, and also from Chambers's *Civil Architecture*.[4] He also made several pages of meticulous tracings from Paley's *Gothic Mouldings*,[5] which he preserved in a carefully bound scrapbook.[6] As well as this, of course, he studied the building press, saving, for instance, from the *Builder* of 12 August 1848 one of two articles by Edmund Sharpe, discussing Rickman's book which set out to categorize the styles of Gothic architecture.[7]

Whether Lane's instruction was as systematic as the scrapbook suggests, or whether that was the result of Waterhouse's own meticulous and tidy mind, it seems that Waterhouse took a carefully analytical view of current theories rather than merely absorbing stylistic rules. In the scrapbook, notes from Chambers and Paley are set down side by side on opposing pages. So far as fashion and preference went, Waterhouse's pupillage was the time when Pugin's writings had made true Gothic the most interesting style;[8] and Ruskin's *Stones of Venice*, appearing in 1849, added further force to this passion. Waterhouse and his friends will have discussed these books as eagerly as any.[9] Out of hours, there was a sketching club where he met friends like Frederic Shields, the Manchester Pre-Raphaelite, and the office contained other lively young architects such as Alfred Darbyshire.[10] But Waterhouse also studied the sober antiquarian publications, and from them, especially Paley, he learnt the principles of the formation of mouldings in both Classical and Gothic architecture.[11]

Waterhouse's scrapbook of these years contains descriptions and cuttings of various buildings, not all by Lane but mostly churches and so mostly in the Gothic style. He describes a new church to be

[4] H. W. Inwood, *Of the Resources of Design in the Architecture of Greece etc., Egypt, and other Countries obtained by the Studies of the Architects of those Countries from Nature* (London, 1834); Sir William Chambers, *Treatise on Civil Architecture*. Waterhouse almost certainly used Gwilt's edition of 1825.

[5] F. A. Paley, *Manual of Gothic Mouldings* (London, 1845).

[6] This bound foolscap volume, begun in 1848, has an elaborate title-page drawn in pen: 'Architectural Fragments, Sketches etc.' headed by a medallion containing a view of a Doric ruin. At the foot is a drawing of the interior of the church of the Holy Sepulchre, Jerusalem. It contains in the early part carefully copied extracts from various works which were no doubt recommended to the student. The later pages are filled with cuttings, including some on his early work up to the time of the Assize Courts commission.

[7] Thomas Rickman, *An Attempt to Discriminate the Styles of Architecture in England* (London, 1817).

[8] A. W. N. Pugin, *Contrasts; or, A Parallel between the Noble Edifices of the Fourteenth and Fifteenth Centuries, and Similar Buildings of the Present Day; Shewing the Present Decay of Taste* (London, 1836); and *The True Principles of Pointed or Christian Architecture* (London, 1841).

[9] Waterhouse possessed the works of both Pugin and Ruskin, but it is not certain when he bought them. We know with certainty that he saw Ruskin's *The Seven Lamps of Architecture* as early as 1851. On a page of one of the two small notebooks of his student days is a copy of an illustration from it of tracery from the upper part of Giotto's campanile in Florence. Office ledger entries between June 1859 and Dec. 1865, under the heading 'Books', include the following technical titles: 'Donaldson's *Specification*, *How to Lay Out a Garden*, *Farm Buildings*, Laxton's *Prices*, Molesworth's *Engineering Formulae*, and (?) Hawkin's *Engineer*'. The list also includes 'Brandon's *Works*; Verdier and Cattois' *Architecture*; Nesfield's *Sketches* [a number of issues]; King's *Continental Architecture*, 2 vols.; *Art of Illuminating*; Johnston's *France*, [several issues]; Viollet-le-Duc's *Dictionnaire Raisonnée*; Viollet-le-Duc's *Mobiliers*; Sharpe's *Parallels*; Dennison's *Church Builder*'. Among other books in private collections which Waterhouse almost certainly bought, but which now contain modern bookplates are: G. E. Street, *An Urgent Plea for the Revival of the True Principles of Architecture in the Public Buildings of Oxford* (1853); Owen Jones, *The Grammar of Ornament* (1856); A. W. N. Pugin, *Contrasts* (1836), *The True Principles of Pointed or Christian Architecture* (1841), and *An Apology for Christian Architecture* (1843); Edmund Sharpe, *Seven Periods of English Architecture Defined and Illustrated* (1871 edn.). Others without bookplates are Bowman and Crowther, *Churches of the Middle Ages*, 2 vols. (1848); R. N. Shaw, *Architectural Sketches from the Continent* (1858); J. J. Fergusson, *The History of the Modern Styles of Architecture* (1862).

[10] A. Darbyshire, *An Architect's Experiences: Professional, Artistic, Theatrical* (Manchester, 1897), gives a picture of this group.

[11] Paley explains the difference between the Greek mouldings, which were based on the effect of light on changing surfaces, and the Gothic, which rely on the effect of cast shadows. He also described methods of copying and measuring all types of mouldings.

erected in Whalley Range, in the decorated style by J. P. Harrison, and the opening of Walters's new Independent Chapel, Cavendish Street, in the Early English style. There are newspaper cuttings of two of Lane's Gothic churches,[12] as well as a copy of a letter which Lane had written to the committee responsible for the erection of St Simon's, Salford, to accompany his competition entry. Lane had defended his choice: 'I have adopted the Early English style of the thirteenth century as being more effective in comparison to its cost than any other.'

Lane was a practical architect, too, a lesson that was not lost on his young pupil. Presumably these years in a fashionable Manchester office also brought the young Waterhouse into contact with various Manchester builders; and his training evidently included attention to practical matters of drainage, ventilation, and so on. The *Builder*, which had begun publication in 1842, provided not only a menu of stylistic criticism and debate but a great deal in the way of reviews and analysis of technical developments; and Waterhouse's interest encompassed those as well. His library included, for instance, Bland's *Experimental Essays on Construction*.[13]

When his apprenticeship was ended Waterhouse began the second stage of his training with an extended continental tour. Not all aspiring architects went abroad so early in their careers; Scott did not travel until he was 30, and Street was several years into his career when he visited Italy in 1855. But foreign travel was recognized as a desirable adjunct at this stage in a career: 'when [the young architect] is out of his pupillage, if he can afford it, he should spend a few months in Italy.'[14] Waterhouse, aged only 23, was very well able to afford it. So in May 1853 he set out for Dieppe (Pl. 6) with his old school friend Thomas Hodgkin, who tells us that he 'was entirely under the influence of Ruskin and communicated [to him] his own passionate admiration for Gothic art and a perfect detestation of that beastly Renaissance'.[15] They aimed to reach the south of France as soon as possible, passing through Rouen and Paris and taking a steamer from Dijon down the Saône to Lyons, then to Avignon taking in Nîmes, Arles, and Orange. Thomas's companion of course required a good deal of time for wayside sketching which made them embarrassingly late arriving to spend the night at the monastery of the Grande Chartreuse, before crossing over into Piedmont, and on to Susa and Turin. Italy was the goal for any disciple of Ruskin at that time; Waterhouse's devotion to France was yet to come. They travelled by steamer and train, by public *calèche*, by diligence at a tedious four miles an hour through the Val d'Aosta, and on foot in a snowstorm over the Saint Bernard pass. Having crossed back into Switzerland they took a steamer down the lake to Geneva (Pl. 7). By now it was June, and Hodgkin was due to meet some of his family in Basle. Waterhouse turned back to Italy in the company of a Manchester acquaintance, George Rooke.

Waterhouse's record of the tour is a bound leather volume with hundreds of small pencil sketches mounted on its pages. It has a carefully designed title page: '*Scraps from France, Switzerland and Italy*', with a drawing of a simple bracketed stone and tile porch seen in Fiesole (Pl. 5). Since every sketch is titled and dated, we can trace his route and his sightseeing precisely. The trip was to last altogether nine months taking him first through Isola Bella, Certosa di Pavia, Milan, Bergamo, Monza, and Brescia to Venice. There he stayed for over two weeks during August: evidently wanting

12 'Architectural Fragments', scrapbook, 48 and 50.

13 W. Bland, *Experimental Essays of the Principles of Construction in Arches, Piers, Buttresses, Etc. Useful to the Practical Builder* (London, 1862).

14 *The Complete Book of Trades, or Parents' Guide and Youths' Instructor* (1842), quoted in Frank Jenkins, *Architect and Patron* (Oxford, 1961), 160.

15 Unpublished autobiographical memoir by Thomas Hodgkin. Thomas had been to France before in 1846, at the age of 15 with his father, uncle, and brother, so no doubt he was able to offer his experience to the unseasoned Waterhouse.

to see at first hand buildings which formed the background to Ruskin's writings. He studied the Doge's Palace, of course, painted the Porta della Carta (Pl. 4), drew a plan of San Marco, and made a pilgrimage to the islands, Murano and Torcello.

From Venice he continued to Padua, Vicenza, and Verona (Pl. 13) and then to Mantua, Modena, and Bologna. At the end of September he was in Florence for a week on Ruskin's trail again. One sketch of a window jamb in the topmost storey of Giotto's campanile of the Duomo bears the note 'for mouldings wanted see Seven Lamps'. From Florence he went to Siena and Fiesole and then on to Lucca and Pisa. Two weeks later he was in Pozzuoli; perhaps he took a boat down the coast to get to Naples and the surrounding places: Pozzuoli, Sorrento, Amalfi, Ravello (Pl. 12), and Paestum. He stayed in Naples, possibly up to three weeks, making few sketches, though completing some watercolours. What sketches there are, are undated, often untitled and lacking in energy. Perhaps he was ill. But by the end of November he was in Rome where he stayed over Christmas and the New Year.[16] Then it was north to Assisi and Perugia, on to Cortona, Arezzo (Pl. 11), Florence again, though only fleetingly, and on to Pisa and Genoa. From there he travelled home through Turin, Basle, and Strasburg stopping only in Basle where he spent five days sketching the Rathaus and the Münster.

The trip was a fortunate luxury for so young an architect, and he used his time to the full. The sketches show that, in spite of the lure of Ruskin, Waterhouse was ready to be catholic in his tastes. Ruskin's towns were the key to the route, but on the way he would look at whatever took his eye. In France he studied Roman buildings as well as the Romanesque churches. He was attracted by the Swiss vernacular, and drew a whole sheet of hillside chalets. He was intrigued by the construction of a timber bridge at Martigny and by the tower of St Maurice at Vallais with its roofs, one central pyramid surrounded by four small ones. He was impressed by the great spaces of the cathedrals, recording in one watercolour the soaring height of the north aisle of Milan cathedral, the chevet glimpsed through the tall pillars with their octagonal drum capitals. In spite of Thomas Hodgkin's assertion, he did look at Italian Renaissance work. He drew the highly ornamented Palazzo della Loggia at Brescia, the Baroque entrance to the Palazzo Brignole at Genoa, and Palladio's loggia to the basilica at Vicenza.

There is no doubt that at this stage he saw Venice, and much else, through the eyes of Ruskin, whose influence shows in the frequent sketches of work in mixed materials done to produce colour and pattern in structural elements. Examples of structural polychromy appear repeatedly, for instance the striped stone and brickwork of the campanile at Vicenza (Pl. 13) or the stone and red and yellow brickwork of the apse of the Duomo on the island of Murano. Waterhouse also noted decorative use of coloured materials: patterns in marble floors and mosaic at Ravello, glass in simple patterns, marble inlay work of pulpits and balconies (Pl. 12). And all the time he was drawing with an eye to constructional methods and the logic of building. There are copious sketches of cornices, with further details of how each was ornamented. His study of mouldings was all embracing; there are Gothic jambs, arches, and vaulting, Renaissance sills and architraves, and the treatment of junctions of different elements. The articulation of the ends of steps on an open staircase and the way a stone skirting mould runs behind the door jamb mould were both noted at Lausanne Cathedral (Pl. 7). His

[16] Although some sketches are included of the fountain of S. Sophia and the Mosque of Sultan Suleiman in Constantinople it has not been possible to reconcile their dates with the time needed to make the return journey from Italy. It can only be assumed that these drawings are copied from others.

eclecticism even embraced the Doric capitals of the temple of Neptune at Paestum in tiny meticulous sections.

In the sketches he produced on this tour we can see the basis of his own system of picturesque design. The massing of spires, the turn of a good staircase were all observed, analysed, and stored as potential solutions to future problems. Later in the tour he seems to have become more interested in the enclosure and linking of space. A tinted drawing done in the basilica of S. Lorenzo fuori le Mure shows the way light from the clerestory windows of the chancel links the arcaded galleries to the side. On his return to Florence he drew two small perspective views of S. Spirito and S. Lorenzo, one looking from aisle to nave, and the second the reverse. Each shows the view through the tall columns supporting the entablature and semicircular arches to the space beyond.

This tour was a crucially important experience; but the great distances travelled and the mass of drawings produced are eloquent testimony to Waterhouse's energy and determination. We are too much beset by images to understand at all readily the overwhelming impact of this first-hand exposure on an untravelled student in the mid-nineteenth century. Brought up in cities shrouded in smoke and industrial gloom, Waterhouse was suddenly confronted with the reality, the light, and colour of the Mediterranean, the texture of its stone marble and brick in their native setting. All the excitement of this discovery is in the drawings, and it seems as though, in these early years, he could hardly draw enough; and he was lucky enough to be able to return soon to his original sources.

On his first visit to France he had been bent on reaching Italy. In 1855 he made a second visit with his brother Theodore, and now everything French seems to have touched him and the closely written pages describing the journey reveal his enjoyment of being among French people and observing French customs and manners. At Amiens, he noted details of iron and lead work and dormers (Pl. 15), was 'much struck with the way in which brick and stone are alternated horizontally in some houses' and found some 'interesting modern work near the cathedral' which was undergoing a complete restoration. 'Its size and beauty failed to overwhelm me as the 1st sight of a French cathedral [Rouen]; but it was refreshing to see the effect produced on my brother.'[17] While drawing stonework details in the cathedral he resolved 'on reaching home to design six caps from nature—the mouldings being same, leafage various'. In Paris for the *Exposition* he spent a whole day at the Palais de l'Industrie. Waterhouse was impressed by the presentation and colouring of drawings: 'The extraordinary care and accuracy as well as artistic treatment in the preparation of their designs' was a lesson which Waterhouse took carefully to heart.

Before leaving Paris he and Theodore took a 'farewell walk amid the gay Champs Elysees with wistful glances at the rotating hobby horses which dignity forbade [their] mounting'. His return to England found him 'much disgusted with English smoke and the poverty of English architecture. We want size, light and shade and colour in our building and in ourselves more good humour and manners.'[18]

The next trip abroad was the family holiday of 1857 spent in Switzerland.[19] They went through Brussels, Namur, and Liège. The cathedral of Aix-la-Chapelle had been '"modernized" almost entirely' but of Cologne Cathedral, then being completed by the building of the western towers, 'any attempt at description would be superfluous'. The whole family attended high mass there. At Bonn

[17] SK 10 Sept. 1855–19 Feb. 1858, 22.

[18] Ibid. 17.

[19] The family seem regularly to have taken holidays together in the early years. In 1856 they were at Nab Cottage, Rydal, and in 1858 at Thorney Howe, Grasmere. 1857 was the first year they had travelled abroad together. SK 13 Aug. 1857–16 Oct. 1859, 1–100, records the trip and includes a written reaction.

too they visited the 'glorious Romanesque Cathedral of light coloured brick and stone'. A few more architectural entries included Freybourg and Basle; but then the holiday began in earnest. Waterhouse filled his sketchbook with pages of rugged peaks and deep valleys and of prose description of the 'glorious peaks and domes' (Pls. 193, 194). He allowed himself two weeks before leaving the family and returning to Manchester on his own, feeling 'more solitary than ever I did on the long tour'. He reminded himself that the 'Rhine towns and Frankfurt [were] worth revisiting but not by steamer'. Switzerland was even then 'spoiled by tourists' and he felt he would not care to go again. France was, he concluded, 'the country after all. Normandy and Brittany—and all the southern provinces I trust I may sometime know.' Visiting Belgium the next year he was less self-indulgent, and systematically missed the paintings, 'being determined to stick to the architecture while I am by myself'.[20]

Waterhouse continued his practice of observing and recording throughout his career, and the fourteen surviving sketchbooks and related drawings provide an invaluable insight into the development of his approach to architecture (Pls. 8, 14).[21] In the early years he made written notes too, describing his travels and recording his reactions. The entries were not entirely architectural. The visit to the Paris *Exposition* of 1855 was carefully planned, probable expenses listed, and a full account kept which included hair oil and braces.[22] Before a trip south from Manchester in January 1856 there is a list of 'Things to be seen in London: Architectural Exhibition; Architectural Museum; Polychromy, Owen Jones, Regent Street; Tues 18th, Rev. Mr. Petit on Utilitarianism in Architecture; W. Burges, Thames Chambers, York Buildings, Adelphi;[23] Thurs. 20th, Howards, Tottenham. Reginald Palgrave, Committee Clerks Room, House of Commons;[24] Geological Museum; All Saints' Church, Margaret Street, Butterfield; and St Stephen's Westminster, Ferrey.'[25]

The drawings and details continue on the model set by the tour volume, although the first two books do include notes on preliminary work for clients. This must have become inconvenient as the practice grew and assistants were employed, and the entries soon become purely for his own use. The first sketchbooks covering the years 1855 to 1859 show how keen Waterhouse was to see the work of leading architects. Pugin, Scott, and Street are the focal points. In 1856 he went to Staffordshire and visited Alton to see Pugin's work and Cheadle to see his church of St Giles of 1841, noting in particular the decorative patterning of stone surfaces.[26] He found the school and cottages by Scott at Ilam, with steep roofs and tile hanging of the upper floors, of particular interest. Three years later he was recording details of Scott's Public Institute at Sandbach (Pls. 9, 10), the brick filling to the tympanum over a window, and the black and red string mould and commenting on the 'Good finials'.[27] Notes from Street's work recur. The first reference is to his 'Brick and Marble Architecture of the Middle Ages',[28] which includes the same stone hood Waterhouse had seen at Fiesole and used for the title-page of his own tour album (Pl. 5). From Reading (where his parents had now moved) he was able to visit Street's churches at Tilehurst and Maidenhead.[29] A visit to the new museum at

[20] Ibid. 193.

[21] See Bibliography, sect. A. The sketchbooks are all in private collections.

[22] SK 10 Sept. 1855–19 Feb. 1858, 3. The cost was £12.

[23] Burges's brother-in-law, R. Popplewell Pullan, was a fellow pupil with Waterhouse in Richard Lane's office. A letter of 20 July 1860 from Waterhouse to A. B. Thompson, the secretary of the Institute of British Architects, suggests that Burges would recommend him as a member.

[24] The purpose of this meeting is not known, though it possibly relates to the Government Offices Competition. It is interesting, however, that Reginald Palgrave's daughter subsequently married Waterhouse's eldest son Paul.

[25] SK 10 Sept. 1855–19 Feb. 1858, 95.

[26] Ibid. 89–90.

[27] SK 13 Aug. 1857–16 Oct. 1859, 225–8.

[28] G. E. Street, *Brick and Marble Architecture in North Italy* (London, 1855).

[29] St Michael, Tilehurst, and All Saints, Maidenhead.

Oxford in 1858 was a necessary pilgrimage (Pl. 9).[30] The architects Deane and Woodward had been closest to the winner in the competition for the new government offices the previous year; and their work at Oxford was being monitored by Ruskin. Waterhouse found the 'upper arcade very pretty', the 'cornice above very simple', and the 'window glass with bull's eyes very good', but the main point of interest was that, after the pleas of Street and Ruskin, Gothic was being adapted for a secular building.

Another useful port of call was the Architectural Exhibition, an annual event since 1849. Waterhouse first records a visit in 1856, when he himself exhibited a perspective of the Chester Street warehouse, which the *Manchester Guardian* noted as having a 'Modesty and simplicity and life like look . . . which makes us hope that Mr. Waterhouse may go forward and do something in his day towards improving Manchester'.[31] The exhibitions were very much a learning experience, an excellent opportunity to see contemporary work, and Waterhouse notes work by a number of less well-known men: a cemetery chapel by George Truefitt, author and contributor to the *Ecclesiologist*, and work by E. B. Lamb, who produced highly original details. The entries show his continuing interest in ornament: 'Remon makes me alter my opinion about flat ornament almost. Perhaps natural imitation [is] in some cases admissable'; 'See Owen Jones on colour, natural forms reduced to conventional forms' and 'Colling most wonderful in ornamental Gothic—see his models'.[32] Waterhouse visited the exhibition again in 1858, noting among other things Burges's design for the Crimea Memorial Church and Street's for the Lille Cathedral Competition.[33] By then, however, Waterhouse was himself established in practice and shortly to become as well known as either Burges or Street.

30 SK 13 Aug. 1857–6 Oct. 1859, 228.

31 *Manchester Guardian* (16 Jan. 1856). According to Andrew Saint, Norman Shaw began to exhibit this year but we find no notes on him. There is a sketch of the elevation of M. D. Wyatt's Portuguese chapel and details of stone gables by Joseph Clarke, a Glasgow workhouse by R. W. Billings, window and door details by E. B. Lamb, and more polychrome details from drawings by William White. Street's work also figures in sketches of buttresses, string moulds, and turrets. The following year Deane and Woodward's design in the competition for new government offices was on show and Waterhouse carefully sketched one of the corner pavilions in which the pitch of the staircase is strongly expressed by the rising windows (SK 13 Aug. 1857–16 Oct. 1859, 216. This is a very rare instance of Waterhouse using ink in the sketchbooks).

32 SK 10 Sept. 1855–19 Feb. 1858, 49.

33 SK 13 Aug. 1857–6 Oct. 1859, 106.

3

Building a Practice: The Client Network

TRAINED in one of the leading offices of Manchester and inspired by his nine-month tour of the Continent, Waterhouse opened his career as an architect from offices in Cross Street Chambers early in 1854. He will have been encouraged by the reception afforded some drawings, which he showed at a conversazione, when he visited his parents at Sneyd Park at the end of February. 'As the first exhibition of, we are told, a very young artist, we may hail with pleasure the architectural drawings of Mr. Waterhouse, jnr., than which we have seldom seen specimens more full of promise.'[1] But the Bristol and Clifton Graphical Society was not the hard world of Manchester business, and topographical drawings were not architecture. It seems likely that his father had helped to set him up, since Cross Street Chambers was a prestigious city centre address and can hardly have been cheap. These first years may also have been slightly less anxious for Waterhouse than for others, since he evidently had some financial reserves.[2] None the less, he had to find clients and prove his ability as a designer.

Manchester may have been a boom town, but any likely commissions would come from the hard-headed businessmen who had made it so. Gothic dreams would have little place if it could not be shown that they were practical, and probably economical as well. So Waterhouse had to begin with a series of humble commissions for alterations to houses or offices, the design of a sideboard, or a new bay window. To start with there was no grand opportunity to display his talents; and one suspects that he was merely being tested out by those men of substance whose commitment to solidity and worth went a great deal further then the appeal of a fashionable façade. Even in modest commissions, however, Waterhouse's thoroughness and the reliability of his service will have impressed them; and in a few years he was building their offices, designing their shops, and providing all that was needed for a mansion or a complete estate.

He began practice in what was, by 1854, a thoroughly traditional way, with surveying work and land valuation as very much a part of the business. Even in the 1860s he was prepared to act as agent

[1] Unidentified newspaper cutting (4 Mar. 1854), report of Bristol and Clifton Graphical Society exhibition in Waterhouse's scrapbook, 'Architectural Fragments'. This is the first cutting in the scrapbook.

[2] Office Ledger, pp. 2, 3, 4, 5. Private collection. The earliest accounts show that by about 1860 he had shares in the Midland Railway, East India Railway Company, and his Pease cousin's Stockton and Darlington Railway Co. It is unlikely that his practice would have provided so much surplus within 6 years, and we can assume that the shares were transferred to him from his father.

for the sale of one of his houses.[3] It is possible too that he benefited, indirectly, from the tail end of speculation of the sort that had been commonplace under the Regency. His master, Richard Lane, had been involved in laying out the suburb of Victoria Park in Manchester, and Waterhouse handled some seven commissions there in the years up to 1871, not counting his own house, Barcombe Cottage **[52]** (Pl. 17).[4]

One of his earliest designs **[3]**, however, was for stables for his father, presumably for Sneyd Park, which must be seen as the family interest at the start of his career.[5] But there were other family commissions too. His uncle, Rogers Waterhouse, asked him to carry out alterations to his house at Mossley Bank, Liverpool, in 1858 **[47]**, and, perhaps equally important, was involved in getting him to draw designs for the proposed Friends' Meeting House in Liverpool **[30]**. Though it was never built, this will have brought him to the notice of Quaker society in a professional capacity. This Quaker connection, through his family and friends, was essential in launching his career. His own marriage, in 1860, was within the Society, to Elizabeth Hodgkin, giving him links to that powerful family with ties in the north-east. He had already overlapped with Thomas Hodgkin, his future brother-in-law, at The Grove School, Tottenham, and travelled with him to France. At The Grove he also met the sons of several other Quaker families who were to be influential in his career: the Gilletts of Banbury, the Heads of Ipswich, with young Edmund Ashworth of Bolton and his brothers, the young Crewdsons, and several of his generation of the Pease dynasty, as well as Barclays, Foxes, Birchalls, Bassetts, and Harrises.[6] Of these families only the Barclays never commissioned work from him, though they were closely connected with his cousins the Backhouses who did. One may assume that most of the Liverpool Friends, such as the Hadwens, would already have known the young man through the father. It seems likely also that he was given initial help by the Quaker architect, P. B. Alley, partner to Richard Lane, who later claimed that it was on his account that Waterhouse was apprenticed to the firm.[7] Alley had connections in Southport and may be the link to one of Waterhouse's earliest extensive commissions, the refurbishment of the Southport Baths **[21]**, completed in 1856. Quaker connections also took him to the Lake District where he designed his earliest complete building, Rothay Holme **[2]** for Elizabeth Head at Ambleside. The Lakes also

[3] MS letter to Mrs Henderson (14 Aug. 1862). OLB. Private collection. The house was Rose Trees near Keswick **[28]**, which had been built for G. H. Henderson, the Recorder of Liverpool. He unfortunately died before the house could be occupied.

[4] See Maurice Spiers, *Victoria Park Manchester* (Manchester, 1976). Richard Lane lived at Park Villa which he had designed in 1834. Waterhouse moved to Barcombe Cottage, Victoria Park, from the Polygon, a desirable part of Moss Side that was rapidly being forsaken by leading industrialists, driven out by increasing pollution. The other inhabitants of the suburb included James Kershaw, R. Barbour, John Slagg, William Hewitt, and T. R. Hetherington among Waterhouse's clients. It was also the address of Ford Madox Brown in Manchester and of the architect Edward Salomons.

[5] There is a hint that Waterhouse may have had an interest in service buildings and farm buildings generally. A long description of Mr Littledale's model farm at Liscard on the Wirral is laboriously copied into his scrapbook 'Architectural Fragments' (1848), 56–74. Private collection. Waterhouse may have been looking at the farm for his father who, in 1848, was on the look-out for land. It is also possible that, as the son of a relatively well-to-do businessman, Waterhouse saw himself as a potential investor in land. He did own 2 farms at Gormellick in Cornwall by the early 1860s (see e.g. OLB II 500. RIBA). It is not known when he acquired these or to what extent he took an interest in the farming. The land was managed for him by a local agent. However, it is equally possible that he studied the Liscard group (was it even by Richard Lane?) as a potential pattern for the development of estates for future clients. He may have reckoned that, in the mid-century agricultural boom, a thorough understanding of such buildings would offer him the same opportunities that workhouses had offered Scott 20 years earlier.

[6] *Grove House, Tottenham School* by a Former Student, in Friends' House Library. It is not entirely clear how actively Waterhouse himself was involved in the Society of Friends. He was not a leading member of the local meeting; but as late as 1857 he was helping in the distribution to the Lancashire Quarterly Meeting of a tract written by his aunt Katherine Backhouse (MS letter to William Satterthwaite (8 Jan. 1857). OLB. Private collection).

[7] MS letter, P. B. Alley to Alfred Darbyshire (24 Jan. 1898). MCL Archives f. 927. 2. Da. 1: 'I could say more but perhaps what I have already said is sufficient to show that which I only wished it to be understood was, that Alfred Waterhouse was practically, really and morally my pupil in the Office of Richard Lane to whom he was indentured as the principal of the office.' I am grateful to J. H. G. Archer for bringing this letter to my attention.

brought him into contact with the Redmaynes of Brathay Hall, whose son became his pupil and assistant and later his brother-in-law. At Brathay **[4]** he undertook the humble job of minor alterations in the house and additions to the kitchen. This was followed by a new lodge **[42]** (Pl. 19), which is a confident composition in the picturesque style long established by Loudon and others.

At this stage Waterhouse was clearly no determined goth. Another early Lakeland house, Fawe Park **[45]** (Pl. 16), is better described as picturesque than as Gothic. This was a cottage for another Quaker, James Bell, a more important client in that he had trained as an architect, was secretary of the Institute of British Architects, and was a Member of Parliament with the ensuing connections. There is every indication that the relationship was friendly and that Bell was well pleased with his architect's readiness to design details, alter the arrangement of rooms, and lower chimneys. As the final carcass of the building cost only in the region of £1,200, it seems that Waterhouse was already efficient in working to a relatively modest scale. The house has two unusual features which are typical of the imaginative and unfettered approach of the mature Waterhouse. First, the plan is unconventional, with the main sitting-room at second-floor level to take advantage of the view over Derwentwater.[8] It was, said Waterhouse, 'magnificent beyond my most sanguine expectations'.[9] The building had a projecting half-timbered open air gallery reached from the sitting-room. This feature, derived from the medieval French houses of the sort illustrated in Shaw's *Sketches from the Continent*, gave the little building a bold picturesque quality.

James Bell, clearly well pleased and wanting to help his architect, arranged for the submission of Waterhouse's perspective of Fawe Park to the Royal Academy where it was exhibited in 1857.[10] Yet Waterhouse still had to win actual clients, and was quite unabashed in offering his services. He wrote to his uncle, William Bevan: 'I understand that a friend or client of thine, Mr. Richd. Wilson, Solr. Kendal is about to build at Ambleside, that further he is a rich well to do gentleman who wld. like a good house: and of course it instantly occurred to me that it wld. be a kindness both to him and myself were we brot together'.[11] Waterhouse was also ready to undertake preliminary surveying for clients who were only contemplating building in a general way. This was much more common as part of an architect's work in the 1840s; and a long letter of 1857 shows how comprehensive his service in this line was. He was asked to examine an estate for sale, and, though he presumably hoped to be the architect of any house that was built, was meticulous in describing the agricultural and sporting possibilities as well as the social amenities:

> Some of the land in the neighbourhood is on gravel and sand, but nearly all the property in question is clay land: it stands however so high that I do not think there would be difficulty in building a dry house anywhere along the ridge . . . the views E. N. and N.W. are very pretty, particularly the latter which has the railway about as far from the house as it is at Bellamont, and backed in this case by the woods of Newbold Park. Over the

[8] This is not the first time he had designed in this way. Rothay Holme had a sitting-room with verandah at first-floor level for the same purpose.

[9] MS letter, AW to James Bell (31 Mar. 1857) OLB. Private collection.

[10] MS letter, AW to James Bell (2 Mar. 1857). OLB. Private collection. See also Royal Academy Exhibition 1857, cat. no. 1094, *Cottage erecting in Cumberland*. This was Waterhouse's first work shown at the Academy. James Bell remained a close friend, and Waterhouse later wrote to him for advice about securing the commmission for Strangeways Gaol (MS letter, AW to James Bell (15 Jan. 1862). OLB. Private collection). The exhibition led to at least one enquiry from a potential client (MS letter, AW to G. Bailey Jones (9 Mar. 1857)). The enquiry clearly referred to the probable cost of building and Waterhouse's charges. Waterhouse pointed out that building costs in the Lake District were increased by extra cartage, but added that his own charges for expenses would be reduced since he was already doing a good deal of work in the area.

[11] MS letter, AW to William Bevan (22 Sept. 1856), OLB. Private collection. The meeting evidently took place and it appears that a commission resulted, although the house is as yet unidentified. An office ledger records a charge of £150. 12*s*. for professional charges against Richard Wilson of Derwent Grange, making a house of about £3,000.

woodlands to the East may be seen Coventry spires 9 miles distant . . . The land when Bates took it was worth 16/- an acre. When he left it was valued at 35/- or 40/- and is still capable of further improvement. It is well adapted for pasturage or for wheat and beans. Some of the finest timber has been cut down. But there is still enough in my opinion to assist greatly in the formation of pleasure grounds. There is a drive thro' the wood in the centre of wh. grows an immense Scotch fir. Bates thinks there will be no difficulty in obtaining a good supply of water. I started a covey of partridge and a fine hare. The stream at the lower extremity of the property is only small. Easenhall consists of about a dozen buildings and cottages, and is a neat well kept little place—Shelton is larger and more picturesque, but inhabited exclusively by farmers or their dependents. Captain Ramsden of Newbold Hall is the nearest gentleman resident, his estate being only separated by the railway from Town Thorn's Wood—Then Lord Denbigh, Sir George Clarke, and Lord Craven of Coombe Abbey are all in the neighborhood [*sic*] and in Rugby I should think there would be pleasant society.[12]

Meanwhile in the Lakes, one commission led to another and Elleray Bank, Windermere **[22]**, was designed for G. G. Cunningham, a local JP, in 1856.[13] This house was also in a Loudonesque style; but the story of its building was not entirely happy. There were endless delays with the builders, to whom Waterhouse wrote: 'No words can express to you the bitter annoyance I feel as architect of Elleray Bank . . . Mr C has written to me *his last remonstrances*. I never was placed in so humiliating a position towards a client.'[14] It was not a position he ever allowed himself to get in again. Part of the problem seems to have been lack of regular or frequent supervision from a distance. For the future Waterhouse was careful to secure reliable Clerks of Works; but on this occasion he was evidently able to mollify the client. Ten years later the Cunninghams called on him again for the design of a stable block **[261]** and for subsequent enlargements to the house.

Within a year or two of setting up in practice Waterhouse was beginning to build up a loyal clientele. They were, it is true, mostly relations, Quakers, or businessmen to whom he was probably introduced through his father; but he built on their satisfaction so that one commission led to another, and delighted clients were sure to recommend him to their friends and relations. A number of early commissions for banks, for instance, were directly linked to the job of designing the banker's mansion. The Heath, at Leighton Buzzard **[178]** (Pl. 116), for Francis Bassett was just such a house, which in turn led to the commission for nearby Aspley Guise Rectory **[312]**, when the vicar married the young Miss Bassett. On another occasion Waterhouse wrote an unsolicited letter offering his services to Gilbert Henderson, the recorder of Liverpool. 'My connexions in Liverpool,' he claimed, 'may not be unknown to you. I beg to refer you to Mr Christr Bushell of Liverpool, Miss Head of Ambleside, Mr Redmayne of Brathay Hall or Mr Cunningham of Birthwaite for whom I have been or am still acting as architect.'[15] Rose Trees **[28]**, the house that resulted, again has an unconventional

[12] MS letter, AW to Edmund Cropper (9 Oct. 1857), OLB. Private collection. Edward Cropper was presumably related to the Revd Cropper, minister of the Cross Street Unitarian Chapel, and could have known Waterhouse through that connection.

[13] The Waterhouse family had connections in Elleray who presumably mentioned his name, though it is possible that Waterhouse got to know him through legal circles or possibly through John Green of Ambleside. A connection with John Green is suggested by the gift of an organ to Rydal Church by G. G. Cunningham some time before 1870. It was John Green who commissioned Waterhouse to restore the finials of that church in 1856, and he is presumably the J. F. Green of Moss Hill, Ambleside, who was a potential client of Waterhouse in 1868. He may possibly be the same Green, a surveyor, who took a survey for Thomas Wrigley, another potential client in Ambleside, in 1861.

[14] MS letter, AW to Ed. Wilson (27 June 1857). OLB. Private collection. There are several other letters testifying to the problems, especially on 27 June to G. G. Cunningham and on 22 May, 23 May, and 20 June to Ed. Wilson. The 'last remonstrances' came only 3 days before Cunningham was due to leave his old house, and yet Elleray Bank was not even glazed.

[15] MS letter, AW to Gilbert Henderson (25 Oct 1856), OLB. Private collection. Hearing that Henderson had bought land next to Fawe Park, Waterhouse wrote directly to him offering his services, with the result that he was commissioned to build Rose Trees. Henderson died before the house was completed, and Waterhouse confessed in a letter to James Bell that it was a

plan with ingenious use made of the different levels to maximize the potential of the views. In common with his other Lake district houses it is built of the local stone (in this case partly quarried in the grounds) which gives the houses a quality of reticence that is entirely appropriate in their surroundings.

These early houses already contradict the view that Waterhouse was only a brash and insensitive Gothicist. In fact it was some years before he ventured into a recognizably Gothic mode, and the town house he built for his cousin Sebastian Waterhouse in Liverpool **[29]** (Pl. 18) was in many ways conventional. It is self-confident to the extent of not repeating the pattern of the existing terrace in Catherine Street, and there is an unusual touch in the introduction of constructional polychromy for the rear façade. For the rest, however, it is a very typical mid-century town house, well built, but fitted with stock items, such as balusters that Waterhouse was soon to abandon in favour of his own designs or items carefully selected to suit his modern Gothic. The importance of the house is in the client, who was not only a cousin, but an estate agent and commission agent for the Commercial Union Insurance Company. Sebastian may well have provided useful contacts with the commercial world.[16] If the conventional exterior was designed to please Sebastian, it showed that his architect was able to work in the prevailing taste of the business world of Liverpool.

In the Quaker stronghold of Darlington, Waterhouse first appeared as the designer of a porter's lodge **[43]** for Compton Birchall, for which he was never paid. He then made alterations to existing Backhouse mansions at Blackwell and Shull **[44, 80]**, advised on drawings for Henry Pease's house at Brinkburn **[98]** and installed a new chimney-piece for J. W. Pease. In the later 1860s and 1870s these contacts were to lead to a handsome and profitable series of commissions (Pls. 20, 100, 201). Yet even by the mid-1850s he was able to attract clients who were willing to entrust him with quite substantial works. For them he developed comfortable and attractive house plans and his own brand of modern domestic Gothic.

Christopher Bushell was undoubtedly the most important of these clients, and his mansion at Hinderton **[13]** (Pl. 26) was Waterhouse's first really substantial commission, as well as being the first of his works to be publicized in the *Builder*.[17] Here again he was patient and meticulous in dealing with all the usual difficulties. In fact there seems to have been no more serious problem than poor workmanship in laying the tiled floor of the hall; but no less than fifty-four letters survive relating to the work and dealing with all sorts of details.[18] He wrote to contractors asking their advice, and carefully justified his design decisions to his client. He went into detail over the various qualities of slate for the roofs,[19] accompanied his client to Sheffield to select iron grilles,[20] and explained his new design for the tower by comparison with Italian Early Christian campaniles.[21] It is also clear that he took advice from Sir Thomas Deane, probably at Christopher Bushell's request, over the design of the entrance porch.[22]

pity to have spoiled the land since the plot was now being sold on. Such remorse did not stop Waterhouse designing the large house, Lingholm, on the next plot in 1871.

[16] The cousins evidently worked closely together. They jointly handled the sale of Rose Trees, for example, for Gilbert Henderson's widow.

[17] *Builder* (15 Jan. 1859), 42–3.

[18] The surviving correspondence covers the period from 18 Sept. 1856, when designs were already drawn and an agreement with the builder was being prepared, to 9 Sept. 1861, when the work was substantially complete.

[19] MS lettter, AW to Christopher Bushell (4 Dec. 1856), OLB. Private collection.

[20] MS letter, AW to Christopher Bushell (29 Nov. 1856), OLB. Private collection.

[21] MS letter, AW to Christopher Bushell (Mar. 1857), OLB. Private collection.

[22] MS letter, AW to Christopher Bushell (9 Sept. 1861), OLB. Private collection.

This sort of close attention was typical and Bushell, like a number of others, returned to Waterhouse for additions to his house, and, much later, was involved, as Vice-President of the Council, in securing for him the commission to build the Liverpool University Buildings (Pls. 151–3). Yet this is not to suggest that Waterhouse took great pains over his work simply with an eye to the future. He was clearly aware that satisfied customers bring in business, but the determination to build works of quality was no cynical calculation. However, he was fortunate that his early clients came from a level of society whose social position and wealth were both advancing, so that one commission was quite likely to lead to another. The legal profession was particularly important to Waterhouse, and one of his early houses in Liverpool, New Heys **[103]** (Pl. 98), was for the lawyer W. G. Bateson. Just as at Hinderton, the correspondence reveals thorough care with the details and constant reference to the client's wishes. And W. G. Bateson was presumably related to William Henry Bateson of Liverpool who became Master of St John's College, Cambridge, from 1857–81; so the commission for New Heys may have brought Waterhouse into contact with yet another sector of the rising middle classes; though it seems unlikely there was any direct connection here with his Cambridge commissions.[23]

Waterhouse's legal connections were also important, not only locally, as we shall see with the Manchester Assize Courts (Pl. 33), but nationally. His brother Theodore and his uncle William Bevan both practised in London, and by the late 1860s he was gaining commissions from senior law officers. The most important of these legal patrons was Roundell Palmer, later Lord Selborne, Solicitor-General 1861–3 and Lord Chancellor 1872–4 and 1880–5, for whom he designed much of the estate at Blackmoor, Hants **[192, 232, 266, 267, 328, 329]**. Once again the commission began in a small way with alterations to Blackmoor Farm House. Once the house was habitable Sir Roundell was able to move in and develop his plans for the whole estate. A new mansion to be built round the old farmhouse was designed in 1866–7 and erected to a new design 1868–73 (Pl. 106). New cottages and a schoolroom were built in 1866–8 and St Matthew's Church followed in 1867–70 with a vicarage (1867) and a new school (1870). It was a very substantial series of commissions and the accounts record no less than thirty-one site visits by Waterhouse. Later Lord Selborne was instrumental in gaining for his architect commissions for St Paul's School **[492]** (Pl. 143) and the City and Guilds' College **[490]** (Pl. 144).

Equally typical of Waterhouse's pattern of patronage was Richard Christy, a barrister of Lincoln's Inn, with strong family connections in commerce. Christy practised law in Manchester from 1857 and held professorships of modern history, political economy, and commercial science at the embryo Owens College. Waterhouse did some work for him at Fairfield in 1859 **[64]**; and he may well have been involved in the selection of Waterhouse as architect for the Assize Courts **[67]** (Pl. 33)[24] and Strangeways Gaol **[113]** (Pl. 37) as well as for Owens College **[89, 321]** (Pl. 147), whose Christy Library **[605]** is one of his last buildings. Having built for one member of the family Waterhouse also got commissions from other Christys—a new wing in Bournemouth **[193]**, a house in Kingston **[202]**, offices in Old Bond Street **[481]**, and even the rebuilding of Britwell Court **[564]** for W. Christie-Miller.

[23] Waterhouse did no work for St John's during Bateson's Mastership, though there was a small commission for unspecified work in 1887.

[24] It must be noted that Richard Christy was not among the 13 members of the building committee recorded on the memorial tablet now preserved in Manchester Crown Court. However, given his position in the city, it is a fair assumption that he was at least aware of the discussions at the time of the competition and able to make informal comments.

In addition to his Quaker and legal connections there was also a considerable overlap with the parliamentary Liberal Party.[25] His friends and patrons here included men such as George Grenfell Glynn, second Lord Wolverton whom Waterhouse came to know through his brother Theodore. Glynn, some six years older than Waterhouse, was Liberal MP for Shaftesbury (1857–73) and his client for Iwerne Minster **[445]** (Pl. 104). Another important client was Julian Goldsmid, Liberal MP and millionaire, whom Waterhouse probably also met through Theodore.[26] He did work for him at three London houses **[299, 419, 448, 513]** (most extensively at 105 Piccadilly), as well as building his country house and a villa at Cannes **[400]**. A. W. Peel, the Liberal MP, was another client. Then Lord Grosvenor was also a Liberal MP until he succeeded his father as Duke of Westminster; and Lord Selborne, though an independent, served in Liberal administrations. Also one should not forget that Gladstone himself was the son of a Liverpool merchant: and Waterhouse was involved with at least two of his cousins, Robert Gladstone who had offices in Cross Street, Manchester, and Murray Gladstone who was one of the donors of St Seiriol's, Penmaenmawr **[224]**.[27] Finally Sir Joseph Whitwell Pease of Darlington was another important client, a Quaker and a good friend who sat as a Liberal MP 1865–1903.

This sort of powerful backing undoubtedly helped Waterhouse, especially in public commissions; but even so it was not simply a case of having friends who would favour him. Joseph Pease cannily tried the young architect out by asking him to design the very modest reading room and teacher's house in the Quaker-dominated resort of Allonby **[92]**. Only when that was satisfactorily completed was he given the series of commissions at Hutton near Guisborough **[176, 272, 367, 368]** (Pls. 108, 112, 119). But Joseph Pease was probably his most constant and most typical client. A Quaker businessman, he had extensive interests in Darlington and Teesside. He was one of the Middlesbrough owners who developed that whole new town, and influenced the decision to ask Waterhouse to design the town hall **[390]**, rather than initiate a competition.[28] When that came to nothing, it was Joseph Pease who gained Waterhouse the commission to design Middlesbrough Grammar School **[389]**. Later still he was involved as a trustee in commissioning Waterhouse to design Guisborough Grammar School **[541]** and he may even have had a hand in developments at Saltburn.

Powerful industrialists such as Joseph Pease made the bridge in Waterhouse's clientele between private and corporate commissions. This overlap between private and corporate patronage had first become common in church architecture when the incumbent or building committee would be the actual client, but the money would be donated by one or more local landlords. Waterhouse had experience of this in the rebuilding of Brasted Church, Kent **[194]**. The donor there was William Tipping, an old friend of his father, for whom Waterhouse was working at Brasted Park as well as designing farm buildings and converting cottages on the estate **[54, 173, 234]**. The commission for

[25] Waterhouse's own political allegiance seems to have been largely Liberal, though he was not an extremist. He was able to intercede with Liverpool Tories for his young friend Hamo Thorneycroft 'though they think me a sad radical' (MS letter, AW to Hamo Thorneycroft (12 July 1881), Leeds City Art Galleries, Henry Moore Centre). Thorneycroft did not in fact get the Liverpool commission though he did later provide the statue of the donor for Waterhouse's Turner Memorial Home in the city.

[26] Both Theodore Waterhouse and Julian Goldsmid studied at University College, London, and will certainly have met at that time. Goldsmid was deeply involved in Theodore's Searle Street Improvement Company; and Lady Goldsmid was involved in supporting the foundation of Girton College, Cambridge.

[27] Gladstone's brother Robertson Gladstone was a mayor of Liverpool; but Gladstone himself is known to have spent holidays at Penmaenmawr as a guest of Samuel Darbyshire, one of the chief donors for the church, so it is likely that he will have seen Waterhouse's church for himself.

[28] *Building News* (20 June 1873), 720.

the local National School flowed from the same source **[76]**. But William Tipping was also a major shareholder and a director of the London and North Western Railway. His influence was at work in the design of the Company's hotel at Lime Street, Liverpool **[274]** (Pl. 203). The commission was won in a limited competition, and it was Tipping who suggested Waterhouse be invited to compete.[29]

Well connected though he was with the professional and commercial middle classes, Waterhouse lacked an easy entrée into the world of aristocratic patronage. Yet, in the third Marquis of Westminster, he found a noble patron almost as wealthy as William Burges had found in Lord Bute. The direct contact appears to have been the second Marquis's commission to design, or redesign, the signboard of the Grosvenor Hotel in Chester **[198]**.[30] As usual, the sources even of the second Marquis's interest were earlier. He had an interest in the church at Penmaenmawr which Waterhouse was building at the time.[31] Whatever the connection Waterhouse was assiduous in his treatment of even such a minor commission as the sign. Almost immediately afterwards he was involved in the proposed rebuilding of the East Gate of Chester as a memorial to the Marquis **[199]**. Then in 1869, as soon as Hugh Lupus Grosvenor succeeded as third Marquis, he began the lavish series of works that resulted in the complete rebuilding of Eaton Hall **[315]** (Pl. 84).[32]

The skills that enabled Waterhouse to expand his circle of interest were, however, the skills of architecture. Undoubtedly the foundation of his career was the strong Quaker presence in the business circle of the north. He was fortunate that by mid-century these wealthy magnates were beginning to forsake the extreme simplicity of the old Quakers and to set themselves up in mansions of solidity and comfort. They were, however, not easily satisfied with poor workmanship or inefficiency; and, without his customary meticulous attention to detail, it is unlikely that Waterhouse would ever have been able to build on the grand foundation that was offered him. Build he did until by 1860 he had moved his drawing office into more spacious rooms that he had just refurbished under the Friends' Institute in Mosely Street **[110]**. By 1865 he, himself, was ensconced even more grandly in rooms over the Royal Insurance Office **[108]** (Pl. 28), a handsome building of his own design in King Street at the heart of commercial Manchester. In terms of that city he had most definitely arrived.

[29] He was also clearly involved in the adoption of Waterhouse's design and estimates (MS letter, AW to W. Tipping (26 Feb. 1867). OLB 753. RIBA Library. Waterhouse suggested he should prepare an elaborate letter setting out the reasons for his estimates which Mr Tipping would then show to the Chairman. In this context it is interesting to note that Waterhouse himself, like many of the Victorian middle classes was a railway shareholder, of the Midland Railway. He designed The Brand at Loughborough for Alfred Ellis, son of the Chairman of the Midland Railway. His collaboration with Tipping, however, evidently worked and resulted in one of Waterhouse's largest commercial buildings.

[30] MS letter, AW to the Rt. Hon. Earl Grosvenor, MP (10 Feb. 1865). OLB 492. RIBA Library. A further possible contact would be through Lord Egerton of Tatton Park who was involved with the extension of Knutsford Gaol on which Waterhouse was working in 1865.

[31] MS letter, AW to the Rt. Hon. Earl Grosvenor, MP (17 Feb. 1865). OLB 542. RIBA Library. The commission for the Grosvenor Hotel sign has an oddly exact parallel in Waterhouse's work for Samuel Brooks, a wealthy Manchester banker, who first commissioned a signboard for his King Street offices (1863) and then the ambitious design for St John's Church, Brooklands (1864) which was continued by his son.

[32] It is just possible that, as a Balliol man, the young Marquis had been impressed by the work Waterhouse was doing at his old college. However, it is hard to see how a collegiate building could influence a client for a private mansion.

4

Some Early Commissions: The First Approach to Styles

Before he could aspire to a major commission, or even to offices in a building of his own designing, Waterhouse had to prove himself by getting work done for his closely connected circle of potential clients. His early buildings represent a fairly standard range of types; and a detailed study of them shows his growing ability to adapt styles for modern needs.

On completing his pupillage Waterhouse continued to use his student scrapbook; but now it was filled with cuttings about his own buildings. The first is a report from the *Manchester Guardian* on the Architectural Exhibition in London. Waterhouse was proud enough of his work to send in his drawing of a set of fireproof offices and stores for the sugar refinery of Messrs Binyon & Fryer **[7]** (Pls. 21, 22).[1] His design was a fairly conventional two-storey base in brick and stone, supporting four storeys of brickwork diapered like the Doge's Palace.[2] This was constructional polychromy of just the sort that Ruskin was preaching; and suggests that Waterhouse was firmly committed to the new belief. It would have made a striking addition to the Manchester streetscape; but the upper, Ruskinian, part appears to have been separately designed, and may even, in 1855, have represented only a potential development, as the warehouse was not completed until 1863 **[154]**.[3] Its hybrid nature did not escape all the critics:

> No. 63, the street front of fireproof offices and stores now erecting in Chester Street, Manchester, by Mr Waterhouse, the ground floor, it is true, has a somewhat nineteenth century aspect, but all above it is very striking; . . . a mass of wall four storeys high, diapered with lozenge patterns of red and white brick, and pierced with small plain round-headed windows, which has considerable breadth of effect.[4]

[1] Thomas Binyon was a Quaker and presumably related to Brightwen Binyon, who was one of Waterhouse's few pupils. It is possible that Binyon failed as a businessman for there were difficulties in getting his bill paid, and subsequent commissions were for the new partnership of Fryer Benson and Forster.

[2] To call the building stylistically conventional is, of course, to ignore the innovation of providing an almost continuous row of windows at first-floor level. In structural terms this was an act of some daring since the front wall had to bear the weight of the 4 additional storeys of the second stage.

[3] One drawing, signed 'A. Waterhouse Archt. 1855', is in the Victoria & Albert Museum (see J. Physick and M. Darby, *Marble Halls* (Victoria & Albert Exh. cat. 1973), 100). A painting in the Manchester City Art Gallery, also signed 'A Waterhouse Archt. 1855', shows only the ground and first floors. The office accounts record substantial alterations and additions to the sugar warehouse in Chester Street for 1863–5, by which time Thomas Binyon's interest had ceased and the firm was Fryer Benson and Forster. It is possible that Binyon was unable to build the whole structure at once and needed an architect who was realist enough to accept a commission in stages.

[4] *Critic* (1 Feb. 1856).

If this wholehearted Venetian Gothicry was only what Waterhouse hoped to build, he was learning that style alone was not enough to result in buildings in the hard-headed world of business, and he was realist enough to see that he must first provide whatever his industrial clients wanted. His Institute at Droylsden **[50]** (Pl. 23) can barely be called Gothic, though it is certainly utilitarian. It already shows a marked Waterhousian character in its straightforward use of patterned brickwork and the bold dormers that he drew from the vernacular buildings of northern France. He did, however, find a better outlet for Gothic in a pair of cemetery chapels at Ince in 1857 **[14]** (Pl. 25).[5] The principal chapel, for the Church of England, was designed in what should have been an ecclesiologically approved 'early English Gothic' style and, according to the local press, presented 'a very unique and appropriate appearance'.[6] Antiquarian ecclesiologists would have been horrified by his heavy treatment, for the little chapel is loaded with a full-size arcaded porch, awkward lancets, and ungainly tracery. The second chapel, a short distance away across the grass, was in what was described as Norman Gothic. Its mouldings and bold chevrons are, if anything, heavier yet, more typical of the 1840s than of the 1850s. It would be easy to dismiss this as the awkwardness of ignorant immaturity; but that is inconsistent with the assurance of Waterhouse's Venetian design of the year before, and hardly fits with his careful study of Gothic at home and on the Continent. The most convincing explanation seems to be that he was determined to adapt Gothic so as to give a duly sombre air to the place of death. He was certainly not bound by formal conventions, for in the Norman chapel he combined two separate chapels, one for Roman Catholics with an altar in the apse, and another for the nonconformists in the 'western' end of the chapel.[7]

It seems as though he already saw Gothic forms as his servant not as his master. It was not a course that would endear him to the church building world; but he was not in the mould of Butterfield, Pearson, or Street. However, by the early 1860s he was winning church commissions. By then the massing of his buildings showed greater assurance; though there were still deliberately quirky and unecclesiological details. At both St Seiriol, Penmaenmawr **[224]**, and St John, Brooklands, in Manchester **[152]** (Pl. 27),[8] Waterhouse turned his back on Ruskinian polychromy in favour of regularly coursed rubble-faced ashlar. St Seiriol's was unusual in having a free-standing tower balancing a north transept. In this it matched Street's better known church of St James the Less, Westminster; but Waterhouse opted for a bolder though less decorative solution for the little slate township on the north Welsh coast. His church was heavily buttressed, with battering on the north, downhill, side, but the bold gables of nave and south aisle were admirably set off by the free-standing south tower. His idiosyncrasy showed in tower buttresses that turned into tourelles, and in the choice of a saddleback roof. St John, on a flat green-field site, was given a conventional layout with broad nave and transepts;[9] but there were once again deliberately individual details in window tracery of

5 The commission was won in competition, though it would seem no professional assessor was employed.

6 Undated cutting in 'Architectural Fragments' scrapbook. Private collection.

7 The decision to combine the chapels was apparently taken as an economy measure by the Burial Board. They reckoned the population at 3,434 Anglicans, 731 Catholics, and 632 nonconformists. The arrangement is certainly unusual, however, and would have worried a more committed goth than Waterhouse.

8 The church of St John, Brooklands, was a direct commission from the Manchester banker Samuel Cunliffe Brooks as a part of his planned development of a new suburb. The Penmaenmawr commission also had links to the Liverpool and Manchester business scene through Murray Gladstone, a major donor, and Vernon Darbyshire.

9 From the exterior the transepts are obscured by the gabled vestries that are set immediately to the east.

plain circles above conventional lancets. It was a pattern hardly ever used by Victorian church builders,[10] and showed again how little Waterhouse felt bound by conventional precedent.

The interiors of both churches were more typically Victorian, being of polychrome brick. St John had the approved deep choir and raised chancel, and was particularly rich with tiled floors and diapered walls. But then this was intended to be the centre-piece of a respectable suburb, and was handsomely funded accordingly. Even so its broad aisleless nave and shallow transepts under the spreading roof seem to have more in common with chapel than with church building. Waterhouse, however, did have as a pupil Ernest Geldart, who was later ordained into the Church of England and became an expert on liturgical fittings.[11] And his own interest in church design was already clear from his student scrapbook. At Brooklands his eagerness produced several alternative schemes, including one with a cruciform plan, central tower, aisled nave with clerestory, and an apsidal chancel.[12] It would probably have outstripped even the developer's handsome budget, but it showed his concern, already standard by 1863, to offer clients choice.

Waterhouse's readiness to break with the conventions of ecclesiological Gothic may indicate that he was never fully committed to church building. Neither churches nor chapels ever formed a major part of his practice, though he was to build important examples of both. In fact the 1860s also brought him opportunities in chapel building as a result of the Congregational Union's bicentenary move for the erection of memorial chapels.[13] He built three chapels in the suburbs of Manchester, which showed his stylistic range. At Ancoats, for his friend Joseph Thompson, Waterhouse designed an imposing building with Gothic windows and a tall tower at the apex of the tapering site **[105]** (Pl. 29). Alas it only survived till 1866, when it was demolished almost unused to make way for a railway extension. At Rusholme he put up a small chapel with a corner tower in the round arched style that he was later to make his own **[164]** (Pl. 30). Both these chapels are interesting for their planning and will be discussed later.

The third of Waterhouse's early chapel commissions was different again, and demonstrated equally clearly his approach to design. Chapels were almost invariably associated with schools, and at Besses o' th' Barn, near Prestwich **[163]** (Pls. 31, 32), he used them to enhance the effect of a relatively modest chapel. Perhaps this was the sort of architectural problem that appealed to him. The chapel itself was a rectangular barnlike building with a west gallery and two smaller galleries on either side of the dais in spaces that were virtually shallow transepts. These were reached from the schoolroom, while a rectangular staircase turret led to the west gallery and also gave some movement to the façade. There being no chancel, the schoolroom was set at right angles across the eastern end of the chapel allowing the whole to tell as a single mass with two great gables to the side.[14] This was the plainest of his three early chapels, of red brick with simple stone banding. Even so it was given idiosyncratic Gothic features. The west window was formed as a rose in simple plate tracery, but set under a heavy pointed arch and clasped tightly between plain butresses. It was a remarkable composition if undeniably effective, and, since it was designed after his acclaimed Assize Courts, it can hardly be seen as poorly managed Gothic. Rather it seems to be another demonstration of

[10] Among the few possible medieval precedents are the church at Lynchmere, Sussex, and the old Deanery at Norwich, both of which have two light windows with a plain circle, and both of which are in the approved early middle pointed style of the 13th cent.

[11] It appears that Ernest Geldart did most of the detailed drawings for St John. In a similar way it was G. T. Redmayne, also an Anglican, who did the drawings for Waterhouse's rebuilding of St Martin at Brasted in Kent.

[12] A drawing survives in a private collection.

[13] The Congregational Union celebrated its bicentenary in 1862.

[14] These were soon after increased to 3 when a further schoolroom was added.

Waterhouse's immediate perception that style should follow after function. In this case the west wall had to contain the depth and thrust of the gallery, so buttresses were essential; and the gallery, which also contained the organ, precluded the making of a major west window.

After 1858 and the Assize Courts, however, commissions had come flowing in. By then Waterhouse was an established architect; and the first development of his style belongs to the years before. After the Chester Street warehouse and the Ince Chapels, came the commission to build Hinderton Hall **[13]** (Pl. 26), where Waterhouse had demonstrated such close attention to his client's wishes and to the management of contractors. The house was a compact block with relatively little movement to the entrance façade; but the roofs and gables that gave the skyline such effect clearly revealed the organization. The tower entrance and stairway stood at the end of a range of lower service rooms, stopped by a cross wing containing the kitchen. The main rooms and family bedrooms, which rose higher, were arranged in an L shape round the tower and part way along the service block. Hinderton was rightly recognized as a successful functional design.[15] Tall and solid in rubble-faced pink sandstone it has undeniable character, but was never a showcase of style. The steep roofs and pointed lights of the staircase mark it out as Gothic, but probably the most Gothic thing about it was the lettering on Waterhouse's elevation.

Style, or the Gothic style in particular, was not, it seems, Waterhouse's first concern. In 1858 he designed his own house, Barcombe Cottage, Fallowfield **[52]** (Pl. 17), into which he moved with his bride, Elizabeth Hodgkin. There, surely, he could let himself go and demonstrate his particular fancy. The fancy was for a rambling but suave version of the Swiss style with low pitched roofs that had been popular in the days of Loudon. Inside there were Ruskinian Gothic touches, his and Bessie's portraits on corbels over the sitting-room fireplace; but the house was emphatically designed as a home first, and as a shape second.

Another early commission was the Bradford Old Bank **[24]** (1867), which Waterhouse designed for Alfred Harris of Bingley and his uncles, the Bradford champions of Ruskin and the Pre-Raphaelites. With its round arched openings, this bank developed the conventional Italianate mode into something approaching the revived Romanesque that Waterhouse was to make his own. He was not prepared to work as the slave of either the old Classical or the new Gothic fashion. However, Gothic was the style of the moment, and where buildings were needed to appear up to date, he was quite prepared to go in for Gothic in a serious way. The little Mechanics Institute at Bingley **[125]** (Pl. 24), for which Waterhouse was also commissioned through Alfred Harris, was, in contrast to the bank, a serious and simple piece of Gothic.

The Royal Insurance Building in Manchester **[108]** (Pl. 28) of 1861 was another important commission that Waterhouse dressed in Gothic. Standing on King Street this new office building would be very visible and must make an impact. Waterhouse adapted the Renaissance palazzo design that was already common for commercial buildings, and designed a similar façade in simple Gothic. He was later to repeat this scheme in Leeds **[336]** and to develop it for banks in Darlington **[175]** (Pl. 20),[16] and Leighton Buzzard **[204]**. But in 1861 he was probably not planning to make his name

[15] The house was published in *Builder* (15 Jan. 1859), 42–3. Its plan was later illustrated by Robert Kerr in *The Gentleman's House* (1864), pl. 29, 443, where it was described as skilfully contrived.

[16] Early sketches of the Darlington bank show a High Renaissance design to the same plan and fenestration; the heavy cornice and window treatment are much as Barry might have executed (drawings in possession of Barclays Bank). This choice of styles as in a competition shows the clients' continuing uncertainty as to what was appropriate. Once J. C. Backhouse was convinced of the appropriateness of the style Waterhouse finally adopted, he maintained it in other branches. The former Backhouse's Bank in Bishop Auckland, by G. G. Hoskins, *c.*1870, is very similar to the Darlington branch.

as a designer of offices. The Royal Insurance building, however, offered interesting design problems, for the custom built office was still a relatively new type,[17] and Waterhouse now had to work out the distinctions between the traditional warehouse and the new commercial office. He had to combine a large public office with space for clerks and managers, and he had to provide unspecified lettable space above that would help to finance the project.[18] His solution was a symmetrical façade in simplified Venetian Gothic. An arched entrance at each end led either into the public office or up to sets of chambers on the floors above. The offices needed as much light as could be managed, so their windows were set in a Gothic arcade that extended across each floor, while a dormered attic added drama to the streetscape. It was a characteristically simple solution, and was so frequently copied that the comparative novelty of Waterhouse's building gets forgotten. At any rate he was sufficiently pleased with it to take rooms there for his own office. He was now well placed to meet important clients, while his draughtsmen, in their rooms at the Friends' Meeting House, had space to produce the hundreds of drawings that were being required for the Assize Courts and Strangeways Gaol.

[17] Robert Thorne in an unpublished lecture to the Urban History Group, 1984, pointed out that I'Anson's block in Moorgate Street of 1837 was one of the very earliest of the type. However, as late as 1842, 67 Fleet Street was rebuilt as an office and 'so arranged that the upper floors and basement may be let off separately for offices, or the whole occupied as a private dwelling'.

[18] The company were uncertain who their tenants might be. In fact with Waterhouse's help the top floor appears to have been let off as one unit as a clubhouse. There is no suggestion that this involved any special design work.

5

National Acclaim: The Assize Courts and Strangeways

In December 1858 Waterhouse bought a copy of a newly published book by G. G. Scott, *Remarks on Secular and Domestic Architecture*.[1] It was a turning-point in the development of his philosophy and was perhaps the reason why his journey in September 1858 was not to France but to Belgium, where he could see a rich store of civic and secular Gothic buildings. As usual his study was intensive and searching.[2] At Bruges he was up at six o'clock at work on the belfry of Les Halles with its 'magnificent proportions' and fancied he would see 'nothing better'. Its tower and octagonal drum formed a combination to which he was often to return. Les Halles at Ypres was disappointing though 'the largest building I ever saw'. At Courtrai the town hall was being rebuilt in its entirety, which Waterhouse found a 'questionable procedure'. At Louvain he was pleased by the early decorated style of the church of the Dominicans and found the polychromatic work of the Antwerp town hall 'very good'. Ghent he saw mostly by moonlight, then Oudenarde and Brussels. In five days he saw nine towns and returned 'convinced of the application of the Gothic style for this purpose [Town Halls] as for others but certainly [preferred] the quiet dignity of Bruges to the elaborate façade of Louvain for instance'.[3]

Whether he was aiming for a specific commission or not this tour stood him in excellent stead. Manchester had recently been designated an assize town; and on his return he found a committee set up and a competition launched for the new Assize Court of the Hundred of Salford. The editor of the *Builder* reckoned it 'one of the remarkable experiences of our time', second only to that for the new government offices in Whitehall,[4] which had aroused such enormous controversy on the issue of style. On that occasion Palmerston had forced Scott to change his Gothic design for the Foreign Office or give up the appointment. That was in the same year as the Assize Courts competition, and all 107 entries paid close attention to style. A fair number even hedged their bets by submitting two designs in alternative styles, usually based on the same plans.[5] Competitors included E. M. Barry, son of the architect of the Houses of Parliament, Cuthbert Brodrick, architect of Leeds Town Hall, Norman Shaw and William Nesfield, who entered a design which the *Building News* considered the

[1] G. G. Scott, *Remarks on Secular and Domestic Architecture* (London, 1858). Waterhouse's own copy, signed and dated by him, survives.

[2] SK 13 Aug. 1857–6 Oct. 1859, 137–94.

[3] Ibid. 174

[4] *Builder* (30 Apr. 1859).

[5] *Building News* (22 Apr. 1859).

most striking Gothic in the collection. Others such as E. B. Lamb ('cleverness and originality') and G. Truefitt ('novelty and quaint unbridled fancy') had been noted by Waterhouse at the Architectural Exhibition. Then there was Thomas Worthington, soon to become his rival in the town. It seems that even at this early date the presentation of the winner's drawings were thought to give him a lead over his rivals, and it did not go unnoticed that in his perspective view he did not confine himself to the use of sepia as required by the rules. Even so the commission of JPs selected the design by Waterhouse **[67]** (Pls. 33, 34), still virtually unknown and only 29 years of age. The Quaker connection may have been important here, for the chairman of the Commission was Edmund Ashworth, a prominent Friend from the Bolton area for whom he had already done work.[6]

On the first day that the designs were on show at the Royal Institution a thousand or so people crowded through the building. The *Building News* in its first report was somewhat grudging in its praise while admitting that the quiet and pleasing character of the winning design did grow on the mind. Its character was considered not so much English as foreign, following the continental use of hipped roofs;[7] but, in the prevailing climate of a search for a national architecture, Waterhouse's design appealed both to practicality and historic nationalism.[8] The apparent economy of construction was in its favour, whereas doubts were expressed as to the possibility of building the highly ornate design of Thomas Allom, the runner up,[9] within the estimated cost. Mr Waterhouse, it was felt, had sacrificed the picturesque to conventional forms, which were medieval but would only be interesting when ancient.

The symmetrical planning, evident in most of the entries, seems to be the result of the requirements set out by the building committee in their conditions for the competition.[10] There were to be two courts, criminal and civil, and rooms for all the ancillary activities from the judges downwards including judges' lodgings. Waterhouse arranged these in descending order of importance around a central public hall, with the Judges' Lodgings as a separate block, domestic in scale but linked to the main building (Fig. 1). This straightforward arrangement allowed the main block to be symmetrical with a grand entrance pavilion as the central feature of the frontage. Behind rose a tall tower, revealing the mass of the building and acting as a foil for the strong horizontals of the façade. The central hall was 100 feet long by 50 wide with immense hammer beams soaring to 75 feet overhead.[11] Wide openings looked down from the first-floor corridors to the busy concourse below.

Waterhouse's composition was remarkably assured for one so young. Not only did he demonstrate skill in handling large buildings, he displayed in full measure the Ruskinian recipe for colour and ornamental carving. The sepia tinting, ordered for the competition, could never have done justice to

[6] Waterhouse had been introduced to Ashworth by Richard Lane's partner, P. B. Alley, himself a Quaker, who claimed to have been Waterhouse's real mentor during his pupillage.

[7] *Building News* (22 Apr. 1859). The difficulty of defining the styles is evident in the variety of expressions used. Gothic may be one of several nationalities or merely foreign. The terminology for the Classical entries is more precise having had more usage.

[8] See Robert Macleod, *Style and Society* (London, 1971), 20.

[9] *Building News* (22 Apr. 1859), 367. The design, though bearing Gothic detail, was a grandiose Classical composition. A long narrow waiting hall with a central drum and dome formed the spine; and blocks of equal height at either end broke well forward, the space between being filled by a low storey and a central grand porch. The courts lay behind surrounded by corridors and offices, all, including the Judges' Lodgings, contained within a rigidly symmetrical plan.

[10] *Builder* (30 Apr. 1859). The criticism in this journal is otherwise confined to the minutiae of the plans and the lack of segregation of the different levels of society and their sanitary accommodation. Just over 10 years later, the same journal considered the finished building a great improvement on the original design, and remarkably beautiful apart from the sculpture which was too small and crude.

[11] The roof, strengthened laterally by timber lattice girders, was a particularly intricate assemblage of timber.

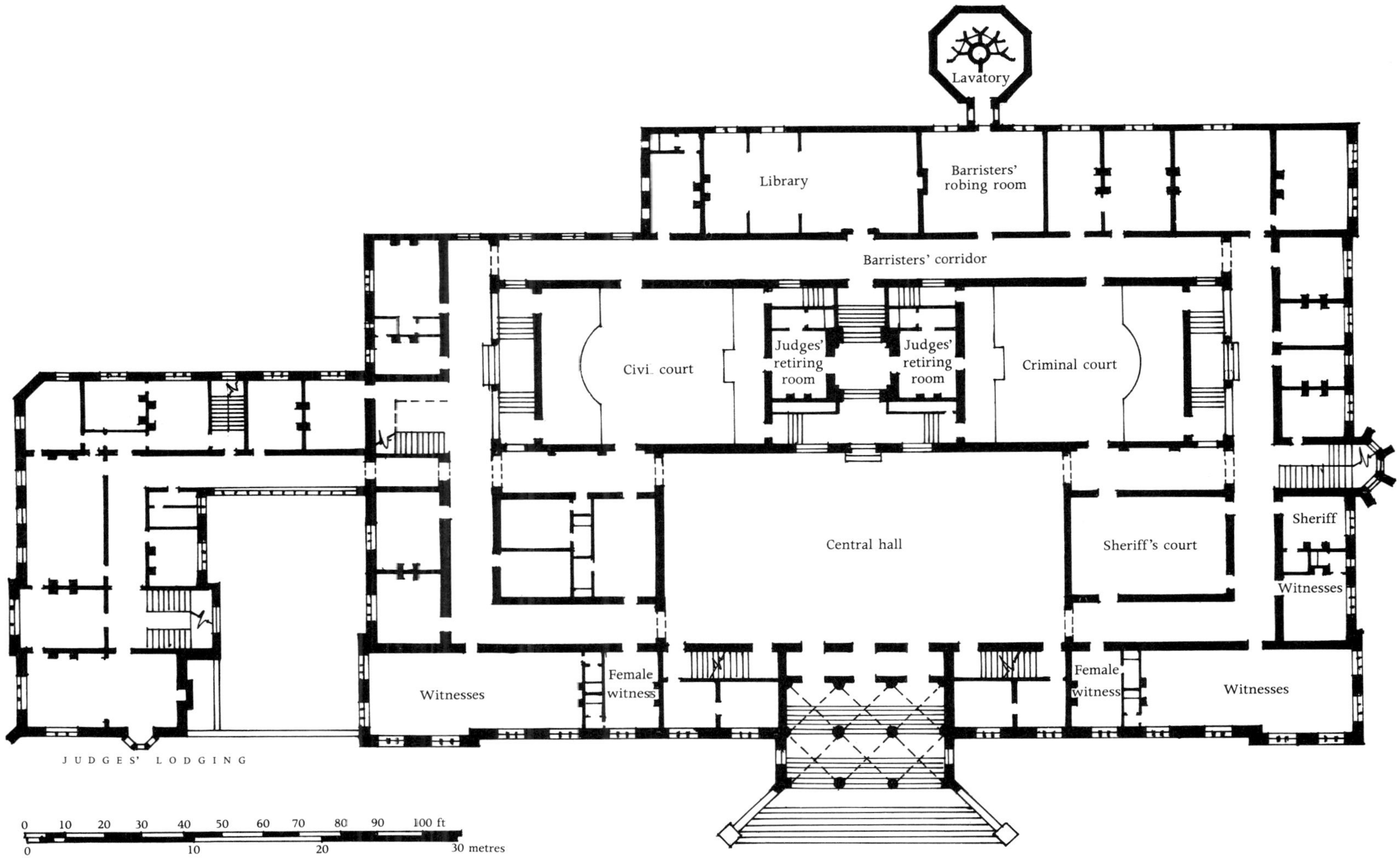

FIG. 1. Manchester Assize Courts (1859) **[67]**, plan of principal floor (RIBA)

Waterhouse's polychromatic system, so he simply disobeyed the instruction.[12] His exterior was to be faced with Darley Dale and greenish blue Forest of Dean stone, to emphasize the horizontals of the decorative banding and string courses and in alternate voussoirs of the window arches. Polished Dalbeattie granite was used for all the external detached shafts and was introduced as decorative chequerwork in the main string course. The interior floor was laid with tiles in chocolate, slate, and drab, and a patterned tile dado ran above a deep stone skirting. The walls were plain but pierced by deeply recessed arched openings flanked with polished shafts and crisp foliate capitals. Most of the ornamental work was done by the O'Shea brothers, fresh from Woodward's Oxford Museum; though in typically cautious fashion Waterhouse demanded that they carve a sample capital before they were entrusted with the work.[13] It is also typical of Waterhouse that the decoration was kept firmly subservient to the composition. Heavy string courses and banding below sill level, an open balustrade at eaves level, and alternate bands of plain and chequer slating on the roof, even the wrought iron cresting on the ridge, all emphasized the horizontality that was carefully stopped by the projecting end pavilions. And Waterhouse showed further independence from Ruskin's Gothic in the repetition of a single window type on each storey.

Probably the richest interior space was the Grand Jury Room in the central pavilion. This had a great medieval style fireplace, panelled dado, and richly patterned walls. Between the exposed timber beams of the ceiling were geometric patterns of plaster; and the whole was decorated by J. G. Crace, whose methods Waterhouse considered brought out the form. He had nothing but 'the highest praise' for Crace's use of colour in architecture.[14] The Judges' Lodgings, by contrast, were composed in a more picturesque manner. An entrance arch led into a courtyard enlivened by an octagonal staircase turret. Inside Waterhouse arranged everything, even designing the tableware himself.[15] He considered the stone screen of the hall to be one of O'Shea's best pieces of work. Both the joinery and furniture were characterized by straight members with chamfered edges, cusped brackets, and geometrical ornament.[16] The functional simplicity is typical of Waterhouse. In an equally practical approach the long drawing-room was in two bays; the division was a structural necessity but at the same time it allowed the room to be used either as one large space or as a smaller more intimate area around the fireplace in the end wall (Pl. 35).[17]

Practicality was certainly one of the key considerations, but Waterhouse was already able to marry it effortlessly with the picturesque. The tall tower, so necessary for the composition, functioned as the main ventilation shaft. When the sheer bulk of the building dictated awkward features like the long barristers' corridor at the rear, this was broken down by the insertion of arches. Faced with difficulties in lighting the Barristers' Library, Waterhouse provided windows high up, and made a feature of them by a frame of open arched ribs springing from the return walls of the bookcase bays (Pl. 36).

[12] A full contemporary description of the building and its decoration is contained in the official *Description of the Hundred of Salford Assize Courts of the County of Lancs* by A. Waterhouse.

[13] Waterhouse also got samples from J. W. Seale and Henry Lane. It is surprising that Thomas Earp, Harry Hems, and the firm of Farmer and Brindley do not seem even to have been considered.

[14] Waterhouse's address to RIBA on the Assize Courts: *Papers Read at the RIBA 1864/5* (1865), 165–76, published in *Building News* (23 June 1865).

[15] Pieces of the tea and dinner services are held in Manchester City Art Gallery.

[16] A few pieces survive in Manchester Town Hall. Much of it is closely related to work from Street's office, e.g. the table designed for Cuddesdon College, Oxford and to designs by A. W. N. and E. W. Pugin. See C. Wainwright and C. Gere *Architect-Designers from Pugin to Mackintosh* (London, 1981), and J. Lever, *Architects' Designs for Furniture* (London, 1982).

[17] This was repeated at Barcombe Cottage, his own house, and in the Mayor's apartments at Manchester Town Hall.

Waterhouse admitted that he had to make a number of changes before the building was complete (Pl. 33).[18] Most were minor adjustments, but one important change was the adoption of simpler arches of an Italian type for the upper windows. He also introduced variations in the window tracery and in the panels beneath to break the monotony of repetition without disturbing its regularity. This approach was characteristic of his work. The most obvious alteration was the removal of the clock turret from the central block. In its first report, the *Builder* noted that this was to be changed because the committee was not in favour.[19] In fact the constructional problems caused by Waterhouse's decision to set it on top of a hipped roof led him to try various solutions before abandoning it in favour of a simple and imposing gable. At its apex he set a statue of Moses and the tablets of the Law; and the final touch was the addition of commemorative statues, for which the sculptor Thomas Woolner joined the brothers O'Shea.[20]

Up to this point in his career Waterhouse's commissions had mainly been small, houses for family and friends, a bank, a warehouse, cemetery chapels, or an educational institute. He had had an introduction to competitions in 1855 with an entry for the Taunton Assize Courts, which was won by G. G. Scott. Now, in a great civic project, he had to address himself to the opinions of the day. After his visit to Belgium he considered Gothic appropriate for this type of building, but he struck out boldly, ignoring the narrow arguments of historicism: 'The designer has not endeavoured slavishly to copy the Gothic of any particular period or country, though the noble civic buildings of the Low Countries and Italy may have given him suggestions.'[21] Later, after the building was complete, he claimed: 'the mouldings and details are thirteenth century in their general character, but whenever I thought that the particular object in view could not be best attained by a strict adherence to precedent, I took the liberty of departing from it.'[22]

He was already prepared to set the practical needs of each building above the demands of precedent, and to ignore emotive calls for English Gothic as the basis for a true national style. This was the key to his success, yet in the Assize Courts he still managed to hold the course set by Pugin's plea for truthfulness, by Ruskin's call for polychromy and traditional craftmanship, and by Street's advocacy of Gothic for secular buildings. Public approval was assured by Ruskin's comment that the building was 'much beyond everything yet done in England on [his] principles'.[23] And once the building was opened in 1865 the architect could feel he was truly a figure on the national scene.

More immediately, the Assize Courts led to the commission to build Strangeways Gaol **[113]** (Pl. 37) and the stipendiary magistrates court **[250]**. The latter provided no real opportunity for urban display since it was tucked away down a side street behind the Assize Courts themselves. Strangeways Gaol, on the other hand, formed a backdrop to his own Assize Courts, and gave as much scope to Waterhouse the practical economist as the courts had given to Waterhouse the artist. The final cost of the gaol was £170,000, £50,000 more than the Assize Courts, so it will have been an

[18] Waterhouse's address to RIBA on the Assize Courts: *Papers Read at the RIBA 1864/5* (1865), 165–76, published in *Building News* (23 June 1865).

[19] *Builder* (30 Apr. 1859).

[20] Little is known of Waterhouse's connection with Thomas Woolner, but some early business letters suggest a fairly close friendship in which Waterhouse felt free to descend on Woolner for occasional nights without warning.

[21] *Manchester Guardian* (19 or 20 Apr. 1859), quoting Waterhouse's written submission.

[22] Waterhouse's address to RIBA on the Assize Courts: *Papers Read at the RIBA 1864/5* (1865), 165–76, published in *Building News* (23 June 1865).

[23] Ruskin's lecture 'Traffic' delivered at Bradford (21 Apr. 1864). Waterhouse's continuing admiration for Ruskin shows in a letter later that year to Alfred T. Jarvis Jun.: 'We had Ruskin in our neighbourhood last night lecturing on Reading. He seemed in good spirits and delighted us all very much. I did not get to speak to him however as he was surrounded by so great a throng of lady admirers' (7 Dec. 1864). OLB 109. RIBA Library.

important job for the practice,[24] which at that time seldom employed as many as eight draughtsmen.

Waterhouse was first approached in December 1861, and promptly took advice from his friend James Bell. Bell he thought, as secretary of the Institute of British Architects, might be able to advise him on fees. 'My dear Friend,' he wrote, 'I have the chance of a prison here: and the question may any day be asked me on what terms I should be willing to undertake the work. Its cost may be £90,000, and I fancy it has been usual with the profession in large works where there is necessarily great repetition and much plain work to except [*sic*] a com^{n} less than the ordinary 5%.'[25] Next he visited the prisons at Reading, Winchester, Holloway, and Wandsworth. Like the two last, Strangeways was to be a large prison with space for some thousand inmates (a separate women's prison was added a few years later), a house for the governor, and all the usual offices. The design was not revolutionary. Waterhouse merely refined the radial plan that had been worked out by Joshua Jebb for Pentonville in the 1840s. Six wings opening off a twelve-sided central hall ensured ease of observation with the accepted system of control and isolation. Waterhouse also took the precaution of having his plans shown to Joshua Jebb, who was by now Surveyor-General of prisons.

Waterhouse's version of Jebb's plan was in a simplified, Gothic style, with a limited vocabulary of detail, as being more economical to construct. Even so Waterhouse managed a degree of architectural display. In keeping with his developing eclecticism, much of the detail was Romanesque, though the main arches of the interior were Gothic and the gatehouse was a version of the typical French chateau gate in banded brick and stone. From this a processional route led between dwarf walls up steps to a terrace, and then into the prison itself. The entry, into the chapel wing which was wider and so more imposing than the others, led up six steps into a vaulted entrance hall. A further set of steps led the prisoner directly up to the first floor of the central hall. This, with its roof supported on giant domed girders, was raised well above the wings to give architectural effect. Finally the heating and ventilation system required a massive shaft,[26] which Waterhouse turned into the tall minaret that is still a feature of the Strangeways townscape.

The real importance of this commission, however, lies not in its architectural character so much as in the opportunity it gave Waterhouse to work out means of economical construction on the grand scale. He had to devise doors, windows, cornices, and archways using the simplest means. Practically everything was achieved in simple brickwork, and he learnt valuable ways of enriching brick structures without the use of elaborate specials. The two court buildings and the gaol together meant that the young architect had covered the whole of a city block, well over an acre in extent, with buildings of his design. The Assize Courts brought him national acclaim as a creative designer, but the gaol will also have impressed potential clients, in the legal profession at least, with his ability as a practical man. The two buildings need to be taken together as the climax of his early career. He was now ready to go on to greater things.

[24] No less than 80 letters survive from Waterhouse's office relating to this commission. Many deal with problems of getting an adequate supply of bricks and with the difficulties caused by a masons' strike. However, given the plainness and repetitive nature of the work the attention given to the job is impressive.

[25] MS letter, AW to James Bell (15 Jan 1862). OLB. Private collection.

[26] The system was installed by Messrs Haden of Trowbridge, Waterhouse's favourite contractors. It is a tribute to its solidity and efficiency that it is still operating today. The brickwork of the horizontal shaft, buried beneath the wings of the prison, is one of the few cases of poor supervision in his work. It was evidently built from both ends simultaneously, but the join in the vault is out by almost 4 in.

6

The Move to London and the Law Courts Competition

In the autumn of 1864 Waterhouse completed negotiations for the purchase of a house in New Cavendish Street, Portland Place. Built at the end of the eighteenth century this Georgian town house is hardly what one would expect of a dedicated goth. Besides Waterhouse had been living in a house of his own design, Barcombe Cottage, in Fallowfield since 1860, and began work on his next house, Foxhill, next door to his parents' home in Reading, in 1867, less than three years after moving to London. Yet his taste was never restricted to the prevailing fashion and even that ardent goth, William Burges, a close friend in the profession, was happy enough in a seventeenth-century house in Buckingham Street.[1]

As a home, 8 (now 61) New Cavendish Street may have been only a stepping-stage; but it was, above all, convenient, within easy reach of Euston for trains to Manchester, and at no great distance from either the City or the RIBA in Conduit Street. It was also a sufficiently fashionable address for a young architect making his debut in London. True he had already 'astonished England' with his Manchester Assize Courts; but metropolitan success, so essential for an architect who was to reach the heights of the profession had yet to be earned. Without it there would be no Foxhill and no Yattendon, though there might very well have been Manchester Town Hall.

It is a mark of Waterhouse's ability that, notwithstanding his northern industrial origins, he did make such a success in the capital. A very few provincial architects, such as E. G. Paley and Bowman and Crowther,[2] had made a national impact through publications without practising in London. Yet in 1860 that was very much the exception and Waterhouse was anxious for a wider theatre than had so far been his in Lancashire and the Lake District. So he set about the move with customary thoroughness. The new house was a deal too large for the family, and, success not being yet assured, they did not set up in style. Evidently they kept no horse as the stables were let off to a Mr Dundas, MP, who lived in Harley Street. The move from Manchester was not, however, a leap in the dark. It

[1] William Burges's sister was married to Richard Popplewell Pullan, who had been articled to Richard Lane at the same time as Waterhouse. Burges's famous dream house in Melbury Road was not begun until 1875, and Burges himself only moved in shortly before his death in 1881.

[2] E. G. Paley (1823–95), whose brother, F. A. Paley, was the author of a number of books on the Gothic style including *Gothic Mouldings* which Waterhouse studied, had a national reputation although he practised largely in the north of England. H. Bowman (1814–81) and J. S. Crowther (1832–93) are best known for their publication, *Churches of the Middle Ages*, but had a nationwide practice, largely for the nonconformist community.

had all the forward planning of cautious Quaker business sense. Waterhouse was in any case familiar with London from his schooldays, and had made frequent visits since his parents had been living in town from 1855–8. This time the ground was very thoroughly prepared.

The 1860s, as Summerson has pointed out, was a time when 'London was more excavated, more cut about, more rebuilt and more extended than at any time in its previous history'.[3] There is no knowing how clear the opportunities appeared to the young architect in his provincial office. But his success with the Manchester Assize Courts had brought him into close contact with the legal profession. His uncle William Bevan already had an established practice as a solicitor in Old Jewry, and indeed had provided early training for Waterhouse's younger brother Theodore. In 1864 that brother had just completed his articles and was shortly to set up in practice on his own at 10 Lincoln's Inn, sharing lodgings with a third brother, Edwin, who was training as an accountant.[4] On the eve of the explosion in investment and mechanized accounting that was fuelling the building boom, and with Theodore's forward-looking interest in the telephone and electric lighting, they must have made a formidable triumvirate. Edwin was able to advise on setting up the new office, but it was Theodore who had the really ambitious plans, involving property speculation linked to major developments.

As early as 1860 Theodore had written to his brother: 'I want to call your attention to a paragraph in today's "Times" relating to the concentration of all the Law Courts at Lincoln's Inn—a thing which is now to be . . . To come to the point, if you work your cards well what is to prevent you having the job for yourself?'[5] There had been nothing like this in the profession since the Government Offices competition a decade before. Now Waterhouse was not only an established practitioner but celebrated for his success with the Manchester Assize Courts, the very type of building that was now in contemplation. Besides he had sufficient contacts in the legal profession to afford him every hope of this great commission. It is a moot point whether it was his contacts or his undoubted success at Manchester that led to his being appointed, in 1865, to advise on the accommodation required for the new Courts; but he resigned hastily on discovering that the appointment would bar him from entering the design competition. However, he was already sufficiently well linked and liked in the legal profession to be designing the alterations to old Blackmoor house **[192]** and cottages on his new estate **[232]** for the Attorney-General, Sir Roundell Palmer,[6] by the middle of 1866. In 1864, however, even that commission was in the future.

Whatever his hopes for a really major building, Waterhouse's move was pinned on something more tangible still, a new bank that he was building in Lombard Street **[179]** (Pl. 39). The commission for Alexander and Cunliffe's Bank came to him towards the end of 1863, work began early in 1864, and the shell was substantially complete by the time he moved to London, so that prospective clients would have a building on a prime site to look at. The bank was in the tradition of the small family banks, many run by Quakers, that were the typical financial institutions of the mid-century. It was a familiar world for Waterhouse, who had already undertaken commissions **[152]** for the banker Samuel Cunliffe Brooks in Manchester,[7] and the senior partner in the London bank,

[3] John Summerson, *The London Building World of the Eighteen Sixties* (London, 1973), 7.

[4] Edwin was later to become the founder of the firm Price Waterhouse. His recently published memoirs give some valuable insights into the business world in which Waterhouse was moving.

[5] TW to AW (18 July 1860). Quoted in *Theodore Waterhouse: A Memoir* (privately printed, 1894).

[6] Roundell Palmer was called to the bar at Lincoln's Inn and served the Liberal Administration as Attorney-General from 1863–6 and as Lord Chancellor from 1872–4 and 1880–5. He was created Viscount Wolmer of Selborne on the occasion of the opening of the Law Courts in 1880.

[7] His precise relationship with Cunliffe's Bank in London has not been traced; but it is assumed that he was at least a relation. The Manchester merchants Cunliffe & Leaf, for whom Waterhouse was to build a warehouse in 1867, were probably also connected.

G. W. Alexander, was also a Quaker. In spite of these links it is not possible to argue that only good connections secured Waterhouse the commission. His experience of office design had begun with the Manchester office of the Royal Insurance Co. **[108]** (Pl. 28) whose London premises were next door to the new bank. He was also already designing banks in Nantwich **[166]**, Darlington **[175]** (Pl. 20), and Leighton Buzzard **[204]** in which he was refining his solution to the problems of combining large banking halls with sets of offices, and of expressing the required sense of worth on the façade.

Alexander and Cunliffe's Bank marks an important development.[8] On its corner site, Waterhouse took advantage of the dynamic changes of direction afforded by an entrance on the angle and two adjacent façades that he refined in several of his later commercial works. In the narrow confines of Lombard Street he was also concerned to get as much light as possible into the offices. He wanted larger windows than would have suited a conventional Gothic façade, and so his elevations are composed in a determinedly eclectic style—a basic Renaissance palazzo with convincingly Gothic detail on all floors. The ground-floor banking hall is separately treated and enriched; while the first and second floors are linked by giant Gothic pilasters supporting a bold round-headed arcade. Smaller second-floor offices fill the level of the entablature, while a gabled attic over the main cornice makes for a picturesque skyline. It is a confidently handled mixture, and the success of the eclecticism suggests a remarkable skill in its designer at this date.[9]

The building also reveals Waterhouse's determination to have a precise control over every detail of his building's appearance. Correspondence with the contractors shows that he knew precisely what he wanted, and was not prepared to accept merely standard workmanship. The carving was in the hands of J. W. Seale, and Waterhouse went to the lengths of stopping the work because the carvers were not following his ideas. 'Too much attention is being given to detail and not nearly eno' given to the mass. The detail is worse than useless destroying altogether the breadth of the work as seen from the street. I wish you could impress your men with the necessity of putting more thought and less work into the stone.'[10] Waterhouse had expressed himself in these terms before at the Assize Courts, where in the end the carving had been the subject of criticism. Now he would no longer tolerate the full Ruskinian approach to carving that left the final result in the hands of the individual mason. What mattered was the overall effect, and mere mechanical ornament, whose scale and depth could be precisely controlled by the architect, was a better bet.

Even with the bank, it is clear that the move to London led Waterhouse to take on several piecemeal commissions not all of which were rewarding. One early job, worth only £200, was for alterations to a dining-room in New Bond Street **[207]** for George Redmayne, his sister's father-in-law. Another small job was superintending alterations to drains at a house in Highgate **[209]** for E. Fry, presumably a connection of the Bristol Quakers. Through Theodore he seems to have been involved in some real estate dealing, advising clients about the purchase of houses in Lancaster Gate, and in a series of subsequent alterations **[239, 240]**. His usual attention to detail paid off here, in that one client was C. P. Stewart, who was later to commission a major country house from him **[439]**. There were also modifications to a house in Wimpole Street **[245]** for R. P. Amphlett, who, besides

[8] Its demolition in 1965 has meant that it has not received the study it deserves. Unfortunately this is one of the schemes for which no drawings survive. Only fragments of the rather rangy carving are left, built into the entrance of the successor Clydesdale Bank.

[9] It has been mistakenly dated to 1880. See C. Dellheim, *The Changing Face of the Past* (Cambridge, 1982).

[10] MS letter, AW to J. W. Seale (11 Apr. 1865). OLB. RIBA Library.

being a neighbour, was one of the Commissioners for the Law Courts competition. Then, also through Theodore, he became involved in the Metropolitan Estates Company. For them he designed a corporate seal **[238]**, and was involved in promoting the private bill that established the Searle Street and Cook's Court Improvement Company, whose buildings in Carey Street he designed **[370, 470]** (Pl. 46).[11] The bank too led to another commission, a house for G. W. Alexander at Stoke Newington **[242]**.[12]

Alexander's Bank and house were not the only London commissions in hand when the new office opened. Even more important was the commission to rebuild the New University Club **[208]** (Pls. 38, 40).[13] The problem, since the club had only been in existence a few years, was to construct a building of appropriate status and with sufficient large rooms within a very restricted budget. Waterhouse was the best man for that approach, and he presented a detailed report on the new club in April 1865. Once again, it shows his typical thorough attention to the realities of the financial constraints. He offered a complete rebuild at an estimated cost of £13,500 or work in two stages of about £7,000 and £7,500, with use of the club maintained throughout. Both stages were finally completed at a total cost of just over £18,000.

There were inevitably problems in grafting a building of a new and larger scale on to the existing urban fabric. The actual building process caused a good deal of discontent among the neighbours, but the end result was a relatively modest yet boldly Gothic structure with a great bay window rising through three storeys as the principal feature of the main façade. However, Waterhouse was eventually prevented from building an oriel window of double height to match at the rear.[14]

The club was an ingenious rather than an elegant building; but it taxed his skill as both planner and designer. And there were other equally public commissions. A large wine and spirit warehouse for H. B. Fearon **[276]** filled part of the northern side of Holborn viaduct, a very visible contribution to the new thoroughfare linking the City with the West End. This warehouse also seems to have been controversial, involving a dozen or more conferences with Horace Jones, the City architect, to sort out the elevation and plans. They were the sort of problems that Waterhouse was adept at coping with. He was always ready to adapt his design if needed, which is probably why he was so well equipped to benefit from the boom in commercial building.

His reason for coming to London, however, cannot have been merely to expand his business. The prospect of a really important commission was the inevitable lure for an ambitious young architect. He had always aimed high and as early as 1856 had considered entering the competition for the Government Offices.[15] It is hardly surprising that, as he set about moving to London, he was busy ensuring he could be a candidate for the great Law Courts competition. He certainly had the right

[11] See The Register No. 1, 285. Accounts from 1862–72. Private collection.

[12] Such a link between a commercial and a private commission was not at all unusual, and houses for bankers formed a considerable part of Waterhouse's early work. In this case it resulted in what was by all accounts an unusual house with an attractive conservatory. Tantalizingly the accounts record stained glass by Heaton Butler and Bayne and carving by Thomas Earp, but the house has gone.

[13] The source of this commission is far from clear; though it may be linked to Theodore's close involvement with University College or to Waterhouse's own success in winning the Cambridge Union Society's building in Feb. 1865. The club had acquired 2 adjoining houses, one facing on to Duke Street and one on to Arlington Street, forming an awkward and irregular plot, and the properties were unsuitable without a good deal of modification.

[14] On this job Waterhouse also had to deal with recalcitrant suppliers of structural ironwork and pacify neighbouring owners who threatened to seek an injunction when the offending girders were finally delivered and dumped in the street obstructing the roadway.

[15] MS letter AW to Alfred Austin, Office of Works (1 Oct. 1856). OLB. Private collection. He sent for the particulars of the competition but there is no evidence from the earliest account book to show whether he submitted a design.

sort of support in the interest of such MP's as Charles Buxton,[16] Frances Powell,[17] and Roundell Palmer.[18] Palmer was involved in preparing the Bill that would bring all the Law Courts together and require the huge building,[19] and, much more important, was one of the five judges for the competition. Of the others, Gladstone already knew Waterhouse both for the Manchester Assize courts and the church at Penmaenmawr **[224]**,[20] though it was always likely that he would favour a design by the High Anglican, G. E. Street. A third member of the panel of judges, William Cowper, First Commissioner of Works, also knew Waterhouse, and could be counted on for support. The other legal judge, Sir Alexander Cockburn, was an unknown quantity, and the Tory, Sir William Stirling-Maxwell, had no links with Waterhouse. However, it was to his Lancashire friends that he turned in the first instance. He had both advice and support from Edmund Ashworth,[21] who had helped secure him the Manchester Assize Courts. Mrs Gaskell even wrote to Ruskin on his behalf,[22] though making it clear that Waterhouse would have deprecated or even forbidden such string pulling.

Given his success at the Manchester Assize Courts it was inevitable that Waterhouse would be one of the architects invited to compete for the London building. He and Street were both extremely keen, and were the only two to stay with the competition through all the changes from the original invitation to six architects in February 1866 until the final eleven began work in July.[23] He was already thinking about his Law Courts design at the time of his move to London.[24]

His design **[278]** (Pl. 41) is a virtuoso display of his characteristic good planning,[25] and is a key step in the development of his modern Gothic. The details are still in the rather angular style that he had used for the Manchester Assize Courts. But the main impact of the building was to be the 'fantastic city'[26] of towers and roofs that would dominate the distant skyline. The great ventilating tower was to rise over 350 feet, and would be supported by other towers, turrets, and steep pitched French roofs in dramatic array. The site was an awkward rectangle with only narrow street perspectives to show off the building. Waterhouse used this difficulty to compose an irregular main façade with its whole eastern half set forward, providing a great deal of useful extra office space. William Burges, whose design was widely believed to be the most successful architecturally, was the only other competitor to use the freedom of the site in this way.

[16] MS letter, AW to Charles Buxton (3 Mar. 1865). OLB 633. RIBA Library. It is clear from Waterhouse's letter that Buxton had already acted on his behalf and spoken to Cowper, the First Commissioner of Works.

[17] MS letter, AW to F. S. Powell, MP (18 Feb. 1865). OLB 556–7. RIBA Library. Powell had advised Waterhouse to write direct to Cowper asking to be among the invited architects, and he also championed Waterhouse in the House.

[18] Roundell Palmer was later a substantial client on his new estate at Blackmoor. In this connection he was a member of the jury and, understandably, no correspondence between him and Waterhouse on the subject of the courts survives.

[19] The eventual fruition of Palmer's plans did not arrive until the Judicature Act of 1873.

[20] See Ch. 3 n. 27.

[21] MS letter, AW to Edmund Ashworth (18 Feb. 1865). OLB 553–4. RIBA Library.

[22] Mrs Gaskell to John Ruskin (Feb. 1865), in J. E. V. Chapple and F. S. Pollard (eds.), *Letters of E. Gaskell* (Manchester, 1966), no. 559. (I am grateful to J. H. G. Archer for bringing this item to my notice.)

[23] The original 6 were: E. M. Barry, P. C. Hardwick, G. G. Scott, G. E. Street, Waterhouse, and T. H. Wyatt. Wyatt and Hardwick declined to enter and were replaced by Raphael Brandon and T. N. Deane. Scott and Barry then withdrew to be replaced by Henry Garling and G. Somers Clarke, with John Gibson and J. P. Seddon as alternates. Somers Clarke declined allowing John Gibson in. Then the number was increased to 12 by the addition of J. P. Seddon, Henry R. Abraham, H. F. Lockwood, and William Burges, while Scott and Barry agreed to rejoin the competition now that the regulations were changed. John Gibson's resignation left the 11 competitors who finally produced designs.

[24] MS letter, AW to Edmund Ashworth (18 Feb. 1865). OLB 553. RIBA Library.

[25] A full description of Waterhouse's design is preserved in the descriptive booklet that he, like the other competitors, submitted with his plans. Alfred Waterhouse, *Courts of Justice Competition: General Description of Design* (London, 1867).

[26] See G. Stamp, *The Great Perspectivists* (London, 1982), 63.

Waterhouse's experience of court design in the competition at Taunton and in Manchester undoubtedly helped him make the building functional, which was always a prime concern. Behind the façade his planning was exemplary (Figs. 2, 3). An outer ring of offices enclosed the whole building. Two internal streets (Pl. 42) isolated the main block,[27] with its twenty-three courtrooms, from the noise of the surrounding area; and the courts were laid out along either side of a huge central hall. But the plan was not merely efficient. Waterhouse had an eye for the dramatic, and here he gave it full rein. The main entry from the Strand was reserved for barristers and parties to the various cases, who climbed straight to the principal floor before crossing the internal street on an open bridge to the southern end of a vaulted transverse hall (Pl. 43). They would then pass along its galleries to the central hall and so make their way up to top-lit courts at a higher level still. The general public were to enter the transverse hall at ground level and could look up to the splendid areas where justice was being enacted literally above their heads. They had no access to the great central hall, but climbed by dark stairs to the public galleries of the courts (Pl. 44).

Waterhouse's spatial organization depended on his monstrous central hall, roughly 300 by 60 feet at first-floor level. Almost all the competitors had found it necessary to insert a central hall, although it was not part of the requirements. Waterhouse was the only one to arrange a second transverse hall at right angles that rose from ground to second-floor level, and met his central floor at the gallery. The two halls were the key to his design, both spatially and in planning. They were the core of his circulation and provided grand and exciting open spaces at the heart of the vast building. Waterhouse used the crossing to split the immense length of the central hall, and simultaneously gave himself the opportunity for more dramatic bridges. The main hall, brilliantly lit by a glazed roof, spanned the transverse hall on one bridge, and was itself divided by higher-level bridges reserved for judges and for barristers. The courts were at a higher level still, flanked by offices that looked either out into the cross halls or into the internal streets.

The whole arrangement was masterly. Although his plan did allow the public into the heart of the building, which the instructions had urged him to avoid, all classes of occupant were carefully managed. He took great pains to explain the means by which he would segregate or bring together the different groups of people who would use the building. 'In the plan I submit a separate entrance is provided for nearly every class which can be considered independent of the rest . . . Judges, Jurors, Witnesses, Registrars, Shorthand Writers, and general public would all have their own special entrances, and might enter and leave the Court without disturbing those around them.'[28] And it was a formidable puzzle to get them safely to their destinations. Waterhouse was quite equal to it, and, for instance, built the public and witnesses' stairs within the same well, their flights interlocking, but with no communication.

The sloping site he used 'so as to obviate the necessity of the Judges passing up or down a single step', providing a separate ramped entry for them direct to the court floor. As he explained, this allowed a further dramatic possibility: 'at the opening of Term the Judges might assemble in their Common Room . . . descend thence by the Grand Staircase at the West end of the Central Hall, and then (as now at Westminster) pass in procession along the latter, and so up the staircases and through

[27] The instructions suggested the concentric layout which became the basis of most of the plans. Barry, Scott, and Waterhouse devised a street between the outer ring of offices and the inner ring of courts. Waterhouse provided a southern, or Common Law, street and a northern, or Equity, street.

[28] Waterhouse, *Courts of Justice Competition*.

the corridors to their respective Courts'.[29] The scarlet robes in procession across the bridge would have made an awesome sight.

It was an enormously complex design, and Waterhouse was well aware that it would not be easy to put its virtues across to a panel of busy men with a heavy parliamentary programme to deal with.[30] He hit upon a new way of advertising his ability, and went to some lengths to provide bound sets of photographs of his designs for every member of the Commission. This was a sharp move, using what was then the very latest in technology. It did, however, involve anxious waits for sunny days on which the photographer could work; and his correspondence reveals his frustration at the delays and his concern whether he would get the albums finished in time. When they discovered how he was stealing a march on them, his competitors hastily put a stop to the project until they could arrange to do the same themselves! The importance of presentation was clearly recognized, and it was an area in which Waterhouse excelled.

The assessors of the competition were asked to judge the entries on their convenience and efficiency, and it was Waterhouse's plan that was preferred by the Bar Committee, who, after all, represented the principal occupants. The architectural press disliked it, critizing his 'thin slices of towers', and was overwhelmingly in favour of Burges's design. However, Waterhouse's planning was recognized as exceptional, and he and Burges, together with E. M. Barry, were reckoned to be the favourites. The two architects who had been added to the panel probably had little liking for Waterhouse on the grounds of his provincial origins;[31] but they were unlikely to favour Burges's continental Gothic either. Eventually, after some months of argument, the panel reached the split decision that called for the elevations of Street's design with Barry's plan.

It was as great a fiasco as the Government Offices competition. The Bar committee had had good reason to like Waterhouse's plan from their experience of his courts at Manchester. There was clear and easy circulation, and he had used his knowledge of the legal process to place the various departments in the most convenient relationships. Barry's plan, on the other hand, set the probate office, which would attract a large daily public, in the very centre of the building. Even the senior law officers who were consulted were unimpressed with Barry, and preferred the work of Waterhouse and Scott, who, like him, had built successful assize courts, at Taunton. True there were suggestions that Waterhouse's glass-roofed hall would have been unbearably noisy in the rain; though his later version of this feature at the Natural History Museum hardly bears the criticism out (Pl. 82). The specialist reports of the fire officer did not single out Waterhouse but T. N. Deane; whereas the specialists in heating and ventilating favoured Scott and Seddon. There were just too many interests at play, artistic and political, for the preferences of the actual users to be paramount.

Theodore, his solicitor brother, seems to have been almost more disappointed than Alfred when in 1867 the commission was finally awarded to Street.[32] However, given the traumas attached to Street's building it is unlikely that Waterhouse's design would have turned out as attractive as it seemed if it had been built. Street had to recast his plan entirely, and paid Waterhouse the

[29] Waterhouse, *Courts of Justice Competition*.

[30] D. B. Brownlee, *The Law Courts: The Architecture of George Edmund Street* (Cambridge, Mass., 1984), has argued persuasively the extent to which the failing Liberal Administration must have been occupied with attempts at the Reform Bill and their problems with the Adullamites.

[31] George Pownall and John Shaw had originally been employed to supply statistical comparisons of the 11 designs. Neither was a particularly gifted architect, and their elevation to the panel was a blow for the Gothic designers, since their taste ran to conventional Renaissance revivals.

[32] MW *Journal*, 209 ' . . . Alfred's designs . . . not being the accepted ones. So many judgements have been given in their favour, that the decision . . . is a great surprise . . . . Theodore was greatly tried.'

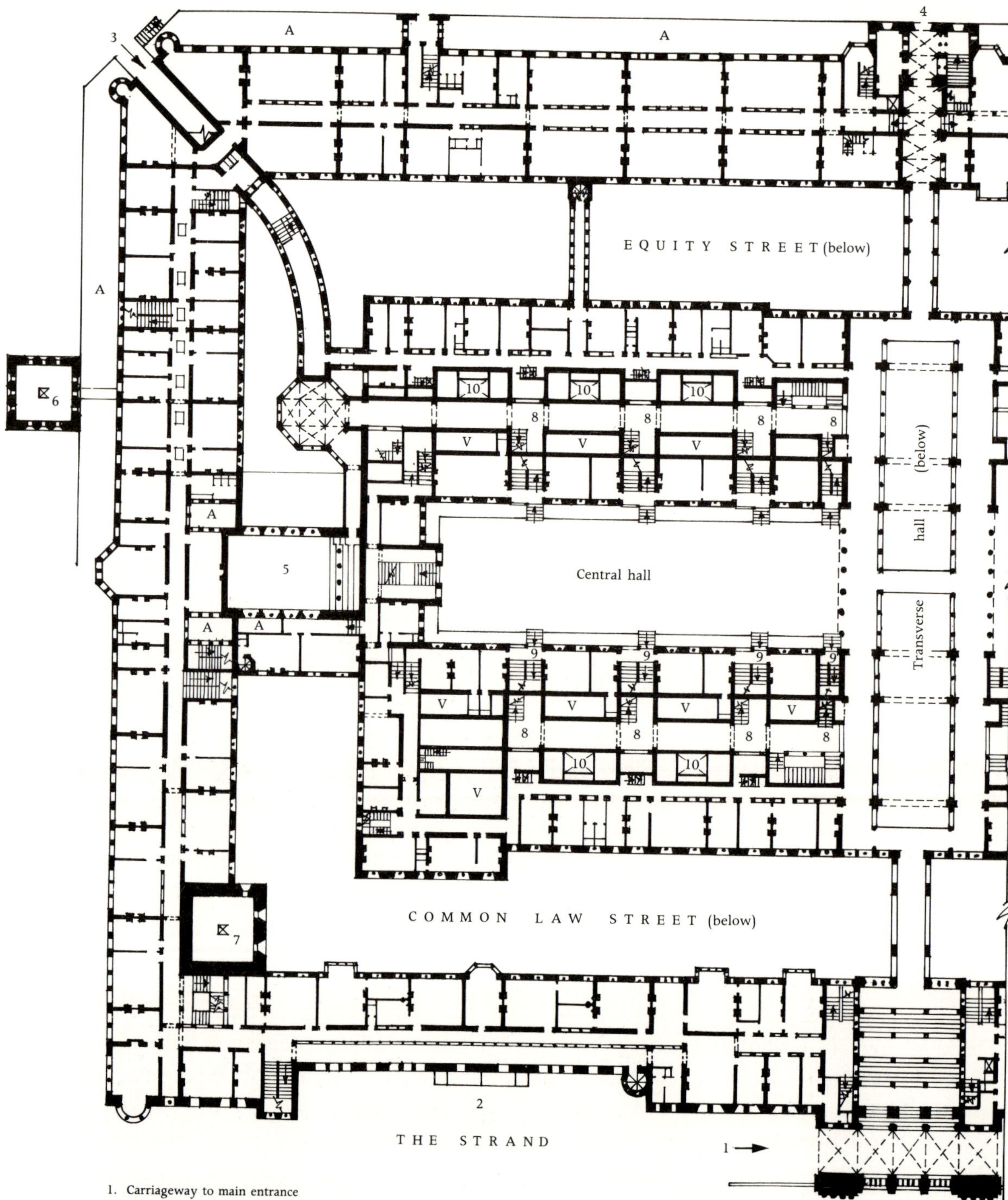

1. Carriageway to main entrance
2. Carriage entrance to Common Law Street (below)
3. Judges' carriageway up to court level
4. Entrance for general public
5. Apellate Court - Gallery level
6. Tower for storage of original wills
7. Tower for storage of will copies
8. Public corridor and stairs to galleries of courts
9. Witnesses' stairs to courts
10. Lifts from ground floor for court exhibits

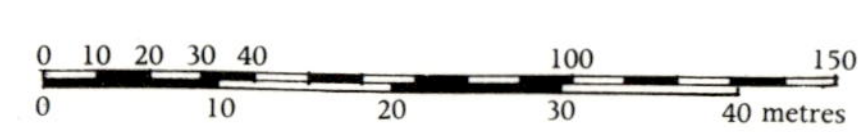

FIG. 2. London, Royal Courts of Justice (1867) [278], half-plan of principal floor

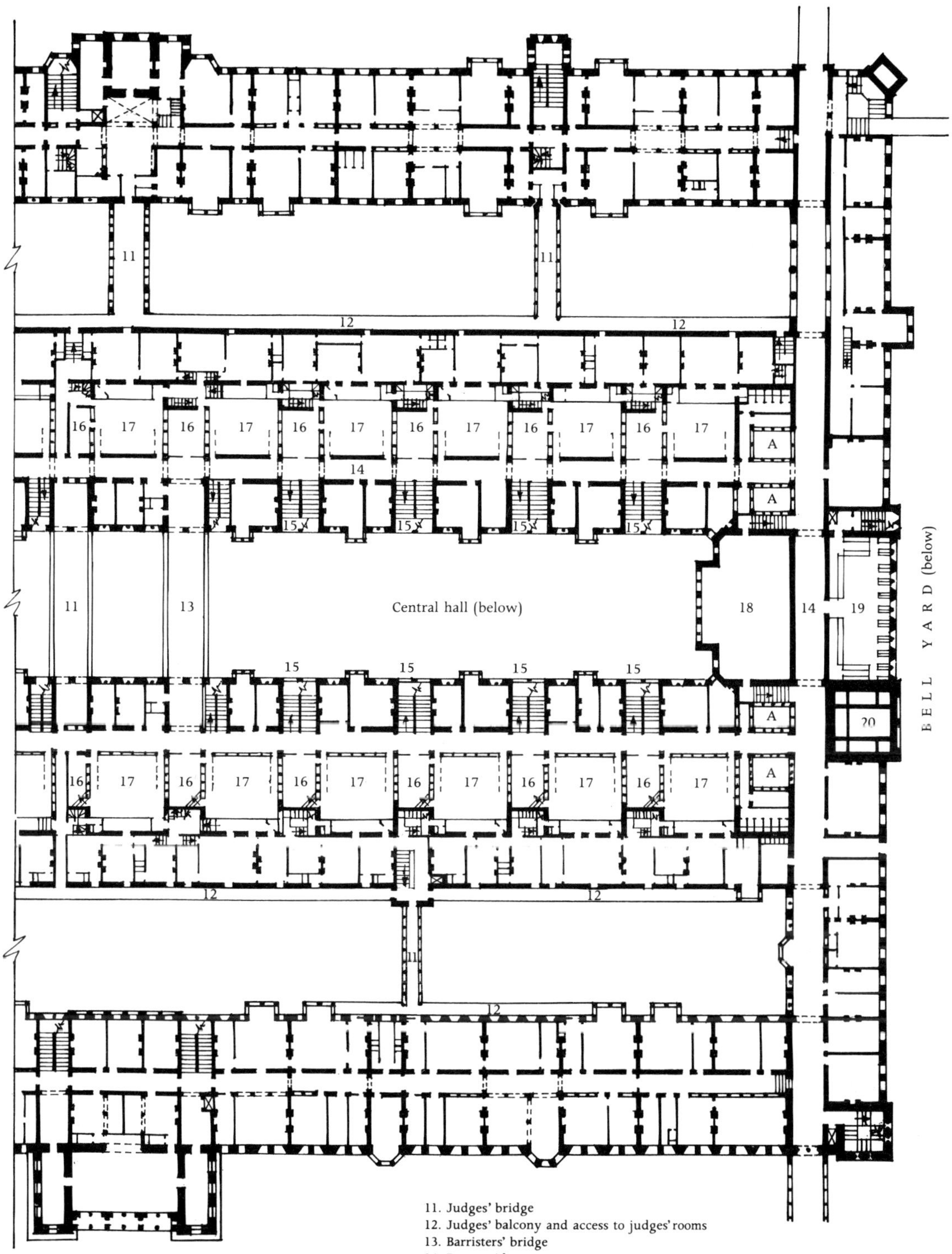

11. Judges' bridge
12. Judges' balcony and access to judges' rooms
13. Barristers' bridge
14. Bar corridor
15. Separate witnesses and public stairs
16. Witnesses' waiting room
17. Court
18. Bar common room
19. Bar refreshment room
20. Smoke tower

A = Area or light well
⊠ = Lift
V = Heating and ventilating voids

FIG. 3. London, Royal Courts of Justice (1867) **[278]**, half-plan of court floor

compliment of adopting all the best features of his. He also had to go through the upheavals of alternative sites. Waterhouse faced exactly the same disruptions over his Natural History Museum, which Austen Layard, as First Commissioner of Works, wanted alongside the new Law Courts as another element in his scheme for an imperial river frontage. Perhaps Waterhouse's skill in handling committees would have secured a more complete building than the one that finally broke Street's health; at least his museum survived better. He had been engaged on preparatory work for that for almost twelve months, and does not seem to have let his failure in the Law Courts competition deter him unduly. And within a few months Street was able to acknowledge his debt to Waterhouse over the planning by selecting his design for Manchester Town Hall.

Meanwhile the Manchester office was flourishing. In the period from June 1865 to April 1866, effectively the first year of his London practice, Waterhouse recorded fees totalling £6,072. The next twelve months were even more successful with a total of £7,776, a perfectly comfortable income even after allowing for the salaries of his staff. During 1865 his brother Theodore discussed with him plans for buying land and building a large development of shops and chambers at New Court, Carey Street. Waterhouse would, of course, be the architect. With the backing of the millionaire Sir Julian Goldsmid, MP, and a mortgage of £176,000 from the Prudential, that project was eventually achieved **[370, 470]** (Pl. 46). His practice was expanding elsewhere too, and at the time of the move to London Waterhouse was designing and building in various parts of Lancashire, in Darlington and Teesside, Bristol, Leeds, Penmaenmawr, Banbury, and Bedfordshire.

Ambition apart, with two major London commissions and work in the offing in Kent and Surrey, the opening of a London office must have seemed imperative. It was in fact opened at least a year in advance of his own move, in chambers in the Royal Insurance Building, 24 New Street, Spring Gardens in Covent Garden,[33] close to the office of George Gilbert Scott. Salary accounts for this office begin in January 1864. It was a small affair, under C. H. Scott who had been with the firm at least since 1859.[34] From the start Scott was assisted by another experienced draughtsman, Evans,[35] who had worked for Waterhouse since 1862. They were joined in March by a third, Bellringer, who had been with the firm a year, and later by the young Ernest Geldart whom Waterhouse had taken on as a pupil.

London was in constant touch with headquarters in Manchester, and every detail of the jobs in hand was referred to Waterhouse. A stream of drawings passed steadily to and fro, though, for safety, he soon insisted on tracings being sent in preference to drawings. When he visited London there were daily reports from Manchester of correspondence and callers. And when he was in Manchester he was busy approving or suggesting alterations to cornices and brackets and, of course, demanding a long series of measurements of the house at 8 New Cavendish Street. The traditional London town house had to be transformed to suit a progressive young architect's home and office. The upper floors were completely refitted, given verandahs and balconies and redecorated as a family home, while the ground floor was turned into the architect's office that it was to remain until 1901.[36]

[33] Some of the office correspondence refers to the office in New Street or Spring Gardens. Confirmation that it was in the Royal Insurance Buildings comes from a letter from Waterhouse to R. B. Dockray with that heading printed (Friends House Library). The connection is interesting given his earlier work for the company and the fact that he also had rooms in the Royal Insurance building in Manchester until 1865.

[34] Waterhouse's salary accounts do not survive before 1859. Payments to Scott in Manchester cease in Aug. 1862 and it may be that the London office opened as early as that.

[35] I am assuming the Evans of the London office was C. Evans. Another draughtsman, F. W. Evans, joined the practice in 1863 but the London salaries do not distinguish the two.

[36] The office then moved to Staple Inn Buildings, newly built by the firm for the Prudential as an investment property.

The most obvious external change was of course the iron verandah and window rails whose authorship is unmistakably Waterhouse's. The correspondence shows that he gave close attention to all the minutiae of the new house.[37] None the less Scott's job as chief draughtsman in the London office was a responsible one. He was clearly acting as site architect for Alexander's Bank and for the alterations at No. 8. In addition the London office seems to have been largely responsible for Bassett's bank in Leighton Buzzard, the stables at least of R. L. Ford's house in Leeds, and houses for Charles Gillett (in Banbury), Lewis Fry (in Bristol), J. H. Fox (in Wellington, Somerset), and James Milne (in Manchester).[38] Evidently by 1867 the London office was running efficiently as the headquarters of the thriving practice. Manchester, in the hands of G. T. Redmayne, was only a branch office. Yet it was in Manchester that the great opportunity finally arose.

[37] Although he wanted the best for his new house, he was careful to exercise sensible economies. The brackets supporting his verandah for instance were the same as those supplied for his work at Strangeways Gaol. MS letter, AW to James Haywood (27 Feb. 1865). OLB 613. RIBA Library.

[38] In the last instance, the fact that the drawings seem to have been largely left to Scott may explain some of the oddity of detail. However, for the most part the drawings were carefully scrutinized by Waterhouse, and he grants permission for drawings to be sent direct to Mr Fox almost as though that were an exception. MS letter, AW to C. H. Scott (16 Dec. 1864). OLB 155. RIBA Library.

Part Two

MODERN GOTHIC MATURITY

7

Manchester Town Hall

THE opportunity to build a major public building was the goal of many an aspiring architect of the mid-nineteenth century. Indeed it was probably the main reason why the widely (and wisely) despised competition system continued to flourish. So Waterhouse's victory with the Manchester Town Hall competition, and the subsequent municipal palace that was built, have a very special place in his work. The building has rightly been called a classic of its age,[1] and long celebrated for its sophisticated planning. It is probably his greatest work; but it is in many ways a building of contradictions which makes it yet more interesting—Gothic in form yet modern in structure and servicing, functionally economical but highly expensive, it successfully combines palatial splendour with a range of simple office spaces.

By the time of the competition Waterhouse was already well established as an architect in London, and hardly needed to risk failure in another contest. Equally, he had already more than made his mark in Manchester with the Assize Courts (Pl. 33), and could have turned his back on that city without discredit. Yet he did enter, and, having won, produced a building of rare distinction, and one that continued to claim his attention even as late as 1895.[2] His greatest contribution to his adoptive city was a commission that occupied more of his, and his draughtsmen's, time than any other, and displays all his most creative talents. However, it needs to be appreciated in the context of his career as a whole; and the roots of this extraordinary masterpiece run deep.

In one sense it is a mistake to think of architects of Waterhouse's period as provincial or metropolitan. Transport had become so easy that a man of energy, like Scott or Waterhouse, could easily manage simultaneous commissions in Manchester and London, Bristol and Bradford, Darlington, Guisborough, Reading, Cheshire, Somerset, and rural Hampshire.[3] Yet it is a fact that local connections were important and Waterhouse was well known and respected in the Manchester of the 1860s. For all his Liverpool origins, family, and business connections, he never quite reached the position there that he achieved in Manchester; and he will always be thought of as a Manchester architect.

[1] J. H. G. Archer (ed.), *Art and Architecture in Victorian Manchester*(Manchester, 1985). Archer's own essay provides a masterly account of the building, and should be read together with his excellent essay on the competition: 'A Civic Achievement: The Building of Manchester Town Hall, Part One: The Commissioning', in *Transactions of the Lancashire and Cheshire Antiquarian Society*, 81 (1982).

[2] MCL, Archives M79/1. 17. Letter AW to the Mayor of Manchester (16 Jan. 1895) in Scrapbook of the Town Hall. Waterhouse in 1895 was still urging the completion of the scheme of decorative mural paintings.

[3] This list is only of the principal locations in which Waterhouse had works in hand in 1866–7, at the time when the Manchester Town Hall competition was coming to fruition.

In another town his connections might have made him the automatic choice as town hall architect. Yet Manchester was too big for one architect to dominate, and the Manchester Town Hall competition was unusual for its fairness. However, Waterhouse's success cannot be divorced from his earlier success at the Assize Courts **[67]**. On that occasion, at least one architect wrote to complain of the influence of interested parties.[4] Such accusations were commonplace, though P. B. Alley's claim to have introduced Waterhouse to the Chairman of the Justices Commission does lend some weight in this instance.[5] Yet the Assize Courts decision was never notorious as a case of bias; and Manchester architects in general do not seem to have held his success against him. None the less, he was relatively young when he collected that commission in the face of 108 other competitors, local as well as national. He beat Thomas Worthington and the partnership of Speakman and Charlesworth,[6] both of whom were to be his rivals, one friendly and one most emphatically not, in the Town Hall competition.

Waterhouse had been careful to ensure that his availability for work in the north was known; and he had been involved with plans for a new Exchange **[215]** (Pl. 45). His design for this, published in the *Building News*,[7] shows several features that are typical of his mature style. The building is Gothic in outline, its ventilation turrets capped with swallowtail battlements of the sort Burges was about to use on his Law Courts tower. The fenestration is of the round arched pattern that he perfected for the Natural History Museum, and the main tower has his beloved steep French chateau roof. The design was probably not worked out in detail; but it has a number of features of interest beyond the optimistic provision of an al fresco promenade at first-floor level.[8]

On such a plot it is, I think, clear that a noble room might be erected say 330 feet long by 80 feet wide containing very nearly 3000 square yards. And down each side of this room there might be a range of arched recesses, lighted from the side streets for Conversation Newspapers &c . . . making a total of nearly 4000 square yards. Above these arched recesses could be a range of windows on either side giving, with the side windows of recesses, abundant light and ventilation to every part of the building without the use of dome lights at all . . . . By such arrangement as I have ventured to suggest the somewhat wedge-shaped character of the Plot would not be discovered at all internally, and I do not think the appearance of the exterior need suffer from this cause.[9]

Waterhouse's report makes clear his concern with planning and light which are precisely the virtues of his town hall design. However, the Exchange Company remained undecided, and he does not appear even to have entered the competition they held in 1866.[10]

[4] *Builder* (7 May 1859), 318. The architect listed 7 of 'the more salient defects' in Waterhouse's design, and concluded : 'When the choice of the committee was announced, the competitors and public both naturally presumed that so speedy a decision proceeded from the manifest and indisputable superiority of the plans selected, and rested content; but an examination of the plan has completely dispelled the illusion and excited the astonishment of all capable of judging the question, more particularly those who have studied the matter carefully to avoid all these defects supposing that they would have naturally influenced the decision of the committee. I have only two conclusions to draw—either the committee were most incompetent, or influenced by interest.'

[5] MS letter, P. B. Alley to Alfred Darbyshire (24 Jan 1898), MCL, Archives Misc. 860/1.

[6] See R. Harper, *Victorian Architectual Competitions* (London, 1983). The other leading Manchester competitor for the Town Hall, E. Salomons, did not compete for the Assize Courts; nor did T. H. Wyatt or J. O. Scott (who was too young), though Cuthbert Brodrick, another of the 8 selected competitiors for the Town Hall, was also among those defeated in the Assize Courts competition.

[7] *Building News*, 12 (1865), 792–3.

[8] To have built this would have been a major coup, and Waterhouse must have felt reasonably sure of his ground for he wrote suggesting an alternative site to the one originally under consideration. He must, I think, have been given some encouragement; for, 3 months later, he was presenting a report on his preferred site (letters AW to Browne Childe Murray and Co. (3 Apr. 1865). OLB 815–16, 817. RIBA Library.

[9] Ibid.

[10] The building was eventually erected to Mills and Murgatroyd's designs in 1869–74.

If the Exchange came to nothing and the Assize Courts were the product of Lancashire influence, Waterhouse's other works still show how much he owed to specifically Mancunian connections. That he was well aware of this is shown in a letter to Joseph Thompson, the Ardwick cotton spinner, on the occasion of his move to London:

> Many thanks for your kind personal remarks. Pray do not think that I have any intention of severing my connection with Manchester or that I shall be unwilling to give my best service to any commissions that may be entrusted to me in the neighbourhood. I propose making my London office my headquarters—but this will in no way interfere with my ability to serve my Manchester friends.[11]

Thompson was a close friend and a leading Congregationalist, for whom Waterhouse had already designed a simple chapel in Altrincham **[231]** and a more elaborate church in Ancoats **[105]** (Pl. 29).[12] Both buildings were under construction at the time, so his client may have needed some reassurance; but it is quite likely that both Waterhouse and Thompson were really thinking of the Town Hall even in 1865.

As a city councillor, Thompson had been added to the New Town Hall subcommittee, that was formed in late 1864. It is inconceivable that Waterhouse was unaware of the prospect of a new building; but it is difficult to be sure what sort of information might have been available to the well informed or connected. Later, in 1866, when he was approached for news by T. N. Deane, he replied cautiously:

> The Manchester Town Hall matter I don't know much about as yet, though some of my friends have asked me to go in for it. I believe as yet they have not settled upon their site! or at any rate its exact boundaries! . . . If I hear anything special about the Manchr. building when next there will let you know.[13]

The letter suggests that he was in the know, for, though the decision to adopt the Albert Square site had been taken in November 1865, the actual purchase of the various properties was not yet completed.

Waterhouse had several friends who were in a position to help him. He was already well known to two, if not three, of the members of the committee set up to consider the question of accommodation in 1863. These were Alderman William Nield, whose house at Bowdon he had altered in 1861 **[95]**,[14] and Robert Neill the building contractor, who had done work for him at the Assize Courts and St John Brooklands, among other places.[15] In addition Waterhouse almost certainly knew Abel Heywood, the driving force behind the whole scheme, through his contacts with the Cross Street Unitarian Church. But it must be presumed Heywood would be more likely to support Worthington, since his old master was Worthington's father. The enlarged New Town Hall sub-committee included not only Joseph Thompson but another potential supporter in John King Jun., Quaker

[11] MS letter, AW to Joseph Thompson (24 Feb. 1865). OLB 587. RIBA Library.

[12] The earliest surviving contact between Waterhouse and Joseph Thompson is a letter of 4 June 1862 (OLB 1. Private collection) in which Waterhouse refers to the completion of the preliminary designs for the Ancoats chapel. A further letter, from the same source, dated 7 May 1862, which explains the plans in more detail, is probably misdated.

[13] MS letter, AW to T. N. Deane (31 Dec. 1866). OLB 347–8. RIBA Library.

[14] High Lawn, the home of William Nield, was in the fashionable suburb of Bowdon near Sale where Waterhouse had a number of commissions. Joseph Thompson also lived in Bowdon, as did J. B. Forster, a partner in the sugar refining firm of Fryer, Benson and Forster (successors to Binyon and Fryer) whose dramatic Venetian warehouse was completed early in 1863.

[15] The third possible contact was Councillor Ogden who may be the Ogden who was secretary to the Manchester Exchange Company.

partner in the firm of Watkin and King, who knew Waterhouse from a commission for alterations to the Friends' Meeting House in Manchester **[110]** and from the conversion of buildings in Rusholme **[220]** and Ancoats **[247]** for the Manchester Free Library Committee.[16] However, it seems to have been Joseph Thompson who was most influential in supporting Waterhouse's entry in the competition.

The first competition closed in August 1867; and Waterhouse wrote to Thompson:

> I have just finished my sketches for your new Town Hall. The Elevations have not been thot. of as they ought to have been, but I hope to have the chance of showing you better things by being one of the selected few. I have just heard that Barry and Street are to carry out the Courts of Justice between them and if so I shall go in for the Town Hall in earnest if permitted to do so.[17]

Waterhouse was now, if not before, very keen to win and Thompson was equally enthusiastic, keeping in touch with him during the selection process, and causing him to write: 'As you may imagine I was greatly interested on your criticism of No. 74 [his own entry in the first competition], only too lenient as, if I know the author, he was much pressed for time, and will think himself fortunate if he be among the first 12.'[18]

Three times during the deliberations Waterhouse thanked him by letter, first for giving support and finally 'for turning the tide in my favour'.[19]

For all the attempts to ensure the competition was a model of fair dealing, other Manchester architects will surely have counted their support fairly carefully before entering. Waterhouse's potential lobby did not mark him out as a clear favourite. If he could count on the support of the three committee members, there were fourteen others with whom he had no connection. He will probably have known that another of the local competitors, Edward Salomons, had been engaged by the Council in negotiations for the site and might therefore count on some support.[20] The very precise competition instructions, ruling out perspectives and forbidding colour in the elevations, made it clear that the corporation as a whole really did have an open mind. Architects were invited to submit an explanatory statement with their designs, but no restrictions were placed on style. This last freedom is a little difficult to understand given the fervour for Gothic in the late 1860s. Certainly there were advocates of the Classical and Free Renaissance styles among the competitors—Edward Salomons was one. However, it is significant that Speakman and Charlesworth described their competition design for the Police Courts, made in 1867 before the Town Hall competition was decided, as Romanesque/Lombardic, 'affording a variation from that in which the Assize Courts and, most probably, the New Town Hall will be erected'.[21]

[16] The contact might well go back further since this could be the same John King who was receiving the acounts for Waterhouse's work at the Victoria Baths in 1856. If so it is likely that King was another of those leading businessmen to whom P. B. Alley, himself with Southport connections, introduced the young architect.

[17] The Archives Department of Manchester City Library has a collection of letters from AW to Alderman Joseph Thompson, hereafter referred to as the Thompson letters. Thompson letters (3 Aug. 1867).

[18] Ibid. (13 Aug. 1867).

[19] Ibid. (28 Mar. 1868).

[20] Manchester City Council. New Town Hall subcommittee minutes (14 Feb. 1867).

[21] The design for the Police Courts was the result of a limited competition arranged when it was clear that the courts could not be accommodated on the proposed Town Hall site. The competition was announced in May 1867 and Waterhouse, Speakman and Charlesworth, Worthington, Salomons, Clegg and Knowles, Mills and Murgatroyd were invited to produce designs. Waterhouse and Mills and Murgatroyd declined to enter. The selection was made in Oct. 1867, without the aid of a professional assessor, the commission going to Thomas Worthington.

The Manchester Town Hall competition was the first to be run in two stages.[22] Between six and twelve competitors from the first competition were to be selected to contribute further designs. The winner would be assured of the commission and each of the finalists would receive £300. When the competition was announced the *Builder* had commended the conditions which, thanks to the care and astuteness of the City Surveyor, J. C. Lynde, were less open to abuse and misrepresentation than most of their day.[23] The instructions for the first round,[24] while giving the accommodation required on four storeys and a basement, had asked for plans of only the ground and principal floors and three elevations (Pl. 48) all in Indian ink or sepia: no perspectives were to be submitted but there was to be included a report on the estimated cost. Little else was defined though it was suggested the main entrance should be from Albert Square and that steps here could break the rigid building line. It was also specified that the building should be faced in ashlar of York or Darley Dale stone. The effects of pollution on the polychrome work of the previous decade, not least on the Assize Courts, could have been the influence here.

In all, 137 sets of drawings from 123 competitors were submitted, and the competition was decided in August and September 1867. One can only speculate as to the arguments over the first selection. George Godwin, editor of the *Builder*, was the assessor, and, though his own best works were Gothic he was remarkably fair minded as to style. He spent two days with the designs, armed presumably with the council's short list of eleven to which a further four had been added at the request of individual councillors. Godwin's selection of eleven designs by nine competitors brought in a further two names. However, one was ruled out for failing to comply with the instructions, and the remaining eight, Waterhouse, Worthington, Speakman and Charlesworth, E. Salomons, J. O. Scott, C. Brodrick, T. H. Wyatt, and William Lee, were instructed to produce further drawings for the second competition. Their designs were withdrawn from the display which was then opened to the public.[25]

The instructions for the second round repeated the original requirements,[26] though it was asked that the Council Chamber be sited towards the Albert Square front of the building. The strict building line could be broken to allow for towers or other architectural features at the extremities of the Albert Square frontage. This time plans were required of all floors and the roof with four elevations and sections (Pls. 50, 61) and the architects could at last enter at least one perspective. A cost of £250,000 was suggested but an estimate was to be submitted. The entrants were at liberty to submit entirely new schemes.

Waterhouse now had the opportunity to give the extra time that he had admitted was necessary to refine his original ideas.[27] The planning concept had been clear from the start; but, as John Archer has pointed out,[28] both the details and the forms were greatly developed. In his letter to Thompson of 3 August Waterhouse had written: 'by putting the hall in the centre and the offices all looking into the streets, I think you may get more than you expected',[29] clearly identifying the strength of his

[22] Archer, 'A Civic Achievement', 15.

[23] *Builder* (30 Mar. 1867). Later the same publication urged that the council should seek professional advice, only hinted at in the instructions to architects, for help in making their selection (24 Aug. 1867). There is no evidence that the council ever intended otherwise.

[24] MCL, Local History Department, *New Town Hall: Instructions to Architects* (first competition).

[25] *Building News* (10 Apr. 1868).

[26] MCL, Local History Department, *New Town Hall: Instructions to Architects* (second competition).

[27] MCL, Thompson letters (13 Aug. 1867). There was also the matter of the tower for which Waterhouse hoped for further hints from Thompson.

[28] Archer, 'A Classic of its Age', in *Art and Architecture in Victorian Manchester*.

[29] MCL, Thompson letters (3 Aug. 1867).

approach as the provision of maximum space and, a necessary corollary, effective circulation. Actually a very similar disposition of rooms was offered by most of the selected architects; so that it was the clarity of the circulation that distinguished them. Waterhouse's rather rushed first design lacked two of the principal virtues of his finished building. The grand corridor at the front was not continuous, and he did not provide stairs at each corner of his ring corridor.[30] The most efficient circulation plan of those selected seems to have been T. H. Wyatt's, where the straight ring-corridors and three corner stairs within the ring correspond precisely to the pattern Waterhouse adopted. It is even possible that he saw these drawings or that Thompson passed on his opinion of them.[31] If he did learn from Wyatt, his own brilliance shows in the elegance of the detailed development of the basic concept (Fig. 4).

In spite of the general opinion that the competition was set out in the fairest possible terms, the *Building News* raised the question of influence and suggested that the race was really between Speakman and Charlesworth, Salomons, Waterhouse, and Worthington in that order.[32] All were Manchester men. As at the time of the first round, the professional advisers were not appointed until after the designs were due in. This may again have been northern astuteness: while the assessors remained unknown the designs could not be modified to suit the judges' taste. The appointment of G. E. Street with Professor Donaldson must have gladdened the heart of Waterhouse who had such regard for Street's work; and Street will, of course, have known his drawing style from the Law Courts competition.

The designs were all, of course, submitted under pseudonyms (Pl. 50); and the judges were asked to make a recommendation based on six points: architectural merit, general arrangement and convenience, daylighting, ventilation and heating, acoustic properties of the Great Hall and Council Chamber, and probable cost in relation to the designer's estimate. The list shows how great was the concern to achieve a building which as well as being a credit to the artistic aspirations of the city was in all practical matters satisfactory. It was in precisely these terms that the report was given concluding that 'after careful adjustment and comparison' Waterhouse was the winner, leading Speakman and Charlesworth, Scott, and Worthington.

The report caused a furore. The entry, 'St. Valentine' (the pseudonym chosen by Waterhouse), they considered best in the second, third, and fourth categories and capable of being built within the proposed cost. On the matter of acoustics they declined to comment.[33] In considering the first of the criteria, that of architectural merit, they had placed Waterhouse fourth after Speakman and Charlesworth, John Oldrid Scott, and Thomas Worthington. This offended not only other architects but the sensibilities of the burgesses who had set out to seek civic splendour. Two local newspapers got hold of the report and published it before the date set for its consideration by the committee, thus giving an accelerated start to the debate.

[30] Waterhouse's first competition plan (DC) shows major staircase at the Cooper Street end and at the Lloyd Street corner of the front; but only a minor staircase at the Princess Street corner. The actual siting of the staircases as well as their form was changed for the second competition.

[31] Waterhouse certainly knew Wyatt, having previously done him small favours in Manchester; and he had visited his new Exchange in Liverpool. MS letters, AW to Giles Redmayne (3 Dec. 1866), to T. H. Wyatt (3 Dec. 1866), and to Redmayne (11 Dec. 1866). OLB 140, 162, and 214–15. RIBA Library. Wyatt had asked Waterhouse to chase up a small debt and he admitted frankly to his brother-in-law, 'I shall be glad to do Wyatt a turn'. The correspondence does not suggest any very cordial friendship, for one letter to Redmayne runs: 'When a swell dates from "Wilton House", I suppose one must do his bidding if it is even to dun for him.' As the amount in question was a mere £2 the episode says more about Wyatt's tight-fistedness, but it does at least demonstrate a more than casual professional acquaintance.

[32] *Building News* (14 Feb. 1868).

[33] The Council Chamber proved to be a disappointment in this respect; the Mayor had curtains hung in the alcoves to try to effect some improvement. Manchester City Council, General Purposes Committee (31 Jan. 1878).

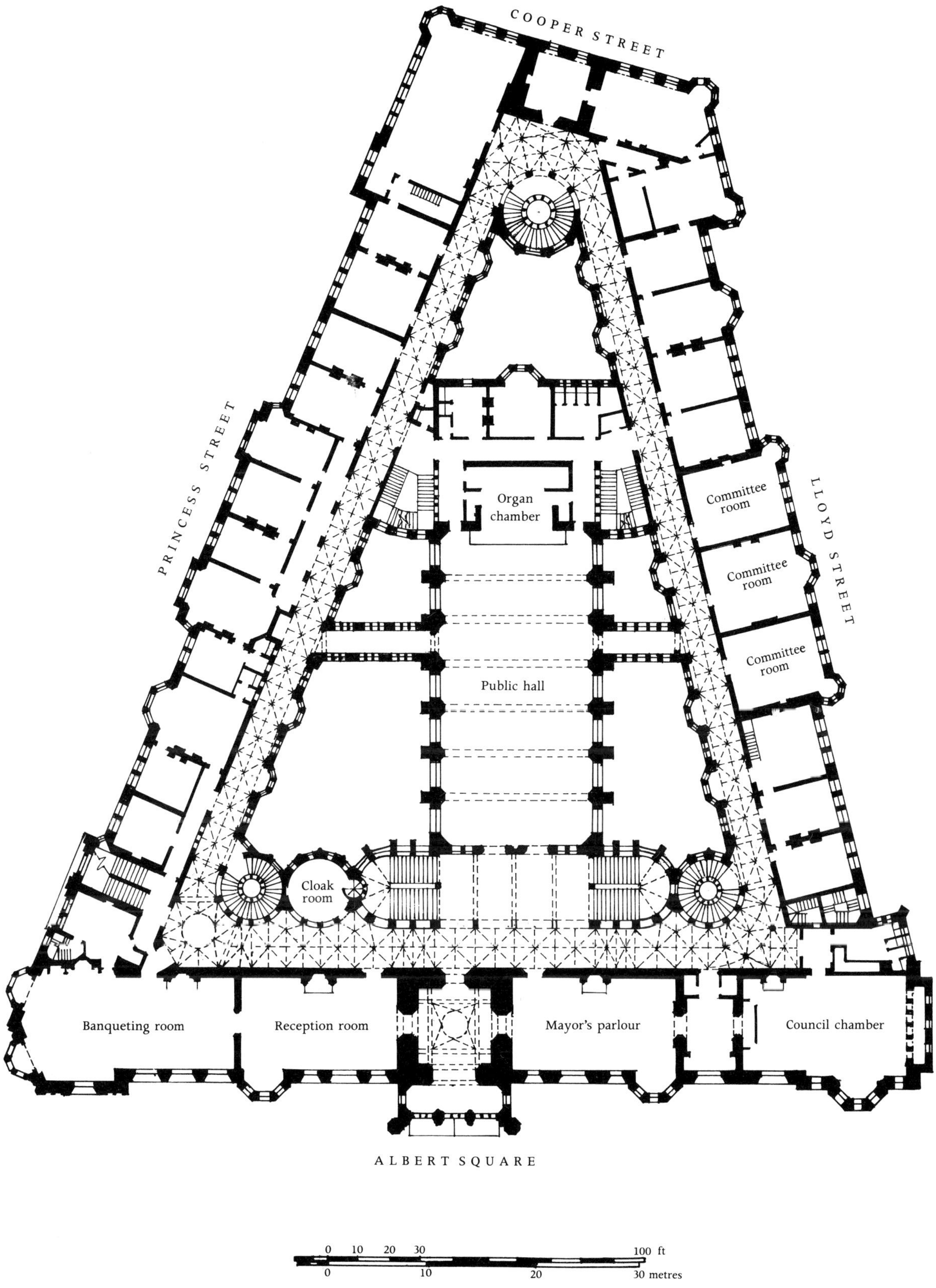

FIG. 4. Manchester Town Hall (1868) **[284]**, plan of principal floor (RIBA)

Donaldson and Street were asked to expand upon their conclusion. They had pointed out defects in the plans of St Valentine and suggested various improvements and alterations but had made no observations in relation to the other plans. Did that mean that in the judges' opinion the rest were not open to any sort of objections? The two replied with a lengthy but confidential report,[34] thus protecting the 'susceptibilities of our professional brethren' and standing back themselves from further involvement in the public debate. In their first report, they said, having answered briefly under the six headings, they had given the grounds for their conclusion, basing them only on Waterhouse's design.

The architectural character of this design is, as we have said, not so good as some of the others, but the plan has such great merits, is so admirably and simply disposed, and so well lighted, that we cannot but feel that it is thoroughly entitled to the first place. The general disposition of the masses of the elevation is very picturesque, and there is much dignity about the treatment of the principal storey towards Albert Square. We are bound to say that in some respects the design appears to us to require additional study and modifications, of which it admits without difficulty.[35]

The council formally approved the recommendation on 1 April 1868, and all the drawings were put on display to the public who could now judge for themselves.

They found six schemes in forms of Gothic and two which were Classical. It seems that there was no doubt that the first four schemes at least were of a very high standard, the *Builder* reckoning the display's quality at least as high as that of the Law Courts competition.[36] The two Classsical designs were by Wyatt and Salomons, Brodrick having changed to Gothic detail over a Classical composition with an immense circular tower and spire. He and Worthington raised the entrance to the level of the principal floor with grand flights of steps. Worthington's, the most romantic of the Gothic entries, was given its best display in the perspective, which exaggerated the massing of oriels, bays, turrets, and spires. Like Waterhouse, he had used such elements to turn the unequal angles at either end of the Albert Square frontage. Only Speakman and Charlesworth set their tower behind the front block; and in general all the plans were similar, which is not surprising in view of the site and the instructions given.

Waterhouse's design, apart from refinements of plan, used the relaxation of the building line to allow the end pavilions of the main façade to break forward. The tower was enlarged, and the façade to Princess Street was radically firmed up, with more attention to the oblique view along the street (Pls. 48, 50, 51). A further tower now marked the Mayor's private entrance, providing a bold accent in the streetscape and masking the changes in floor level imposed by the extra height of the receptiom rooms. With the Council Chamber now moved to the Albert Square front,[37] Waterhouse redesigned the Cooper Street frontage with a substantial tower 'serving to give as varied a skyline as possible'.[38] The displaced committee rooms made a central block for the Lloyd Street elevation.

The storm about the award continued with pressure for publication of the confidential explanation. Waterhouse stayed aloof from the row but the *Building News* published an acrimonious

[34] This report has not been found. See Archer, 'A Civic Achievement', 25.

[35] *Building News* (13 Mar. 1868). Street and Donaldson's report to the New Town Hall Committee.

[36] *Builder* (11 Apr. 1868).

[37] This position was not to Waterhouse's liking; see AW RIBA Presidential address to students Jan. 1890, published in the *Builder* (25 Jan. 1890), but the instructions for the second round requested it.

[38] MCL, Local History Department, AW [St. Valentine] Written statement accompanying design submission.

and bitter comment on his design for which they could find merit only in the plan.[39] However, the council had already authorized the General Purposes Committee to 'proceed with full power to make the necessary arrangements with the Architect and to take all other steps required to secure the prosecution without delay, of the necessary works'. On 22 April Waterhouse attended the New Town Hall Committee to discuss his plans, the modifications suggested by Street and Donaldson, and the programme of building. The contracts were to be let in stages, first the foundations and basement, then the rest of the carcase, succeeded by the tower, spire, and finishing. This division of the work enabled Waterhouse to revise and reconsider at every stage and indeed to leave a good deal of the interior detail to the last stages.

Waterhouse agreed that the assessors had 'pointed out defects which perhaps were too patent to have escaped even this author'.[40] The architectural character that they had found wanting could, they suggested, be gained by the modification and improvement the architect would doubtless apply: the scheme needed better openings, more light and greater dignity for the great entrance, and work to the clock tower and angle features of the main façade. They considered that the internal courts being so spacious should have a better appearance, 'Good, solid, simple, but really architectural character'.[41] Waterhouse and the council did not see this as a criticism but the implementation immediately affected the cost which eventually rose by four times to almost one million pounds, a fact of which the city was to be immensely proud.

During the rest of 1868 and into 1869 Waterhouse continued to work on the main elevation and the tower which, above all, gave him the greatest concern.[42] His first design, of a pointed dome on a drum for the clock, he abandoned in favour of an octagonal clock drum set between corner turrets and surmounted by a stone spire; but the relationship between tower and drum was unsatisfactory. He made another attempt at a dome,[43] but eventually, in 1869, incorporated the clock in the uppermost storey of a tower more than twice the height of the building itself. This in turn was capped by a light and lacy octagonal lantern with a blunt spire supported by flying buttresses. That was not the end of the problem of the tower; and the last change was made as late as July 1875, when the tower was actually under construction. Waterhouse stopped the work and asked to be allowed to build the tower 16 feet higher at an added cost of £450 to £500,[44] an addition to which the council happily agreed (Pl. 47).

[39] *Building News* (10 and 17 Apr. 1868). The final prick was their accusation that Waterhouse's plan could not have been so successful had he not 'encroached somewhat on the space allowed for the buildings, and thus [obtained] one advantage over those who have complied with the conditions'. This hurt so much that he made a rare entry into the correspondence coloumns quoting the two sentences of the conditions which had given the architects leave to alter the lengths of the projecting portions and areas in Princess Street and Lloyd Street to suit the design of the competitors, *Building News* (17 Apr. 1868). A sharp reply came from Charlesworth: His interpretation was different and he claimed entries which had broken this rule had been rejected from the first round. The full instruction was printed out and reveals its ambiguity, *Building News* (24 Apr. 1868). Waterhouse, as ever courteous, replied smoothing Charlesworth's ruffled feathers and pointing out that he had not been the only one to interpret the instruction in the way he had.

[40] AW on Manchester Town Hall, *RIBA Sessional Reports*, 12 (1876–7), 117 ff.

[41] *Building News* (13 Mar. 1868). Street and Donaldson's report to the New Town Hall Committee.

[42] The 1869 drawing (DC) shows a full development of the order hinted at in the first attempt while the uncertainty about the tower remains: glue marks on the drawing show that yet another sketch was overlaid at some stage. The agony of finding a satisfactory design for the upper stage of the tower and the spire is proved by the number and variety of other perspective sketches in Waterhouse's own hand: an octagonal drum incorporating the clock faces with and without gables, one version including flying buttresses to the corner turrets, yet another having the spire surmounted by a lantern, the spire itself varying in its detail from plain, to crockets on the angles, from simple banding to elaborate panelling or filigree tracery. Other versions with the clock faces in the upper stage of the tower lead to the final outcome.

[43] Perspective of 1868 at Manchester University School of Architecture.

[44] A small sketch shows that the comparative heights of the main tower as originally planned and the Cooper Street tower, when seen from the ground, were very little different. The increase in height was to correct the effect of foreshortening.

In developing his design, Waterhouse was following Ruskinian principles of organization and growth, rather than simply adding ornament. However, this was a building in which he moved steadily away from Ruskin towards a more practical modern Gothic of his own devising. Detailed problems of plan and elevation were worked out together. The unequal angles of the site broke down the symmetry of the main elevation, and each end pavilion, while relating to the other, was differently designed, with the windows grouped to match the rooms they served. The main entrance was given added importance by increasing its size and adding a filigree gable over it. A three-storey bay increased the scale and weight of the Council Chamber corner (Pl. 59), while for the other corner pavilion Waterhouse developed a Ruskinian 'solemn frown of projection' in the uppermost storey,[45] so that the window heads beneath appeared almost as machicolations. Finally, the intermediate bays were reduced by one storey so that one continuous band of upper windows, with parapet above and string band below, tied each wing to the centre.[46] Within the building the importance of the main corridors was increased by vaulting, which also provided a fireproof ceiling (Pl. 52); and the addition of the recesses in each alternate bay alleviated their length. The internal courts were enlivened with bridges, a suggestion from Street and Donaldson.[47] The resulting motley of scales of medieval quality was combined with simple detail and lively surface interest (Col. Pl. 4; Pls. 47, 60).

The progress of the work was recorded in Waterhouse's monthly reports to the New Town Hall subcommittee. From time to time further subcommittees were appointed to deal with particular items, such as the decorations, which included the famous set of twelve murals for the hall commissioned from Ford Madox Brown. There were of course difficulties in that the client was the council, two stages up through the General Purposes Committee so that, while a great deal was delegated to the New Town Hall subcommittee, which met once a month, there was considerable delay in any major decisions.[48] None the less revised plans were approved by June 1868, and tenders out in July for the foundations. By early 1873 the shell was nearing completion and the New Town Hall subcommittee began to involve itself even more actively in the detail.

Waterhouse, who was running the job from London, now needed the help of his friend Thompson. He discussed the schemes for statuary with him, and sought his advice about explaining to the committee the implications of the latest tenders.[49] A year later he was writing: 'I hope you will be at the next meeting. Harwood and the chairman talked of selecting the gas fittings themselves';[50] and a week after that: 'When you go over [to Bradford] Mr. Salt (is he not Sir Titus?) will get the Town Clerk to let all the tunes off one after another. It is important TC. should grant us this

[45] John Ruskin 'The Lamp of Power', *The Seven Lamps of Architecture* (London, 1849), 69.

[46] The scale of the principal floors is thrown up by the contrast between these close-set windows and the large areas of blank wall beneath which, like the rest of the building, carry the rigidity given by the broad and narrow masonry courses of unvarying dimensions. This is a common method of coursing in which the narrow courses tie into the brickwork. The regularity is a hallmark of Waterhouse.

[47] AW on Manchester Town Hall, *RIBA Sessional Papers*, 12 (1876–7), 117 ff.

[48] Indecision and changes of mind were frequent too, an inevitable consequence of the cumbersome structure. The choice of site had taken more than 3 years, the present site being abandoned once during the process and an attempt made to increase it and improve the shape. A further attempt to square up the site was made even in Feb. 1868 as the final competition designs were about to be submitted (*Building News*, 14 Feb. 1868). In the spring of 1876 it was realized that the new building should properly be called the City Hall. After 3 months and as many debates so it became, but after a 6 months' attempt to get used to the idea the council reverted to the familiar old title, maintaining the anomaly to this day. The most notable example of vacillation, though, is the story of the mural decorations (the full story of the murals in the Great Hall is told in 'Ford Madox Brown and the Manchester Murals' by Julian Treuherz, included in *Art and Architecture in Victorian Manchester*, ed. John H. G. Archer). Those in the Great Hall were not completed until 1892, and the schemes for the reception rooms were never undertaken.

[49] MCL, Thompson letters (30 June 1873).

[50] Ibid. (23 July 1874).

favour.'[51] Thompson also played a large role in planning the themes for the decoration throughout the building. Subjects for sculpture, arms of families, suggestions for subjects for the mural panels were all requested by Waterhouse, who must have hoped to avoid tiresome discussions with the committee.[52] At the end of 1877, when the question of the subject-matter of the mural schemes was taking its protracted course, he wrote to Thompson: 'The more I talk to the painters [artists] the more confused I become. Those were palmy days when we . . . had only to get the money and then put Guppins and Sweltz to work.'[53] But Thompson clearly had considerable influence when matters concerning the cost of the project were looming; and on the death of the Chairman of the Finance Committee, 'dear old Willert a good friend to both of us', Waterhouse's hopes that Thompson would succeed him were, fortunately, fulfilled.[54]

Nevertheless it was not themes of decoration that most concerned the architect but the building itself, and he was meticulous about his responsibilities; he even had reports and certificates sent to him while on holiday abroad.[55] By the spring of 1877, however, ten years after the initial competition, the building was far enough advanced for a formal opening to be contemplated. Visits by the public were stopped in order to speed up the work; but even after the proud opening ceremony on 13 September 1877 the building was not finally finished.[56]

Today it is remarkably little altered. The main entrance is the start of a processional route beneath the tower, past the vaulted waiting hall, to a pair of matched staircases dramatically lit by a wall of tall traceried windows (Pls. 53, 61). Corbelled out into the main stairway is a spiral staircase with open balustrade, up which town hall servants could be seen taking coats and hats to an upper cloakroom. Finally at the head of the stairs a top lit landing, separated by stone screens, affords a breathing space before entering the great hall. It is one of the most magical promenades of revived Gothic architecture.

The hall, the centre of the whole edifice, is magnificent (Pls. 54, 56). One end is filled with a great organ by Cavaillé Coll of Paris, then considered the best organ builder in the world. Overhead soars a rich timber roof, its panels painted with heraldic symbols of the countries with which Manchester is connected in trade. Windows of 'white glass tenderly relieved by subdued colour'[57] fill each side above the mural panels, giving light without spoiling the colour of the paintings. Wrought iron tie beams support elaborate iron gasoliers, and further light comes from great sconces between the windows. Opposite the hall, the grand corridor opens into a suite of rooms overlooking Albert Square (Pls. 49, 57). Each has an elaborately panelled ceiling, colourfully painted except in the Banqueting Room, where colour is confined to gilding on natural oak. All but the Council Chamber (Pl. 55) have a high oak dado, each room a different pattern; but, above, the walls remain plain

[51] Ibid. (27 July 1874). Waterhouse also persuaded them to visit Worcester Cathedral to hear the carillon there.

[52] Thompson's position on the decorations subcommittee set up in 1876 ensured his influence. Minutes of the New Town Hall subcommittee (19 Jan. 1876).

[53] These were the 2 Belgian artists whom Waterhouse had selected and recommended both for their manner of painting and their method of working direct on to plaster. The murals had involved him in trips to the Continent, numerous interviews with artists and reports to committees, all without payment, such was his conviction that the opportunity to complete the schemes should not be lost. But the matter was taken out of the architect's hands; clearly the committee felt local artists were more appropriate.

[54] MCL, Thompson letters (25 Mar. 1879).

[55] The arrangement was not without its flaws. While in Italy one spring he changed his route and missed the certificates his manager Willey had sent out for him to sign. MS letter from Willey, Waterhouse's manager (15 Mar. 1876).

[56] Details of the opening ceremony and banquet were recorded in W. E. A. Axon, *An Architectural and General Description of the Town Hall, Manchester, &c.* (Manchester and London, 1878).

[57] AW on Manchester Town Hall, *RIBA Sessional Papers*, 12 (1876–7), 117 ff.

awaiting the murals that never came. Waterhouse's failure to carry the council with him over the mural scheme has been a loss to the Town Hall and to Manchester.

At each corner of the building the vaulting turns around spiral stairs that not only provide vertical circulation but also serve as shafts to distribute hot air from the basement. These stairs are the key to the working of the whole building. Set close to the office entrances, they link easily to all floors of the building. Their circular form allows landings at any point, so the extra height of the reception rooms and the resulting break in the floor levels above goes almost unnoticed.

All the offices face out on to the streets, while the long passages draw ample light from the courtyards. On the two grand floors the corridors are faced with terracotta and roofed with painted concrete vaults.[58] They are all enlivened by views into the courts, picturesque glimpses of bridges, roofs, great chimney stacks, and wall surfaces patterned in glazed tiles, white, grey, and red. The two long side corridors gain a rhythm from the added recesses in alternate bays and, to improve the circulation, intermediate stairs are slipped into the tapering angles against the hall. These are uncompromisingly supported on exposed iron girders, a frank expression of modern structure that would probably have distressed Ruskin. However, as far as Waterhouse was concerned, the building needed also to be a functional, and the business part of the building is not disguised. Instead the Dennet fireproof system of wrought iron beams and concrete arches is left visible and allowed to dictate the division of the committee rooms and offices. Only in the roofs of the spiral stairs does he let himself go again. They are capped with exhilarating umbrella-like constructions of timber.

It was a mammoth commission and dauntingly complex. But Waterhouse was the man to resolve the complexity into a clear and logical plan. Stylistically his building owes as much to the nineteenth century as to medieval sources. It is emphatically modern Gothic, using up-to-date techniques of fireproofing and steam powered lathes for polished shafts. The choice of terracotta for wall facing was thoroughly modern, and a new departure for Waterhouse. Even the Shap granite for one of the stairs was from a newly opened quarry. Above all Manchester Town Hall is a building where the modern function has been allowed to dictate the form, although the Gothic approach gives the picturesque associative values that the Victorians prized so highly. Waterhouse, by his close control of all aspects, created a vital unity, not by uniform treatment of every part but by a hierarchical appropriateness which makes the whole building comprehensible. Civic display and administrative function welded closely together.

[58] On the upper floors only simple arches were added to relieve the monotony.

8

The Oxbridge Commissions

MANCHESTER Town Hall may be the grandest achievement of Waterhouse's maturity; but among the other results of the national acclaim that followed his success with the Manchester Assize Courts was a series of commissions in the Universities of Oxford and Cambridge. In fact he had work in one or other town almost continuously from 1865 until his retirement.[1] These commissions are all works of his maturity, in concept at least, and so deserve to be considered together. Even to an architect who was by now successful they were important, because commissions in the old collegiate universities offered the same sort of prestige as major public commissions. The position of the two universities was virtually that of a corporate aristocracy. Following the reforms of the mid-century and after Prince Albert's Chancellorship at Cambridge, these ancient institutions were blooming; and the expansion of numbers offered major opportunities throughout the second half of the century. Waterhouse did as much as any other architect in Cambridge; but we should begin at Oxford with his modern Gothic work for Balliol.

His first commission at Balliol dates from 1866 **[254]** (Pl. 63), when both the Law Courts and Manchester Town Hall were only future possibilities. It is the only example of Waterhouse's work, other than the Assize Courts, discussed by Eastlake, who commented that at Balliol he had 'kept up with the stream of advancing taste without losing that individuality of design which every true artist wishes to retain'.[2] Probably the college was drawn to approach Waterhouse on the strength of his Assize Courts work (Pl. 33). But, as a new building type, the Balliol buildings are important in his career; and the complicated story of their genesis is also interesting as an indication of his ability to face problems.

The leading figure was Benjamin Jowett, energetic and outspoken, almost openly aiming for the mastership which he achieved in 1870, and held until 1893. In 1866 the college had barely recovered from the furore caused by his contribution to the theological bombshell, *Essays and Reviews*; and they had been through an earlier, though more local, controversy in 1843 when Pugin was commissioned to design a replacement for the Broad Street Buildings, and a new Chapel.[3] In an

[1] Girton was the longest running commission, beginning in 1872 and continuing until Waterhouse's retirement, and afterwards, under his son Paul, and subsequently his grandson Michael, until 1931.

[2] C. L. Eastlake, *The History of the Gothic Revival* (London, 1872). Facsimile edn. with introduction by J. Mordaunt Crook (Leicester, 1970), 361. Ten other works by Waterhouse appear in the list of selected examples, as corrected and expanded by Professor Crook, including Manchester Town Hall and Eaton Hall where works were in progress at the time of publication.

[3] See H. M. Colvin, *Unbuilt Oxford* (New Haven, Conn., 1983).

Oxford still bound fast to the Established Church,[4] the work of this ardent Roman Catholic convert was too much to swallow. In the event Salvin was called in for a number of minor works; and the new Chapel was entrusted to the safely Anglican William Butterfield in 1856. This left unsolved the major and longstanding problem of decaying rooms which Basevi, in 1841, had proposed rebuilding in a formal Classical style. Pugin had been equally radical in suggesting the reconstruction (and refurnishing) of the whole college as a medieval enclave. But the college had to wait until 1865, when a handsome donation, the result of a certain amount of fishing by the Master, Scott, allowed them to proceed.

The College Minutes of 13 April 1866 record that 'Mr Alfred Waterhouse be approached to furnish plans and designs, the College reserving the right of rejecting them'.[5] Jowett's choice of a northern industrial architect of Quaker stock was a major change. The donor, Miss Brackenbury, had named Henry Woodyer as architect, but 'left the choice . . . open, and, it being found that those of high standing objected to furnish competitive designs, the Society with much difficulty and after great diversity of opinion, finally directed' the Master to write to Mr Waterhouse.[6] Butterfield may well have wondered why he was not considered, though it is probable that the discussions leading to the foundation of Keble College were already linking his name with that building.[7] At any rate Waterhouse was fortunate in his more eminent colleagues' suspicion of competition; and Butterfield was not to everyone's taste, having 'lately erected some buildings for Merton College which very few fellows are found to admire'.[8]

Waterhouse's approach needs to be compared with Pugin's attractively presented dream college. Significantly the commission was 'that plans and designs be procured for a restoration of the buildings from the Master's House (inclusive) to the Chapel; the work to commence at the West end, and to be carried on so far as the available funds will allow'.[9] As Miss Brackenbury had only promised £6,000 at this stage, and the eventual cost of the lodge and Brackenbury Buildings was to be £30,000, this was a wise precaution. The cautious approach is the key to his design. Pugin had drawn a more or less symmetrical front about the fifteenth-century tower, a concept which Waterhouse respected; indeed he borrowed the drawings to help him make his perspective. Yet he kept only the position and approximate width of the old tower, and assembled round it a set of buildings that is symmetrical only in barest outline, allowing the plan to dictate the forms and making full use of the freedom of the Gothic style (Pl. 63).

First came the Master's Lodge, conceived and built as a separate unit, yet incorporating existing rooms in the old Hall range. Though Waterhouse was rebuilding from scratch, he was clearly conscious of value in the old work. 'The oriel window of the dining-room in the Master's Lodge has been, of course, vigorously preserved, indeed a feeling of the absolute necessity of its preservation has dictated the arrangement of much of the present plan.'[10] It dictated a new window facing Broad

[4] The Oxford University Act, abolishing religious tests at matriculation, was passed in 1854. Cambridge followed suit 2 years later.

[5] Minutes of College Council (13 Apr. 1866). Balliol College Archives.

[6] MS letters, Revd J. H. B. Scott (Master) to W. H. Rosper (23 Feb. 1866 and 14 Apr. 1866). Balliol College Archives.

[7] Butterfield's plans for Keble College were laid out at a public meeting on 28 Apr. 1866.

[8] MS letter, Scott to Rosper (23 Feb. 1866). Scott also notes in his letter of 14 April that Waterhouse 'has gained a high reputation by building the Assize Courts etc. at Manchester.' Balliol College Archives.

[9] Minutes of College Council (13 Apr. 1866). Balliol College Archives.

[10] Waterhouse, writing in the *Architect* (2 Jan. 1869). He did not carry his conservationist approach throughout the building. He accepted the need to move one wall of the dining-room for structural reasons, cutting off one row of lozenges of the plaster ceiling, though retaining the cornice.

Street, so that this portion of the lodge was set back from the street making the main body of the house read as a completely separate building.

The main buildings formed an L-shaped range with a tower in the centre of the front.[11] Waterhouse followed the logic of the commission, and the separate sections were designed and built from the west (Pl. 64). The first part stopped one bay short of the tower and was divided from the rest by a solid wall. However, both eaves and roof line were maintained so that the block appeared as part of a single building with windows freely placed to allow a lofty lecture theatre (now the Bursary) with two floors of rooms above.[12] The rest of the block had four floors, the top one fitted into the roof and lit by dormers. It was a heavy development of the site and made the front quad cramped and rather dark, an inevitable result of the pressure for accommodation in a rapidly expanding university. Even so Waterhouse was concerned to keep the bulk of the building down, and, when asked to add extra rooms during the process of building, was careful to do so without adding dormers to the Broad Street front and only one or two to the side facing Trinity College.[13] He was, he said, anxious to avoid damaging the architectural effect of the buildings. He had already surrendered a steep pyramidal roof of the kind he favoured for the tower in deference to the Fellows' taste but gave his building a clear dramatic presence by placing a large tourelle at the south-west corner balanced by a large gable and oriel window at the south-east.

The local detail was all Gothic of a thirteenth-century French variety, as one would expect from a serious architect of the 1860s. And though the Bath stone was adopted as an economy measure, it allowed deep crisp mouldings, which fitted well in a range of considerable size. Its front measured almost 250 feet and took the opportunity of tidying up the irregular street line. Waterhouse gave it a pleasing unity by string courses, which were picked up in the Master's Lodge; while the disposition of the four oriel windows and the slight break forward in two dormers added an essential element of variety. The building was hard edged and eminently high Victorian, but the way in which it accommodated the requirements of site and space as well as the historical precedents marked it out as a work of quality. It emphatically establishes Lethaby's definition of Waterhouse as a 'hard' goth, one who thought Gothic and built accordingly.

Waterhouse also built satisfactorily, as far as his clients were concerned, and immediately on completion of the new buildings was commissioned to produce further designs for the college, many of them adapting buildings by Salvin. In 1872 he designed an organ case for the Chapel. Then there were new rooms on the site of the stables and refurbishment of the Fisher building. In 1873, when the college, now with Jowett as Master, decided to build a new Hall, they again commissioned Waterhouse **[392]**.[14] Jowett thought his new Hall 'really beautiful'; but Ruskin scathingly dismissed it as a 'dull sort of Church'. It was a complex mix of first-floor hall with buttery and teaching rooms beneath, and it was probably conceived as the centre-piece of more extended development. However,

[11] Conservation seems to have dictated the rather narrow dimensions of the tower. At one point Waterhouse was asked to reuse the old groining, a suggestion that he firmly rejected. However, the very suggestion strengthens the belief that the width of his tower, if not its depth, was the same as its predecessor.

[12] The first designs show all the building to the west of the tower designed as 3-storey accommodation. Balliol College Archives.

[13] MS letter, AW to Revd Dr Scott (1868). Balliol College Archives.

[14] This work was followed by the conversion of the old Hall into a library. In 1881 Waterhouse designed the college barge (I am indebted to Edward Hibbert for this information); and he finished his connection with Balliol with a design for the case of the hall organ that Jowett presented to his college in 1885. The year before, Waterhouse's eldest son, Paul, had graduated from the college.

any such plans were never realized and the Hall is now less easy to appreciate, shorn of the block of rooms that adjoined it to the east, and with its windows partly blocked.

Balliol's Broad Street frontage was one of the more important commissions in Waterhouse's career; but his earliest university commission was a debating hall for the Cambridge Union Society in 1865 **[197]** (Pl. 62).[15] This was clearly similar to Deane and Woodward's Oxford Union, which he visited before drawing his own design. Yet, practical as ever, he turned his back on such exaggerated Gothic and produced a simple but attractive building with decoration pared to a minimum. Details such as the flying flight of stairs to the smoking-room, and access to the caretaker's house through the roof space of the debating hall seem quirky, and probably result from the attempt to get the maximum accommodation for the minimum price. The building certainly was economical, costing a little under £9,000; but the list of subcontractors, which included Farmer and Brindley for stone carving, W. Godwin for tiling, and Lavers and Barraud for stained glass, showed that this was no pinch-penny job; and there was even a complete set of furniture designed by the architect. The society was so satisfied that, twenty years later, they commissioned Waterhouse to double the size of the club with a new wing in a richer version of the same style (Pl. 69).

In the interim he was called in at Oxford to increase the size of Deane and Woodward's building **[464]**.[16] After trying various means of extending the original block, he persuaded the society to consider a free-standing building.[17] His design showed a progressive rejection of Gothic forms and detailing (Pls. 66, 67). The final structure, set apart from the main building, did not attempt to follow its style. Instead he offered a chamfered box of brick and yellow terracotta with an attractive and, for him rare, deep coved cornice. The Gothic here is so minimal that the building seems to deserve a new style label, Old English.[18] Stylistically the building has more in common with later country houses by Devey, Shaw, and Philip Webb than with the spiky cavern next door in which Rosetti and Morris's paintings were already fading into obscurity. But as always Waterhouse put practicality before mere style.

Coming so soon after Balliol Hall, the smoothness with which he is able to use this new style is impressive;[19] but he had already shown the range of his eclecticism in Cambridge with the new buildings for Gonville and Caius College **[268]** (Pl. 70). He was invited to produce designs in December 1866,[20] when the Union building was little more than a shell, and actually before he was

[15] The commission was won in a limited competition early in 1865 when Waterhouse was in touch with Henry Peto of Trinity College. The details of Waterhouse's connection with Trinity have yet to be researched; but they resulted in one of his most unusual Gothic works—a memorial brass for the Tractarian Revd J. W. Beaumont in the Chapel dated 1868. Waterhouse may also have had some connection with St John's College who owned the Union site. The Master W. H. Bateson appears to be the brother of W. G. Bateson of Liverpool for whom Waterhouse had already built a house, New Heys, in Liverpool.

[16] Whether there was any connection with his work at Balliol is not clear; though it is interesting that the Hall was completed the year before, and 6 of the 9 Union Presidents from 1870–9 were Balliol men. More significant, however, is the society's decision not to re-employ Woodward's old firm (Benjamin Woodward himself had died in 1861 and Thomas Deane in 1871; but the practice was being continued by T. N. Deane) but to commission the more practically minded Waterhouse.

[17] Designs were also drawn for an additional wing along the side of the site or facing New Inn Hall Street. This block was to contain writing rooms and a coffee room; but it was abandoned, presumably on grounds of cost. Those spaces were eventually provided, just where Waterhouse had suggested in a sort of revived Tudor–Gothic by Mills & Thorpe in 1910–11.

[18] Old English was not a term ever used by Waterhouse, but it recognizes the domestic qualities of works such as the Oxford Union, and in particular the Tudor origins of his mullioned and transomed windows, as well as the general solidity of his interior woodwork. The term is used to describe his furniture by Jill Lever in *Architects' Designs for Furniture* (London, 1982).

[19] Waterhouse had already been experimenting with this mode in houses such as that for James Milne in Cheadle **[174]** in the early 1860s.

[20] MS letter, AW to Revd J. Lamb, Master (18 Dec. 1866). OLB 267. RIBA Library.

approached by Balliol. So keen was he to accept, that he travelled down to Cambridge on the very day after the opening of the exhibition of designs for the Law Courts competition. Pevsner dismisses Tree Court as 'pretentious in a characteristically High Victorian way, utterly unconscious of the character of the architecture into which it should fit'.[21] The criticism overstates the case and takes little account of what Waterhouse was asked to do. He had to provide some sixty sets of rooms on a very small site (it was the problem of rapid expansion again); and it was inevitable that he would have to go to four storeys. As at Balliol the top floor was fitted into the attic; and an aerial view reveals, from the width of the lead flat, just how much space he managed to fit behind his dormers. However, there was no real disguising the bulk of this building, and Waterhouse accordingly opted for a dramatic contrast to avoid the difficulty of finding a single style that would accord with both Gibbs' Senate House and Great St Mary's Church. He selected French Renaissance, slipping convincingly into an idiom which he had not previously tried, and using its lavish ornament to produce a fantastic skyline of dormers topped by a bold tower, which, if not universally loved, is none the less a striking feature of the Cambridge townscape.[22] Architecturally speaking, the tower should be the main entrance to the college, instead of the old Gate of Humility. This was a radical change for the college, and Waterhouse might have been criticized for sweeping away interesting old work. Early plans show that he intended to retain the old gate; but as 'it had been long ago either restored or faced with Roman cement, till there was scarcely a vestige of ancient work left',[23] its removal to the Master's Garden was sanctioned. However, the traditional entrance to the College was, and remains, from Trinity Street through a new Gate of Humility. Waterhouse provided an alternative design with a tower over the old entrance; but clearly preferred the architectural effect of his dramatic gesture on King's Parade.[24]

If the building is somewhat deadened by mechanical ornament, that is no more than to say the building is clearly high Victorian. Waterhouse's buildings were mostly new structures, the product of, and designed for, a machine age, and, as such, disdaining the gentling irregularities of hand craftsmanship. Increasingly Waterhouse was rejecting Ruskinian doctrine in favour of a system that gave more precise control over the texture of his buildings. Criticism of Caius centres on the exterior; it is generally admitted that the court works well from within (Pl. 71). Its bulk is nicely set off by a rich pattern of fenestration and a corner staircase of some grandeur. Originally there was also the unusual feature of a handsome wall with stone piers (at first conceived as a dwarf wall with railings between piers) to maintain the privacy of the Fellows' garden. Waterhouse's work was extended to include the rebuilding or refacing of half of the third side of the court, culminating in an added apse to the Chapel (Pl. 68). The interior of this is an imaginative grafting on to the seventeenth-century original. The new apse is entered through a smooth arch in grey stone clearly inspired by Brunelleschi's side chapels at S. Spirito, but discreetly decorated with nineteenth-century Gothic foliage.

In his Oxbridge commissions Waterhouse was continually brought up against tradition and the presence of other work of quality. The fact that he was re-employed at Caius suggests that again he

[21] N. Pevsner, *Buildings of England: Cambridge* (2nd edn., Harmondsworth, 1970), 81.

[22] A sketch of Feb. 1867 in Caius College archives shows that in the course of the design the gables were elaborated while the decoration of the tower was simplified and made rather less in tune with Dr Caius's exuberant Gate of Honour.

[23] J. Venn, *Biographical History of Gonville and Caius College*, III Cambridge, 1901), 147–8. The present state of the remains suggests that there was a good deal of justice in this claim.

[24] MS letter, AW to Revd Dr Lamb, Master (26 Feb. 1867). Caius College Archives. 'I cannot but very much hope that you will allow me to amplify scheme no I so as to make a more effective thing of the tower'.

satisfied his clients. And the range of styles and materials that he used is an indication of the flexibility of his approach. At Caius in 1883 he abandoned stone for brick with stone dressings in his new Lecture Theatre **[502]** that would be seen alongside Salvin's brick and stone Hall block. At Jesus in 1868 he not only used brick and stone but adopted a very similar pattern of fenestration to the original buildings **[290]**;[25] and his new gate piers were based closely on the existing seventeenth-century ones by Robert Grumbold. It is true that he did produce two designs for his new buildings at Jesus, one of which was traditionally Gothic with steep pitched roofs. But whether the decision to choose a low pitch and parapet matching the old buildings was his or his clients', the fact remains that he produced a design that handles the Tudor pattern with confident ease. And his reconstruction of the Hall roof and bay window, though identifiable as nineteenth-century work, is an exemplary piece of restoration.

A further Cambridge commission was at Trinity Hall where, in yet another idiom, he produced an attractive little stone building **[332]** (Pl. 65). He used tile hanging and much breaking up of the surface to afford interest on the cramped site. This building too could really be called Old English, adding one more colour to his palette of eclecticism within the one town.[26]

He changed styles again when he was called in at Pembroke College **[333]** (Pl. 72), where he was asked to examine the buildings and to report on 'the best way of providing for the College a group of buildings, as efficient, convenient, and architecturally effective as the site was capable of'.[27] A number of the college buildings were in a sad state, and the college was desperately short of space. The temptation must have been to sweep it all away and start afresh, especially since the college development fund had been accumulating for eighty years and now stood at some £70,000. In his first plan, Waterhouse did virtually begin from scratch; and his most outrageous suggestion was the complete rebuilding of the Chapel, an early work by Wren.[28] Though a larger chapel was clearly needed there does not seem to have been much difficulty in persuading him to drop this idea; but in the flat Cambridgeshire countryside it is hard not to sympathize with his desire to set a campanile on to his new chapel. 'I believe also,' he wrote, 'the Campanile will be a most effective feature from Trumpington Street, as well as from the First and Third Courts. It might be made sufficiently high to be the most conspicuous tower in Cambridge which suffers much from the lack of lofty architectural features.'[29]

He did not get his tower, but the college immediately accepted his suggestion of relocating the Master's Lodge. For this he designed a little mansion in the plushest version of his domestic Gothic with a variety of gables, bays, and oriels tied together by richly banded stonework and string courses. The design was sufficiently impressive to be fully illustrated in the French architectural press;[30] and the Master was evidently well pleased with his new status. Waterhouse also put up a block of

[25] His work there included rebuilding of chimneys for which he adopted precisely the pattern of the existing work. Sketches in DC refer only to chimneys for the Master's Lodge. It is, of course, possible that every chimney in the college was rebuilt by Waterhouse in which case he must be given credit for the use of matching bricks in different buildings instead.

[26] At Trinity Hall the new building stands on the site of the old College Gate, the masonry of which Waterhouse retained and rebuilt into the rear entrance of the college.

[27] MS report, AW to Master (14 Apr. 1870). Pembroke College Archives.

[28] Waterhouse's report did specifically recognize that the demolition of the Chapel would be contentious and offered the alternative of its retention while recommending that the decision be deferred until the other buildings had been decided. The Chapel was eventually extended by G. G. Scott Jun. who slapped a stone bay in a matching style on to the east end of Wren's brick building.

[29] MS report, AW to Master (14 Apr. 1870). Pembroke College Archives.

[30] E. E. Viollet-le-duc, *Habitations modernes* (Paris, 1875), ii. 162–4.

undergraduate rooms in bright red brick with stone dressings (Pl. 72). The vertical aspect of the steep pitched turret roof was as near as he got to his campanile, and the block's picturesque asymmetry makes a most effective feature from Trumpington Street. None the less, it was widely criticized when completed in 1874 for introducing a new material and French motifs to the town. This is a far cry from Jowett's confidence that in choosing Waterhouse he would 'avoid non-English styles and fancies'.[31]

The new rooms were soon followed by a commission to enlarge the Hall **[415]**; and, when it was finally decided that the Hall must be rebuilt, the opportunity was taken to build a separate new Library instead of placing it over the Hall (Pls. 73, 74). The Library is a decent enough building and attractively fitted inside, but it has a less happy mixture of Gothic and French Renaissance motifs. It may be that by 1875, when it was designed, Waterhouse was losing interest in Pembroke. His replacement Hall, on the other hand, was a convincing exercise in the Gothic of the fourteenth century, and he went to some lengths to make the work archaeologically acceptable.[32] Yet, perhaps because the college had by now been a building site for a full five years, the Fellows were becoming increasingly restive. One wrote angrily to the Master of 'the feeling that we have not been fairly dealt with by our architect', and complained that Waterhouse did 'not seem even to have the time, if he had the will, to devote the proper attention' to the work.[33]

Certainly Waterhouse was busy in the 1870s and he may well have found the conservative approach of the Fellows restrictive compared to his other commissions. In the end discontent reached such a peak that Waterhouse offered to resign.[34] As yet the plans for the Hall had not been finalized so he was not allowed to back out; but ill feeling remained, and in July 1878, shortly after completion of the Hall, his services were summarily dispensed with. He was far too well established for the dismissal to damage his career, though the ostensible reason, his lack of concern for the old buildings, has done much to damage his reputation. He was being asked to widen, lengthen, and re-roof the Hall and, as he had already publicly pointed out, all that remained of the old Hall was probably the core of the walls. Even so, he claimed, he would not have demolished them if he had been able to certify their stability.[35] Now, as he rather sadly wrote to the college 'the exterior of the hall on its principal front has never yet been properly seen by anyone'.[36] Yet, even though he had carefully reused old materials where possible, the replacement was not the old Hall, and was not valued. It was brutally modified in the 1920s and his attempt to re-create a convincing late medieval hall was obliterated.

He has been accused of being the last architect to be guided by respect for the character of old work; he was also the only architect to refuse to enter the limited competition to replace William Wilkins's screen with a range of undergraduate rooms abutting King's College Chapel. Scott, Street, and Burges were all happy to oblige but fortunately the Fellows there thought better of the plan.[37] In

[31] MS letter, Benjamin Jowett to Thomas Woolner. Bodleian Library.

[32] MS letter, AW to Master (8 May 1875): 'the difficulty has been to get authority for the exact form of window tracery prevalent at the date (1360). I went to Winchester yesterday to make drawings of the Bishop Eddington work (he died in 1366) but the windows there are of 4 lights not of 3. At Wells, same date, they were of 2 lights—so at Westminster Abbots Hall and at Penhurst'. Pembroke College Archives.

[33] MS letter, Budd to Master (19 Feb. 1875). Pembroke College Archives.

[34] MS letter, AW to Master (31 Mar. 1875). Pembroke College Archives.

[35] Letter to *The Times* (29 Mar. 1875).

[36] MS letter, AW to the Fellows (17 Feb. 1879). Pembroke College Archives.

[37] King's College Archives. See A. Doig, *The Architectural Drawings Collection of King's College, Cambridge* (Amersham, 1979).

this light the condemnation of Waterhouse at Pembroke does seem rather unfair. The real point, however, is that his new buildings were designed with a confidence that meant them to be modern and to tell by contrast with their neighbours.

The surviving collection of his works in central Cambridge is a remarkably comprehensive display of his ability to adopt a variety of stylistic dresses to provide buildings of individual character for a variety of fiercely independent corporations. They suffer, if that is the right word, from being emphatically of their own age rather than anonymous approximations to another. If that is to admit that their age was in general more concerned with go-ahead modernity, Waterhouse was also ready to accept its values. He did exactly that at Girton where he was lucky enough to collect the commission for a brand new college on a green-field site.

In some respects the story of Girton **[346]** (Pl. 75) belongs with Waterhouse's work for the northern universities, for when first built it was emphatically not a part of Cambridge University. None the less, it is important here because it showed the way that he and his client, Emily Davies, set about designing a collegiate building from scratch. The struggle for women's education is a story in its own right; but the decision to found the college that eventually became Girton seems to date from a meeting of schoolmistresses in Manchester in 1866. Emily Davies was the moving force, and was already thinking in specific terms, for the same year she wrote to her friend Anna Richardson that her new college was:

> to be as beautiful as the Assize Courts at Manchester, with gardens and grounds and everything that is good for body and soul and spirit. I don't think I told you how intensely we enjoyed the beauty of the Assize Courts. I have seen no modern building to be compared with it, and the delight one felt in it made one realize how much one's happiness may be influenced by external objects.[38]

A year later she wrote again 'I hope we shall get Waterhouse, I am anxious that the building shall be as beautiful as we can make it'.[39] That, however, was two years before the Girton site was bought, when Emily Davies clearly did not want to be in, or near, Cambridge. There was strong opposition in the university to this radical new departure. The whole concept of educating women was far from having majority support in 1870;[40] and Miss Davies preferred to site the new college in Hitchin. It would be easy for sympathetic scholars to travel out by train to give lectures; but the substantial body of opponents in the Regent House could have no jurisdiction.

In 1869 Waterhouse was asked to report on four sites in Hitchin **[316]**; and the next year did some minor designing to adapt the mansion that was being bought for the college **[335]**.[41] In turning to him, Emily Davies was not only in thrall to the Assize Courts. The movement for the education of women already included a number of his friends and connections. His brother-in-law, Thomas Hodgkin, was involved in the early financial arrangements for the college. Supporters of the venture included Mrs Kershaw, widow of the Manchester cotton spinner and Liberal MP whose tomb Waterhouse had recently designed **[181]** (Col. Pl. 8); and the general committee included Quaker connections such as Russell Gurney, MP, and his wife, and others, such as Lady Goldsmid, whose

[38] MS letter, Emily Davies to Anna Richardson (1866). Girton College Archives.

[39] MS letter, Emily Davies to Anna Richardson (1867). Girton College Archives.

[40] Waterhouse's own cousin, Nicholas Waterhouse, seems to have been a firm opponent of women's education (Letters of Nicholas Waterhouse, Picton Library, Liverpool).

[41] There was even talk of an iron building for the college, though whether this was to be designed by Waterhouse is uncertain.

husband was such a close friend of Theodore Waterhouse. So Waterhouse was, in many ways, the natural choice for the new college, when, early in 1871, Miss Davies took the decision to move from Hitchin to a rural site at Girton, a mere three miles from Cambridge, but still conveniently outside the city and free from its or the university's jurisdiction.[42] Now that she had a site, however, Emily Davies was looking for precisely the same qualities as Pembroke in her new building—efficiency, convenience, and architectural effect—but with an overriding fourth concern, for economy. She also had firm ideas about designing for healthful living which Waterhouse had to accommodate. On the open windy site at Girton he designed a building that fulfilled her conditions. For economy he planned a building that could be extended in stages; and building took place in a whole series of campaigns from 1872 – 99 **[346, 414, 459, 501, 531, 621]**. The rooms were laid out along corridors, instead of the traditional staircases, which made for efficiency. There was careful attention to lighting, making the rooms pleasant to live in; and the addition of laboratories, and even a swimming bath, made Girton a remarkably complete little academic community.

The bright red brick and hot terracotta are striking, though perhaps no more assertive than the contemporary new college at Keble. Some would call the building grim. It is certainly spare; but then the constraints on cost meant that there was little scope for ornament. Still, there was a positive attempt to achieve interest by varying the shapes of dormers and subtle banding of the roof tiles. There was also a ration of picturesque features; an open loggia, a spiral staircase turret, and an ingenious balcony for the Mistress's rooms (Pl. 76). The composition culminates in the tall gate tower of the 1880s, which is a convincing reinterpretation of the traditional college gate tower such as he had carefully retained at Balliol. However, at Girton, Waterhouse was not tramelled by exact precedents and the sources for most of the details are to be found in a host of anonymous buildings. Some dramatic features, such as the Hall gable, are virtually repeated from his own work, and at least one detail reproduces a sketch he made of work by C. F. Hayward shown at the Architectural Exhibition in 1858. But the overall impression is of a consistent building, rambling and perhaps rather severe, that exactly reflects the earnest nature of the institution. The style is again more nearly Old English than Gothic. Its characteristics are the straightforward expression of function in traditional shapes.

Taken together, the Oxbridge works show how, at the peak of his career, Waterhouse had the courage to develop his own versions of styles, and to select and adapt forms to suit each particular commission. If Manchester Town Hall demonstrated his ability to devise a rich modern Gothic, these collegiate buildings show how far and how freely he could range in choosing precedents. Each building is given its own individual character; yet there is a consistently thought out relationship with tradition and older structures. Their straightforward solidity marks them all out as his work; yet none is in the round arched mode for which he is so well known. That was a style he was developing, at much the same time, for the Natural History Museum in London.

[42] Girton was so far from Cambridge that the captain of the city's fire brigade solemnly declared that, if the college was built there, he would be unable to undertake to deal with any fires in it. In a typical reaction the Girton girls set up their own fire brigade.

9

The Natural History Museum

THE building is not a Classical one, although it has Classical traditions in its balanced symmetry . . . Nor is it a Romanesque building, notwithstanding the varying recurrence of the round arch. It is not a Gothic building, though having steep gables and an arrangement of roofing which are eminently Gothic in *motif* as they are effective and picturesque in outline and grouping. It is, in short, a Victorian building and no other, designed upon principles which have informed the great works of all time, but adapted to the wants, using the materials and employing the methods of the age in which we live.[1]

The *Magazine of Art*'s encomium nicely summarizes Waterhouse's approach to design in this, the greatest of his London commissions. The Natural History Museum **[243]** (Pls. 78, 79, 80), with its Gothic towers, cathedral portal, and dramatic halls was the third great masterpiece of his career. Its stylistic ambiguity makes it an odd monument for one whose reputation so far was as a goth. But at Oxford and Cambridge he had shown himself ready to take a boldly cavalier approach to mere style. Now he devised a new style, a sort of Romanesque revival, that was ideally suited to the new material he selected; and he was instantly established as the leading figure in the use of terracotta. This museum was effectively the first complete building on the South Kensington site.[2] When it was opened in April 1881, the Science School was the only building of the South Kensington Museums to have a street frontage,[3] and the principal entrance to the Museum of Science and Art was set back amid a clutter of other structures and concealed by its partially completed library block. In fact the Natural History Museum itself was by no means as complete as intended. The story of its building reveals a commission as perilous as any in Waterhouse's career, and its success was no easy achievement.

Like many of his great commissions, the museum was a long time in the building. Waterhouse was first approached in February 1866, when he was still relatively new on the London scene. At that stage he had not even been formally invited to submit designs for the new Law Courts **[278]**, which was the scheme uppermost in his mind.[4] He was asked to accept the commission as executant

[1] *Magazine of Art*, 4 (1881), 36, quoted in 'The Natural History Museum' by John Olley and Caroline Wilson in D. Cruikshank (ed.), *Timeless Architecture*, I (London, 1985).

[2] For a detailed study of the museums area of South Kensington, and of Waterhouse's work at the Natural History Museum, see *Survey of London*, XXXVIII. *The Museums Area of South Kensington and Westminster* (London, 1975). For the Natural History Museum alone, M. Girouard, *Alfred Waterhouse and the Natural History Museum* (New Haven, Conn., 1981) and Olley and Wilson, 'The Natural History Museum' both provide excellent accounts.

[3] Now the Huxley Building of the Victoria & Albert Museum, which itself was not known by that name until 1901.

[4] Girouard suggests (*Alfred Waterhouse*) that Waterhouse's ability and his tact in the fiasco of the request to draw up specifications for the Law Courts Commission may have impressed First Commissioner Cowper and influenced him to give Waterhouse the Museum commission.

architect of the late Captain Fowke's designs for a museum, that had first been conceived by Richard Owen as long ago as 1859. It comes as a bit of a surprise that an ambitious and well-connected young architect should accept an apparently secondary role. He may have hoped to replace Fowkes's confection with something of his own, or, more likely, to have the opportunity of demonstrating his efficiency to the politicians who would commission the Law Courts. At all events he can have had little time or energy for thinking out a creative design.

A change of government in June 1866 left the whole proposal in limbo, and one suspects that the busy Waterhouse was more relieved than anything else. When the idea was revived in late 1867, he had failed in the Law Courts, and had not yet won the Manchester Town Hall. Here at least was another chance of a major public building, even if not entirely his own. It was as well that the basic arrangement of the museum had been fixed by Owen and the design worked out, for he was busy working up his drawings for the second Manchester competition. Nevertheless by March 1868 he had succeeded in the formidable task of virtually redesigning Fowkes's building as something entirely new. Once again a change of government forced delays. The new First Commissioner had visions of a grand metropolitan development, and wanted the new museum transported to the Embankment. Perhaps the delay was just as well, for he could adapt his design to the new curved site by doing little more that bend the façade, while he pressed on with designs for Manchester. Still it must have been trying to see this golden prospect continually receding.

A large part of the success of Waterhouse's design, when it did come, must be due to his ability to handle committees and to charm a wide variety of interested parties, as well as to the provision of adequate flexibility within the basic concept. Like Barry at the Houses of Parliament, he had to learn how to cope with government departments, changes of government, and the occasional swift departure of key civil servants. By early 1870 the old site was in favour again, but there was a new First Commissioner. This was Acton Smee Ayrton, who ought to have been favourably impressed by Waterhouse's reputation for economy. He was to hold the post for the entire period of building; but he nearly wrecked the scheme by demanding a reduction in the budget from £500,000 to £330,000. However, detailed design work could now go ahead. Waterhouse was not worried about the cost limit —he was facing exactly the same problem at Balliol. He merely designed the museum to be built in two stages. The main façade, great hall, and single-storey study galleries were all essential, and should come within the budget; so the side wings and rear could be left for the present. They were to be left for ever. Building costs were rising, and, in 1872 when the work went to tender, he had to ask for a further £20,000 even to build the reduced version of his museum. Even so things did not go smoothly, and the final cost rose to £412,000. In view of the complexity of the building that is not altogether surprising; but Waterhouse was attacked in Parliament. Hunt, the surveyor to the Office of Works, called his work a scandal and a shame and urged proceedings against him for 'serious dereliction of duty' and 'negligence or indifference to his moral, or, as I humbly think, his legal obligation'.[5] Waterhouse had for too long been used to producing workable designs and submitting detailed accounts that satisfied economy-minded practical men for that attack to be more than so much hot air.

The building has all the clarity of his usual planning skills in working out Owen's original conception. Its cathedral portico leads (Col. Pl. 3) into the vast main hall, designed to contain an

[5] Quoted in *The Museums Area of South Kensington and Westminster*, 215.

index collection which would display in resumé the whole of God's creation (Pls. 81, 82). Beyond was a further hall for British natural history, while galleries for displaying more detailed exhibits were stacked on three floors facing the Cromwell Road. Owen had wanted a single-storey museum; but restrictions of site and cost forced him to agree to galleries on three floors, thus allowing his architects some scope for architectural display. Behind the façade single-storey study galleries were laid out, alternately wide, for public access, and narrow, for private study.[6] These were reached by short bridge corridors from the main ground-floor galleries leading back to a long east–west corridor that opened into all the galleries in each wing. Finally the gently sloping site allowed the whole museum to be raised up on a floor of well lit basement workrooms, looking out along the façade on to a sunken garden.

In addition to its simple basic layout the museum was to be thoroughly practical. Waterhouse's main concern was lighting, for there was to be no gas light in the galleries.[7] The index halls and the study galleries were all top lit, and the façade galleries drew ample daylight from the long rows of windows on either side.[8] Ventilation was more of a problem. He was unable, possibly because of political interest, to make use of his favourite heating contractors, Hadens of Trowbridge. Instead the contract went to F. P. Stevens & Son, whose system required a complicated arrangement of flues to draw hot air up from the basement, through the galleries, and so out through the ventilating turrets on the roof. Their system was to a great extent a failure, and needed almost immediate modification. But at least Waterhouse was able to defend his picturesque skyline on the grounds of practicality,[9] and the great chimney in the north-eastern tower still functions as the flue for area heating for the whole group of museums.

What marks this building out, however, as central in Waterhouse's career is the use of terracotta.[10] The walls are of load-bearing brick, but all the façades are faced in terracotta in buff or pale blue tints. The iron columns of the galleries are sheathed in the same terracotta, and only on the rear walls is the brown stock brick left visible. It was a remarkably bold decision to face the whole of so large a building in this new material. The choice led to endless problems;[11] but it afforded him, as no other material could, the opportunity to concoct a rich and varied scheme of decoration inside and out. The modelled figures could directly reflect the museum's own contents, with extinct species on the eastern, geological, galleries (Pl. 77) and living on the western, zoological, wing. The arches of the index museum were alive with monkeys as a reminder of man's evolution, while, originally, Adam himself stood on top of the main gable.[12] Birds, beasts, and fishes were on every surface. They

[6] These galleries were designed to be linked by double-sided cases filling the arcade, so that students could easily remove specimens for study. The sceme had the sort of logic that appealed to Owen; but in the course of construction his departmental curators demanded various alterations to suit their own particular wants, and the arcades were bricked up. In the western galleries the basic arrangement has been obscured by later alterations. In the east wing, the last of the galleries to retain the original layout has recently been refurbished.

[7] At a late stage it was decided, in keeping with South Kensington policy, to open the museum in the evenings for the working classes. This required the insertion of gas lighting, but only to the 2 index galleries to which, presumably for reasons of supervision, the artisans were to be restricted.

[8] By setting the index hall lighting in the lower slope of the roof, and by using lightly tinted glass, Waterhouse hoped to avoid the problem of reflection from the glass cases. Problems of over-exposure of exhibits led to the gradual use of curtains and other screens in various galleries, so that now a good many spaces are more gloomy than he intended.

[9] Waterhouse had had fortunate support for his skyline when the Fire Brigade demanded large tanks at roof level because of the danger of large numbers of exhibits preserved in spirit. This allowed him to persuade Ayrton to restore funds previously cut for an extra storey on his east and west towers.

[10] This aspect of the commission is discussed in more detail in Ch. 21.

[11] See Ch. 21.

[12] In one version Waterhouse designed a flat parapet, allowing Adam and Eve to stand side by side. In the event architectural considerations triumphed over fairness, and Eve vanished. Adam's fall from his perch during the Second World War was a belated piece of poetic justice.

peered from a series of splendid panels set into the Portland stone gate piers, they swam round the gallery columns, and stood beside the staircase. Perhaps the most dramatic ones were the free-standing creatures that stared down from the first floor or stood proudly on the parapet between each gable (Pls. 81, 164, 165, 166).

The design and manufacture were an extraordinary feat. Each creature was drawn by Waterhouse himself, who must have revelled in this concession to the artist in him.[13] The drawings were checked by Owen, and then modelled by M. Dujardin of Farmer & Brindley, before being sent to Tamworth for casting. Yet even here he did not lose sight of the principal object of creating a fine building. The animals were carefully and regularly spaced, so that they read as decoration first and foremost. And they were backed up by other decoration, geometric and foliate patterns on clustered capitals, window heads, and string courses with richly stepped corbels and complex arrangements of double and treble colonnettes in the balustrades (Pls. 83, 167, 168). As if this was not enough, Waterhouse complemented his moulded decoration with contrasted colours of terracotta, using the more expensive blue sparingly to enliven arches and in decorative bands made rich by alternating simple shapes.

The Natural History Museum is an extraordinary *tour de force*, brought off in the face of government parsimony.[14] It can be criticized today as being too much architecture and too little museum. Even a contemporary critic felt that walls of mottled fawn and dun colour were hardly the best background for exhibits of dun coloured bone![15] But the index hall still seems a suitably dramatic space for the great dinosaurs that inhabit it. The cathedral proportions are grand, but the terracotta menagerie and the flying staircase inject an air of vitality into what might easily have become a sterile academic storehouse. There was plenty of scope for varied arrangement of the different galleries to suit the needs of the different keepers. Waterhouse's design was flexible enough to allow him to abandon the strict logic of his alternating single-storey galleries as necessary; and even during construction he accepted a series of modifications suggested by different keepers. The joinery was not all designed in his office, but it followed his intentions closely enough; and the collection of spaces worked well for almost a hundred years. Now one can hardly separate the obvious enjoyment of the throngs of visitors in the building from their delight and wonder at the exhibits. On that basis alone one would be justified in calling this a masterpiece.

In the 1870s Waterhouse linked it with Manchester Town Hall and Eaton Hall as one of his three greatest works; but in 1893, when he was asked to send two exhibits to the Chicago World's Fair, it was the Natural History Museum that accompanied his more recent drawing of the National Liberal Club.[16] It was not by any means his most expensive commission, but it is probably the one in which he is most clearly himself. He transformed Fowkes's Florentine Renaissance design into his own Romanesque revival, retaining only the pattern of the first-floor windows. He used his understanding

[13] His sketches filled 2 large volumes, and are preserved in the Natural History Museum Library. The extent of Owen's involvement as adviser and corrector of the drawings is a story that has yet to be told. Presumably, however, he had a considerable say in the selection of the species.

[14] One might add that there were also pressures from the different keepers, who formed a quarrelsome and disparate collection of scholars.

[15] *Builders* (May 1881).

[16] More recently the perspective of the Natural History Museum, along with one of the sketches for the terra cotta ornament, was selected for inclusion in HRH The Prince of Wales's exhibition, 'A Vision of Britain'. The 2 images were chosen to illustrate 2 of the Prince's 10 principles, Art and Hierarchy, applied to buildings of the South Kensington area. Waterhouse is almost the only late Victorian architect to have drawings included in that exhibition.

of Gothic to give a picturesque skyline to the immense façade, and he let his imagination run riot with the decoration. Yet underlying all this is the sensible practicality of its clear plan and washable surfaces. Had he not already won the RIBA's gold medal for Manchester Town Hall, he would surely have deserved it for this.

10

Eaton Hall

To win a commission from the richest aristocrat in the land for a rebuilding of his principal seat to the tune of upwards of £600,000 ought to be the ultimate accolade and an unparalleled opportunity. Yet Eaton Hall **[315]**, the immense pile which Waterhouse put together for Hugh Lupus Grosvenor, third Marquis and later Duke of Westminster, has not quite made for Waterhouse the reputation that Cardiff Castle has for Burges or even Highclere for Barry. One reason is, of course, that the house has been demolished and we are left only with the fragments. These are substantial and comprise the Chapel, stable block, parrot house, laundries, a lodge, and several cottages; and there are also the drawings. Yet even so it is difficult to recover the true flavour of this stylistically complex and rambling palace.

Waterhouse's other greatest works, Manchester Town Hall **[284]**, the Natural History Museum **[243]**, and the National Liberal Club **[514]**, all have a stylistic consistency that was not so clearly part of Eaton Hall. All three are also public buildings, and their architectural qualities are not those of a private commission. The design of Eaton ought perhaps to have been more easy to control: after all Waterhouse had only to satisfy the Duke and his wife. There is some slight evidence that he was not entirely successful in this,[1] though the Duke was sufficiently pleased to retain his services from 1869 to 1899, which hardly suggests real displeasure. More likely, it was the very length of this commission that was the root of the trouble.

When he was first commissioned in 1869 Waterhouse had no great reputation as a country house designer. True, he had already designed a number of middle-sized mansions for successful magnates; and his house for the Liberal MP, J. W. Pease, at Hutton Hall (Pl. 108) was well under way, as was Lord Selborne's new mansion at Blackmoor (Pl. 106). However, it is unlikely that the social position of Roundell Palmer, who was not created Lord Selborne until 1872, still less that of Joseph Pease, would have caused the Duke to be even aware of their houses.[2] The most likely supposition is that he

[1] Gervas Huxley, *Victorian Duke* (Oxford, 1967) 134, records that the Duke 'was heard to say that if his grandson and heir had any sense he would pull it down and build another house'. This biography, of course, reflects the prejudices of the first half of the 20th cent. The recollections of Lady Lindsay of Dowhill (formerly Duchess of Westminster) are almost entirely of comfort, splendour, and quantities of flowers, though she does recall the long passage (M. Waterson (ed.), *The Country House Remembered* (London, 1985)). In any case the 1st Duke was canny enough to employ several different architects in and around Eaton. E. R. Robson was commissioned to design St John's School, Chester, in 1882, Bodley to design St Mary's Church, Eccleston, in 1899; and from the start 2 local architects, T. M. Lockwood and John Douglas were given a number of smaller commissions. Had Waterhouse really failed to give satisfaction he could easily have been replaced.

[2] Equally Waterhouse's better known local house, Hinderton Hall, built for the Liverpool magnate, Christopher Bushell, can hardly have offered any precedent for what was to be done at Eaton.

was appointed on the strength of his reputation as the winner of the Manchester Town Hall competition. The Duke was to be housed by the best-known architect of the day. Besides which, Waterhouse had already satisfied the Grosvenors in one or two minor commissions.[3] None the less, it is significant that he did not go on to design further mansions for the aristocracy, and only found clients for large houses among the newer peerage and industrialists such as Lord Wolverton, Sir Julian Goldsmid, or Charles P. Stewart. Maybe he found himself temperamentally out of tune with the old style aristocrats and their patronage. His master at Eaton was, indeed, unusual in his class, seeing himself as 'not so much a private millionaire as the head of a great public institution or trust'.[4] This serious-minded view of what a great nobleman should be accords more easily with Waterhouse's own background and inclinations.

However, the extent of the Eaton commission, and the vast number of alternative designs proffered, suggest very strongly that Waterhouse accepted the work with enthusiasm. If the finished product did not, in some eyes, fulfil its potential as a complete work of art, that is because of the nature of the commission itself. There were genuine problems for both architect and patron; and it is important to remember that the commission was for rebuilding, not for a new creation (Pls. 84, 87). The original Eaton Hall was a substantial seventeenth-century house by Samwell, which had been extensively rebuilt and Gothicized first by Porden and his assistant Gummow, then by William Burn.[5] The main state rooms had been retained throughout, and much of Waterhouse's difficulty arose from the decision to keep them as the centre of ducal ceremonial. Besides there seems to have been no clear decision from the start to embark on an almost complete rebuild.[6] The work took place over a thirty-year period, so that for almost the whole of his time as head of the family,[7] some part or other of the Duke's principal seat was a building site.[8] This alone makes it difficult to assess the architect's success.

Waterhouse was thus not asked to design a palace for a duke, but to refurbish and extend an existing one; and there was a good deal of modification as the work went ahead. The first job was the replacement of draughty cast iron window frames in Porden's state rooms.[9] By all accounts this was essential for the barest comfort. There were plans, too, to replace the Great Library; but some of the earliest drawings are only for rebuilding the laundry. These show a good deal of the original work retained in a project that was apparently complete by 1874. Then in 1877 plans were drawn for a completely new laundry, on a site much further from the Hall; that was completed about 1880.[10]

[3] A new sign for the Grosvenor Hotel, Chester, in 1865 and a new reredos for Aldford Church in 1870, commissioned by the 2nd Marquis.

[4] Gervas Huxley quoted in M. Girouard, *The Victorian Country House* (New Haven, Conn., 2nd edn. 1979), 2.

[5] *Country Life*, 2 (1897), 182 ff.; also 9 (1901), 496 ff.; and 47 (1920), 724 ff. The main parade rooms and some of their plaster work were retained in Porden's extensive rebuilding. Gummow had the commission as site architect, but the extent of his individual contribution, if any, is uncertain. Burn was responsible for a good deal of work in the wings and the Chapel and Library.

[6] The main contract, to George Smith for a total of over £580,000, was under way by May 1870. The survival of survey drawings, dated 10 July 1871 (DC) relating to basement windows, suggests that full plans were not drawn from the start but that the work was essentially conceived piecemeal.

[7] Hugh Lupus Grosvenor was born in 1825, inherited the title of Marquis in 1869, was created 1st Duke of Westminster in 1874, and died in 1899.

[8] The first certificate was paid in May 1870, the last in Aug. 1893; and the cycle was set to begin all over again with the refurbishing of the kitchen in 1898–9.

[9] Detail drawings and correspondence survive from 1869 for the insertion of wooden sashes and shutters (DC). It is evident from notes on the drawings that a good deal of the original stone was to be retained, though a number of mouldings had to be replaced or recut. The same series of drawings shows the first proposals for the upper floors of the main front, where the Porden/Burn windows were retained but given square heads. The masonry on the main front, which did not conform to Waterhouse's preferred alternate coursing, suggests that much of the old facing was also retained (see n. 22).

[10] Over 1,000 drawings for Eaton Hall survive, almost all in

Such chopping and changing may have improved the final outcome, but it was a nightmare for the designer. The first major extension to the house was the new Library wing to the south of the old house. A new billiard room wing (containing bachelors' bedrooms) was ingeniously joined to the north-west corner of the old house. Then new floors were added over the parade rooms, the rooms themselves were refitted, but not rebuilt, and the bulk of the west façade was retained. Meanwhile a new and distinct kitchen wing and a separate private wing were being added to the north and north-east, and a new grand staircase to link the Library wing to the old mansion.

Such a commission would be taxing enough; and needed all Waterhouse's skill to make a coherent whole out of such disparate elements. The solution has all the virtues of clear planning (at least on the ground floor) for which he is justly famous (Fig. 5), and nicely accommodates many of the preferences of Victorian country house planning that had been recently set out by Kerr.[11] There had to be a formal approach, but also easy access to both gardens and stables. He had to provide rooms of state for ducal entertaining, but business rooms must be separated from pleasure. The Duke's duty to entertain required numerous guest rooms which were fitted in on all sorts of levels, even on the ground floor in the kitchen block and behind, or rather beneath, the grand staircase. There was also the need to segregate children from grown-ups, bachelors from married couples, guests from family, and, above all, servants from gentry. There had to be extensive provision for cooking, but the kitchens must be as convenient for the state dining-room as the private wing. Visitors' servants had to have access to their gentlemen and ladies; and there was the whole gamut of household services to be provided for. Fortunately many of these could be in separate buildings. Finally there had to be adequate plumbing, and the house boasted thirty-three WCs and eight bathrooms.[12] Waterhouse linked all the new units by a single corridor running roughly 330 feet from the south to north and opened out the Saloon and, later, the entrance hall to form a cross axis (Pl. 93). This grand corridor ran from the Library and grand stairs past the main parade rooms, bisected the kitchen block, and continued north towards the steward's office where another corridor ran across the northern end from the Chapel and stables to the private wing which looked out over the gardens of the main block.

This layout has the same bold logic that underlies the plan of Manchester Town Hall and is a tribute to its designer. But only the basement and ground floor were on a continuous level. The Great Dining-Room, Saloon and Great Drawing-Room rose through one and a half storeys, while the entrance hall and organ lobby at the south end of the grand corridor were of double height, leaving the first floor seriously interrupted. Confusion was further compounded by the Library wing being almost one and a half storeys in height, so that its first floor was about 5 feet above the first floor of the main block. There were further changes of level in the billiard room wing and kitchen block.[13] The consequent variations of level on the second and third floors were yet more complex. Guest rooms over the Great Drawing-Room were reached by a half-flight from the first floor of the Library wing: those over the Great Dining-Room were reached by four steps from the half-landing of the north-east

the RIBA Drawings Collection, and further work remains to be done in analysing the office numbers to determine the precise design sequence, particularly as most of the drawings are undated. Plans for the first laundry have job numbers in the sixties indicating an early stage in the work. Drawings for the new laundry are signed and dated 6 June 1877 (DC).

[11] Robert Kerr, *The Gentleman's House or How to Plan English Residences from the Parsonage to the Palace* (London, 1864).

[12] In spite of this apparently lavish provision, some sets of guest rooms had no easy access to a bathroom and in one case a gentlemen's dressing-room had to double as a bathroom.

[13] If the private wing and Mr Allen's house are counted as separate levels, being unconnected with other wings at first-floor level, no less than 10 different levels are included on the first-floor plan of 1885 (DC).

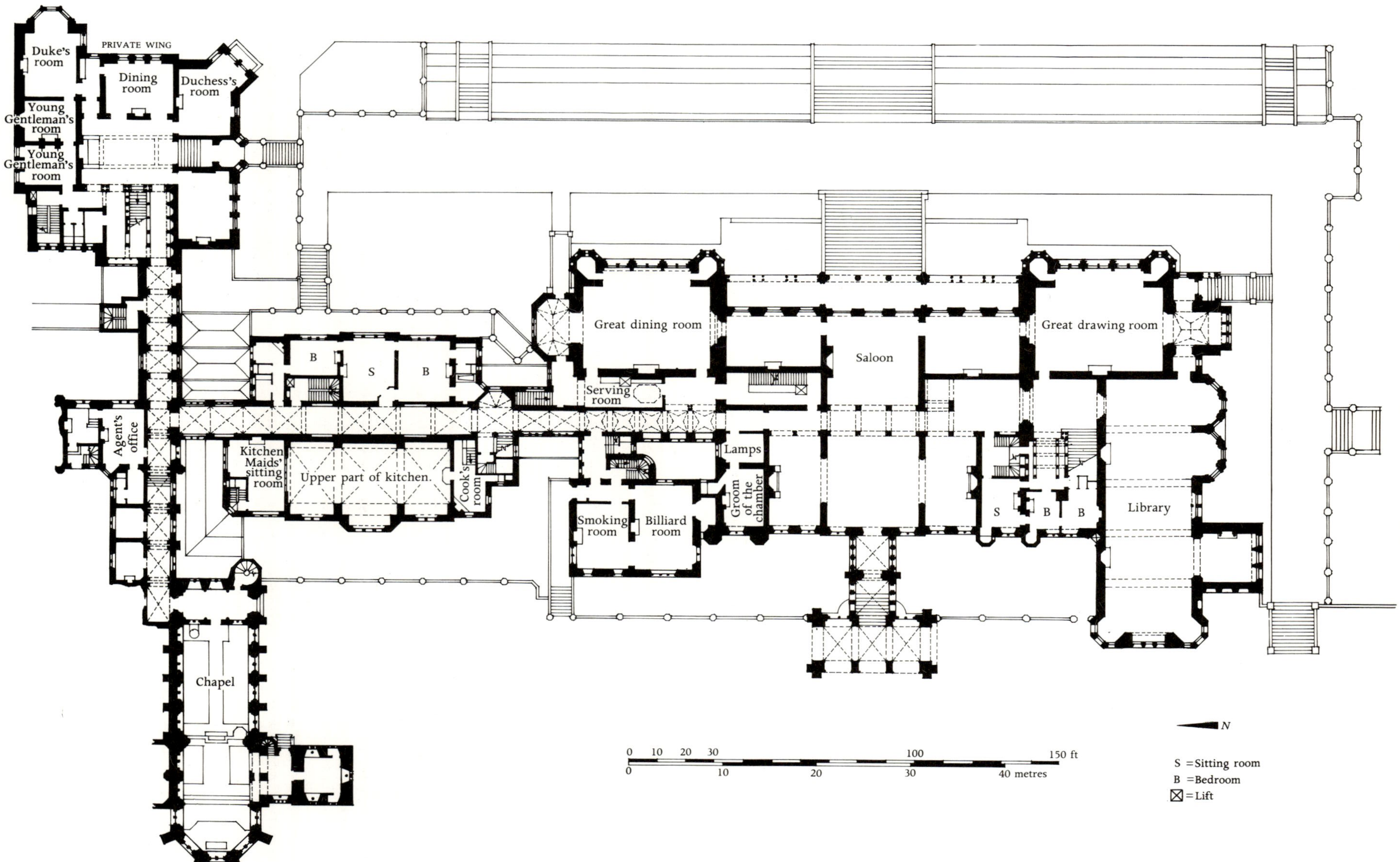

FIG. 5. Eaton Hall (1869) **[315]**, plan of ground floor (RIBA)

stairs. The second floor of the main block was for further guest bedrooms and included nurseries for visiting children. The second floor of the Library wing was for servants but connected to the main block by a dramatic bridge corridor through the roof timbers of the great staircase.[14] Third-floor servants' accommodation was in four entirely unconnected blocks over the Great Dining-Room, Saloon, Great Drawing-Room, and over the west façade.[15]

It was an extraordinarily complex pile in its upper floors (Fig. 6), and, since the commission was handled in separate parts, it was only with considerable ingenuity that Waterhouse made it function at all. But the very complexity of the circulation above the parade level may well have made the house confusing to guests and socially unfortunate in that servants and gentry had to share several routes.[16] Only the private wing, a separate and completely new structure, had the clear social separation required by the Victorians; and the discomfort of his guests may well have been in the Duke's mind when he told his son that it would be better to pull the whole thing down and begin again from scratch.[17] It may indeed have been a mistake to require the retention of so much of the earlier building. If so it was the client's error, and one can only admire the ingenuity with which Waterhouse handled the increasingly complex problems of a variety of different levels and structurally distinct blocks, while at the same time maintaining a unified façade.

The building, as it stood before Waterhouse arrived, was illustrated in Eastlake's *History of the Gothic Revival*[18] but it can hardly, in the 1870s, have been seen as reflecting credit on an educated nobleman. It was in the frivolous ignorant Gothick of the first half of the nineteenth century and as such an affront to serious-minded Victorian stylists. However, even by 1870, the passion for true Gothic was almost past, and it was as well for the Duke that he engaged a modern goth rather than an antiquarian. The end result owes a good deal more to French chateaux than to the approved centuries of medievalism.

Waterhouse's designs for the Library wing illustrate this development (Pls. 85, 86). His sketches were all drawn in late 1869 or 1870,[19] and the earliest show that he considered retaining the

[14] This entered the main block at a half-landing between the second and third floors.

[15] Some idea of the complexity of the whole palace is gained from the following summary of the 1885 survey plan: the basement, excluding kitchen and scullery, contained rooms 1–58; the ground floor, excluding hall, Great Dining-Room, Ante Dining-Room, Saloon, Ante Drawing-Room, and Great Drawing-Room, contained rooms 59–81; the first floor contained rooms 82–120; the second floor contained rooms 121–69; the third floor contained rooms 170–88. This analysis excludes all WCs, closets, etc., and all circulation space.

[16] Vertical circulation for this complex jigsaw was provided by a grand stone staircase, which Waterhouse added at the southern end of his great corridor forming a link between the old house and his new Library wing. These stone stairs reached only to the first floor. There were two further stairs used by the gentry. The iron staircase, a service staircase from the basement, rose to the second floor and provided the only access to second-floor rooms over the Great Dining-Room. The red stairs rose the full height of the building to the third floor, and appear to have been retained from the old house (though most probably rebuilt in their upper flights). The red stairs adjoin the grand staircase, but lie within the original mansion. They cannot have been intended to act merely as a service staircase since they are the principal means of access to rooms on the second floor. I assume that they were the secondary staircase of the Burn/Porden mansion. A ground-floor plan (DC) shows the staircase and parts of the old house in blue ink, the convention used by Waterhouse to indicate existing structures. The steps are as shown in the final survey plans of 1885, though those on the upper flights follow a different pattern to accommodate the bridge corridor. A photograph taken by Edward Hubberd shortly before the demolition showing the stairway from second to third floor at the centre of the east front indicates that the Burn/Porden work in that area was left untouched. A further staircase served the second floor of the Library wing, and 3 further staircases (at least one of which was part of the original house) ran up to the eastern sections of the third floor. The billiard room block had its own staircase! In fact 2 staircases, since the WCs were set on a lower first-floor level, almost a mezzanine. The bachelors' bedrooms had a separate staircase and had no access to the lower part of the first floor.

[17] Huxley, *Victorian Duke*, 134; 'Few would regret the destruction of a building which, in today's opinion at least, was so lacking in architectural merit.' Huxley's description of the private wing was, however, thoroughly complimentary, emphasizing its convenience.

[18] C. L. Eastlake, *The History of the Gothic Revival* (London, 1872). Facsimile edn. with introduction by J. Mordaunt Crook (Leicester, 1970), 77, where it is listed simply as the work of Porden with no mention of Burn.

[19] An early drawing for the library wing (DC) dated 23 Dec.

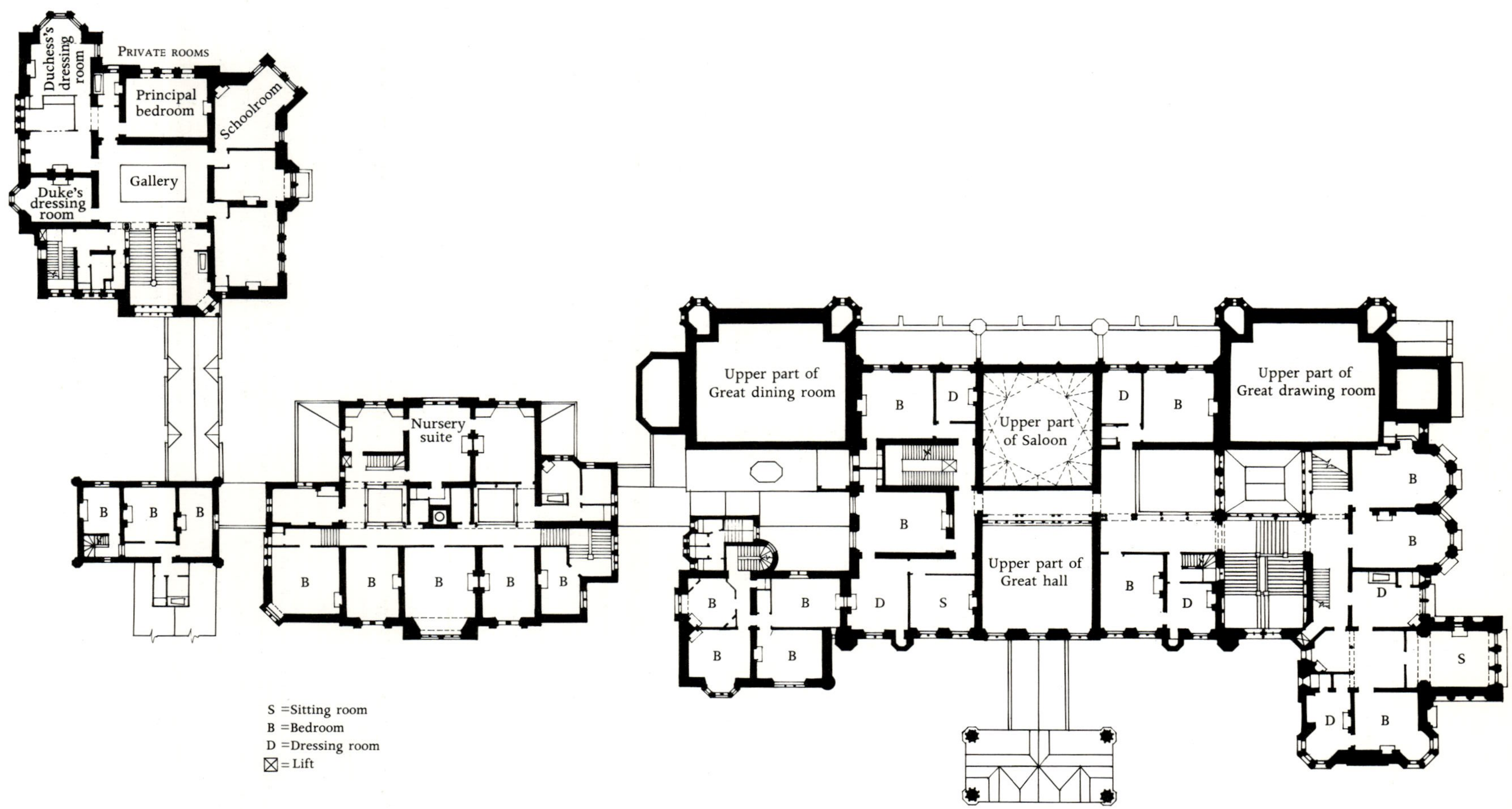

FIG. 6. Eaton Hall (1869) **[315]**, plan of first floor (RIBA)

castellated skyline and ogee pinnacles of the old house by Porden, as remodelled by Burn. At that point he was planning a low block of two bays to balance the proposed billiard room wing with the Library itself set some 70 feet south of the old mansion.[20] This was to be enriched with a dramatic tower 120 feet high topped by a 60 foot spire in high Gothic at its north-western corner.[21] The south façade is worked out in some detail with a picturesque corner turret, traceried windows, and elaborate blind tracery in the gables. The second design is noticeably more restrained in its Gothic, though the layout is not altered and some features, like the Gothic angle turret, are merely transferred to the inner block. However, the thinking on the roof line of the main block is much advanced, displaying a clear intention to replace the whole of the Burn/Porden skyline, at least to the west. The Library wing itself now has two towers, though more modest. The outline already appears more determinedly French and the two-stage tower is reminiscent of the almost contemporary one by Godwin at Beauvale. There are still, however, bay windows with cusped lights and a romantic arcaded loggia at the level of the parade rooms. It is a much more lavish conception than the Duke eventually commissioned and balances an equally lavish French thirteenth-century chapel with a tall saddleback west tower. This version was redrawn in a sketch elevation, and simplified at the same time as the drawings for the central façade were worked up almost to the design that was built. The accompanying south elevation shows yet further simplification. A further version represents the plan eventually adopted, except that neither the south-east nor the south-west tower is shown. Possibly the decision to move the Library northwards represented a sufficient saving to allow the towers, essential features of the southern end of the building, to be reinstated.

Waterhouse's preference for steeply pitched roofs, tourelles, and corner turrets shows clearly; and there is his usual attention to the skyline, with carefully disposed gables and dormer windows (Pl. 87). The choice of a Gothic style was probably inevitable given the decision to retain the bulk of the old structure.[22] However, it was equally certain that it would be impossible to retain the original skyline and provide unity if the new wings were designed in fashionable, and functional, modern Gothic. Waterhouse was committed to modern Gothic and opted for unity and the replacement of the original skyline. Most of the old windows were enlarged and there was a good deal of decorative carving, by Farmer and Brindley, to the window mouldings. To balance the complex massing, however, Waterhouse chose mostly plainer shapes. Almost all the windows are square headed or blunt lancets set in square frames. The grand rooms have cusped lights or shouldered arches, which are still set in bold square frames. Tracery and fully Gothic arcading are reserved almost exclusively for the *porte-cochère* and east loggia, though the whole mansion is threaded on a vaulted Gothic grand corridor.[23] Yet everywhere Waterhouse wisely kept local detail strictly subservient to the broad picturesque concept.

1870, shows the basement plan as built. By that date all the sketch schemes had been examined and rejected, though a note on the drawing shows that the design was not yet finally accepted. 'This tower may not possibly be executed' refers to the south-west tower which was in fact built as the dominant feature of the block.

[20] In that position it must have approximated to the site of the original Library added by Burn.

[21] Waterhouse was determined to have his vertical axis. A drawing of 1873, after the library spire had been ruled out, shows the chapel spire rising fully 185 feet. As built it rises almost to that height.

[22] A plan and elevation of the east façade (DC) shows that original masonry was left to eaves height, but that the second-floor windows were considerably enlarged. A comparison of photographs of the ashlar in the central block and the Waterhouse additions reveal that the early 19th-cent. walling was left in the whole of the central block. Waterhouse used his characteristic alternate deep and shallow courses for the stone facing of the new work.

[23] The problems caused by retaining the old work, and by designing a series of almost independent blocks, surfaced clearly at the junctions of the various sections, where string courses had to be ingeniously adjusted and differing floor levels concealed.

Those who experienced the long approach up the Belgrave avenue can testify to the success of the vision. Steep roofs and turrets were silhouetted against the sky in a way that no photograph can adequately recall. Eaton was a romantic palace of an age when the chivalric ideal was much to the fore. But it was no Gothic dream, rather it was a luxurious and up-to-date mansion, and so its stylistic roots are in the Loire chateaux of the sixteenth century, not in the thirteenth.

The pressure of fashion was more insistent in the interior, however. Considerable sums were spent on the decoration, which was of the very finest throughout. It included a specially commissioned set of paintings by H. Stacey Marks of the Canterbury Pilgrims and Gertrude Jekyll's largest commission for tapestry in the shape of panels for the Library which were made at the Royal School of Art Needlework. There were stained glass windows, made by Heaton, Butler, and Bayne, of armorial bearings and a set of six illustrating Tennyson's *Idylls of the King*. There were carvings, such as the Caxton panel over the library doorway. There were clocks and an anemometer as well as firedogs and built-in furniture, all designed by Waterhouse. Tiles came from Godwin's, W. B. Simpson, and William de Morgan; and there were inlaid floors, patterned ceilings, and panelling by the yard. The only thing that Eaton lacked was a complete set of Waterhouse furniture; but that was not needed. The Duke's father had bought four rooms worth of Gothic furniture from the Great Exhibition.

Waterhouse did design most of the fittings (Pls. 91, 185), often in styles that were unfamiliar to him. The rich alabaster frame of the main door, for instance, was based on a cast in the South Kensington Museums.[24] But there were endless changes. Fine oak panelling installed in the saloon in the 1870s was ripped out in the 1880s and replaced by coloured marble to match the hall. The new design was a long way from honest Gothic. However, the effect was undeniably sumptuous and appropriate to the entrance of a nobleman's palace. It works curiously well with the standard Gothic arcading of the grand corridor, where detached marble shafts pick up the richness of the marble mosaic floor, itself inspired by Venetian work at Murano or Torcello (Pls. 93, 191). Of course Waterhouse had never been exclusively committed to Gothic, even to his own version of modern Gothic, and was prepared to adapt, as required. His high Gothic design for a drawing-room fireplace (Pl. 184) was set aside in favour of a more lush and colourful Renaissance-inspired piece. Part of the problem was that some of the old decoration was retained.[25] The Great Drawing-Room contained a white marble fireplace made in Rome in 1869 by Steinhauser for the second Marquis (Pl. 89). In the Great Dining-Room was an antique fireplace by Montorsoli from a palazzo in Genoa. Waterhouse carefully adjusted his designs to incorporate these pieces, and even attempted a suitably florid Renaissance clock to match the Montorsoli fireplace.[26]

The only room at Eaton which was characteristically his was the library (Pls. 88, 185). This sumptuous apartment, upwards of 30 feet by 90 feet, was a new construction so he had complete freedom. In addition to Gertrude Jekyll's medievalizing tapestries, there were two fine walnut fireplaces, their panels beautifully inlaid in boxwood and mother of pearl. The ceiling was heavily

[24] An unidentified photograph of a similar doorway, supplied by the South Kensington Museums photographic service, was found with the drawings for Eaton Hall. Waterhouse's drawings for this feature, with alternative sketches, are in the RIBA Drawings Collection.

[25] A detail drawing for joiners' work in the new panelling is annotated with instructions for modifying the design so that it relates to the old dado (DC). Photographs taken before demolition record that the old dado was indeed retained in the Great Drawing-Room, though not in the Great Dining-Room. The old flooring was left in all the main rooms of the east front.

[26] The Renaissance version is not a very happy composition, though it is only a rough sketch. Waterhouse's alternative design for a Gothic clock in a wrought-iron case is altogether more convincing (DC).

beamed and panelled with moulded plaster stencilled and gilt; and the walls were lined with walnut panelling above the bookshelves. It would be better described as Old English rather than Gothic, but its chief characteristic was an impression of substance and quality, features which were possibly more important than any commitment to style on the part of Waterhouse or his client.

The fitted decorations were quite sufficiently sumptuous and varied to offer the artist full rein and though the contents of the house are now sadly scattered, the Chapel remains as an indication of the original richness of the whole (Pl. 94). The design was worked out with Waterhouse's customary thoroughness, backed by the Duke's bottomless purse. Dr Stainer, the well-known organist and composer of church music, was personally consulted about the carillon for the Chapel. He was even sent to Louvain to examine the bells there. The interior of the Chapel is chiefly remarkable for the stained glass and mosaic panels designed by Waterhouse's friend, the Manchester Pre-Raphaelite, F. J. Shields.[27] William Morris was consulted about the material, glass or marble mosaic, and recommended the latter; and Shields was paid £1,300 (between 1878 and 1884) for cartoons for sixteen panels illustrating the Te Deum with nearly one hundred figures.[28] Yet even these panels are subordinate to the overall impression of the rich materials, of the cosmati-work floor, of the heavy alabaster of the pulpit and low choir screen, of the wrought iron communion rail (probably by Skidmore), and of the pew ends carved with pomegranates and sunflowers. Behind the Chapel's Gothic exterior an extraordinary eclecticism was at work. The strength of Waterhouse's detail gives a determined unity; but the use of Byzantine patterns and sunflowers and pomegranates reveals the new direction in which taste was moving.

The selection of style in the service buildings was less constrained by existing work and so is, perhaps, more interesting. The early drawings for the laundry, which retained part of the old structure, show banded stonework with blunt lancet windows and a simple cinquefoil in the shallow gable. By the time the new laundry was designed in 1877, the livery has altered completely and the building is in red brick with a half-timbered upper storey.[29] This new style, the local Cheshire vernacular, was already popular with John Douglas, the Duke's estate architect, but it was not Waterhouse's first choice. A sketch 'prospect of ye stables at Eaton' shows the main stable court entirely in Gothic.[30] The buildings at the north end of the stable court were put up first and are indeed Gothic with an open arcade and steep-roofed turrets. This block faces the Gothic chapel; but it is flanked by an east wing and slightly later west range both of which use bold half-timbering for the upper storey (Pl. 95). The same style is used to dramatic effect in the outer façade of the north range, beyond the carriage court. There, boldly jettied upper storeys face across the carriage entrance towards a group of neat half-timbered coachmen's cottages. Although the west wing was

[27] Frederic Shields had already been named by Waterhouse as one of the two artists to handle the commission for murals in Manchester Town Hall. That work was not yet begun, and was in any case ceded to Ford Madox Brown alone. Shields, however, was also at work on designs for stained glass windows for William Houldsworth's chapel at Coodham, another commission that he received through Waterhouse. Some of these windows are now in St Elisabeth's Church at Reddish.

[28] These are in the windows to the south and in the apse or in mosaic panels of the blind north wall to the stable court.

[29] Drawings for both old and new laundry buildings are in the RIBA Drawings Collection. A fragment of the old laundry, which was converted into a game larder, survives just east of the stables.

[30] The design of Aug. 1873 (DC) has apparently not survived and exists only in a photograph. This version is worked up in the complete presentation of the whole scheme dated Aug. 1873. Construction of the stables, however, did not begin until 1877 and was largely over by 1879; and the structures of the first campaign combine Gothic and half-timbering. The finished sketch of the west elevation also combines the two styles, using Waterhouse's simplified Gothic for the refacing of the riding school at the south end, and to make a transition from the high Gothic of the Chapel which adjoins the stables to the south. The revised version, with all the half-timbering and the altered west wall to the riding school, is dated Jan. 1879. The two-style sketch carries a note that the Gothic gable should be 'as others half timbered'.

built after the north range, and the junction is admittedly awkward, it does not seem likely that the Duke or his architect changed their minds about the style half way. The Duke could after all have afforded to demolish anything that was really offensive stylistically. The variety is more likely to be intended both for interest and as a deliberate expression of the separate sections. The bold Gothic arcade of the north range would face the Duke as he entered the court from his Gothic corridor, and, appropriately, has a degree of solemn splendour. The two side ranges, in a more humble mode, would add a pleasing variety, linked by the continuation of the same red brick and Manley stone dressings for the ground floor.

Such a design concept would be quite in keeping with the eclectic tastes of the time, especially at a point when high Gothic was falling out of fashion for other than ecclesiastical work. Gothic, however, was entirely suited to the Chapel, which survives as a dramatic and convincing essay on thirteenth-century French sources. For buildings that did not adjoin the main edifice, Waterhouse felt quite free to use different styles. The north lodge for instance is designed as a free-standing fragment of a French chateau; while the parrot house is an accurate exercise in classicism, a circular temple of honey coloured terracotta, complete with caryatids atop its interior columns (Pls. 90, 92). Without stronger proof of the Duke's dislike of the style, rather than the inconvenience, it seems preferable to consider Eaton Hall and its outbuildings as a flamboyant demonstration of high Victorian eclecticism. The general inclination for French sources gave a unity of appearance to the chateau, while the local vernacular was a suitable inspiration for the more earthy buildings of the stable court. In short Eaton Hall displayed all the traditional values that might be expected to appeal to an English nobleman. It also afforded sufficient variety, both within, in the shapes of rooms and the detail of their decoration, and without, in the massing, outline, and materials used. It was intended to be an imposing edifice, boasting almost 190 rooms. Even the private wing was a substantial mansion in its own right (Pls. 96, 97). But it was also up to date, fitted with an ample sufficiency of WCs, a series of lifts for goods, and with a truly monstrous double-height kitchen, 25 feet by 55 feet, complete with a first-floor gallery from which the cook could survey his minions. The planning owed everything to mid-Victorian aspirations of comfort.[31] Its style was rooted in tradition. It was essentially a creation of its age, marred perhaps by the contortions required to keep the old work, but a masterpiece none the less.

Because it was so much less determinedly Gothic than Manchester Town Hall, and because it was so much less consistent than the Natural History Museum, it is particularly important that Eaton be remembered as the third of Waterhouse's great commissions. If the Duke himself was the littlest bit swamped by the gargantuan splendour of it all, we should not forget that it was he who felt impelled to spend so lavishly in demonstrating his position. Waterhouse's genius was in sustaining and controlling a recognizable uniformity of splendour throughout so complex and long drawn out a commission. What he achieved here was the very opposite of the historicist approach of earlier country house builders such as Salvin or Pugin. His immediate readiness to embrace stylistic compromise and make something of it was suited to an age in which extremism was becoming less fashionable, yet in which status and tradition were still paramount.

[31] Among the up-to-date facilities installed at the Duke's insistence was gas for lighting and cooking (for which Waterhouse designed the special gas works). After 1887 there was electric light, and even, from 1896, a light railway with its own station. A fire engine house had been included from the start. Bathrooms were installed in addition to WCs, but fortunately for the architect he was not required to provide one for each main room. A further indication of the comfort and status of the mansion is the size of the staff. There were 35 indoor servants and 40 gardeners and on the estate as a whole some 300 employees.

11

Residences for the Rising Gentry

DURING thirty-five years of practice Waterhouse built, or substantially altered, some ninety houses for clients of varying means. Most, even the country houses, are relatively modest in their display, more like Ken Hill than Bestwood Lodge.[1] All, or almost all, share his regular commitment to comfort and solidity as the prime consideration. Just occasionally, as at Buckhold near Pangbourne, he could be swayed by an eccentric client into producing an uncomfortable plan. For the most part his well-to-do clients had similar needs, and he was able to work out a flexible plan that could accommodate varying tastes and varying purses, as he adapted his recipe for different sites and materials.

Waterhouse was one of the many architects providing for the increasingly wealthy upper middle classes. He built little for the nobility, and few of his country houses were on established estates. To start with his houses were either country cottages or suburban family houses for industrialists. Most of these were in the expanding suburbs of great cities, or in villages where they were far enough from the smoke and grime yet close enough for the daily journey to mill or warehouse. Some, like Wood Green **[172]**, built for Charles Gillett the Banbury banker, or Belmont in Cheadle **[174]** for James Milne, were relatively modest affairs. The later 1860s and 1870s brought an increasing number of commissions for larger mansions, often for the bankers and industrialists whose business premises he was also designing. And, as those industrialists grew richer and bought themselves into the landed gentry, so they came to him for alterations to the grand houses they had bought or for new country houses as the centre-pieces of their new estates. One was the substantial Gothic mansion with picturesque tower that he raised in Kirby Lonsdale **[295]** for Alfred Harris, the industrialist and man of taste who had introduced Ruskin and the Pre-Raphaelites to Bradford. At the peak of his career he designed several such works, places like Hutton Hall **[176]** (Pl. 108, Fig. 7) for Joseph, later Sir Joseph, Whitwell Pease, Iwerne Minster **[445]** (Pl. 104) for the new Lord Wolverton, friend of the Prince of Wales, or C. P. Stewart's Silwood Park near Ascot **[439]**. But there were still many commissions for more modest houses, like the pair in Victoria Drive, Saltburn, of 1874 **[408, 409]**. At least half a dozen such commissions were in hand in any one year from 1860 to 1879. After 1880, presumably because Waterhouse's reputation was so firmly established by then as an institutional designer, he was given less of these commissions. His style was, in any case, becoming out of tune with the times by then, and it is significant that the Chairman of the Prudential Assurance Company,

[1] Ken Hill, Snettisham, Norfolk, by J. J. Stevenson, 1879–80, for Edward Green, a Yorkshire businessman. Bestwood Lodge, Nottingham, by S. S. Teulon, 1862–4, for the Duke of St Albans.

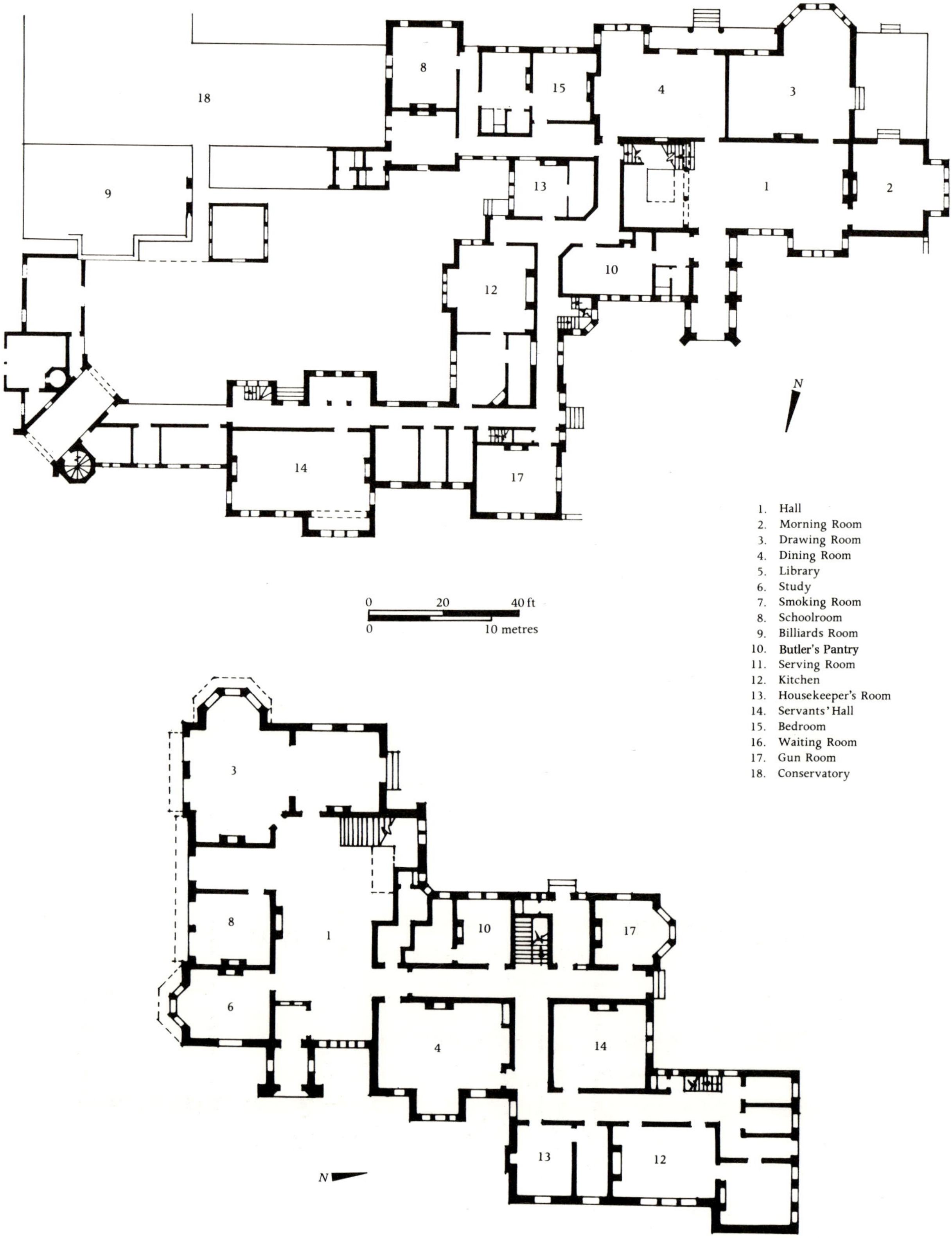

FIG. 7. Simplified plans of (*above*) Hutton Hall (1864) **[176]** and (*below*) Easneye Park (1866) **[258]** (*continued opposite*)

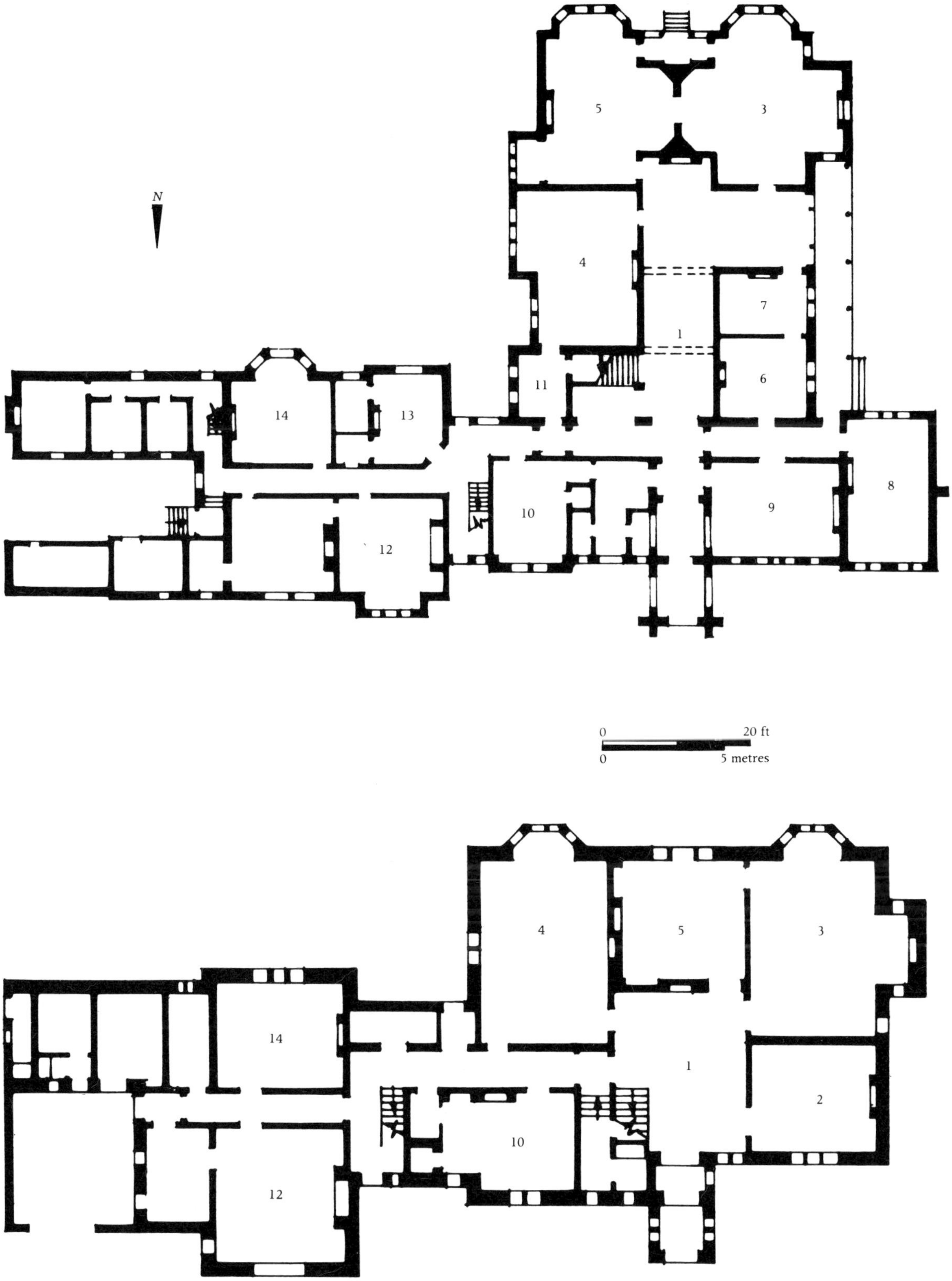

FIG. 7. (*contd*) Simplified plans of (*above*) Allerton Priory (1866) **[237]** and (*below*) The Brand, Loughborough (1875) **[421]** (*continued overleaf*)

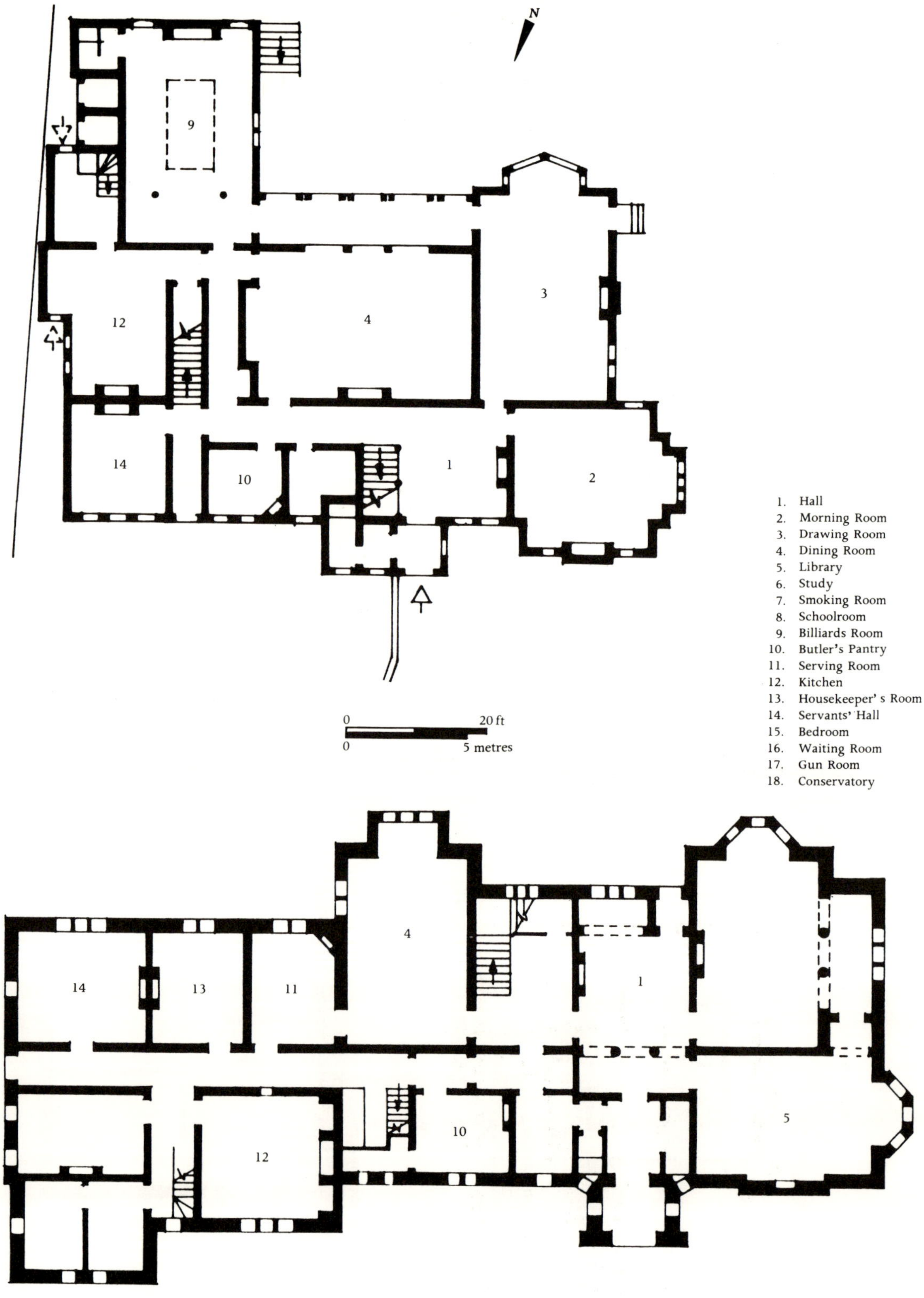

FIG. 7. (*contd*) Simplified plans of (*above*) West Bank House, Manchester (1867) **[281]** and (*below*) Coldhayes (1869) **[323]** (*continued opposite*)

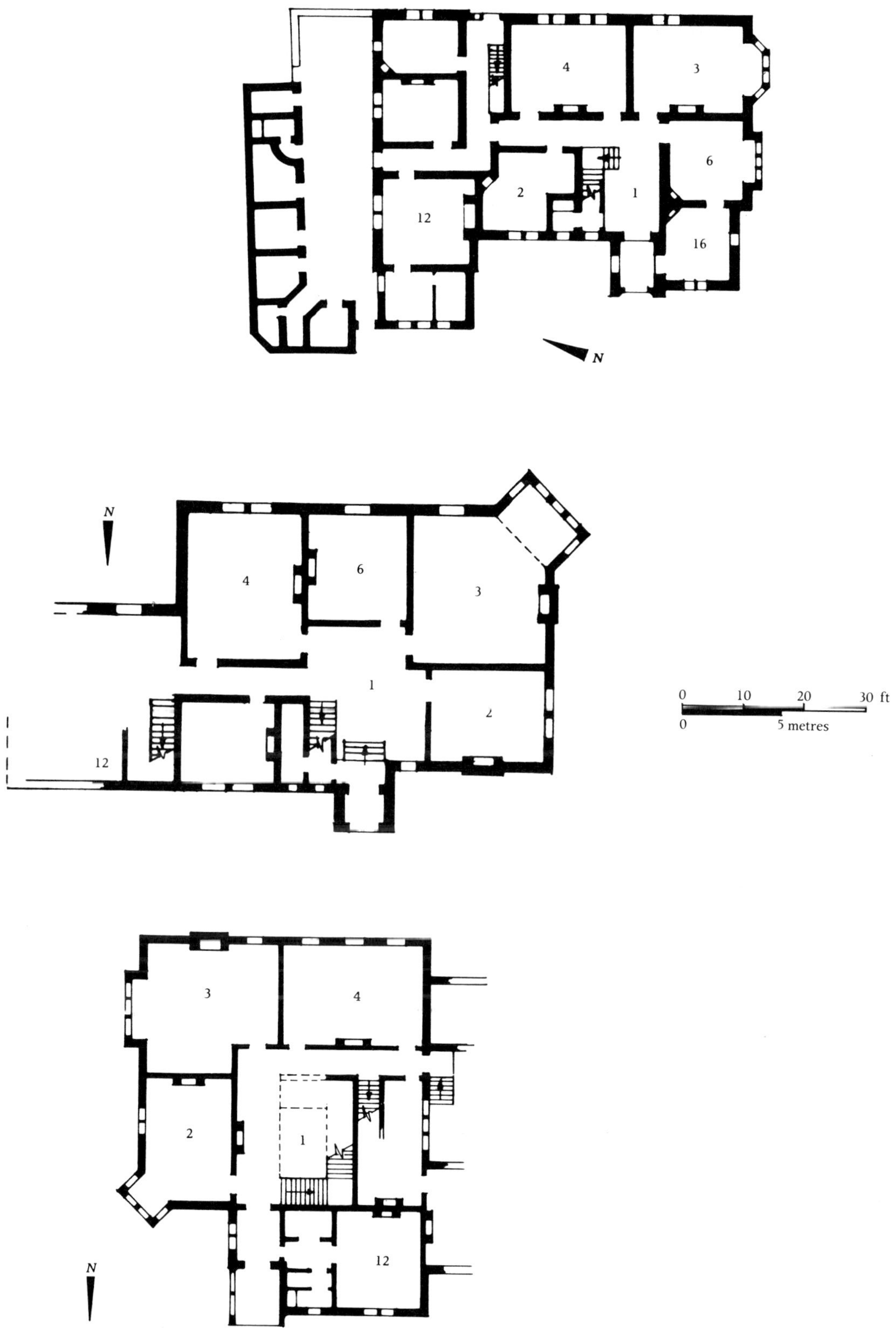

FIG. 7. (*contd*) Simplified plans of (*top*) Bakewell Parsonage (1869) **[313]**, (*middle*) Merle Bank, Bowdon (1869) **[314]**, (*bottom*) T. R. Hetherington's house, Manchester (1871) **[358]**

for whom he built so much and so well, turned not to Waterhouse for his house at Tigbourne Court (1899) but to Lutyens.

Waterhouse, as a house designer, belonged to an earlier generation. His first houses had their stylistic roots in John Claudius Loudon and early Victorian taste.[2] Hinderton Hall **[13]** (Pl. 26) was the first house in which he really developed his own design theories, using Gothic-picturesque shapes to allow him to group his rooms freely but logically round a double-height hall. And it is the logic of his houses that is always their prime virtue. They were, after all, designed as comfortable and substantial homes where the pattern of respectable living could be easily enjoyed. Many were for Quaker friends and relatives.[3] To show the worth of their owners they needed to be as solid and as spacious as possible, but extravagant display was frowned on. Individuality and taste were necessary, but were the final bloom rather than the root.

In the early modest commissions it was often difficult to achieve much architectural effect. The little house he designed for the retiring Recorder of Liverpool on the shores of Derwentwater **[28]**, however, though conventional enough from the exterior, was ingeniously laid out with different levels to make the best of the views while remaining thoroughly compact. At Lymm near Manchester, he built for the brothers Dewhurst (of Sylko fame), a pair of semi-detached houses which together have the appearance of a single mansion **[48]**. It was an acceptable ruse at the time, allowing a pair of houses with a shared driveway a substantially well-to-do appearance belying the modest accommodation of each unit. Then there were alterations to existing houses in Manchester suburban villages, like Bowden or Pendlebury, in Cheadle or in Aigburth near Liverpool. New Heys **[103]** (Pl. 98), in the village of Allerton just outside Liverpool, was an early example of the Waterhouse suburban mansion. It was a compact block, with Gothic detailing and steep pitched roofs. There Waterhouse streamlined the arrangement he had already tried at Hinderton. The house stands at the top of a sloping site; but, as became his normal practice, he made no dramatic use of the slope beyond providing an attractive south verandah above the basement rooms. The porch on the northern side led directly to a hall out of which rose the stairs. These were lit by a great window at the half-landing, which was a feature of the façade next to the porch. The main rooms, as at Hinderton, were laid out in an L around this core, but this time on the east and south instead of the west and south sides.

This was the basic concept that Waterhouse adopted whenever possible (Fig. 7). Where possible he preferred a site which was flat or could be easily terraced. The block was an asymmetric double pile, though it could be three rooms deep in larger houses, closely linked to an expandable service wing. This provided a logical and convenient arrangement with the main rooms around a central space and the domestic wing to one side. The system could be varied to take account of orientation or aspect, or to allow for the exigencies of the site. It was a practical approach; and having devised a plan that worked so satisfactorily Waterhouse saw no point in changing it merely to be original or different. Whatever the size of his houses, the preferred system showed through.

[2] J. C. Loudon's best-known book, *An Encyclopaedia of Cottage Farm and Villa Architecture*, originally published in 1833, had been issued in a new edition at about the time Waterhouse began his articles.

[3] Quaker connections among the enterprising magnates of Darlington brought a number of clients. Alfred Backhouse engaged him to build Pilmore Hall, and his nephew The Grange, at Hurworth, nearby. Backhouse also later commissioned a summer house in the Hamsterley Forest, Dryderdale. In 1864 Joseph Whitwell Pease commissioned Hutton Hall, Guisborough, the centre of a new estate, while other members of his family required houses on the edge of Darlington at Uplands and Hummersknott.

Foxhill in Whiteknights Park, Reading **[285]** (Pl. 117, Fig. 8), shows the system at its best. The house belongs to the middle rank in size and luxury, but here presumably was Waterhouse's ideal because this was his own house. The site slopes gently down to the shores of a lake.[4] The drive approached the house on the north side, where porch, vestibule, stairs, and cloakroom saved the sunny side for the principal rooms. The generous hall, with its large fireplace, also functioned as a living room, and a morning room looked out west to the garden. An ample stairway rose beside the entrance, lit by a great window overlooking the drive. The drawing-room, dining-room, and conservatory spread along the garden front taking in the sun and the view over the lake. On the floor above were the principal bedrooms with the nursery accommodation on the second floor. From the foot of the stairs a passage led to the service wing, with a north-facing kitchen, and, beyond the pantries, a door into the yard at the eastern end of the block. The kitchen yard was closed by the coachhouse and stable with groom's quarters straddling the entrance arch. The servants lived over the kitchen area.[5] Comfort and practicality emphatically came first here; but the house was made picturesque with diapered brickwork, steep gables, and modest Gothic detailing.

The basic pattern could be reversed to allow the kitchens to be set on the other side of the entry; and the service wing was often set apart by a break in the façade. Sometimes it was set forward, sometimes back, but it was always expressed by a lower roof; and often it contained three storeys compared with the two of the principal block. Waterhouse usually had sound reasons for reversing the order of the blocks; though occasionally it seems to have been done simply for variety. At Saltburn **[408, 409]**, for instance, he built two houses on adjoining plots, sharing the same aspect, with identical but opposing plans. At Dryderdale **[349]** (Pl. 107) he turned the axis of the main rooms to catch the view eastwards down the valley that was the principal reason for building the house in the first place. It meant lengthening the side façade, though the entrance was left on the north.

The same basic logic could be expanded or contracted to suit individual purses. At Hurworth Grange **[381]** (Pl. 100), for the junior partner in a banking enterprise, he used the method to produce one of his most typical houses.[6] Rather larger than Foxhill, it had a service wing forward of the house, with a projecting spiral service staircase, that made an enclosed entrance court. A north porch led into the usual hall linked to the staircase with the main rooms facing south and east. The short south façade was enlivened with a great ribbed chimney stack, such as Shaw used later at Banstead Wood,[7] which actually embraced the window of the drawing-room. To the east was a large projecting bay, octagonal at ground level rising to a square upper bay with boldly chamfered corners.[8] The

[4] The lake existed before Foxhill. Whiteknights Park was bought by Sir Isaac Lyons Goldsmid in 1847. The house then standing was later leased by Waterhouse's father and other plots were acquired leasehold including those for Foxhill and Erlegh Park. Papers giving the detailed history of the estate, which is now the campus of Reading University, are held in the Reading University Archives.

[5] These stables and quarters are not grand in size but such as were then required by a businessman and his family, the forerunner of the two-car garage. Houses of comparable size to Foxhill may or may not have their stables forming part of the house group. At Coldhayes (1869), near Steep, Hants, there is a separate courtyard which breaks forward of the house and has an entrance arch skewed across the corner. The house in Reading for Alfred Palmer, East Thorpe (1880), has a kitchen court at the north end and a separate stable block just beyond the house to the south. Here a different orientation is prescribed by the narrow frontage which allows the bulk of the house to shield the garden from the busy suburban road.

[6] Hurworth was altered by Waterhouse in 1886–7 by the addition of a billiard room wing at right angles to the garden front.

[7] Banstead Wood, near Epsom, Surrey, 1884–90, by R. N. Shaw for the Hon. Francis Baring.

[8] This whole bay is virtually identical to a contemporary one at Girton, a rare example of Waterhouse reusing a major design element *in toto*. It is not clear which was drawn first.

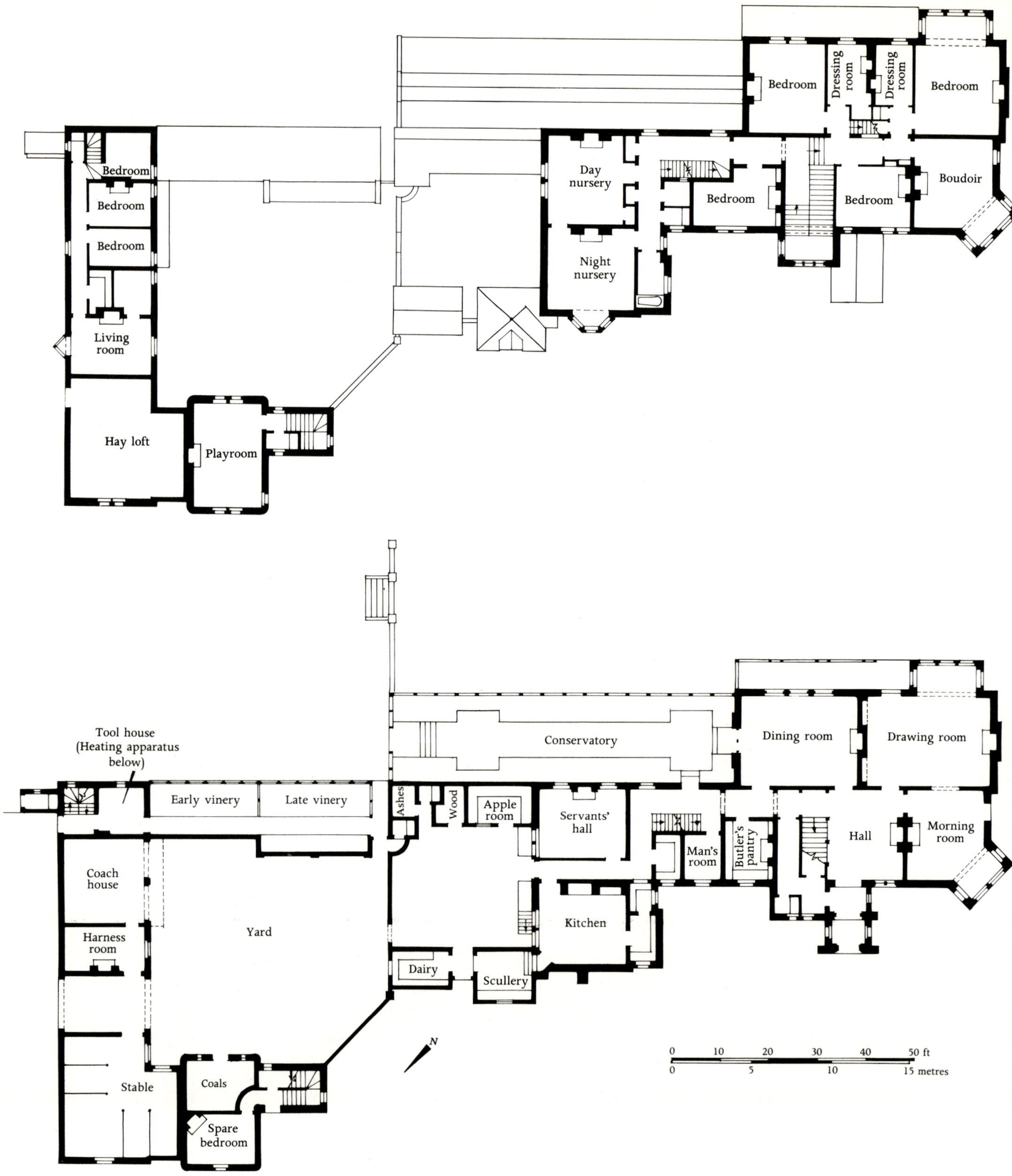

FIG. 8. Reading, Foxhill (1867) **[285]**, plans of ground and first floor

decoration was achieved through inventive bricklaying, with only minimal stonework giving an effect of solid well-mannered reticence.

The same plan, but with the main axis running east–west, was used for the extensive mansion Waterhouse built for Lord Wolverton at Iwerne **[445]** (Pl. 104). The Brand **[421]** (Fig. 7), at Loughborough, for one of the principal shareholders of the Midland Railway, and Abinger Hall **[326]** (Pl. 113), for T. H. Farrer, were on a less extravagant scale but still imposing. While the former parsonage at Bakewell **[313]** (Fig. 7) was considerably more modest but just as complete.

All these houses were in Waterhouse's modified Gothic manner which owed a good deal to Butterfield's simple rectories. He liked to introduce the occasional romantic Gothic feature, for instance, an oriel window on the first floor supported on a buttress or an attached shaft. Another favourite for larger houses was a dwarf wall with an archway, which he would introduce either to screen the service court or to frame the entrance to the gardens. Generally, however, the architectural effects were achieved by steep roofs[9] and tall chimneys, by the use of banded stone or brickwork, and by the massing of gables and dormers, rather than by additive ornament, carving, dramatic buttresses, or turrets. The main garden front was usually composed with two gables or a pair of bays, which Waterhouse almost invariably placed asymmetrically. There was often a further gable or bay on the short side, and the entrance porch or the staircase might be announced by a similar feature. The lower roof of the service wing established the hierarchy of the house; but both roofs were frequently broken up by the insertion of picturesque dormers. With the addition of verandahs and pentice roofs over the main windows, the whole effect was of cosiness and comfort before splendour.

Within his basic concept Waterhouse could accommodate all the normal spaces of a Victorian household.[10] Dining-room, drawing-room, and library were the three basic reception rooms, and a conservatory was almost *de rigueur*. However, he could also break with convention in order to create effect. At Merle Bank, in Bowdon **[314]** (Pl. 102, Fig. 7), for instance, the morning room lay beside a rather narrow entrace hall. So he designed the two spaces to open into one through a timber arcade.[11] He also varied the shapes and sizes of the rooms themselves. Deep bay windows, often on two adjoining sides of a room, added interest to the space and helped to break down the sense of formality. Two projecting rooms could be linked by a verandah; and within the general composition it was easy to include nurseries and schoolrooms, if needed, either in the main block or as part of the service wing. Billiard rooms were often a later addition, and set in a projecting wing.

However, Waterhouse did not arrive at the complete concept all at once, and he was always ready to make substantial adaptations for special circumstances. For instance two houses of the early 1860s, Prince Park **[132]**, in Liverpool, and Benwelldene **[184]**, for his brother-in-law Thomas Hodgkin, in Newcastle, had single-storey octagonal central halls, which obscured the clarity of the plan while adding greater complexity to the routes through the house. Benwelldene also had the large kitchen and nursery wing set at right angles to the main block, giving the children fine views down over the gardens to the Tyne. West Bank House **[281]** (Pl. 101, Fig. 7) at Heaton Mersey in Manchester, also

[9] Waterhouse regularly exaggerated the pitch of his roofs. Some are as steep as 70%; but a steep outer pitch was usually combined with a shallower inner slope leaving plenty of space for generous attics.

[10] For a full discussion of rooms and their uses see Jill Franklin, *The Gentleman's Country House and its Plan, 1834–1914* (London, 1981), and Mark Girouard, *Life in the English Country House* (New Haven, Conn., 1978).

[11] Merlebank also uses the steep fall of the ground to provide a billiard room and an entertaining room facing the garden. Originally the house stood in 4 acres of grounds, and, as at New Heys, Waterhouse set it close to the road with views across its own property. Now, with the grounds sold off and houses close on either side, it looks considerably less impressive than in its original state.

had its service wing at right angles; but here it was increased to three storeys under a single roof, so that it could be squeezed in against the boundary of the site, allowing the house spacious views south and west to the Mersey valley across a moderate sized suburban garden.

West Bank House was also untypical in its use of low pitched roofs and big eaves. This style, with Swiss or Italian roofs, was already old fashioned by the 1860s; yet Waterhouse used it as late as 1880 at Crimplesham Hall **[476]** (Pl. 99), where it led Pevsner to doubt the date and suggest 1850–60.[12] Waterhouse was certainly not concerned to be in the forefront of fashion, and was quite happy to abide by his clients' wishes. This Loudonesque style was still popular in the mid-century; and he had selected it for his own house, Barcombe Cottage **[52]** (Pl. 17). He felt quite free to combine it with various Gothic details such as arched doorways, carved capitals, and oriel windows. Hummersknott **[147]** (Pl. 109) in Darlington, for Arthur Pease, was another attractive house in this style, where the low roofs accentuated the spread of the building and added to its air of spaciousness. Yet it also had a fully Gothic oriel and Gothic capitals in the hall.

At Hummersknott, Waterhouse also introduced a modification to his basic plan by completely surrounding his hall, now top lit, with rooms and setting the stairs at one end. The Heath at Leighton Buzzard **[178]** (Pls. 114, 116), and Mossley Park in Liverpool **[296]** followed a very similar arrangement. But these were among the larger of his middle-ranking suburban mansions. Generally he reserved large central halls, reminiscent of Paxton at Mentmore or Burn at Taplow Court,[13] for his more substantial houses. For Thomas Hetherington, in Victoria Park, Manchester **[358]** (Pls. 110, 120, Fig. 7),[14] he adapted this grand feature even further, the hall rising through three storeys. Its stairs rose only as far as the first floor, and those to the upper floors were set behind the hall with a linking arcade. Yet such radical departures in smaller houses were rare; and at Coldhayes **[323]** (Pl. 111, Fig. 7), in Hampshire, where he set the stairs on the south front, he achieved a less satisfactory design. The staircase window disturbed the balance of the façade, there was less space for the rooms to enjoy the view, and the north facing hall, divided from stairs and passage, was uncomfortably dark.[15] Another variation was Arnsbrae **[508]**, a house of the 1880s near Alloa, where entry, stairs, and long narrow hall formed a single rectangle round which parade and service rooms were added in a complete U. The stairs rose along one side of the hall, which was top lit, to a bridge linking front and back bedrooms. A first-floor passage looked into the hall through arched openings, and gave access to separate stairs to the upper floors.

Waterhouse's larger mansions are more varied, though most have elements of his basic planning scheme, and all are logical and easy to use. Blackmoor House **[192]** (Pl. 106), which he built as the centre-piece of Roundell Palmer's new estate, was one of his grander commissions.[16] It is more typical of his method of managing commissions than of his approach to house design. Palmer began by modernizing an eighteenth-century farmhouse that still forms much of the core of the house. Later Waterhouse managed skilfully to graft on to this a further large mansion on his normal plan, making an unusually long garden front.[17] The rambling additive nature of the commission gave

[12] N. Pevsner, *The Buildings of England: North West and South Norfolk* (Harmondsworth 1962), 126.

[13] Mentmore near Leighton Buzzard, 1850–5 by Paxton and Stokes for Meyer Amschel de Rothschild. Taplow Court, Buckinghamshire, remodelled 1855 by William Burn for C. P. Grenfell who had a fortune from Cornish mining.

[14] Now part of the Xaverian College.

[15] The dividing arcade has now been removed. It consisted of pillars of polished black marble, which suggests that Waterhouse may have been trying to make a virtue out of its dimness.

[16] See *Country Life*, 156 (1974).

[17] Much of the old house forms the service wing with extensive outbuildings around courts to the rear. These can be reached through a picturesque archway flanking the entrance porch.

plenty of scope for picturesque features, which made the rather stern modern Gothic pile thoroughly attractive.

Waterhouse was, however, much more himself at Crookhey Hall **[401]**, in Cumberland, a building on a fresh site for a wealthy American. There he gave the entrance a *porte-cochère* and a tall tower. The basic plan was expanded to include a top-lit central hall, amply large enough to play its country house role as an area for recreation and entertainment for the whole family and as a meeting-ground for family and servants.[18] Waterhouse gave it a splendid inglenook fireplace round which the staircase climbed to divide at a half-landing for the lower service wing or to return to the corner of the hall gallery. The exterior, in hard local stone, is even plainer than was usual for Waterhouse, though he created a picturesque silhouette by his usual mixture of gables and dormers. At Silwood Park **[439]**, in Berkshire, he used a very similar plan to richer effect in red brick and stone, adding a corner turret and various bay windows as well as an open-topped tower in the style of E. W. Godwin.[19]

Hutton Hall **[176]** and Easneye Park **[258]** were the most typical of his grand mansions, in that both were built from scratch for newly rich magnates as the centre-piece of new country properties. Both had elements of the basic design, yet both also showed significant departures from Waterhouse's preferred plan. Hutton (Pl. 108, Fig. 7) was a tall mansion of brick and stone with a two-storey main wing joined at one corner by an L-shaped service wing of three storeys rising to the same height.[20] The entrance was on the north side from the open court formed by the two wings, which allowed easy access for both gentry and their visiting servants and baggage. As usual the main rooms were set on two sides of the hall facing south and west. The hall itself occupied almost the whole of the north front, but it was only a single storey with top-lit stairs at its eastern end behind a stone arcade. Above were both north- and south-facing bedrooms. The service wing and a later billiard room, game larder, and other stores surrounded a sizeable kitchen court[21]; but Hutton had perhaps thirty living-in staff, as well as rooms for visiting ladies' maids, nannies, and children, so that the main rooms occupied only about a third of the house.

At Easneye (Pl. 103, Fig. 7) the proportion was half and half, a much more usual division; but there were more complexities. An early scheme with a north entrance and a range of axially symmetrical rooms (probably the client's original choice) was abandoned in favour of a south entrance linked by an intricate series of stepped bays to a service wing forward of the main house. Movement through the house was conceived in a series of diagonals. The great hall ran from the front to beyond the service wing, and was lit by staircase windows in the far corner. As at Hutton it was only a single-storey hall, though a double-height hall was provided above, possibly for billiards.[22] By all accounts the client for Easneye, Thomas Fowell Buxton, had definite ideas; and it was part of Waterhouse's success that he was able to accommodate such people. His readiness to offer more than one design was very much part of this too. The whole of the first design for Blackmoor **[192]** had been set aside in favour of retaining the old building. The same happened with the reconstruction of Heythrop Hall **[350]** for Thomas Brassey. The old house had been gutted by fire, and Waterhouse spent almost 400 days designing a replacement. This whole effort was set aside in favour of retaining

[18] For a fuller description of the use of these halls see Girouard, *The Victorian Country House*.

[19] See Godwin's Beavale Lodge, 1871.

[20] The nurseries and schoolroom were set into the corner adjoining both the main rooms and the service wing, but virtually self-contained with their own staircase. They looked out on to the gardens forming a continuation of the main garden front.

[21] As originally designed in 1864, the kitchen court was incomplete; but the servants' wing was doubled in size and the billiard room built almost immediately upon completion in 1870.

[22] See Franklin, *The Gentleman's Country House and Its Plan*, 162.

the old structure.[23] At Easenhall, near Rugby **[348]** (Pl. 118),[24] for Washington Jackson, another American client, Waterhouse set aside a whole design for a new house in favour of one that is quite untypical. The final design was for a rectangular three-storey block of Palladian proportions with a Classical centre-piece and some French Renaissance detailing. He gave it variety by the range of bay windows, two of single storey to the façade, two of two storeys to the west, and one full height one to each of the south and east façades, the latter placed asymmetrically.[25] The plan too was quite unusual. A *porte-cochère* leads to a panelled entrance hall which in turn opens into a long gallery running through the heart of the house to the bay at the centre of the garden front. On one side he placed three rooms designed to be used *en suite*, and on the other two large rooms and a staircase that opened directly off the gallery. The service wing was a linked building to one side.

Waterhouse had used a similar plan at Allerton Priory **[237]** (Pl. 115, Fig. 7), where the long hall bisects the house and is set at right angles to the service passage. However, Allerton, in spite of the typically Waterhousian Gothic of many of its parts, also includes a number of features that were probably specific requirements of the client.[26] Waterhouse was prepared to modify or even abandon his own ideas for a client's wishes; though this flexibility could produce awkward results. At Buckhold **[517]**, near Pangbourne, he was forced into the complete abandonment of his set plan while using his preferred modern Gothic in the wildest eclectic manner. The result is an ungainly structure with half-timbering, tile hanging, and a large corner turret, combined with projecting upper bays heavily corbelled out in orange terracotta. The interior has painted tile panels (a medium Waterhouse used only once elsewhere, in the children's ward of University College Hospital) to enliven the hall. Apparently the client, Dr Watney, was a difficult man who preferred to live in isolation at one end of the house with his manservant. He caused such problems that, in order to get agreement to anything reasonable, Waterhouse would suggest that he did not want or need it. For the most part, however, his houses were solid comfortable homes. Mostly in his version of domestic Gothic, they were given individuality by his choice of materials and by the richly varied vocabulary of oriels, dormers, bays, steep roofs, and bold chimneys.

Once he had finished with the house, Waterhouse was equally keen to see that it had proper surroundings. Lodges, gatepiers, and boundary walls often bear his unmistakable stamp, and he even selected the lamp-posts for the driveways.[27] Many of his houses still possess terraces and steps, clearly part of the whole design, bounded by low walls of brick or stone to match the house itself, the walls punctuated by sturdy bollards (Fig. 15). To what extent he designed the landscaping beyond is less clear, though most of the planting seems to have followed the trend of the 1870s for natural and

[23] The original house was by Thomas Archer (H. M. Colvin, *A Biographical Dictionary of British Architects 1600–1840* (London, 1978). Waterhouse merely filled the lower portion of Archer's windows with stone panels to accommodate Victorian taste, and provided bold new chimneys. The interior was virtually rebuilt with a top-lit central hall and fine stone staircase rising behind a heavy Italianate arcade. Waterhouse's own tastes show most clearly in the high quality of the joinery.

[24] Town Thorns is built on the estate at Easenhall surveyed by Waterhouse in 1857 for Edward Cropper (see p. 20). It is referred to in the practice accounts only as Easenhall Hall, but takes its name from a wood mentioned in the earlier survey. As recommended by Waterhouse in 1857, the house is built on the crest of a low ridge with fine views to the north, south, and west.

[25] It appears that a free-standing tower also added to the picturesque asymmetry. This has now been demolished, but may, in any case, have been an extension of the 1880s.

[26] The client, John Grant Morris, was a former mayor of Liverpool who apparently retired from public life to Allerton. This retirement may have been determined by the needs of a disabled child; and, certainly, the house contains an elaborately fitted pair of rooms where the nurseries ought to be. Among the other unusual features was a smoking-room in the tower reached over the great central lead flat.

[27] His own design for the lamp-posts surrounding Manchester Town Hall survives. However, it is unlikely that he did more than select from catalogues for these private commissions. None the less, the few survivors are remarkably similar to those designed for the Town Hall.

informal groupings of shrubs and trees. A number of drawings, some with planting details, do exist for gardens for other types of buildings, such as Jesus, Caius, and Pembroke Colleges. One for the Natural History Museum shows a tight layout of geometric beds; but most, including those for Mowden Hall, the house at Buckhurst Hill, and for Caius College simply indicate species of trees and shrubs in serpentine borders (Pl. 162). A large number of houses appear in the accounts with 'garden works'; but these may have covered only the terracing near the house, which was more or less standard.[28] Correspondence, however, shows that Waterhouse consulted with landscape designers: Edward Kemp for Rose Trees at Keswick **[28]**, New Heys in Liverpool **[103]**, and Erlegh Park, Reading **[71]**; also, in the Manchester area, John Kellett for the houses in Victoria Park and at Northenden **[282]** and Merle Bank, Bowdon **[314]**; and, in the later 1860s, Robert Marnock, who was involved in the Regent's Park Gardens. In 1869 when Waterhouse was building Bakewell Parsonage **[313]**, Edward Milner, a pupil of Paxton at nearby Chatsworth, was called in by either the architect or the client. Later Milner was working at Iwerne Minster **[445]**.[29]

Then there might be a whole host of subsidiary buildings. Waterhouse designed a few timber boathouses, none of which survive, and a host of conservatories. Hutton Hall **[176]** was provided with one, of course; but the extensions of 1870 included a further great conservatory of three aisles with stone arcading. For the most part these too have disappeared.[30] Further from the house were walled gardens, where the gentry side was often provided with attractive wrought iron gates, as at Hutton, or occasionally fountains, as at Heythrop **[350]**. And Waterhouse regularly produced designs for greenhouses and vineries.

In his service buildings Waterhouse invariably respected the hierarchy of the whole establishment; but he could none the less achieve an effect with the simplest means. At Hutton **[176]** and Iwerne **[445]**, for instance, he provided towers over the stable entry. At Iwerne he built the stables of bright red brick, where the house is of stone, and devised a striking roof. At Coldhayes **[323]** he achieved a similar effect more simply by setting the groom's lodgings in a raised block, set at an angle across the corner of the stable yard with an arch beneath. His lodges needed to be less plain, and they were often equipped with elaborate barge boards and deep porches made with turned timber supports of the sort that he favoured for his furniture. None the less Waterhouse was committed to achieving grandeur only with economy in such structures. The substantial south lodge at Easneye **[258]**, for instance, consisted of two dwellings set back to back in a cruciform plan about a single chimney.[31] The upper floors are jettied and tile hung, which adds to the picturesque effect while being an economy in construction.

Where the house was the centre of an estate, there might be substantial extra work in village reading rooms, cottages, and farm buildings (Pls. 112, 119). These were regularly built in a sort of architectural livery, and, on most estates where he was involved, conform closely to Waterhouse's style. He was certainly involved in this way on his own estate at Yattendon **[452, 474, 494, 617]**, and

[28] Waterhouse may have gleaned some suggestions from *How to Lay Out a Garden*, one of the books whose title appears in the ledger entry of book purchases.

[29] While Milner was at work at Iwerne, Waterhouse took the opportunity of getting him to advise on the layout of his own country house at Yattendon.

[30] At Pierremont (1873) **[380]** in Darlington his stone conservatory survives as a roofless enclosure. The timber conservatories at Adel Grange (1864) **[177]** and Grinkle Park (1881) **[489]**, now converted to a ballroom, are more typical but rare survivors. Silwood Park (1876) **[439]** still has its long conservatory in brick and stone, screening the gardens from the carriage drive.

[31] Drawings for this building, signed by Waterhouse, survive in the archives of the All Nations Christian College. They are among relatively few drawings for service buildings that survive; though the accounts make it clear that he did regularly undertake such work.

at Blackmoor **[232, 266, 267, 328, 329]** and Hutton **[272, 367, 368]**.[32] More often, however, it seems that clients were prepared to leave the building of cottages to local builders, and Waterhouse appears to have been quite happy to see his design details copied or adapted by others. At Easneye **[258]**, for instance, the home farm and several cottages appear on stylistic grounds to be from the Waterhouse office; yet there are no records of their design in the office accounts. They are probably the work of a local designer, possibly the clerk of works, continuing to design in the mode established by Waterhouse.[33] There was a similar arrangement at Buckhold **[517]** where farm buildings in a distinctly Waterhousian style are known to be by local men, some to designs of Leslie, an architect of Reading, and others by Wheeler, a local contractor.[34]

The Waterhouse recipe for estates and houses was never in the forefront of fashion. Instead his designs can be seen as standing for traditional values, and his houses were more concerned with worth, solidity, comfort, and convenience than with the switchbacks of taste. When Hermann Muthesius made his tour of English houses at the turn of the century,[35] he did not include a single example by Waterhouse. By then his traditional forms were too obviously old fashioned. None the less, his own summary, based on his extensive experience, highlights qualities that are uncannily close to those admired by Muthesius:

> The domestic buildings of this country are without rivals elsewhere . . . not only in sanitary appliances is this superiority apparent, but the arrangements of plan to promote comfort and the air of homeliness which pervades the vast majority of English houses, have, I believe, never been approached in former times or in any other country.[36]

That Salvin had been able to achieve this in a castle, Waterhouse found commendable, but he himself achieved it in a practical and logical manner.[37] His houses were designed for the comfort and convenience of those who were to live and work in them rather than for effect. His basic theory, that once the plan was right the elevation should follow,[38] is supported by the expression of the function of each part. His planning and functionalism together gave a consistency to the massing of his houses, which, like the plan, was capable of expansion or contraction according to the status and purse of the client. In providing such a complete and appropriate service Waterhouse was an eminently safe architect for the rising gentry to employ.

[32] Many of his designs for cottages are not separately shown in the accounts. This is invariably the case with lodge cottages, and, since it was his own estate, appears to be the case with most of the minor works at Yattendon.

[33] This system of passing on the work on subsidiary buildings is attested at Backhouses' Bank at Darlington, where the drawings for the manager's house and stables, signed by G. G. Hoskins, the clerk of works, still survive in the archives of Barclays Bank.

[34] Signed drawings are in the possession of St Andrew's School, Buckhold, Pangbourne.

[35] See Herman Muthesius, *Das Englische Haus* (Berlin, 1904); and abridged trans. edn., ed. D. Sharp (Oxford, 1987).

[36] AW, Inaugural Presidential Address, 1888, published in the *Builder* (10 Nov. 1888), 336.

[37] Letter from AW to Lord Tollemache in 1878 quoted by Mark Girouard in *The Victorian Country House*.

[38] AW, President's Address to Students, 1889, published in the *Builder* (2 Feb. 1889), 90.

12

Home and Friends: The Squire of Yattendon

THE year 1878 saw Waterhouse at the pinnacle of his career. His two largest buildings, Manchester Town Hall **[284]** and the Natural History Museum **[243]**, were nearing completion, the extensive work at Eaton Hall **[315]** was in hand, and the first building of many for the Prudential Assurance Company **[431]** was under construction. The fruits of success were coming his way too. He became an Associate of the Royal Academy and President of the Manchester Society of Architects. He was awarded a *rappel* to the *Grand Prix* which had been presented to him at the Paris *Exposition* of 1867, and, perhaps most prized of all, he won the Royal Gold Medal of the Royal Institute of British Architects.

Not long after moving to London and setting up office and home in New Cavendish Street **[180]**, he had bought a plot of land in Whiteknights Park, Reading, and there he built his out-of-town home, Foxhill **[285]** (Pl. 117).[1] The London house was still used, though only for entertaining; Foxhill was the family home. Indeed Whiteknights Park was the centre of the Waterhouse clan, and Waterhouse's move was simply a reuniting of the wider family. His parents had moved there in 1858, and were living in Whiteknights House **[116]** with his unmarried sister Maria. His brother Theodore had already bought land at Bowsey Hill, in nearby Twyford, and there Waterhouse designed him a small house **[378, 425]**.

By 1878, however, Waterhouse himself could afford to buy a regular country estate **[453]**; he had, in fact, already purchased the land.

It was, I think, in the summer Vacation of 75 that Paul and I . . . rode over to Yattendon to see Mr Howard's Estate. We went by way of Theale, Bradfield and Ashampstead Common, were lost in the woods and were perhaps not altogether in a condition to enjoy the place when at length we found it . . . At any rate I thought no more of Yattendon for a twelvemonth or more, but instead allowed my thoughts and feet to run towards Devonshire and became much enamoured of the Chagford country (where the Shilstone Farm was offered to me) and of Hawkmore near Lustleigh, a pleasant place I should have bought had it not been for the mismanagement of my Exeter agent. I believe it was a little before this time that Wood's Farm at Streatley was again offered to me and only lost through the inaction of my London agent! Both of these flashes in the pan somewhat annoyed me at the time, but I have had reason since to believe that I may be grateful for these disappointments.

[1] Waterhouse's first 2 children, Paul and Monica, were born at Barcombe Cottage in Manchester. Florence and Alfred Maurice were born in New Cavendish Street. Only the youngest, Amyas Theodore, was born at Foxhill. But even Paul was only 8 when they moved in, and it was at Foxhill that the family was brought up.

Last autumn [1876] when the beechwoods were clothed in gold I visited Ealing and on my way home found myself for the 2nd time on Yattendon hill. Its contrast with Ealing struck me as pleasant.

A week or two afterwards B [Bessie, his wife] drove me over to Ealing, on coming through Yattendon we went up into the park and I was cheered to find her decided preference for the place over Ealing . . .

This visit determined me to speak to the owner . . . Mr H told me what the place yielded and offered it to me for £30,000 including timber . . .

On a dismally cold day (snow on the ground) somewhere about Xmas Edwin [Waterhouse's brother] Nics. W [his cousin] Howard H and I walked over there from Pangbourne, visited the hill . . . after lunch I concluded my bargain with the squire . . . The squire treated us after lunch to some liquor called . . . 'The Brown'. He was very amusing this latter, and we trudged home through the snow and the wood paths very merrily.[2]

Waterhouse now had an estate with farms, cottages, shops, and a Georgian manor house. The next step was to set about creating a suitable setting for himself as the new squire. During the first year he installed an agent, arranged to buy up further cottages in the village to complete the estate, and set about its management with typical thoroughness. His delight in the whole venture is evident:

On the 23rd of April [1877] I went with Theodore to Yattendon. We tramped in the east wind and drizzle from Goring to Aldworth, inspected the wonderful alabaster effigies of the de la Beche family in the church, and approached Yattendon from the north . . . In the afternoon the weather cleared, the view from the hill was lovely and Theodore sat him down while I marked wood with Green. The brougham met us at Pangbourne and we had a most exquisite illumination of purple buds of beech and elm on our way home in the glow of the sunset.[3]

But the site for his future house was not easily settled. It was picturesque country, and Waterhouse was keen to make the most of it:

2 Sept 1877. Sunday took a walk with Theodore and Rotton up to Kiln farm from whence is a wonderful view. We believe we could see the Hog's back at Guildford, and of course all the Basingstoke and Wiltshire hills . . . The hill top of Yattendon . . . is the eye of the picture. A dense forest seems to clothe the hills about to one's feet, behind are the fir hills of Bucklebury and beyond the blue hills of Highclere.[4]

The house was to be set so as to catch as much of the view as possible, and became known as the House on the Hill (Pl. 121, Fig. 9). But there were grounds to be thought of as well, and Waterhouse asked Edward Milner, the landscape designer, to help choose the exact site for the house on the knoll, 400 feet above sea level.[5] Once the position was settled Milner made his suggestions for further planting. The first order for trees contained over 3,000 items with, as might be expected for the times, a high proportion of evergreens and rhododendrons. The plan was for close planting and glades, with more planting than clearing. A coach house and laundry, kitchen gardens, a small lake, a rose garden, and a tennis court were all included in the scheme. By the autumn of 1878 all this was progressing and the rose garden being laid out by James Fair 'one of Mr. Milner's lieutenants after

[2] AW, Journal, a manuscript account of the purchase of Yattendon estate, 1877, in possession of the Waterhouse family. This is one of very few private documents relating to Waterhouse's life, and its preservation indicates how important this step was to him. The final price of the estate was £28,000, the last £500 being decided by the toss of a coin, in the seller's favour.

[3] Ibid.

[4] Ibid.

[5] Edward Milner (*fl.* 1850–84), see Ch. 11.

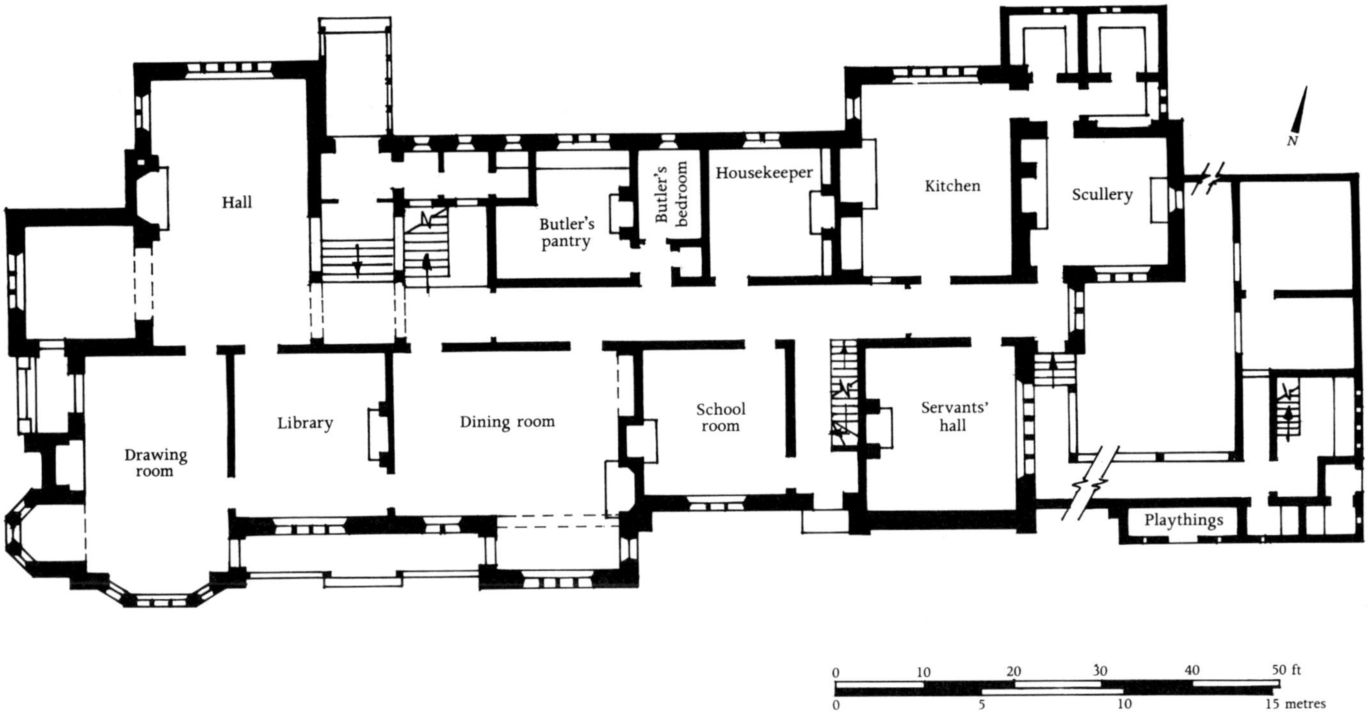

FIG. 9. Yattendon Court (1877) **[453]**, plan of ground floor

the third visit from the general himself'. The house itself was up to the ground floor.[6] It did not include a conservatory but did have in the basement a Turkish bath lined with casts of the Parthenon frieze.

Once ensconced there as new lord of the manor and patron of the living, Waterhouse had, in terms of Victorian competitive society, finally made it. Here was a home where he could entertain his wide circle of friends and enjoy to the full the fruits of his labours. No other English architect of the time managed to set himself up as landed gentry in quite that way; but the step was a measure at least of his financial success. Certainly he came from a well-to-do family accustomed to the comforts of middle-class Victorian life, but there was a contrast between this new life and his and Bessie's background. Like many of his clients, such as Leatham of Miserden **[436]**, and Ellis of The Brand **[421]**, Waterhouse and his wife had come from strongly Quaker families. She was particularly devout. As early as 1864, five of Alfred's younger brothers and sisters, all at that time unmarried, had been received into the Church of England;[7] but Alfred, then father of two children, had remained doubtful about the best way to bring them up.[8] He was in the next few years very much influenced by the Christian Socialist, F. D. Maurice, to the extent that he named his next son after him.[9] Finally in 1877, Alfred, Bessie, and Paul, who was by that time at Eton, were baptized, and the four younger children a few months later.[10]

[6] During the digging of the foundations a hoard of Bronze Age implements was found (AW Journal). The collection was placed on loan to the Newbury Museum in 1970.

[7] MW *Journal*, 188.

[8] Ibid. 191.

[9] Thompson letters (6 Jan. 1891). Thompson had writtten a letter of condolence after Maurice's sudden death at the age of 22. In his reply Waterhouse explains the origin of his son's name and his own admiration for F. D. Maurice saying also that his son was very like his namesake 'in the way in which he lived bringing the ground he trod on as near to heaven as he could'.

[10] MW *Journal*, 300 and 313.

Coupled with his purchase of the estate and the shouldering of its responsibilities, this move completed Waterhouse's migration into the Establishment. It was a social adjustment not uncommon at the time, but Waterhouse retained from nonconformity a liberal and free-thinking attitude. His concern for the welfare of those for whom he was responsible arises from genuine care which was long remembered both in the village and in the practice. The family carried out their duties as principal landowners and landlords with love and thoroughness. Cottages were restored and new ones built, the school enlarged, the church repaired and a vestry and porch added, a temperance reading room was built, and a co-operative shop established. Following the example, perhaps, of his client Lord Selborne at Blackmoor, Waterhouse put much of this work in hand while the house was being built.

Elizabeth's role as lady of the manor was increasingly important. She came from a very gifted family whose strict principles, while allowing the development of an interest in literature, poetry, and art, precluded music and drama. Now she was freer to develop her widespread talents. She was a writer of prose and poetry,[11] and wrote and illustrated stories for her children. A devout woman, she compiled and later published a book of prayers for daily household use and two volumes of the homilies prepared for the women's meetings of the village;[12] but her serious interest in literature took her into a much wider field which included modern English and American authors.[13] Like many ladies of her time, she was an accomplished water-colourist; and she and Waterhouse frequently sketched together. She could also paint the doors of a cupboard designed by her husband with all the skill shown by many a better-known artist.[14] Her designs for embroidery and for copper and brass work have very much the flavour of Walter Crane.[15]

Once they were established at Yattendon, Elizabeth organized evening craft classes for the village men and boys in a manner that was fully in keeping with the burgeoning Arts and Crafts Movement. The enterprise, known as the Yattendon Guild, was similar to G. F. Watts's terracotta workshops at Compton, and built up a reputation for good design. The Guild exhibited regularly, and at the Home Arts and Industries exhibition of 1899 its work was favourably noted (Pl. 122). 'We know, for instance that we may expect good metal *repoussé* from Keswick and Five-mile-town and from Mrs Waterhouse's pupils at Yattendon; from these also sound joinery and wood carving.'[16] The *Studio*'s report of the exhibition illustrated nine items from Yattendon (in a total of twenty-one illustrations), all of *repoussé* metalwork. The bulk of the designing was done by Elizabeth herself, though her son Amyas and her daughter Monica were also credited, and it is entirely possible that Waterhouse advised as well. Such educative work was not uncommon, but the Yattendon Guild was more successful than many and is typical of the concern for good design and sound craftsmanship that had brought Waterhouse his success.

[11] Her *Verses*, written over some years, were first published by Thomas Hawkins of Newbury in 1897 and reissued in enlarged form by Methuen in 1912. Her published prose writings are slender but potent: *The Island of Anarchy: A Fragment of History in the 20th Century*, published by Miss Langley Lovejoy's Library (Reading, 1887); *The Brotherhood of Rest*, published by E. Langley, (Reading), and Hamilton, Adams & Co., (London, 1889). Langley also published *A Book of Simple Prayers*, arranged for daily household prayers from the manuscript book Bessie used in her own home.

[12] Two series of these talks were published by Methuen later in her life: *With the Simple-Hearted* (1904) and *The House by the Cherry Tree* (1911).

[13] Elizabeth compiled 2 anthologies: *A Little Book of Life and Death* (London, 1902), which ran to 10 editions, and *Companions of the Way* (London, 1908). A further very small collection taken from the anthologies appeared as *Thoughts of a Tertiary*, published by Methuen in 1908. The lists of acknowledgements show the wide scope of her reading and her intimate knowledge of contemporary authors.

[14] A corner cupboard designed by Waterhouse with a painted door panel by Elizabeth is the property of the V. & A. Illustrated in E. Aslin, *The Aesthetic Movement* (London, 1969).

[15] A copy of *The Bases of Design* by Walter Crane, published by G. Bell in 1898, is autographed 'E. Waterhouse'.

[16] *Studio*, 17 (1899), 99.

The tastes and interests of husband and wife brought them into artistic and literary circles in London closely associated with the aesthetic movement. Through the spring of each year the house in New Cavendish Street was the setting for weekly dinner parties, 'Mrs. Waterhouse's dim, delicious Thursdays', in candle-lit rooms, where the guests included Hamo Thorneycroft the sculptor, Frank Dicksee, Alma Tadema and Edmund Gosse, whose wives were sisters, Mr and Mrs Humphrey Ward, James Bryce and his sister Julia Gaskell, Benjamin Jowett the Master of Balliol, as well as old family friends.[17] Ford Madox Brown, who painted the Manchester Town Hall murals, was another friend, and had taught Ellen Epps, who became Gosse's wife, and Theresa Thorneycroft, the sculptor's sister. Brown had also taught at the Working Men's College founded by F. D. Maurice, which provided a further link for Waterhouse, who admired Maurice enormously. Frederic Shields, the stained-glass artist, moved from Manchester to London and seems to have been more than a business acquaintance, having known Waterhouse since his pupillage in Manchester.[18]

There is curiously little evidence of friendship with architectural contemporaries. In view of his natural charm and friendliness this lack of information may be due to Waterhouse's own reticence. At his death the *Building News* described him as having a 'genial, cheery, and yet modest and unassuming demeanour' which had 'won him a wide circle of friends within and without the profession'.[19] Norman Shaw was certainly a friend, for he had designed a cradle for one of his children.[20] Together with Shaw and Street, Waterhouse was visitor to the Royal Academy Schools, and the Academy itself was another source of friends. William Burges was a friend from early days, but his death in 1881 meant that he never visited Yattendon. There were obviously numerous other professional acquaintances. He had, for instance, been a guest at the annual two-day outing of FABS, the Foreign Architectural Book Society, in 1867 for a visit to Penshurst Place and Hever Castle. This select but light-hearted group, limited to fifteen in number, included among its members Burges, Street, T. H. Wyatt, and Frederick Leighton, the artist; but it did not have room for Waterhouse as a member until 1881.[21] With them he was able to relax and talk architecture, but he seems to have kept his work strictly separate from his life at home at Yattendon.

Waterhouse seems to have been most at home with friends from the Royal Academy. Hamo Thorneycroft records a visit by many Academy members to a play at Bushey by Herkomer, *An Idyll*: 'We all came home in the train with Waterhouse who was as nice as ever. Dicksee too was charming and amusing. . . . Walter Crane . . . always rather earnest and anxious.'[22] Waterhouse had taken a keen interest in the young Thorneycroft, and wrote to Edmund Gosse, another younger friend, of his delight when Hamo was elected an Associate of the Royal Academy. 'He has won at a canter . . . Nobody can now accuse the Academy of not appreciating deserving youth.'[23] Hamo spent the Christmas of 1882 at Yattendon and wrote to Agatha Cox, later his wife:

[17] *The Centenary of Elizabeth Waterhouse, 1834–1934*, by L. V. Houldsworth (Elizabeth's niece and daughter of Thomas Hodgkin). Reprinted from *Friends' Quarterly Examiner* (July 1934).

[18] A copy of Bunyan's *Pilgrim's Progress*, illustrated by Frederic Shields and published in 1864, is inscribed by the illustrator 'To Alfred Waterhouse Esquire RA with sincere affection' (Waterhouse became RA in 1885). Waterhouse had introduced Shields to W. H. Houldsworth (the client for St Elisabeth, Reddish) for stained glass work at Coodham. This was early in Shields's career and he confessed to having had to learn a lot for this commission. See Ernestine Mills, *Life and Letters of Frederic Shields* (London, 1912).

[19] *Building News* (25 Aug. 1905).

[20] This cradle is now in the V. & A.; Andrew Saint gives the date as 1867. Andrew Saint, *Richard Norman Shaw* (New Haven, Conn., 1976), pl. 72. The Waterhouse's third child, Florence, was born in 1866 and Maurice in 1868.

[21] J. Mourdant Crook, *William Burges and the High Victorian Dream* (London, 1981), 73. Details of the Society's membership are recorded in a printed list preserved in the RIBA Drawings Collection.

[22] Henry Moore Centre for the Study of Sculpture, Leeds City Art Galleries, Thorneycroft Papers, J4 (5 Apr. 1889).

[23] Leeds University, Brotherton Collection, Gosse correspondence. See also Ann Thwaite, *Edmund Gosse: A Literary Landscape* (Oxford, 1984), 187.

Being a well ordered house the family and the guests retire at the hour of eleven, so I am writing to you in my room seated before a jolly fire, anxious to tell you of the two days' sport, for we have been cover shooting and you will be glad to hear no one was 'peppered' except 98 head of game. This is such a charming house to be in that I feel I am never weary of talking of it. Life is made so pleasant that it becomes precious! Intellectually and physically one feels the better for every hour one is here. Everyone is courteous, kind and intelligent, and conversation so bright and happy . . .

After a charming dinner and a *menu intellectuelle* (but is menu m. or f. ?) we had music and then a chat around the great hall fire, like all fires here, on the hearth.

The bedrooms have names such as these: 'Valour', 'Hope', 'Courage'. I am in 'Honour',—Is it not a nice idea? Much better than the 'blue', 'red', or 'yellow' room or 1, 2, 3.

It is pleasant to find that both the Carol singers and the Mummers are still in existence here . . . .[24]

The life of the house was intertwined with that of the village. It would have been Elizabeth who ensured the Christmas entertainment just as she arranged theatricals at home and in the Reading Room **[452]**.[25] On a later visit Thorneycroft stayed at the house on the Sunday evening while his host and hostess attended church for the second time that day; he noted in his journal: 'They have to be constant attendants as *Manor people*.'[26]

Thorneycroft was very much a member of the Yattendon circle; but Waterhouse also helped him in his career as a sculptor. His introduction to the Duke of Westminster led to a commission for a marble Artemis for Eaton Hall **[315]**. Waterhouse was also, presumably, responsible for getting him the commission for the Dean Colet memorial at St Paul's School, Hammersmith **[492]**, and the Turner statue at the Turner Memorial Home in Liverpool **[497]**.[27] On Thorneycroft's failing to win the competition for a statue for St George's Hall, Liverpool, Waterhouse suggested he should consider producing smaller pieces suitable for less pretentious homes, like his Artemis panel, which had been put in the drawing-room ceiling of the new house at Yattendon.[28]

Edmund Gosse, the writer, was a close friend of Thorneycroft and also a regular visitor to Yattendon. To judge from his misbehaviour at family prayers he evidently found the Waterhouses rather a serious couple. He had apparently attended prayers with the whole household, but seated himself close to the fire so that whenever he stood up his head and shoulders disappeared under the chimney hood. The parlourmaids were reduced to stitches, but his hostess was not amused. 'I think, Mr Gosse, we should get on better if you were to come out of the chimney.'[29] But he and his hostess were drawn together by their common interest in literature; and hospitality was extended freely to the Gosses, as it must have been to many others. Gosse was, in fact, their first visitor in the new house as they made the final move from Foxhill over Easter 1881.[30] While the new house at Yattendon, The Court, was being built the Waterhouses used the manor, lending it to the impecunious Gosses for a holiday in 1878.[31]

[24] MS letter, Hamo Thorneycroft to Agatha Cox (23 Dec. 1882), Thorneycroft Papers, Henry Moore Centre for the Study of Sculpture, Leeds City Art Gallery.

[25] MW *Journal*.

[26] Henry Moore Centre for the Study of Sculpture, Leeds City Art Galleries, Thorneycroft Papers, J5 (15 June 1890).

[27] MS letter, AW to Hamo Thorneycroft: 'Mrs Turner expresses herself as delighted with your work now that it is in place.' Henry Moore Centre for the Study of Sculpture, Leeds City Art Galleries, Thorneycroft Papers, C673.

[28] Henry Moore Centre for the Study of Sculpture, Leeds City Art Galleries, Thorneycroft Papers, C669, 677, and 673.

[29] Elfrida Manning, *Marble and Bronze* (London, 1982).

[30] AW Journal.

[31] Thwaite, *Edmund Gosse*, 174.

The manor was finally let to another literary friend, Robert Bridges, who later married their daughter Monica. He was a privileged visitor to Yattendon, even though he used to put his boots up on the mantleshelf.[32] Bessie, at any rate, admired his work, and engaged him to help in her literary interests. The *Yattendon Hymnal*, which contained only hymns of genuine literary merit and was issued in several editions from 1895, was the product of their co-operation.

Waterhouse's ability to help the younger generation of writers and sculptors would have been a natural outcome of the status he had achieved in both society and the profession; it was also a confirmation of his position as squire of Yattendon and as a member of the literary and artistic circles of the time. The fact that he made his home at Yattendon the focus of his own artistic circle suggests how important he felt this aspect was, involving, as it did, his wife, Bessie, and introducing his children to a civilized and intellectual circle. This part of his life was deliberately divorced from his architectural business; yet his existence as squire of Yattendon was the direct product of the success of that business. At Yattendon he was able to reap the rewards of his architectural labours.

[32] The story may be apocryphal. In any case it relates to his behaviour as tenant of the manor, where a chipped marble overmantel bears witness.

Part Three

THE LATER DECADES

13

The Prudential Commissions

By the mid-1870s Waterhouse was one of the most popular and successful architects in the country, a position he retained well into the 1880s. Indeed it was almost as if, whenever an architect was sought for a substantial building, Waterhouse's name was sure to be put forward. However, his experiences at the Natural History Museum, even at Eaton and the Manchester Town Hall, must have taught him how fraught with difficulty major commissions from institutions or government bodies could be. Commercial pressures might be even worse. Yet it is for his commercial commissions, particularly those for the Prudential Assurance Company, that Waterhouse is best known. In a series of twenty-seven works he stamped the presence of the company and his style on cities all over the kingdom. Pink and spiky, often in hard hot terracotta, their survivors stand witness to the extraordinary growth of that company. They are also monuments to the development of office building.

Companies no longer wanted simply an urban mansion with space for a few clerks. They needed to combine large public offices with space for substantial numbers of employees in different departments. Hierarchies of managers had to be appropriately housed; and there was usually a sumptuous boardroom to which directors would travel weekly from their country homes. Here was an architectural challenge in itself; but the increase in size meant that whole buildings could be designed, whose entire façade would express the presence and confidence of a particular company. The old era of modification, that led to façades such as the Equity and Law building in Lincoln's Inn Fields **[338]**, was passing. In the late 1870s the young Prudential Assurance Company adopted a policy of combining property speculation with the construction of custom built offices.[1] The company minute books do not suggest any conscious decision to adopt a new image. Rather, the policy seems to have been to a degree opportunistic and, at least to begin with, conditional on their being satisfied with their architect. Pressure of business in London was considerable, and their building on Ludgate Hill had only been intended for about fifty clerks; now there were ten times that number. The company was not happy with their existing architect or surveyor;[2] so they employed Chatfield Clarke to draw plans for rebuilding their offices on Ludgate Hill. Even he was not entirely

[1] The Prudential Assurance Company was founded in 1848, but had grown remarkably rapidly, and by 1870 was already busily swallowing other smaller offices. See *A Century of Service: The Story of the Prudential 1848–1948* (London, 1948).

[2] Prudential Assurance Company Board Minutes, 5 (18 Feb. 1875), 22: 'The Resident Director brought forward the question of the employment of an architect in place of Mr. Moseley at whose hands he considered the interests of the company had not been so carefully attended to as he considered they might have been. Resolved: that no additional supervision be placed in the hands of Mr. Moseley when the present alterations are complete.'

up to scratch.[3] They were unhappy with his apparent extravagance, and, though the work went ahead, he is not mentioned again. But the company was going through a radical modernization. Their first typewriter was bought in 1875, and dating machines were added soon after. They were in the forefront of the new development, women's employment;[4] and the decision had already been taken 'that if an eligible site sufficiently large can be obtained in a position where land has not yet reached such high prices it would be advisable to secure it and build offices with a view to extension as required'.[5]

The site they found was the beginning of the estate at Furnival's Inn now covered with the Prudential Assurance Chief Office building. The block was not available complete and was only purchased over a period of some years. Nevertheless there was enough ground to build the new office block. The corner plot facing Holborn and Brooke Street was bought, with a straggling extension down Furnival's Inn Yard that finally reached to Greville Street at the rear of the block (Pl. 123). Holborn viaduct had only been opened in 1869, linking the City to the West End; and the Prudential building was one of the first prestige structures at the boundary of the City, setting a new scale of urban grandeur in an area that was clearly destined for improvement.

The new building **[431]** was to be 'good [but] not apparently costly',[6] with an entrance to Holborn and a separate door beneath the tower for lady clerks. The public offices were to be at the front with space for the 500 clerks running all the way back to Greville Street. The office 100 metres race could have been run entirely within doors.[7] Managers occupied the first-floor front, with the boardroom at second-floor level marked by extra height and an impressive oriel. Plans were drawn by June 1876 and altered elevations approved the same month,[8] and work went ahead swiftly.[9] Waterhouse was regularly pressed to ensure that the buildings were completed on time;[10] but his efficient office was well adapted to succeed in this in spite of delays over the delivery of terracotta and design changes as the buildings progressed.[11]

[3] Ibid. (25 Nov. 1875), 199. Chatfield Clarke attended the meeting with plans and elevations for rebuilding 64 Ludgate Hill. The Board 'approved the general plan of the building as proposed by him but suggested that he should rearrange the elevations upon a less ornate and more economical principle, which he promised to do and retired'. It is also interesting that about this time Chatfield Clarke was employing A. H. Mackmurdo.

[4] The employment of lady clerks in large numbers was still relatively new. The approach of the Prudential reveals their attitude to the women in their employ and may have had some bearing on the quality of office accommodation they required. 'Considering that the social status of the lady clerks forms a great element in the success of the system it is: Resolved: that lady clerks be selected from the following classes: Officers of the Army and Navy, Clergymen, Merchants, Bankers and Wholesale Dealers, Members of the Stock Exchange, Professional Men viz. Medical Men, Lawyers, Artists and purely Literary Men, Managers and Secretaries of Companies and Chief Officers of same, Clerks in the Houses of Parliament and any exceptional cases in harmony with the above principle.' Prudential Assurance Company Board Minutes, 5 (12 Oct. 1872) Standing Orders.

[5] Ibid. (4 Nov. 1875), 186. It was also specifically stated that the offices for the different branches of business should be kept together, so a major new building was clearly in prospect. Henry Harben, the Resident Director, repeated that he had seen 'a plot of ground which, having a good frontage and address ran back on to a small street of poor houses, and which it appeared, might suit the Company's purpose'. He and the Chairman were empowered to proceed with the purchase.

[6] Undated MS notes of an early briefing meeting (now in the RIBA Drawings Collection) record this instruction as well as the general principles of the layout. The front office and Boardroom were 'the only rooms which are to be costly'.

[7] This possibility was recorded after the building was opened in *Ibis*, the Prudential house magazine for 1879. It would not have been a very straight route.

[8] Prudential Assurance Company Board Minutes, 5 (8 June 1876), 334 (22 June 1876), 344. The first mention of Waterhouse's name is when his chief clerk, J. Willey, attended to present estimates.

[9] The first Board meeting was held in the new offices on 29 May 1879; but even this was not achieved without some problems. The first clerk of works, employed on Waterhouse's recommendation, had to be sacked for neglect.

[10] Prudential Assurance Company Board Minutes, 5 (25 Jan. 1877), 505 and (1 Feb. 1877), 511. See also 6 (7 Mar. 1878), 355.

[11] Ibid. (6 June 1878), 439. Waterhouse recommended the adoption of a galvanized iron lattice for the lead flats in place of the teak proposed at a stage when the carcase was virtually complete. The Directors had already considered designs for office fittings (of their own drafting) and the commission to design the furniture was given to Waterhouse in Oct. 1878. The building was complete and occupied by the end of May 1879. The use of terracotta seems inevitably to have led to delays

The virtues of the new headquarters were appreciated by the female staff, who commented on the provision of plenty of light and air in the office, plenty of space, pegs, lockers, and looking glasses in the cloakroom, and a luncheon room like 'the saloon of a fine steamer with its leaning side and windows through which may be observed the pallisading of the roof looking like the upper deck'. After luncheon the clerks could promenade on the roof 'where being so high, we are not inconvenienced by the smoke from the adjacent chimneys'.[12]

The building press also saw the advantages of the new building as street architecture.

> We have here then a building which presents some commendable features, and which may challenge, without exaggerated pretentions, many costlier buildings in the same thoroughfare of brick and stone. It marks a more sensible use of the material [terracotta] than we have been accustomed to see since Mr Christian gave us the front of the Economic Assurance Office in Bridge Street, Blackfriars. We have generally had to complain of a redundant use of enrichment, a mechanical monotony in the setting and distribution, that has greatly injured the reputation of the material as one of the most natural we possess, and one that will resist the elements besides acids and alkalis.[13]

In 142 Holborn Bars can be seen the first of some twenty-seven commissions Waterhouse received from the company between 1877 and 1901,[14] the beginning of a relationship in which Waterhouse came to be regarded more as a trusted servant than a supplier of designs. On several occasions he was even allowed to negotiate on behalf of the company over the acquisition of sites.[15]

Once the company was satisfied with its new head office, it lost no time in putting Waterhouse to work again. His second building for the company, a speculative development of shops and offices in Brooke Street **[480]**, dated from 1880. The contract for excavations was let three weeks before the elevations of the proposed buildings were submitted.[16] The company was also able to combine its interests as a property developer with the expansion of the insurance market. It began to buy sites in

though in most cases Waterhouse managed to avoid their affecting the project to any serious extent. This was probably the result of very close attention to details in the production of drawings for which he used a 1″ scale and 'shrinkage' scale. In 1901 the contractors for the new Chief Office did, however, successfully claim an allowance of £1,500 for losses incurred through delays in the delivery of terracotta and iron. The case was clearly exceptional. (Prudential Assurance Company Board Minutes 27 (28 Nov. 1901), 490; 28 (16 Jan. 1902), 24.

[12] These facilities were reserved for lady clerks only. There were no such perks for men. The glowing description is from an article in the house magazine *Ibis* (1879), 288–9. The authoress claimed: 'to say that we are one and all perfectly charmed with our new home is but feebly to express what we really feel . . . there seems to be absolutely nothing wanting to make us perfectly good and supremely happy.' She almost suggests it was a pleasure to climb the 70 steps to the general office. There was no lift for clerks.

[13] *Building News* (8 Apr. 1878), 353.

[14] The list of commissions is as follows: 1877 Holborn first stage; 1880 30–9 Brooke Street; 1884 Holborn second stage; 1885 Liverpool; 1885 Holborn third stage; 1886 Portsmouth; 1886 Manchester; 1888 Glasgow; 1889 Birmingham; 1890 Leeds; 1891 Cardiff; 1891 Newcastle; 1892 Leicester; 1893 Bradford; 1893 Nottingham; 1895 Sheffield; 1895 Dundee; 1895 Edinburgh; 1895 Holborn fourth stage; 1898 Huddersfield; 1898 Oldham; 1899 Bristol; 1899 Plymouth; 1901 Hull; 1901 Southampton; 1901 Staple Inn Buildings. There is also a Prudential building at Bolton that approximates to Waterhouse's style. It appears, however, that in that case a local architect was working to his general direction. The relationship between the practice and the company continued even after Waterhouse's death, with Paul designing several Divisional Offices for them such as that in Aberdeen.

[15] For instance he was confident enough to commit the company, without previous consultation, to a £500 payment to settle an incipient dispute that was threatening to hold up the building of the Leeds offices. Prudential Assurance Company Board Minutes, 18 (30 Oct. 1890), 215, record that 'the agreement be approved and Mr. Waterhouse be thanked for his prompt action in the matter'. On another occasion Waterhouse was authorized to negotiate for the company over the sale or exchange of ground adjoining Staple Inn, which the Birkbeck Bank wanted for their new building (ibid. 19 (2 July 1891), 154). He also negotiated for the company the purchase of the vaults under the roadway fronting their Leicester site (ibid. 20 (10 Mar. 1892), 73). And Waterhouse was responsible for the agreement with Gamage's architect over rights of light when the new Chief Office buildings were about to go up along Leather Lane (ibid. 25 (14 July 1898), 321).

[16] Ibid. 8 (27 May 1880), 54 and (17 June 1880), 74. This development was directly across the road from the head office, and included flats above the office chambers. Some of these were actually taken by the new lady clerks.

good locations in various towns as they became available and then built offices on them which were partly occupied by the company and partly by its tenants.[17]

Waterhouse's skill was to combine for the company a variety of different accommodation on sites that were often tightly built round (Pl. 161). The usual arrangement was a large public office for the Prudential, backed by some clerical space, and sets of managers' offices, which might altogether occupy two floors. Then there were smaller sets of chambers, reached by independent stairs, that might house a whole range of businesses. At Leeds **[559]** (Col. Pl. 7), no less than twenty separate businesses are recorded, including an office for the Leeds Fireclay Company, suppliers of the bulk of Waterhouse's faience.[18] The web of business connections was every bit as complex as the family ties of the Quakers; and the Prudential appears to have had a link with one of the early catering chains. Ye Mecca leased restaurants in the Prudential buildings at Nottingham **[584]**, Leeds **[559]**, and Birmingham **[554]**. These basement eating houses were important for the hundreds of Mr Pooters in the offices above; but they also required careful thought on the part of the architect to create large and reasonably attractive spaces within the structure. There was no standard design, partly because the sites varied, but also because the tenants did. At one stage the Leeds building housed an art gallery; at Bradford **[581]** there was a Turkish bath (with all the screens and fittings designed by Waterhouse), while Portsmouth **[538]** even included a private hotel.

While all this was going on in the provinces, the company was expanding its London property until by 1900 it occupied the whole city block (Pl. 125). It acquired Staple Inn, on the south side of Holborn, which Waterhouse 'restored' by stripping off its eighteenth-century plaster façade and re-creating its series of medieval windows **[543]**. Next door the company acquired land on which to erect Staple Inn Buildings **[645]**, one of Waterhouse's last works, and destined to become the home of his practice along with other Prudential tenants. However, the principal office, the flagship for the company, posed real problems. The site was purchased piecemeal, and the company's requirements changed over the quarter-century, as did the building trade. None the less, Waterhouse managed to create out of the disparate jumble of buildings a mammoth office block with a powerful façade and a dramatic series of interior spaces.

The first expansion was along Brooke Street **[522]**, where the existing design was simply extended, although there were problems with the new structural system.[19] Then a block along Greville Street was acquired and built over **[523]**; but the intention was clearly to spread over the whole block.

[17] The property developer's approach is epitomized by the case of the Bristol office. Waterhouse's building there dates from 1899; but the negotiations for the site began in Aug. 1890 (ibid. 18, 122). The site was acquired in Nov. of the same year (ibid. 260) complete with the leases on the existing houses, which were allowed to stand (evidently with little repair) until the leases expired.

[18] A detailed picture is provided by the survival (Prudential Assurance Company MSS, now in the RIBA Drawings Collection) of a list of tenants in the Leeds building. The list is not dated but is of *c.*1891 and was prepared for the letter box: 'Front Block Ground Floor: Commercial Union Assurance Co. 1st Floor: Clough Ford & Co. Accountants: Leeds Fireclay Co. 2nd Floor: Lakeman & Oliver, Accountants: Rowley Walter, Mining Engineer: W. C. Hull, Architect. 3rd Floor: R. H. Fowler, Private Residence. 4th Floor: Joseph Shaw, Coal Factor: Tronsdale & Co., Locksmith. Back Block 1st Floor: Middlebrook & Hopkins, Solicitors. 2nd Floor: John Peacock King, Architect: Rudolph Hauptmann, Mosaic Manufacturer: E. Firth, Woollen Manufacturer: Young & Co., Wine & Spirit Merchant: W. Spinks, Civil Engineer. 3rd Floor: P. Carlin, Typewriter Co.: Markham & Holmes, Artists: G. F. Fitton, Colliery Supply Co.: Murfin & Durbury, Artists: R. Midgeley, Butter and Egg Dealer. N.B. There may be 3 tenants in the Basement whoes [*sic*] entrance will be in the front block.' The basement is shown on the contract drawings as reserved for an art gallery, but in the event the lease seems to have been taken by Ye Mecca restaurant.

[19] The original block of 1877 was constructed with an internal frame of cast iron stanchions and iron girder floors. The fireproof muniment block of the same period was of iron stanchions and shallow brick vaults. Between them Waterhouse had, in 1880, to insert an extension built with Lindsay's patent steel decking. The difficulty seems to have come in matching the depth of the floors. Load-bearing external brick walls were used thoughout.

Difficulties in completing the site meant that the final commission was built in four stages **[602]**;[20] and there are a series of minor differences in the decorative detailing. But the principal concept was the retention of the great court of old Furnival's Inn. This was reached by an arched entrance under the tower leading to a vaulted passageway and a small courtyard. Then, beneath a dramatic semicircular bridge, the whole court was suddenly revealed (Pl. 127). A five-storey office block, richly ornamented in glazed tiles and terracotta, filled the whole of its northern side (Pls. 124, 126).

The principal front was given over to the main public offices (Pls. 128, 129), the managers and the directors (Pl. 183), while all behind was left for the army of clerical staff (Pls. 130, 131, 132). The extra height of the rooms along the façade made for some pretty complex stairways, with landings at all sorts of levels, and the junction with the original block was never entirely satisfactory.[21] Inside, the old arrangement of a second-floor Boardroom was maintained; but it was now linked to a large Shareholders' Hall, made splendid by a six-bay hammerbeam roof.[22] Rooms, stairs, and passages were all decorated with glazed brick and faience or panelling. Everything was of the highest quality, but arranged according to a strict hierarchy. The Boardroom had oak panelling and fireplaces of alabaster. Managers had marble fireplaces and deep mahogany panelling. The grand staircase was aglow with honey-coloured faience and marble mosaic floors (Col. Pl. 5; Pls. 169, 190); but there were simpler ascents for managerial and clerical staff. In the main public offices there were faience clad arcades and long mahogany counters topped by brass grilles, all designed by the Waterhouse office. The decoration thinned as one penetrated deeper into the building, until, beyond the public gaze, steel flooring was plainly exposed and only the dramatic arches of the brick structure gave interest to these areas.[23]

Despite its extent, the building, with its series of blocks around different courts, retains a sense of variety. In spite of its long period of growth and the deliberate modifications of its patterns, Waterhouse and his office managed to impart an underlying unity. That the building and its furnishings were only completed after his death, and yet conform so closely to his style, merely indicates the lasting power of Waterhouse's design. The Prudential buildings certainly helped to create his later reputation. Taken as a whole they have been seen as one of the first examples of buildings in a recognizable house style. Yet there is little direct evidence that the Prudential set out to acquire any such image, and there were always a number of Prudential offices that were not housed in Waterhouse buildings.[24]

Waterhouse's attention to skyline and broad handling of ornament gave all his Prudential

[20] The last unit, Ridler's Hotel on the corner of Leather Lane, was not acquired until 1898, after plans for the Holborn façade had been drawn and when the northern side of the courtyard had already been built.

[21] The original block on the corner of Holborn and Brooke Street was demolished in 1930 and replaced with a clever copy of Waterhouse's 1897 façade by Joseph. The front section of Joseph's work still remains.

[22] This space has unfortunately been divided horizontally into 3 floors of offices, destroying the roof and effectively marring the original circulation.

[23] The pattern of brick arches supporting steel decking was used on the blocks along Greville Street. With the advent of ash concrete floors in the late 1880s, the system was abandoned, and the Leather Lane block lacked this feature. Most of the arched sections have recently been demolished.

[24] Prudential Assurance Company Board Minutes (26 Feb. 1891), 456, lists the district offices as: Manchester, Liverpool, Portsmouth, Leeds, Dundee, Edinburgh, Cardiff, Birmingham, Nottingham, Bradford, Hull, Sheffield, Newcastle, Huddersfield, Bolton, Oldham, Southampton, Glasgow, Swansea, Barrow, Middlesbrough, Wakefield, Falkirk, Southport, Belfast, St Helen's, Halifax, Brighton, Torquay, Ipswich, Paisley, Sunderland, Bury, Exeter, Rochdale, Preston, Castleford, Blackburn, Greenock, and Dublin. Only the first 18 towns had Waterhouse buildings, and several of those were not even designed in 1891. And the list is not exhaustive as offices in Northampton, Coatbridge, Wednesbury, Walsall, Wolverhampton, and Oxford are all mentioned in the next volume of the company minutes. By 1897 there were as many as 81 district offices.

buildings a certain unity; but there is more variety than appears at first sight. The buildings all have a polished granite base. Hard red terracotta occurs on several offices, notably Manchester **[535]**, Liverpool **[521]**, and the London headquarters **[602]**. Bradford **[581]** uses a paler tone and Leeds **[559]** and Leicester **[575]** have red brick and buff terracotta. Glasgow **[548]** and Newcastle **[572]** use brick and stone while the Edinburgh **[596]** and Dundee **[595]** offices are faced with stone throughout.[25] Waterhouse was clearly given a certain amount of freedom to adapt his basic design and could be relied on to produce what was wanted. As its sole architect he was able to give the company's buildings a distinctive character; though the similarity of treatment of 1 Old Bond Street **[481]**, which he designed for Christy's, the hatters, suggests that the design was a successful recipe for office development rather than a specific Prudential livery. However, the advantages of a clear corporate image must have become apparent, and the company did take a close interest in the appearance of its buildings. It was probably not committed to terracotta in the way Waterhouse was; but it was responsible for the choice of the bright red. The company specifically praised the pink stone and red brick of Glasgow **[548]**, objected to the buff terracotta at Leeds **[559]**, and for Staple Inn Buildings **[645]** insisted on the substitution of red terracotta for buff.[26] The material was chiefly sanctioned by its economy, and the impression is very much that Waterhouse satisfied the Prudential by his ability to produce effective designs swiftly,[27] to use modern materials sensibly, and to see that the buildings were finished at close to the original estimate. The extent of the company's confidence in him is indicated by the way it remitted him the £1,655 that should have been deducted from his commission for the Edinburgh building because a wrong tender had been accepted in error.[28] Indeed when a stroke forced his retirement from practice the Board wrote to Paul Waterhouse of 'their regret at the proposed severance of his father's connection with the Company, and to express their appreciation and admiration of Mr. Waterhouse as an architect and the personal esteem in which they had held him during an association extending over many years'.[29]

[25] The contract drawings show the Edinburgh building as brick and stone. Prudential Assurance Company Board Minutes, 23 (8 Aug. 1895), 347, record that 'The Secretary read the Correspondence which had passed in reference to the objection of the Town Council to the use of brick in this building and he was instructed to inform the Architects that the Board would agree to substitute Dumfries stone in place of brick.'

[26] Ibid. 26 (5 July 1900), 445. The brick and stone of the Glasgow offices clearly pleased the board. When the Newcastle office was being designed 'the plans were considered and the Secretary was directed to instruct Mr Waterhouse that the Board would wish the same material used as that which had been adopted for the Glasgow offices, viz: pink stone and red brick' (ibid. 19 (30 July 1891), 201). Later, after the Secretary had visited the construction site of the Leeds offices he 'mentioned that the terracotta used in the building was of a buff colour, and he was instructed to remind Mr Waterhouse that the Board preferred red terracotta' (ibid. 20 (23 June 1892), 232). Significantly the surviving model of the first design for the Chief Office extension is banded in buff and dark red whereas the final building is all bright red.

[27] The design of the Cardiff offices is instructive. Negotiations for the site began in Apr. 1889 'it being understood that the Company's architect, Mr Waterhouse, should be unrestricted in the plan and elevation of the building' (ibid. 16 (11 Apr. 1889), 361). Plans were not drawn immediately but Waterhouse's assistant attended a Board Meeting with them in June 1890 (ibid. 17, 516–17) when modifications were requested. These were ready by 26th June when Waterhouse himself attended (ibid. 18, 21). Further modifications were requested and passed at a Board Meeting a week later (ibid. (3 July 1890), 34). Evidently Waterhouse's office was both efficient enough and flexible enough to give high priority to a major client.

[28] Ibid. 23 (31 Oct. 1895), 424; 24 (18 Mar. 1897), 400; (25 Mar. 1897), 401. The company showed no such mercy when the stone-polishing contractors, Farmer & Brindley, found themselves in a similar position.

[29] Ibid. 28 (10 Apr. 1902), 141.

14

The National Liberal Club

ONE building of Waterhouse's later period deserves special consideration, the National Liberal Club **[514]** (Pl. 133, Fig. 14). It was the building he himself selected, along with the Natural History Museum **[243]**, to exhibit at the Chicago World's Fair of 1893; and he obviously thought highly of it. Since his brother Theodore was the club's solicitor, and he himself had so many Liberal connections,[1] this commission might be seen as the reward of his political affiliations—a prestige work to display his powers to the full. Once again, however, the story is a good deal more complex than that, and the success of the National Liberal Club as a building was achieved in the face of considerable difficulties. It is a masterly display of his planning skill, though that aspect will be treated separately;[2] but it is also a fine contribution to the embankment skyline, and its interiors are sumptuous.

Waterhouse was not in fact the architect originally intended, though it was as well for the club that he did become involved. The club was only founded in 1882, with temporary premises in Trafalgar Square. However, as the centre of political power, it needed a new and distinctive home. Fortunately in 1883 a large tract of Crown land fronting the river was being touted for development. The hopeful developer was Jonathan Carr, the originator of Bedford Park, who planned to build high-class flats in a sumptuous pile of red brick and terracotta.[3] Although he did not join the club until 1884, he did include space for a large clubhouse on his early plans, and there cannot have been many contenders. Presumably he saw the powerful National Liberal Club as a useful financial foundation for the rest of his plans.

The National Liberal Club, however, was keen to get a site, and an embankment frontage would be ideal. It demanded the northern, rather than the southern, end, so that its club should have frontages to both Whitehall Court and Whitehall Avenue; but otherwise it went along with Carr's plans. However, he grew dissatisfied with his first architect, and the successors failed to satisfy the Crown's surveyor with the quality of their elevations.[4] This dissatisfaction will certainly have reached the ears of prominent Liberals, who had just formed a limited company with a share capital of

[1] Waterhouse was not politically active himself; but many of his clients were Liberals. Joseph Whitwell Pease and Thomas Fowell Buxton were both Liberal party members, as were James Bell, MP, and Julian Goldsmid, MP; and Roundell Palmer served in Liberal administrations. Even the Duke of Westminster had sat as a Liberal MP until he inherited his title.

[2] See Ch. 20.

[3] Jonathan Carr's plans for Whitehall Court led to the spectacular bankruptcy that ruined him. The story is told by Andrew Saint, 'Whatever Happened to Jonathan Carr?' in the *London Journal*, 12/1 (Summer 1986), Commemorative volume for Francis Sheppard.

[4] C. J. Phipps was succeeded by Archer and Green to whose designs Whitehall Court was eventually built.

£200,000 and the sole purpose of building a clubhouse.[5] The directors immediately declared that 'in any elevation approved, the Club *must* have a distinctive character';[6] and, since the company's solicitor was none other than Theodore Waterhouse, it seems likely that they already had their architect in mind.

Waterhouse was commissioned within a fortnight of the company's launch,[7] and can have had no worries about the club's decided preference for an Italian Classic design rather than anything Gothic, Queen Anne, or Elizabethan. The style was an unusual one for him, but well within his eclectic capacity. In any case, he had the help of his son Paul who, in a reversal of his father's expertise, combined a real interest in Classical forms with his Gothic training. The first task was to decide what was wanted, and to arrange the series of grand rooms on the awkward corner site. Waterhouse managed to combine the splendour of the major appartments with the largest number of members' bedrooms of any club in London (Pls. 134, 135; Fig. 14). In this the National Liberal Club was effectively a new development. Its members could no longer dine at their club of an evening and then trundle happily home in the small hours. They lived increasingly further from town, commuting by train from the country, or even belonging to a provincial centre altogether. So their club was designed as a sort of compromise between a clubhouse and a hotel,[8] with 100 members' bedrooms and staff accommodation to match.

This mongrel function had to be clothed in suitable dress. The Crown Estates stipulated Portland stone, which cannot have pleased Waterhouse,[9] although they had earlier been prepared to allow red brick and terracotta for a proposed American hotel on the site. There were further problems too, in that the façade of the club, while retaining its individuality, was supposed to accord with Whitehall Court next door. Yet Archer and Green, the architects of that building, had not even begun to build, and could only offer drawn proposals as a guide. Still the work went ahead, initially with the same contractor that Jonathan Carr was using.[10] Waterhouse and the National Liberal Club seem to have been acting cautiously in letting the foundation contract separately. By the time this was completed in mid-1885, it was clear that Carr's plans were in serious financial trouble; and the new contract for the superstructure was let to George Shaw, who had just completed a new spirit room at the Natural History Museum for Waterhouse.

Waterhouse treated the preferred Italian Classic in the same free way that he usually handled Gothic. The acute angle was turned by a tall tower with Romanesque pilasters rising to the eaves (Pl. 133).[11] The windows of its octagon were elongated versions of the ones he had put in the Natural History Museum, and its roof owed a good deal to Chambord. At the tower's foot crouched a cheery little gazebo to set off the scale.[12] The mass of the building is cleverly broken by bay windows rising

[5] The Company Prospectus, dated 10 June 1884, is preserved in the University of Bristol Library, National Liberal Club Papers.

[6] National Liberal Club Buildings Co., Board Minutes (14 June 1884), quoted in Saint 'Whatever Happened to Jonathan Carr?' The Board of Directors, incidentally, included 2 men for whose families Waterhouse had already done work: Sir Thomas Brassey (Heythrop Hall, 1871) and Leatham Bright (Miserden, 1876).

[7] On 23 June 1884. See University of Bristol Library, National Liberal Club Papers, Board Minutes (23 June 1884).

[8] Financial pressures have recently forced the club to sell off the upper floors to be permanently run as a hotel.

[9] Portland stone was a material he rarely used, though he approved of its qualities for urban buildings. The National Liberal Club is the only building of his to be entirely clad in this material.

[10] This was Henry Lovatt, who had already worked for Waterhouse at Steventon Manor and the City and Guilds College. He later had 4 more contracts from Waterhouse, but was given no major work.

[11] In the original perspective published in *Building News* (30 Jan. 1885), the turret windows were given round Romanesque heads. These were altered to square in the interests of visual unity.

[12] Family tradition has it that the gazebo is all the work of Paul Waterhouse. If so, it is a remarkable tribute to the closeness with which father and son worked, for it makes an important contribution to the design as a whole.

through three floors and by a series of alternating dormers and gables for the seventh floor in the roof. There are clear horizontal divisions, too, that reflect the organization. A single-storey row of billiard rooms flanked the embankment gardens. These were top lit, allowing their roof to double as a broad terrace reached from the main dining-room or the coffee room. More public rooms lay above, divided from the bedrooms by a major cornice. Two floors of substantial bedrooms offered the chance of either a loggia within the bay or a balcony on top. Then came the eaves line, and the roof was two-storeys high with wide gables linked by an open arcade.

The building was certainly most successful on the embankment side, where members had magnificent views over the gardens and the river. The other sides were less successful, and the entrance, a small-scale exercise in Baroque forms, was rather dwarfed by the white cliff of stonework above it. But Waterhouse would not be the first architect to concentrate his efforts on a show front; and this was, in any case, a building to be enjoyed from inside. The structure was of load-bearing external walls with steel columns and steel floors of the same patent type that he was using in the Prudential. The walls, columns, and even one of the ceilings, were all clad in faience in a variety of his subdued tones, green and honey-gold, pale blue, buff, and brown, adopted 'without overelaboration of ornament or crudity of colouring'.[13] In the very middle of the building rose a great staircase of white Sicilian marble with a balustrade of alabaster and Pavonazetto marble supporting paired columns of Breccia Africano (Pl. 136).[14] The secondary staircase in the tower was equally exciting. Curving flights about an open well link landings on all seven floors, and there was a wrought iron balustrade in his typical manner.

Nor was this all, for the glittering caves that are the public rooms were largely furnished to designs by Waterhouse (Pls. 134, 135, 137). The chairs were of the same type that he had devised a decade before for Manchester Town Hall. Solid and yet designed for easy manufacture,[15] they fit as easily into this Classical magnificence as into the Gothic splendours of the north. The whole edifice provided comfortably plush surroundings for the dominant political grouping of the day. Waterhouse served them well, and could be justly proud of a building that, until recently, has been much undervalued.

[13] *Builder* (21 May 1887), 783.

[14] The staircase was destroyed by bombing during the Second World War. The replacement uses an identical alabaster balustrade (presumably incorporating any old balusters that survived) but the Breccia marble columns were not replaced. The change certainly makes the staircase lighter though less dramatic. It is also worth noting that the bomb, a direct hit, did no serious damage beyond the stairwell, which is proof of the the solidity of the structure.

[15] They were made by the firm of Marris and Norton. This was not Waterhouse's favourite furniture maker, but they had already had work from him at Hove Town Hall and at the Gartside warehouse in Manchester. Their National Liberal Club chairs are slightly lighter than those made by Doveston Bird and Hull for Manchester and more elegant, in keeping with the later decade.

15

Two Late Churches

ALTHOUGH Waterhouse's reputation was never as a church builder, two of his late churches need to be examined because they stand apart from the commercial success and the Establishment architecture of the Prudential **[602]** and the National Liberal Club **[514]**. St Elisabeth's, Reddish **[507]**, and the Lyndhurst Road Congregational Church **[505]** in London are among his best buildings, and, both being commissions of the 1880s, show his continuing power and originality. The Congregational Church is a fairly determined example of his round-arched style, but with a complex and original plan. St Elisabeth's displays much more concern for style, and is as complete as any Tractarian church could be. To design so successfully for such very different forms of worship at the same time says a good deal for his imagination, and for the care with which he listened to the client's needs. True he had already built some half-dozen Anglican churches,[1] and about the same number of chapels;[2] but each had been treated as an individual case, and there was no sense of design by formula.

His earliest churches in Manchester **[152]** and Penmaenmawr **[224]** had been striking, but hardly elegant. In his maturity he had produced a convincingly English parish church in rebuilding St Mary's, Twyford **[440]** (though the effect is slightly marred as the spire was never built). Probably his best church so far was St Matthew's, Blackmoor **[266]**, which he designed for Roundell Palmer. That had been a structure of some individuality. The overall impression is of a traditional English church, though the tiled pyramid roof on the west tower is unusual. The tracery of the windows is similar to that used in Manchester Town Hall **[284]**, and derives from the classic late thirteenth-century work at St Mary, Temple Balsall.[3] The interior is fully equipped for the revivalist liturgy of the mid-century Anglican church. However, Waterhouse may have regarded St Matthew's as only partly complete. It was built with nave and chancel only; the arcade that would open into a north aisle was also constructed and is expressed in the outer wall, though the aisle was never built.

[1] St John, Brooklands (1863) **[152]**; St Seiriol, Penmaenmawr (1865) **[224]**; St Martin, Brasted (1865) **[194]**; St Matthew, Blackmoor (1867) **[266]**; St Mary, Twyford (1876) **[440]**; St Bartholomew, Reading (1877) **[450]**. St Mary's and St Martin's were in fact rebuildings, but in both cases the building is to all intents and purposes by Waterhouse.

[2] Ince Cemetery Chapels (1856) **[14]**; Ancoats Congregational Chapel (1861) **[105]**; Rusholme Congregational Chapel (1863) **[164]**; Besses o' th' Barn Congregational Chapel (1863) **[163]**; Ashton-in-Makerfield Congregational Chapel (1865) **[191]**; Caversham Methodist Chapel (1865 and 1876) **[225]**, **[437]**; Reading School Chapel (Anglican)(1873) **[394]**.

[3] St. Mary, Temple Balsall, was almost totally rebuilt by G. G. Scott in 1849. However, the church was illustrated in H. Bowman and J. S. Crowther, *Churches of the Middle Ages* (London, 1848). The page in Waterhouse's copy with the illustration of the windows is heavily marked with compass points, indicating that he had made several tracings from it.

In the 1880s his individuality in church design is just as evident, even when he might easily be expected to be producing work in a more mechanical way. The only thing these churches have in common is that they are of brick, the one with terracotta dressings, the other with stone.

At St Elisabeth's **[507]** (Pl. 138) Waterhouse was working for W. H. Houldsworth who owned the local mill and was the principal local employer. A man of considerable wealth and piety, he had already retired to a large country house in Scotland, to which Waterhouse had added a chapel and music room **[369]**. At the same time he set about improving Reddish with a school **[373]**, a working men's club **[388]**, and parsonage **[407]**, also designed by Waterhouse. The superb church was to be the centre-piece of this development.

Waterhouse decided against the more usual stone for the church interiors, and turned his back on the conventional Gothic to create a majestic pile in red brick Romanesque. The plan is quite conventional, of nave and aisles, with apses to the chancel and lady chapel. But the striking feature is the tower, set outside the south aisle at the level of the chancel arch. It rises over a south-east porch and is dramatically linked to the nave by a bridge and dizzy flight of steps to a dormer in the ridge of the roof. Waterhouse was evidently determined to have a good tower for effect, and tried several possibilities.[4] The first plans show a square east end, transepts, and a western tower. Next he tried a large tower at the south-west corner, linked by a porch. A further scheme shows a porch, but the tower is gone. Presumably cost was the problem, and later plans are annotated with references to 'the tower, which, however, is not to be built'. Whatever the pressures in making the final decision, Waterhouse intended to make the most of what he got. A note on one drawing read: 'Memo to Mr Cooper [senior draughtsman] See Block plan and position of surrounding streets. On former I have indicated the position of the tower I think. I want it to exhibit itself along at any rate *one* street.'[5] The tower he eventually built was only modest in height, but its architectural effect is thrilling. The great nave roof gives the building a solid barnlike feeling that is reinforced by the plain stone banding and the simple round-headed windows. The little tower rises clear of the church, well above the aisle roof, gaining its effect with remarkable economy.

The air of solidity of the exterior was continued in the interior, where a muscular arcade of only four bays was supported by beefy columns of polished granite (Pl. 139). But Waterhouse did not finish there. He fitted out the church lavishly, designing everything down to the altar cross, communion vessels, frontals, and pulpit falls (Pl. 140).[6] Only the stained glass panels in the clerestory windows were not his; and they were designed by his old friend Frederic Shields.[7]

The impression was rich without being fussy. The wide arcade gave a great feeling of space to the nave, while the aisles were dim. The eye was led straight to the chancel, where the patron's affluence was displayed in a chancel screen of alabaster and Devonshire marble columns supporting a frieze and the figures of the rood. The chancel was vaulted in stone and brick, and wrought iron screens separated the side chapel and the organ. Marble and alabaster were used again for the reredos and sedilia, and the sanctuary walls were panelled in marble of soft green and grey. Finally, above the panelling Waterhouse ran a blind arcade of round arches, the shadowy depths emphasizing the richness below. All his student excitement in the Byzantine splendour of St Mark's was recalled and redefined in this Mancunian suburb.

[4] The various schemes are preserved in the RIBA Drawings Collection.

[5] Drawings in RIBA Drawings Collection.

[6] A series of fully coloured drawings for the altar frontals and other fittings survive in the RIBA Drawings Collection.

[7] The windows were originally intended for Houldsworth's private chapel at Coodham. At about this time Shields was also working on the panels for the chapel at Eaton Hall.

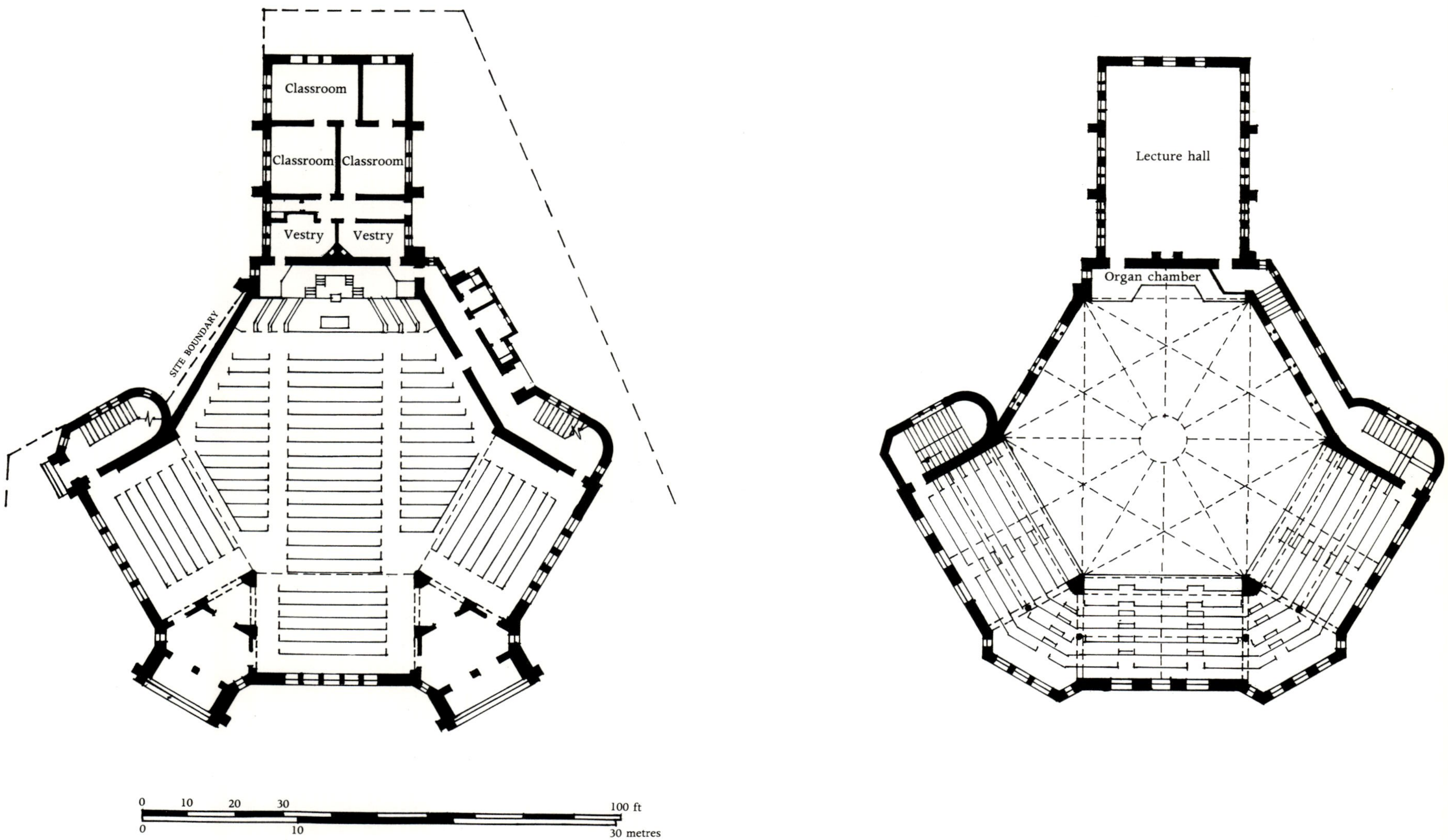

FIG. 10. London, Lyndhurst Road Congregational Church (1883) **[505]**, plan (RIBA)

Study of those Venetian buildings and other Romanesque sources was brought out here with surprising freshness. But this was no exercise in narrow historicism. The chancel screen, in particular, showed his originality excellently. The flowing leaves of the capitals rose gracefully from the round column, recalling Ruskin's studies of St Mark's; but the leafy panels of the frieze were entirely his own. Yet the whole interior was stylistically less uniform than the exterior, a free mixture of Gothic Romanesque and early Italian forms. Waterhouse's consistency in handling draws it all together, however, and makes of this amalgam a powerful unity that is strangely moving.

The richness and romance of St Elisabeth's are in marked contrast to the Lyndhurst Road chapel **[505]** (Pl. 141), where the effect was achieved entirely by the complex geometry of interlocking shapes. The building stands as a grimly determined pile of blue brick and dark red terracotta, beneath a great wigwam of roof. From the outside three great gables project, with porches under lesser gables tucked tightly in between. The effect is to obscure the basic shape and the orientation, which can only be discovered by passing through the porches and experiencing the space from within (Fig. 10).

Waterhouse devised a route which involved several changes of direction. He ended his porches with two doors set at an angle, so the worshipper had to turn to enter the church. Each door led directly into one corner of the worship room, so that one had to turn again to face the dais. Then moving forward under the gallery one reached the main space. It was only a short distance, but the twists and turns and changes of height created a rich spatial experience. In typical fashion, Waterhouse achieved this with simple geometrical forms. His central space was a hexagon, with organ and dais filling one end, and lit from the adjacent south-facing side. The three sides of the hexagon opposite the dais were simply extended at full height to create a huge triple gallery, fulfilling the prime function of such a chapel, to seat as many as possible within full view of the pulpit. Stairs at the outer ends gave access to the gallery which was linked into one space by wide triangular passages over the porches. Vestries and a lecture hall filled a two-storey extension behind the dais, and were linked to the church by passages along the remaining side of the hexagon. The concept was brilliantly simple and original. It combined all the traditional requirements of a congregation, in their traditional relationship to each other, with a complex and subtle plan.

Two such different churches would be a considerable feat for any architect. That Waterhouse produced them at the same time and when he was already past the peak of his career, shows that even in the 1880s there was no dimming of his power. He continued to produce schemes where planning, assembling of forms, and manipulation of space are as exciting as ever till the late 1890s. And there were major commissions yet to come for colleges and schools, for hospitals, and for the commercial world.

16

Collegiate Building of the Later Decades

WATERHOUSE'S position as architect to the three northern universities, Liverpool, Leeds, and Manchester, gives him a very special position in the development of collegiate building. The commission for the Owens College that was to become Manchester University actually goes back to 1860 **[89]**.[1] However, in all three universities Waterhouse was at work for most of the 1880s and 1890s, so the schemes need to be considered with his later educational works. These included two major commissions in London, both now demolished.

The City and Guilds of London Institute **[490]** (Pl. 144) was Waterhouse's second contribution to South Kensington. The provision of a technical college to train draughtsmen and technicians was close to the original ideas of 'Albertopolis'. It had its origins in the need for well-educated artisans and clerks, and resulted from an unusual effort of co-operation on the part of the City Livery Companies, led by the Mercers' Company and the Drapers, who had, till then, controlled their own systems of apprenticeship. However, in the late 1870s, with the spread of examinations and growing stress on technical education, a new unified institution was launched. Its aim was to produce teachers as well as superior workmen, foremen, and even to provide technical education for managers and the sons of factory owners. In spite of considerable misgivings about its site in South Kensington, the City institutions decided to back the new college, when it became clear that no government funding would be available for a new Science School.

As an educational development the City and Guilds of London Institute **[490]** was enormously important; and most of the leading scientists of the day, such as Huxley and Roscoe, were consulted about its design and equipment. However, as a building it never offered the spatial opportunities of the Natural History Museum **[243]** (Pl. 82), and was demolished in 1962 without much protest. Nevertheless it showed several characteristic features of Waterhouse's work. The original report stated categorically that 'regard should be had rather to what is wanted on the inside, than what will look well from the outside'.[2] The original allowance of £35,000 was only increased to £50,000 by the time the architect was selected, so it was clear that the committee meant what they said. Even if the position of Roundell Palmer, Lord Selborne, on the committee made Waterhouse's appointment

[1] Waterhouse was commissioned to design alterations and the fitting out of a building as a temporary home for the college. The Leeds commission began in the same way in 1878; likewise Liverpool University started with the conversion of the old asylum buildings, but not until 1881.

[2] Livery Company's Report on Technical Education, 1878, quoted in F. H. W. Shepperd (ed.), *Survey of London*, xxxviii. *The Museums Area of South Kensington and Westminster* (London, 1975), 238.

likely, his reputation for economical building was probably a strong recommendation.[3] In the first months of 1881 he produced designs for a college to house 200 students, which could be built for £66,000. There is some evidence that he simplified his original design before submitting it.[4] With a few modifications the tender was reduced slightly, and work went ahead on the structure that was opened in 1884. In the four years it took to build costs rose to over £80,000,[5] and the work almost outstripped the fundraising. But, for that relatively tight price, he produced a simple but effective structure that was well received. The *Builder* praised its common sense and everyday aspect, and spoke of commendable eclecticism.[6]

Typically, the building's quality lay in the simple plan, and its character was derived from the pattern of repeated units with only minimal decoration. It was designed to contain a large number of classrooms, which were simply stacked in a pile of three floors, and an attic with a main faience clad staircase in the centre, and a single straight corridor on each floor. A rear wing at one end housed the larger lecture rooms and a secondary staircase. As at the Manchester Assize Courts the considerable bulk of the pile was reduced by a strong cornice and string courses, with the second-floor windows designed as a continuous arcade. Like the Assize Courts **[67]** and the Natural History Museum **[243]**, the composition was strictly symmetrical with an imposing centre-piece topped by turrets and an elongated gable carrying a clock and bellcote. Similar gables emphasized the end pavilions, but were pierced by triplets of round-headed windows under a single arch. The dormers of the attic floor were linked by traceried arches, a motif he was developing just at this time for the National Liberal Club **[514]** (Pl. 133), and which he was to use again on the Christie Library of Manchester University **[605]**. Waterhouse used his straightforward approach to give ample light to the classrooms. Indeed the provision of light was the main virtue of the building, and a sensible one at that. The long ranges of windows were virtually uninterrupted, even running round the central pavilion.

Here was another variation on Romanesque revival. There was a great deal less elaboration and less ornament than at the Natural History Museum down the road; but the free eclecticism gave this building its own richness and character. The materials of the façade, red brick and red terracotta, also marked it out as different from the other structures of the South Kensington site. It was typical of what came to be imagined as the universal Waterhouse style; but in this case most of the forms were closer to the Dutch Renaissance than to his favourite Gothic. The mixture lightened the straightforward Romanesque openings; but there was a slightly uneasy marriage of Waterhouse's simplicity and the finicky ornament that was almost *de rigueur* in the South Kensington style. On grounds of expense, moulded ornament was kept to a minimum; but on this site he could hardly escape some influence of that style. In fact, what modelling there was, was designed by J. Gamble, formerly of the South Kensington Museums' staff.

The respectful reception of this building probably reflects its conventionality, for it broke no new ground either in terms of style or structure. No one, however, criticized it as unsuitable functionally, although it was one of the largest assemblages of science laboratories built up to that time. The final fitting out did take some years, but the building was soon adapted as the Engineering Department of the new Imperial College of Science and Technology, and functioned as such until its demolition.

3 The short list also included Shaw and Street.

4 The *Survey of London*, 239, refers to a letter in which Henry Cole praises Waterhouse for designing a *porte-cochère* for the proposed building. That at least never arrived.

5 According to Waterhouse's office records the figure was £88,120. The *Survey of London* gives a much lower figure of £73,356. The fact that the fittings were not complete at the time of the opening, or indeed for several years after, may account for the discrepancy.

6 *Builder* (5 Jan. 1884).

St. Paul's School in Hammersmith **[492]** (Pl. 143), also demolished in the late 1960s, showed Waterhouse as very much more himself and less fettered by surrounding works. The old school in the shadow of the cathedral had needed new accommodation for years. The arrival of a new high master, F. W. Walker, fresh from raising Manchester Grammar School to a premier position,[7] heralded the start of immediate improvements. The school was governed by the powerful Mercers' Company, who were anxious, and ready to pay, for radical improvement; so an entirely new spacious building for 1,000 scholars was proposed. Scott, E. M. Barry, and Waterhouse were considered as architects. However, since Lord Selborne, a leading member of the Mercers' Company, was again involved, and since F. W. Walker will have been familiar with Waterhouse's work in Manchester,[8] the choice was probably easily made.

For the new school he provided a tall E-shaped block on the northern third of the site, with playing fields to the south. The hall and two classroom wings projected southwards, and, like the City and Guilds Institute, the main façade consisted of a long symmetrical block broken by a central entrance and two end pavilions. But there the similarity ended. At St Paul's there were secondary features in the shape of smaller gables between the centre and each wing, and, most important of all, the style was firmly the Gothic of the fifteenth century. Using a traditional educational dress, Waterhouse was able to build a good deal of variety into the design. Windows on the first two floors were grouped in an alternating rhythm, with segmental arches to enrich the first floor. The narrow lights of the Gothic windows allowed the second floor to be treated as a single arcade with groups of glazed arches alternating with blind ones. The gables were fully decorated with arcading and, a system of which Waterhouse was particularly fond, ribs rising through the gable line to end in pinnacles. There was the inevitable clock in the central gable, and a great *flèche* over the hall behind.

The whole thing was dressed in dark plum-coloured brick and deep pink terracotta, giving what the school magazine described as 'fascinating stateliness' to the new edifice.[9] The mechanical nature of the material and repeated shapes was typical of Waterhouse, sensible and economical; but it was given a vital touch of excitement by the high master's house. This was built in front of the school, of matching materials, but only two storeys. The caretaker's accommodation was in a canted wing reaching to the street, and the whole building was linked to the school by an open arcaded cloister. A matching circular gardener's hut set off the other side of the tree lined forecourt, which was almost immediately graced by a fine bronze statue of the founder, Dean Colet, by Waterhouse's friend and *protégé* Hamo Thorneycroft.

St. Paul's School was a remarkably complete and serious essay in modern mechanical Gothic. It was, however, a style that was soon to give way to the revived Baroque, and was already being challenged by the new buildings of the London School Board erected in a domestic Dutch Renaissance style. Gothic, however, was the traditional style for educational buildings, and Waterhouse had already adopted it for Owens College in Manchester. He was to maintain it there till the end of the century.

At Owens College **[321]** (Pls. 145, 147), later to become Manchester University, the first new buildings were put up between 1869 and 1875. They belonged to the period when Gothic was still

[7] The Manchester Grammar School buildings on Long Millgate, of 1873, are attributed to Waterhouse by Pevsner (*Buildings of England*). This seems unlikely on stylistic grounds, and no office documentation exists.

[8] Apart from his major works in the town, including his first work for Owens College, Waterhouse was actually at work with restorations and a new schoolroom for Chetham's Hospital School, on the site adjoining Manchester Grammar School, in 1876.

[9] Editorial in the *Pauline* (Oct. 1883).

firmly in vogue, and so are, perhaps, more convincing than St Paul's. Like St Paul's the first block was an E-plan, but this time with the wings to the front. Between the wings and the central pavilion Waterhouse added two half-octagon bays containing the entrances and stairways;[10] and the whole was capped with steep pitched roofs enlivened by dormers and patterned tiling. This first block, built to a much tighter budget than the contemporary Town Hall, was a splendid demonstration of his modern approach to Gothic. It was stone faced, but surface decoration was strictly limited, so that the building gained interest from massing alone, and that in turn was the direct result of a simple plan. Waterhouse's usual ingenuity was shown in the way large teaching rooms were fitted into the roof space, with the trusses concealed in the walls. Typically, the whole rhythm was dictated by the pattern of the Dennet fireproof bays of the structure.[11] It was a thoroughly practical approach, and practicality was the rule in detailed solutions too. The small dormers of the rear elevation, for instance, were linked by glazed panels, giving almost a curtain of glass to what might otherwise be poorly lit attics (Pl. 146).

The first buildings were swiftly followed by a medical school **[499]**, with a typically picturesque roof line, but even less in the way of decorative Gothic detailing. This was the economical pattern that he followed for the whole of the rest of the site. The general layout was fixed in 1869 and scarcely departed from, but the buildings behind the original college were built in an increasingly stripped Gothic that scarcely merits the term at all. For the engineering laboratory **[536]** (and its extensions), for the extensions of the medical school **[569]**, and for the Schorlemmer Laboratory **[590]** the watchword was practicality. They are all characterized by large windows and airy spaces on a simple grid, and yellow brick replaces stone as a facing material.

Waterhouse reserved grandeur for the main façade where he built an imposing block, as he had always planned, along Oxford Road closing the court (Pl. 149) **[515]**. Here the buildings are of stone, and there is a dramatic tower with steep pitched roof. Beside it the staircase to the Council Room is given the drama of a spiral staircase corbelled out into its space, just like the one he had added to Manchester Town Hall **[284]** fifteen years before in 1869. But the college buildings do not simply repeat old themes. The Christie Library **[605]** still shows originality in its roof line of gables linked by open arches. Its Gothic is almost a new style, combining features from the arts and crafts tradition with some of the slender elegance that Basil Champneys was putting into the John Rylands Library at the same time.[12] Yet the Christie Library matches the firmness and solidity of the original college buildings,[13] and lacks the slightly etiolated quality of Champneys's work. Waterhouse had moved away from the almost Puginian Gothic of 1869; but the overall effect was of remarkable consistency.

He showed much the same consistency at the Yorkshire College **[460]** (Pl. 150), later Leeds University, where he was at work from the mid-1870s until his retirement.[14] There he used red brick

[10] Internally these have the added picturesque effect of a handsome stone arcade dividing the staircases from the passage.

[11] This rhythm can be most clearly seen at the rear.

[12] The fact that Waterhouse had recently been passed over in favour of Champneys as architect for Mansfield College, Oxford, may have added an edge to the rivalry in this case. At Oxford Waterhouse may have hoped for the commission because his old friend Joseph Thompson was on the Council of the new college. However, his design was described as 'a tired version of Gothic' and Champneys was appointed in 1887. Waterhouse's drawings are preserved in the college.

[13] The last buildings to be completed were the hall and museum. Both of these appear to be largely the work of Paul Waterhouse. The hall with its splendid hammerbeam roof was not commissioned until 1898, and was finished several years after Waterhouse's retirement.

[14] Waterhouse's selection at Leeds will have been partly due to his connections with J. R. Ford, a member of the college Council, for whose father he had built Adel Grange. However, a deciding factor may have been the voice of Professor Thorpe, newly appointed from Glasgow, who arrived at the Yorkshire College meeting with the plans of Owens College in his hand.

and native Yorkshire stone in a variety of Gothic forms,[15] providing 'buildings admirably adapted to their varied purposes, with a character of excellent taste, and at a very moderate cost for works of such magnitude'.[16] The cost needed to be moderate, for, of the anticipated £60,000, only £20,000 was raised for the first buildings. The scheme might never have got off the ground but for the support of the Clothworkers' Company, alarmed at the threat to their trade demonstrated by the Paris International Exhibition of 1867. However, Clothworkers' Court made a start, with lecture rooms in Waterhouse's domestic Gothic mode of steeply pitched roofs and decorated dormers. Behind them were the textile weaving sheds, extended in 1885 to end in a small but pretty tower **[510]**.[17] But funding was a continuous problem, and Waterhouse was warned that one block 'cannot be completed unless the utmost care be used in keeping down the expenses of the elevation'.[18] He did keep them down, and, to judge by the drawings, the resulting Clavering Road block was a piece of building rather than architecture. It was swept away unmourned in the early 1930s.[19]

None the less, Waterhouse did not deal with this commission without thought. The site was an awkward butterfly shape, which he reduced to one long rectangular court, linked to two further irregular spaces (Fig. 11). A great hall with an open cloister beneath was to divide Clothworkers, Court from the irregular quad behind, and the other awkward space was to be glazed over to form a museum with arcades joining it to the Clavering Road block. Waterhouse even planned a tall tower at the corner of the site facing down the hill towards the city. However, money was so slow coming in that the hall was delayed, and only finally built in a much altered form (Pl. 148) **[574]**. The museum got no further than the provision of uncarved corbels for its roof in the rear of the Baines wing **[504]**; and the tower disappeared without trace after its brief appearance, in 1880, in a drawing on the appeal brochure.[20]

By the mid-1880s things were on a firmer footing, and the Baines memorial wing had a little tower of its own and some ration of display in the Gothic detail of its doorway.[21] Practicality was still the watchword, however, with plenty of large square-headed mullioned and transomed windows to light the lecture halls, and a huge double aisled chemistry laboratory filling most of the second floor.

> A moderate use was made—in those places where there would be no undue interference with light—of arched window heads, of cusps, of transoms, but in the main the windows, though mullioned, present large glass surfaces, and, though an effect of vertical dignity is gained by the main tower, the turrets of the Hall and the high-pitched roofs of the frontal range of buildings, all these features were subordinated to the practical needs of the College.[22]

By 1892, when the Great Hall was finally commissioned, the three northern colleges, in Manchester,

[15] Waterhouse had obviously suggested terracotta as an economy (Yorkshire College Council Minutes, 1 (7 June 1878), 371. Leeds University Archives). The decision to abandon terracotta was not taken until 1878 (ibid. (12 Feb. 1878), 329 (7 June 1878), 371. Leeds University Archives).

[16] Yorkshire College, Annual Report (1881–2) 13, quoted in P. H. J. H. Gosden and A. J. Taylor (eds.) *Studies in the History of a University 1874–1974* (Leeds, 1975). It is very interesting to find (Yorkshire College, Court Proceedings (3 Mar. 1880)) that in spite of their own financial difficulties, the College made a donation of £315 to the Girton Building Fund. It is an indication of the way the Yorkshire College saw itself as a pioneer venture in the forefront of education.

[17] By the time of the extension there was evidently some more scope for display. When the walls of the weaving shed were raised by 4 feet to accommodate better looms it was simply ordered that 'the new iron pillars required by the above alteration should be more ornamental than the present ones' (Yorkshire College Council Minutes, 2, Committees (10 June 1879) 65*b*. Leeds University Archives).

[18] Ibid. (22 Aug. 1881), 232*b*. Leeds University Archives.

[19] It was replaced by the heavy Classical Brotherton Library, part of which absorbed the site.

[20] 'The Yorkshire College', appeal in *Leeds Mercury* (2 Dec. 1880).

[21] A memorial to the recently deceased Chairman of the College Council, Sir Edward Baines.

[22] Leeds University brochure (1908) written probably by the Bursar. Quoted in Gosden and Taylor (eds.), *Studies in the History of a University 1874–1974*, 151.

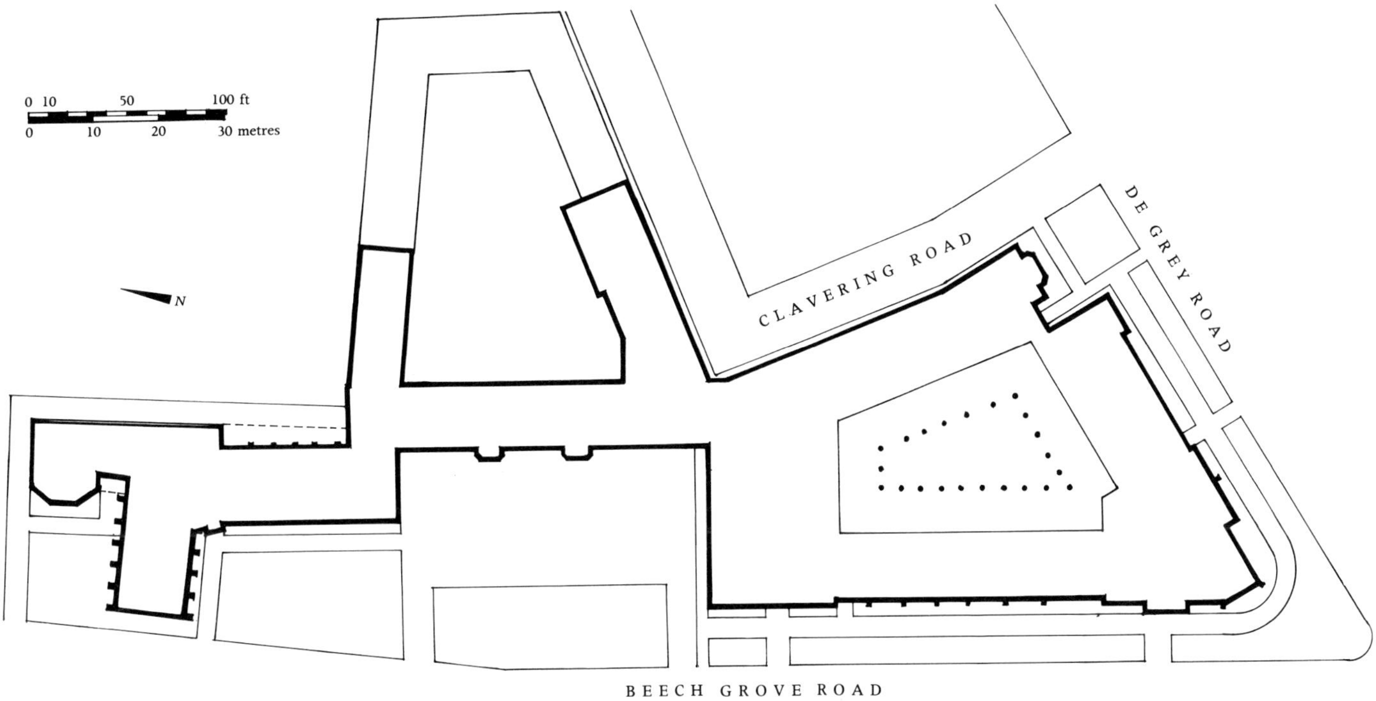

FIG. 11. Leeds, Yorkshire College **[460]**, site plan *c.*1878 (RIBA)

Leeds, and Liverpool, were linked as The Victoria University. The Leeds hall and its attendant art rooms were clearly intended for show (Col. Pl. 4; Pl. 148). Waterhouse turned the axis through ninety degrees, dividing his rectangular court. He then made an imposing façade with a pair of turrets, complete with tourelles and steep roofs. The wall between them was filled with an arcaded gable and a huge perpendicular traceried window, with massive foliate ornament in its arch. It was of almost cathedral proportions, and emphatically stamped the presence of the new institution on what was then still a smart residential area of town. Thereafter the new buildings were mostly very modest structures tucked away into the rear quadrangle. Extensions to the dyehouse, a new engine house, new woollen department, leather industries department, and a new cloakroom and cycle shed were all commissioned in 1898 **[622, 623, 624]**. There was to be no more grand building. Not long after his father's retirement, Paul Waterhouse was approached about continuing work round the main façade on to Headingly Road, and wrote: 'I think my father will be much pleased to hear of these possibilities of the completion of his design.'[23] But the idea came to nothing. The collegiate Gothic for which Waterhouse was so well known was finally going out of fashion.

Leeds, however, was not the latest of his collegiate commissions. In 1881 he was called in, just as he had been at Owens College in Manchester, to modify old buildings for the university college in Liverpool **[488]**. This commission developed into what is probably the most extensive group of buildings in his best-known style.[24] Once the college moved to its new site on Brownlow Hill there

[23] MS letter, Paul Waterhouse to W. J. Husband (29 Mar. 1903). Leeds University Archives.

[24] A detailed history of the main building is provided by A. R. Allan, 'The Victoria Building: Its Furnishing and Use', in *University of Liverpool Recorder*, 83 (Apr. 1980). A general history of the university is T. Kelly, *For the Advancement of Learning* (Liverpool, 1981).

were a whole series of commissions. First came the chemistry laboratories **[512]**, then the Walker engineering laboratory **[542]** and the Victoria Building and Jubilee Tower **[549]** (Pl. 151). There was an enlarged medical school **[599]**, the new Gossage chemistry laboratories **[598]**, and the Thompson Yates laboratories **[600]**. And the whole thing was linked to the Royal Infirmary **[532]** with its medical school and nurses' home.

Waterhouse used his favourite mixture of brick and terracotta throughout. He set the grander Gothic buildings on the street façade, leaving those behind as only minimally Gothic. Even so he did not abandon opportunities to create exciting spaces. One awkward alley-way was opened out by giving one building a semicircular end. This contained quadrant-shaped teaching rooms and a quadrant staircase with flights curving up through four floors. It was architectural drama combined with practical sense, and all achieved at relatively low cost. Later Sir Charles Reilly was to dismiss the building as 'pastry cook's architecture';[25] but the Victoria Building at least showed Waterhouse's ability to make something out of a complicated commission on a low budget (Pls. 151–3). It was the centre of the university, and contained its most public spaces, so the decision to concentrate architectural display here was obviously sensible. It occupied a corner site, but the tower was set away from the corner with an arched entry beneath marking the end of the building. The tower was flanked by the grand staircase, which rose only to the first floor, and was signalled by a stepped arcade of lancet windows. The corner of the building was designed as a subtle exercise in cutting back about a small turret. At ground level the plan is almost square, with a rich postern door. Above, one wall was curved, echoing the apse of the great hall within; while the top floor developed into a complete semicircle behind the turret, revealing the shape of the top-lit main lecture theatre.

The main spaces of the interior consisted of a hall and library. The latter, now shorn of its galleries and book bays, has become rather an uncomfortable space. But the double-height hall on the ground floor, which acts as a collection point for all the circulation routes of the building, still retains its drama as a space (Pl. 152). It was faced in Waterhouse's favourite buff and blue terracotta, set off by treacly brown faience for the dado and fireplace. The mixture presents a startling originality, and it is as typical of the man as the splendid space it decorates.

Each of these five great educational commissions of Waterhouse's later years has its own undeniable character. None was perhaps quite as exciting as the Manchester Assize Courts or Manchester Town Hall, but economy and practicality were rather more to the fore in the late works. They all show his skill in planning, picturesque layout, and constructive ornament developed to the full. The complexity of each of the university commissions produced fascinating problems of organization. Fortunately the clients were also conscious of the need to make some architectural impact; and that Waterhouse was happy to supply. His mixture of practicality with traditional forms was nicely suited to new institutions that needed to demonstrate their status at once.

[25] Sir Charles Reilly, quoted in Kelly, *For the Advancement of Learning*.

17

Other Late Commissions

TOWARDS the end of his career Waterhouse was still collecting commissions for large and small buildings from leading institutions; but his stylistic conservatism was becoming increasingly apparent. None the less, examples like the Liverpool Royal Infirmary display all his habitual skills of planning, massing, and economy.[1] Not that that building was cheap. It cost over £120,000. Coming late in his career the hospital represents a thorough maturing of his thoughts.[2]

The Liverpool Royal Infirmary **[532]** (Pls. 155, 156) was of grey brick with neo-Romanesque detailing picked out in red Ruabon terracotta, matching the university and medical school behind. The interior was faced with white and grey glazed bricks or pale coloured faience, except for the chapel which was a riot of turquoise faience and tiling (Col. Pl. 4; Pl. 171). The plan was simple and fairly typical of hospital building of the day. An administrative block fronted the street; behind it a spine corridor gave access to six blocks of Nightingale wards (women on the first floor, men on the second). Yet there were subtle modifications. The placing of sluices at the ends of the wards had been pioneered by Scott as long before as 1863, at Leeds General Infirmary. Here, as at St Mary's Hospital in Manchester **[571]**, Waterhouse gave them a sense of character by the great semicircular archway that linked them and provided an open loggia at the outer ends of the wards. Two of the ward blocks were circular, which gave an element of variety to the composition, as well as accommodating a short-lived medical fashion.[3] The kitchen was on the top floor of the administration block, with access to the wards by way of an airy bridge, a rather *ad hoc* solution, but one that hardly detracted from the composition, since it was practically invisible from outside. The Liverpool Royal Infirmary was a building of considerable character, though of only moderate splendour; above all it was clean, light, and functional—an admirable recipe for a Victorian institutional structure.

From the 1870s Waterhouse's office handled a succession of such commissions until it often seemed as though he was the inevitable choice as architect. This probably ignores the fact that in the last quarter of the century there was a very considerable amount of institutional building. He was simply fortunate to receive several of the premier commissions in a rich field. However, what he did

[1] Even the chapel, a necessary adjunct, was filled with glass merely by Odell's, not with stained glass from any of Waterhouse's normal suppliers.

[2] Waterhouse had been in touch with hospital building from his early years. He had worked previously at the Cheadle Lunatic Asylum (1863 *et seq.*) **[146]** and at the old Infirmary at Manchester (1861 *et seq.*) **[111]**. Waterhouse also entered the competition for the new Macclesfield Infirmary in 1865. He withdrew because he was dissatisfied with the conditions of the competition. He probably suspected that outsiders were not to have a chance; and in the event the infirmary was built to the designs of the local architect, J. Stevens (1867–8).

[3] For an account of circular wards see J. Taylor, 'Circular Hospital Wards: Professor John Marshall's Concept and its Exploration by the Medical Profession in the 1880s', *Medical History*, 32 (1989), 426 ff.

build showed that he generally deserved his good fortune. The Liverpool Royal Infirmary was eminently functional, but it was also a refinement of what, by the late 1880s, was already a tried and tested layout. His other major hospital, University College Hospital in London **[588]** (Pl. 154) was altogether more adventurous. Waterhouse had been invited to report on the old hospital as early as 1877.[4] The new hospital was to be on a cramped site, roughly square; and the initial instructions envisaged a gridiron plan with a central spine corridor, and six wings, three on each side, dividing the site into four tiny courtyards. It was to rise only three storeys, which would have kept it close to the scale of the existing houses in Gower Street. Waterhouse abandoned this idea in favour of an X-plan rising five storeys,[5] that provided far better light and air on this inner-city site (Pl. 163). The base was a semi-basement covering the whole block and including the outpatients' department, waiting rooms, and casualty rooms. The outpatients' waiting hall rose through two floors into the ground floor of the block above. It also had its own entrance with an arcaded stairway down, which was richly decorated with mosaic floors and marble veneers. Above the podium rose a central tower containing lifts, stairs, and operating theatres, linked to four wings. Three of these contained the wards, the fourth the nurses' home.[6]

It was probably the first vertically planned hospital in the world, and as such is the ancestor of many modern structures. On each floor of the ward wings was a single twenty-four-bed ward; but this was laid out in a cruciform shape, allowing an element of privacy in the short arms. As at Liverpool, Waterhouse had gone into the business of ventilating thoroughly, and the whole hospital was ventilated on the plenum system. There were intake vents under each window, and the vitiated air was drawn out by vents in the fireplace flue at the centre of each ward. As usual the sluices were set at the outer end of each wing; and here again Waterhouse concentrated on ventilation to cut down infection. The sluices could only be reached across open bridges from the wards, allowing the fresh air to blow freely through.[7] From each bridge an iron fire escape staircase wound down to the ground. It was a remarkable concept allowing a surprising amount of space, plenty of ventilation, and good light in the wards. There was also excellent natural light in the operating theatres, which were all designed with a complete glazed wall facing south-west.[8]

The building's bright red terracotta is criticized as insensitive to Wilkins's University College opposite. It is certainly a contrast; but the two buildings face one another, which makes it impossible to see both together; so perhaps Waterhouse may be forgiven. None the less, it did introduce a new scale; and, for all the material's popularity at the time, the terracotta is certainly an aggressive element in the townscape. In trying to introduce a large new hospital into this largely residential late Georgian area, it was characteristic of Waterhouse, and of his age, to opt for a bold contrast.

The Refuge building in Manchester **[570]** (Pl. 157, Fig. 12), on the other hand, seems entirely at home with its five floors of brick and terracotta. But that is because Waterhouse was setting a new scale for a whole quarter of the town. First Paul Waterhouse, and later the company's own architect, merely repeated his recipe in adjoining buildings, while the rest of Whitworth Street was built up in

[4] UCH General Committee Minutes (July 1877), quoted in J. Merriden, 'University College Hospital', unpublished Open University BA diss., 1982.

[5] The fifth floor was set entirely in the roof to reduce the bulk. It appears that the X-plan may have been suggested by the hospital secretary, Newton H. Dixon.

[6] The floor levels of the nurses' home, lacking the huge spaces and elaborate servicing of the wards, were adjusted to allow an extra floor in that wing.

[7] Even at the time the open bridges caused problems with draughts and cold, and they were soon fitted with extra doors. Subsequently they have all been enclosed.

[8] It is interesting to note that, as late as 1897, since this was a teaching hospital, the operating theatres were designed with tiered space around the operating space and galleries to accommodate the spectators.

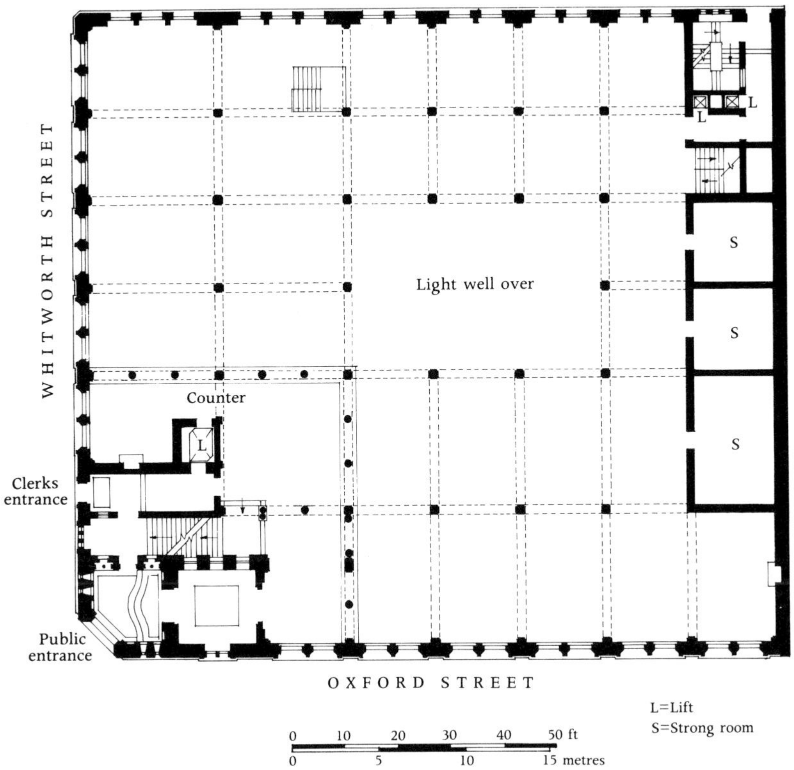

FIG. 12. Manchester, Refuge Assurance Building (1891) **[570]**, plan of ground floor

the same confident Edwardian scale. Until the arrival of his Town Hall there had been few buildings in the city on the scale he adopted for the Refuge. By the time the Refuge was built the whole pattern of business had changed; and new corporations needed new offices for very large numbers of clerks. There was no alternative but to build high.

Waterhouse chose his preferred materials, brick and terracotta, so that his building should have some chance of remaining in its pristine condition. His perspective shows that he also intended his preferred mix of red brick and buff terracotta;[9] but this was abandoned in favour of a deep plum-coloured clay. He also opted for free Renaissance detailing with broken segmental pediments at second-floor level to lift the eye out of the canyon of the street. The bulk of his effect was put into the picturesque skyline above the cornice. First came a broadly scalloped parapet, then each façade was held between high gables, not unlike those of the City and Guilds Institute **[490]**. Finally, over the corner was an octagonal turret with a decoratively roofed spire.[10]

The corner was important for commercial as well as compositional reasons; so Waterhouse did not simply carry his turret down to ground level. The whole of the corner was given over to the widest possible door with little room for an elaborate frame. Instead ground and first floors were linked with a simple frame that turned into a decorative gable on the angle bearing the company's crest. This in turn became a second-floor balcony at the base of the turret. It would have been an imaginative

[9] This perspective formerly used to hang in the Boardroom of the Refuge building. Its present whereabouts are unknown.

[10] This was replaced by Paul Waterhouse with a squat Baroque cap to match that on his tower.

enough piece of designing in two dimensions, but it also functions as the junction of the two façades and as an advertisement for the main entrance. This leads through a vestibule, rich with faience in brown, honey gold, and cream, with blue touches (Pl. 170), into a large pillared public office, with the Boardroom above (Col. Pl. 6). The predominant colour is pale cream, which must have seemed warm and clean after the sooty air outside.

In its basic form the building matched some of his larger district offices for the Prudential, and an unthinking visitor might be forgiven for thinking it just that. It had a similar approach to design. However, the colour of the materials and the detailing of the windows went a long way towards distinguishing one insurance company from the other. Waterhouse went further when he built the Liverpool offices of the Pearl Insurance **[613]**, and clad them in stone. But there he also used a tower that was very much a version of his National Liberal Club one **[514]**. However, one should not be surprised that a busy architect did occasionally adapt solutions that had already been worked out for similar problems on other sites.

Two more late buildings, however, show that Waterhouse continued to design with unabated skill into the 1890s, even if his stylistic taste became increasingly out of date. Fosters' Bank, in Cambridge **[565]** (Pl. 158; Fig. 13),[11] showed all his old verve as he entered on his sixtieth year. The site was an irregular and incomplete wedge with a frontage to two streets. The main façade contained offices and consultation rooms, with the entrance to the banking hall under a tall turret at one side. The banking hall itself was set to the back of the site, and lit partly from above and partly from the rear. What made it particularly exciting was its octagonal shape. This neatly filled almost the whole of the site, leaving only a few awkward corners that were used up for lightwells and WCs. The strong room, in order to provide an axis for the main space, was designed to fill one segment of the aisle of the octagon. The whole space was, of course, clad in faience in smoky green and pale tints. This little building provides a great deal of drama and excitement in what was always a straightforward commercial building. It is also a brilliantly ingenious solution to an awkward site.

Five years later Waterhouse could adopt a different, more pragmatic approach for the much more prestigious commission of the Royal Institution of Chartered Surveyors **[615]**.[12] For them he produced a surprising little block of red brick and stone, decked out with jolly Franco-Flemish and English Renaissance detail. Yet it was very far from being an ill thought out asssemblage. The façade block was formal, rectangular with a central gable over the entrance; but the site was deep, and a good deal of accommodation was required. Waterhouse's original plan showed typical attention to practicality, with a central top-lit staircase. However, he was asked to move this to one side, where he used its window to break up a long façade that he had designed in regular blocks. The result was a relaxed free composition that wandered back to finish in the curved end wall of the lecture theatre with its massive Baroque chimney stack. Such a disruption might have been disastrous; but Waterhouse had been adept for years at managing commissions that were to be built piecemeal. That experience had taught him the need to design large buildings with groups of rooms. Here the lecture theatre was clearly one block, and the library and forestry museum in the façade another. What had been a range of offices along the side was simply redesigned as an office-and-staircase block. There was surprisingly little damage to the overall design; and, now that the very long side façade is open to view, the sense of variety may be seen as a gain. What is clear, however, is that Waterhouse's

[11] Now Lloyds Bank in Sidney Street.

[12] The detailed story of this building is told by Jill Lever in 'Charter Centenary', supplement to the *RICS Journal* (Sept. 1981).

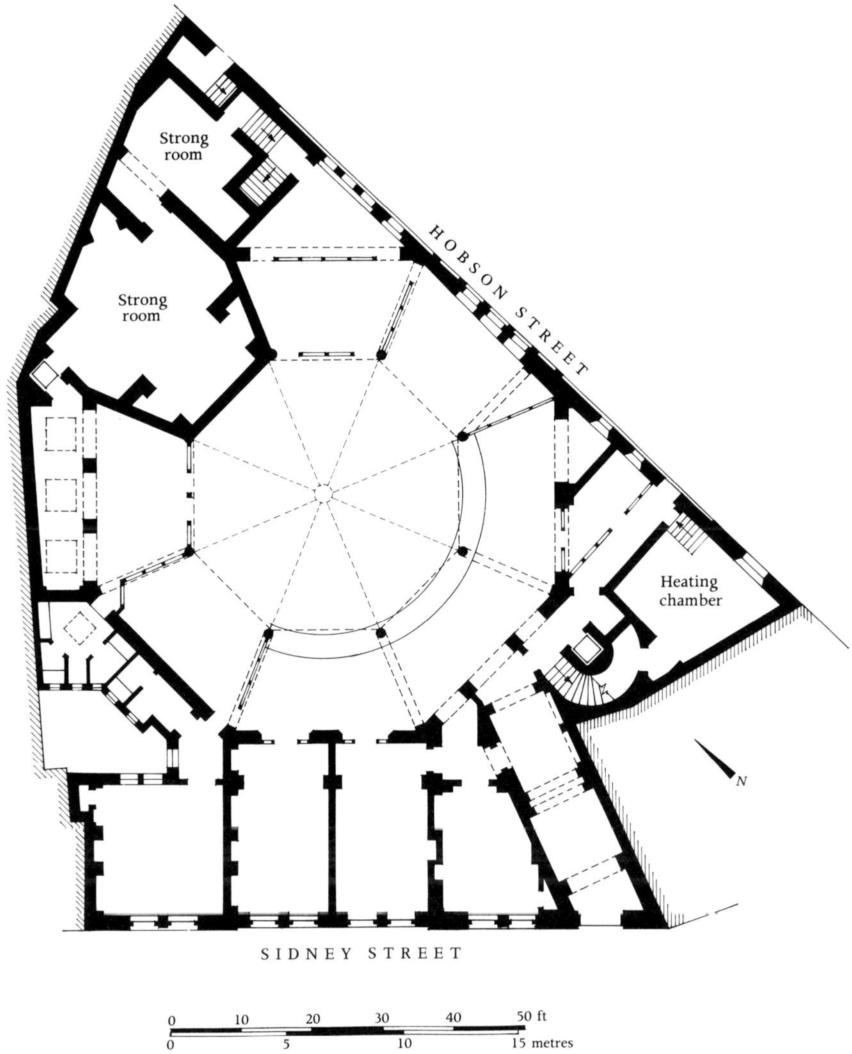

FIG. 13. Cambridge, Fosters' Bank (1891) **[565]**, plan of ground floor

flexibility as a designer had not diminished. Although his eclectic decorative tastes were, by 1895, *retardataire*, his ability and his adaptability as a composer of buildings allowed him to continue to please his clients even when they were as demanding as the Surveyors. In a practice lasting almost half a century and embracing over 600 commissions it was another remarkable achievement.

Part Four

THEMES IN WATERHOUSE'S CAREER

18

The Business of Architecture: The Office, Contractors, and Suppliers

WHEN Waterhouse began his articles with Richard Lane the popular image of the architect was of Mr Pecksniff, adept at taking fees from pupils but of whose 'architectural doings nothing was clearly known, except that he had never designed or built anything'.[1] Dickens's acid portrait shows a man morally and professionally unfit to practise but profoundly aware of the importance of his status. By 1886, 'when the Forsytes assembled at old Jolyon's to celebrate the engagement of June to Philip Bosinney'[2], Galsworthy's architect, both the actual status of the profession and the picture had changed. Bosinney is never accused of architectural incompetence and is as attractive as his client is repellant. Yet the unhappy story of Robin Hill is another of the best-known literary vignettes of an architect at work. Waterhouse was a good deal more successful than Philip Webb, who has been suggested as one of the originals of Bosinney,[3] and he survived the onslaught of his clients better than Bosinney, or indeed Webb. For all his geniality and genius, Bosinney signally failed to satisfy his hard-headed client:

> On the one hand the architect's devotion to his idea, to the image of a house which he had created and believed in, had made him nervous of being stopped or forced to use make-shifts; on the other Soames' not less true and whole-hearted devotion to the very best article that could be obtained for the money, had rendered him averse to believing that things worth thirteen shillings could not be bought for twelve.[4]

Galsworthy draws a trenchantly realistic portrait of the established upper middle classes of the latter part of the nineteenth century who made up the bulk of Waterhouse's clients; and Bossiney's *impasse* was one that Waterhouse managed to solve, or at least survive, so that he was immensely successful in winning and keeping clients, and acquired a well-deserved reputation for respecting their budgets.

Having won commissions, Waterhouse was also a master at producing clear and effective designs for the builders. He certainly had great energy to carry out the volume of work he undertook during forty-five years of practice. Yet the regularity of holidays, frequently abroad, suggests that he paced

[1] Charles Dickens, *Martin Chuzzlewit* (London, 1843–4), ch. 2.

[2] John Galsworthy, *The Forsyte Saga* (London, 1922; repr. 1970), p. xiv.

[3] William Amos, *The Originals* (London, 1985), 58, claims that the character of Bossiney is based on Edward Garnett, his publisher's reader; but that Webb was the prototype for the architect. It is worth noting that Philip Webb (1831–1915) is an almost exact contemporary of Waterhouse.

[4] John Galsworthy, *The Man of Property* (London, repr. 1970), part II, ch. 1, 91.

himself, unlike 'poor Street' of whom he was very critical for 'not knowing how to take a little leisure when he ought to have had it'.[5] Equally important, he ran an efficient office on which he could rely. Significantly most of his senior draughtsmen stayed with him for many years. He was to continue in single-handed practice until he took his son Paul as a partner in 1891. Like Butterfield and Burges, Waterhouse worked in isolation from such master–pupil hierarchies as Burn (who took in Norman Shaw and Nesfield) or Scott (who trained Street, Jackson, Bodley, and William White).

In his office there was a clear hierarchy from the lads, Popplewell and Entwistle, who began at a wage of 5*s*. a week, to the senior draughtsmen, such as C. H. Scott, assistants, such as G. T. Redmayne, and chief clerk John Willey, all of whom earned £3 a week or more.[6] The staff, together with clerks and office boy, seems seldom to have numbered more than about fifteen (Pl. 159). Until almost the end of 1859 the office employed only three draughtsmen. Once he had won the commission for the Assize Courts it is no surprise that Waterhouse took on more men. In 1860 the average number of staff was five, and by 1862 he was occasionally employing eighteen. His Manchester office reached a peak of twelve in 1863. After 1865 the norm of around fifteen seems to have been maintained, and a photograph of 1900 shows only thirteen including the office boy. Surviving records from the start of his practice suggest that extras were taken on for limited periods to help with specific jobs; and this system was probably continued in later years.[7]

It is clear that Waterhouse trained his draughtsmen thoroughly in the ways he preferred. As a result he could delegate a good deal of routine designing, detailing of cornices, windows, and so on, secure in the knowledge that his men would produce Waterhouse work. For instance, when he was commuting between London and Manchester, he wrote to the chief draughtsman in London, Scott, about Alexander's bank **[179]**: 'I am very anxious about the light. Do secure every inch you can';[8] but he offered no instruction how to do so. This could only be because he trusted Scott to work within the systems he had laid down.

One of the more taxing jobs, for which drawings were done in the London office in Waterhouse's absence, was the restoration, or rather rebuilding, of Brasted Church **[194]**. It was a job that occupied a good deal of his personal attention; yet several of the surviving drawings were done by Ernest Geldart, of whom he wrote to Scott: 'I am very much pleased with Ernest's drawings, they speak of diligence and care on his part, and of good tuition on yours.'[9] However, it is clear from other correspondence that there were limits to freedom and initiative, and the preferred system was to refer questions to Waterhouse. When the work on Pembroke College hall **[415]** was going ahead, his chief clerk, Willey, wrote to the Master: 'We have devised this method of carrying out the suggestions made by you in a recent note, in Mr Waterhouse's absence. He may perhaps see a better way of doing it when he comes to look into the matter personally.'[10] His draughtsmen remained very much draughtsmen, so anonymous in fact that they very seldom initialled drawings and it is difficult to identify their individual contributions. There are certainly occasions, too, where the designing is

[5] MS letter, AW to Hamo Thorneycroft (16 Dec 1881). Thorneycroft Papers, Henry Moore Centre for the Study of Sculpture, Leeds City Art Galleries.

[6] Their wages vary from week to week, so, presumably, overtime was paid. The starting salary for draughtsmen seems to have been £1 a week, and the senior draughtsmen were regularly paid £3. This was a respectable salary in the 1860s.

[7] This information has been largely culled from the first office account book, The Ledger, in which names and earnings are recorded week by week for the Manchester office. The London office salaries are recorded only for the first few months of 1864.

[8] MS letter, AW to C. H. Scott (15 Dec. 1864). OLB. RIBA Library.

[9] MS letter, AW to C. H. Scott (9 Mar. 1865) OLB. RIBA Library.

[10] MS letter J. Willey to the Master of Pembroke (11 Dec. 1874). Pembroke College Archives.

routine; and there are even a few jobs in which Waterhouse does not seem to have been personally involved at all.[11]

A well-organized system, however, was essential to cope with the multiplicity of jobs. The chief clerk, Willey, had been responsible for helping to set up the office systems when the London office was opened. He sought advice from Waterhouse's accountant brother Edwin: [12]

> upon one or two new features which I think it desirable to introduce into the bookkeeping of this establishment. First we have no Day Book—no book in which charges appear as a matter of course as soon as our clients become liable for them. Each draughtsman keeps a register of his work which is abstracted monthly and yearly . . . There is also a Journal in which the charges to clients are entered. But I believe no rule whatever has been followed in entering these charges. I think the system is chiefly defective as in the Time Abstracts there is no provision for describing what the labour was expended upon—nor for showing whether it was expended upon one work only or upon several. Indeed it is Mr W's belief that many charges for small works given to him by clients for whom perhaps large works were at the time in hand have never been made.[13]

Clearly there was room for improvement, and the system set up then has left us Certificate Registers, recording the payments to contractors, a Day Book, recording charges due to clients and other details, and a Register of draughtsmen's time.[14] Obviously Willey was in charge of day-to-day organization in the office and would oversee the distribution of work, booking in and out drawings. In designing, the general system was for Waterhouse to work out the sketches;[15] a group of draughtsmen would then be set to work up details with occasional supervision, about a day's worth a month, from the master or occasionally from a senior draughtsman such as Scott, C. Evans, or Cooper.[16]

Senior draughtsmen would have a general responsibility for one or more jobs. Cooper, for instance, was responsible for much of the work at Backhouse's Bank **[175]**, Strangeways Gaol **[113]**, Allerton Priory **[237]**, Foxhill **[285]**, the Natural History Museum **[243]**, and Eaton Hall **[315]**. Evans seems to have been responsible for the New University Club **[208]**, Balliol **[254]**, Caius **[268]**, and Pembroke Colleges **[333, 415]**. Scott was principal draughtsman for the Law Courts scheme **[278]**,[17] Lime Street Hotel **[274]**, the Fearon Warehouse **[276]**, and Heythrop Hall **[350]**, though he

[11] e.g. the design of a summer house for Alfred Fryer (1862) **[134]**, the design of a bookcase for his brother Edwin Waterhouse (1874) **[413]**, alterations for the Manchester Royal Infirmary (new stables) (1865) **[218]**, or, more significantly, the tombstone for Joseph Thompson in Ardwick cemetery (1866) **[248]**.

[12] Edwin Waterhouse, founder of the accountancy partnership Price Waterhouse & Co., had trained in the firm of Coleman, Turquand, Youngs & Co., and in 1864 had just set up in business on his own. He was working from chambers at 11 Old Jewry.

[13] MS letter, J. Willey to Edwin Waterhouse (22 Feb. 1865). OLB 564–5. RIBA Library.

[14] These survive, along with an earlier certificate ledger begun in 1864, 3 volumes of office correspondence, and a scattering of other notes and letters (many still in the possession of clients' descendants). In addition there was double entry bookkeeping and presumably the monthly summaries of designers' time and contractors' certificates were continued. Also, it is clear from the correspondence that by the mid-1860s Waterhouse was using a standard form of contract as well as standardized report forms for his clerks of works. It all affords a rare and detailed picture of the way that work was managed in what was now a busy and progressive office.

[15] It seems likely that Waterhouse's early work on any scheme would not be charged in the Register. The existence of occasional pages in the sketchbooks with notes and plans for projects that were later built suggests that the very first ideas were not infrequently sketched out there, and may well have been discussed with clients before any proper drawings were made. Among the Prudential drawings, now in the RIBA Drawings Collection, are also one or two examples of rough notes in Waterhouse's hand recording decisions reached at meetings when his design brief was agreed. One such was even written on the back of a Midland Railway dinner menu as he travelled home.

[16] Waterhouse felt that his senior draughtsmen were sufficiently close for him to remember them in his will. They were evidently the core of his designing enterprise.

[17] The Register No. 1, 117–18, records the full details of the time charged to this account. The total is 1,047.5 days, of which Waterhouse himself charged 124. The draughtsmen involved were: Scott (121.5 days), Cooper, Ernest Geldart, Harrison, G. Steane, C. Evans, King, Thornicraft, Hodgson, Turner, Wilson, and Popplewell. The accounts run from July 1865, when Waterhouse was first consulted. Designing began in

also had a hand in the Natural History Museum **[243]**. Supervision was also entrusted to the assistants, such as Giles Redmayne,[18] who became architects and occasionally took over commissions in their own right. However there were never many such in the office,[19] and Waterhouse would regularly check and correct drawings himself often working alone in the office long after the staff had left.[20]

Assistants would do all the minor tasks such as ruling up sheets for brick courses (a necessary chore before the advent of the standard brick) to inking in designs and making tracings. A few draughtsmen had special skills, such as perspective work. They would rule up perspectives for the master, and were even allowed to sign some.[21] In later years Waterhouse employed the specialist T. Raffles Davison to draw perspective versions of his work in addition to those he drew himself. There was a regular call for these, not only to show clients, but for illustration in the building press.

For a large commission covering several years, there was a great deal of drawing to do, and it is no surprise to find a considerable team involved; though a close examination of the accounts does not bear out the myth that Waterhouse simply managed a huge drawing factory. Instead it is clear that there was a considerable turnover of junior draughtsmen. Some were employed for rather less than a month; so it looks as though he was able to draw on something approaching the 'lump' among draughtsmen. In all, the names of eighty-eight draughtsmen are known from the period up to 1875; but of these twenty-nine worked for him for substantially less than a year, charging at most seventy-three days designing time and often much less. A further twenty-five draughtsmen were only employed for a year or two, and even then apparently not full-time.[22] Of the remaining thirty-three names, excluding Willey the chief clerk, only five are known to have worked right through the period 1865–75, and they, of course, included the senior draughtsmen.

Under the supervision of one of the seniors, a team would be assembled to work on each commission. At Manchester Town Hall, for instance, a total of forty draughtsmen were involved at

earnest in Feb. 1866 and the competition closed on 16 Jan. 1867. Thereafter Waterhouse found it necessary to charge some 30 days work to this job, and the account closes in July 1867. Compensation for all this was the government premium of £800.

[18] It appears that church work was generally given to Redmayne or to Ernest Geldart, whose status were respectively qualified assistant and pupil. Geldart actually signed some of the drawings for St Martin, Brasted, and Redmayne some for St John, Brooklands, a rare privilege. But by then Redmayne was Waterhouse's brother-in-law, and several commissions were taken over by him to be completed independently. Even under Waterhouse he had a good deal of freedom, and appears to have been responsible without detailed guidance for the tomb in Ardwick cemetery that was commissioned in 1866 by Waterhouse's close friend and patron Joseph Thompson.

[19] It has often been remarked that few independent architects trained in Waterhouse's office. In this respect he was quite unlike Scott, and even refused some applications to train with him. (See for instance, letter, AW to William Young (22 Jan. 1867). OLB 520 RIBA Library. This, I think must have been the William Young who made his name with the building of Glasgow City Chambers.) The names of Waterhouse's pupils, Redmayne, Geldart, Brightwen Binyon, or, later, Sylvester Sullivan are hardly of the first rank. Redmayne's work in Manchester is not without quality, and shows that he learnt his master's ways quite effectively. Binyon was presumably a relative of Thomas Binyon, an early Quaker patron, and built a respectable practice for himself in Ipswich with municipal commissions as far afield as Sunderland and Swindon. Ernest Geldart left architecture for the Church and became an authority on liturgical fittings. Most of Sullivan's work is in the London area.

[20] L. Sylvester Sullivan, a pupil in the practice during the 1890s, confirmed, to Alfred's grandson Michael, that Waterhouse examined drawings before they left the office, often going round the drawing boards late at night after having been away during the day.

[21] An ink perspective of St Paul's School, Hammersmith, bears G. W. Nicolay's name. The splendid perspective of the Assize Courts is signed by J. Battye who started as a draughtsman (Pl. 33).

[22] Some of these may have been pupils serving part of their apprenticeship with Waterhouse. The Willink who was employed between 1873 and 1875 could possibly be the W. E. Willink who was President of the Liverpool Architectural Society in 1896. And it is tempting to link the I'Anson, who put in small amounts of time on 19 different commissions in 1867 and 1868, with the well-known London architectural family. It is even possible that the Hansom, who worked for almost a whole year on the designs for the Law Courts, was related to J. A. Hansom who, after the Birmingham Town Hall fiasco and his ensuing bankruptcy, was back in architectural practice in the 1860s.

one point or another, though a team of under twenty was more common. There were never more than eighteen employed on that job in any one year; and of those only three or four would be heavily involved, the rest contributing a day here or there as needed. In fact of the forty only three worked on the Town Hall every year from 1868–75.[23] A broadly similar pattern can be established for most of his major jobs. At the Natural History Museum a total of twenty-four assistants are recorded between 1870 and 1876, not one of whom worked on the building every year; and the bulk of the designing was done by only four men.[24] Again, at Eaton Hall, between 1869 and 1875, twenty-seven assistants are recorded, but only four of them worked continually.[25]

The product of this system was a huge number of drawings.[26] The extraordinary quantity of work has left stories of the master having drawings held up for his signature as he hurried to the cab,[27] but the impressive feature of this large number of drawings is their consistency and their clarity (Pls. 56, 62, 67, 83, 119, 161, 172). With the exception of sketches, they were all carefully numbered, one number for the sequence in the office production, and one for the sequence in a particular job. Tracings were given duplicate numbers and added letters to indicate how many tracings were taken. From about 1858 the drawing style was also standardized. Waterhouse abandoned the pale colours and fine line that he had learnt from Richard Lane (Pls. 17, 18) in favour of a stronger line and deeper tints, a buff red for brick, yellow for stone, brown for timber, and blue for metal. The tints changed remarkably little over four decades,[28] though the scheme had to be modified slightly to accommodate terracotta. Finished tracings were also tinted, a practice that was continued even after the blueprint came into the office in about 1890.[29] Without such a detailed system for producing drawings, it is doubtful whether Waterhouse could have maintained any sort of consistency as a designer. His method resulted in some extraordinarily fine technical drawings. Significantly his work was among the few examples that R. P. Spiers chose to illustrate in his book on architectural draughtsmanship.[30]

In addition to his clerks and draughtsmen, Waterhouse also employed his own specialist quantity surveyor. This was Michael Robinson, who was with him from 1860 until at least 1875, when the records stop. His pay was equivalent to the senior draughtsmen, and he may also have been involved

[23] These were Thornicraft, Windsor, and Robinson, the quantity surveyor, who would be involved in all jobs as a matter of course. The Register No. 1 215–16, 289–90, 313–18, 367–8, 364, covers the period Mar. 1867–July 1875. Full accounts for the remainder of the period unfortunately do not survive. The days charged are: 1st competition 82.5 days, of which Waterhouse worked 8.5; 2nd competition 303.5 days, of which Waterhouse worked 21.5; 15.5 days were spent on a perspective for the Royal Academy; Apr. 1868–July 1875 5,833.5 days were charged. Thornicraft worked 683 days, Windsor 700.5, and Robinson 726.5. Others involved were Haigh, Lamb, Harrison, Spice, Hardcastle, Cooper, Wilson, Nicolay, Gordon, Evans, Willink, Watson, Burgess, Tatham, Lee, Brace, Spurrell, Deane, Geldart, Maitland, Foster, Marnock, Steane, Thomas, Gibbs, Scott, Binyon, Froude, Hendry, Reynolds, Dorman, Cotton, Peabody, Speechly, I'Anson, Atkins, Popplewell.

[24] The Register No. 1, 156–8 and 361–3. The bulk of the work was done by Steane and Thornicraft (1870–2) and Cooper and Hardcastle (1873–6).

[25] Ibid. 305–8, 329–34. The vast bulk of the work was done by Cooper, with regular help from Thornicraft and Nicolay. C. Evans was regularly working half a day a month on this scheme, which may well mean that he was only supervising. In 1874–5 he was involved on a more regular basis.

[26] The RIBA Drawings Collection holds around 20,000 drawings, which have yet to be catalogued in detail. There are also smaller holdings elsewhere, and large numbers of schemes for which no drawings survive.

[27] Paul Waterhouse and his family lived in the upper part of the New Cavendish Street house, and their son Michael as a small boy often observed, from the top of the stairs, the scene below of the secretary holding up letters to be signed and assistants holding up drawings as his grandfather hurried to a carriage waiting to take him to the station.

[28] In the late 1860s Waterhouse took to even stronger versions of the same colours, and continued to use those until the late 1890s, when, presumably under the influence of his son Paul, there was a return to paler tones. It may be that paler tones were easier for mechanical reproduction.

[29] Blueprints of the Bristol Prudential, for instance, from 1899 have carefully added colour, and are attractive and unusual.

[30] R. P. Spiers, *Architectural Drawing* (2nd edn., 1905). The other drawings illustrated are by Rowland Plumbe.

in draughtsmanship since his work is often charged like the other staff. He was involved in at least one hundred commissions; though all but a handful of the jobs he surveyed are in the north, so he was presumably based in the Manchester office. Another surveyor was T. M. Rickman, son of the Quaker architect who had known the Waterhouse family before Alfred was born. T. M. Rickman II was apparently in independent practice, but was regularly retained by Waterhouse from the early 1860s for jobs as diverse as measuring extras for some stables in Reigate **[185]**, advice on the girders for the New University Club **[208]**, taking out quantities for the Natural History Museum **[243]**, or surveying Emmanuel College, Cambridge **[345]**. The survey drawings he did of Leather Lane in advance of the new Prudential building **[602]** are a model of clarity.

Waterhouse's attention to detail did not stop when the drawings left his office. He was concerned to see that what he had designed was built precisely, and so endeavoured to find reliable clerks of works. J. Battye began as a draughtsman in the office, worked as the contractors' setter out at Manchester Assize Courts, and was still working for Waterhouse as clerk of works at the Yorkshire College and at Liverpool University. Of course, as a clerk of works, he was tied to a particular job and might not be assured of continuous employment with his former master. But Waterhouse was ready to help where he could, recommending him to a fellow architect as 'a very careful young fellow once in my employ'.[31] There were others like Henry Littler who supervised the building works at the Assize Courts, Strangeways Gaol, and later at Manchester Town Hall.[32]

Only occasionally did Waterhouse have problems. There was a row with the clerk of works at Backhouse's Bank at Darlington **[175]**, which obviously touched a point of professional pride. The clerk of works was G. G. Hoskins who practised as an architect and was later mayor of Darlington. He appears to have resented having to work to the strict system devised by Waterhouse, who was clearly dissatisfied with the imprecise nature of his reports and lack of scrutiny of the contractor's accounts leading to delays to payments. As Waterhouse wrote:

> Dear Sir, I have received your letter of the 11th, and am sorry to think from the tone of it that you are under the impression that I wish, to some extent at least, to impose upon you labour and responsibility which does not legitimately attach to your office. So far is this from my wish that I shall be quite willing to relieve you of the trouble of making your *frequent* (surveys) of the Works, and shall be content if you will furnish a report only so often as the Builders request Certificates . . . then perhaps when the builders do make their applications for certificates—you will kindly (since their applications are addressed to you) glance over their statements and make any remarks that may suggest themselves to you for my guidance.[33]

The row had been brewing for some time and there was similar trouble with Pilmore Hall **[100]** that Hoskins was also supervising. The week before, Willey had been driven to almost his only break from impersonal business style in his report from Manchester to Waterhouse in London: 'I enclose a letter from the Great Darlington Frog. and hope it will enable you to take the whole case before Mr Backhouse. Unless indeed you think it better the said frog should collapse (like his illustrious ancestor) by the very violence of his own conceit.'[34]

[31] MS letter, AW to Henry Woodhouse (10 July 1862). OLB 445. This was for a post as an architectural draughtsman.

[32] Henry Tift is another example. He was first recommended as clerk of works at Balliol but replaced by another man who had already worked for Waterhouse. However, having proved himself, Waterhouse saw that he was re-employed and he worked at Caius, Jesus, Pembroke, and Trinity Hall in Cambridge.

[33] MS letter, AW to G. G. Hoskins (16 Jan. 1865). OLB 296. RIBA Library.

[34] MS letter, J. Willey to AW (12 Jan. 1865). OLB 296. RIBA Library.

As usual, however, Waterhouse managed to smooth things over. Both commissions were successfully completed and Hoskins's pride assuaged. Apparently he even took over some parts of the bank commission himself;[35] and the experience of working for so efficient a master may have helped Hoskins in the manual he published on *The Clerk of Works*.[36] Much later, in a competition of 1882, Waterhouse selected Hoskins's design for Middlesbrough Town Hall and offered suggestions for improving it. Its debt to his style is fairly apparent.

The final stage in achieving precisely what he had designed lay in Waterhouse's choice of contractors. Poor workmanship at Rothay Holme in 1854 **[2]** seems to have taught him at a very early stage the need to select contractors carefully and to provide proper on-site supervision.[37] Increasingly his building work was handled by a small number of contractors such as Parnell's of Rugby or Holland and Hannen whom he knew and trusted. Parnell's were the main contractors for sixteen of his buildings between 1870 and the early 1890s,[38] and Holland and Hannen built thirteen.[39] However, the big names were not always available at the right price. Parnell's, for instance, tendered for at least seven other buildings without success. The larger firms were particularly suited for major institutional commissions; but Waterhouse was well aware of the advantages of smaller contractors with a strong local base, and used them for jobs such as Kircassock House at Lurgan **[236]** or the expensive remodelling of Goldney House, Bristol **[196]**. His Manchester training meant that he was better informed about Lancashire contractors, who undertook most of his work in the north.[40] But there were others such as Stephens & Bastow, with a base in Reading, to whom he turned for a good deal of work in that area. Only occasionally was he at all at a loss for a contractor, and that was early in his career.[41]

Waterhouse also built up relations with a small group of subcontractors and suppliers whom he considered reliable. In his early years, naturally, he experimented. For instance, he went first for stained glass to Lavers and Barraud, who had a reputation for patterned, rather than figured, glass of the sort that the nonconformist world preferrred. The firm of Clayton and Bell, much better known in Anglican circles, he employed only twice.[42] Pugin's favourite supplier, Hardman, he never commissioned for glass, though the firm did make the fine memorial brass he designed for Trinity College Chapel in 1868 **[291]**. However, once he had found a congenial supplier, he preferred to stay with them; and the vast majority of his stained glass was supplied by Heaton Butler & Bayne, who

[35] Drawings for the manager's house and stables dated 16 Apr. 1867 (in possession of Barclays Bank) bear Hoskins's signature. If the whole design is by Hoskins there must have been some delay, for details of the manager's house are referred to in Waterhouse's letter of 16 Jan. 1865. In any case the buildings are remarkably close to Waterhouse's style.

[36] G. G. Hoskins, *The Clerk of Works* (London, 1876).

[37] The main drains seem to have been badly laid and there was serious trouble with a chimney. Waterhouse roundly blames the contractor who, it appears, had gone out of business shortly afterwards. OLB 240, 246, 247, 249, 250, 252. Private collection.

[38] These were: Cambridge Union extension (1878); Reading Grammar School (1870); Knutsford Market Hall (1870); Christ Church Parsonage, Reading (1871); Reading Municipal Buildings (1872); Upcroft Lodge, Bowsey Hill (1872); Hurworth Grange (1873); Pilmore Hall (1876); Oxford Union Debating Hall (1878); East Thorpe, Reading (1880); Lyndhurst Road Church (1880); St Paul's School (1881); Girton College (1885); Putney Hill House (1887); Hillmorton Paddox, Rugby (1889); Springfield, Maidstone (1890).

[39] These were: 49 Grosvenor Square, renovations (1868); 45 Grosvenor Place (1875); Prudential, Holborn (1877); Woodbines, Kingston, additions (1878); 1 Old Bond Street (1880); Prudential, Brooke Street (1885); Prudential, Greville Street (1885); Staple Inn, renovation (1887); Pangbourne Coffee Shop (1890); 21 St James' Place (1891); Britwell Court (1895); Prudential, Furnival's Inn (1895); Staple Inn Buildings (1901).

[40] He regularly worked with such as Samuel Bramall, Thomas Clay, Mark Foggett, Robert Neill, or Joseph Thompson (not to be confused with Alderman Joseph Thompson of Pin Mill Brow, Waterhouse's friend and patron).

[41] There is evidence in the office correspondence that he had difficulty in getting sufficient competitive tenders for his church St Seiriol at Penmaenmawr.

[42] Clayton and Bell supplied glass for Balliol College Hall (1877) **[392]**, and for the Council Room in University College Hall (1871–2) **[355]**.

appear to have been the only firm willing to supply the pale pink and grey glass that became his hallmark.

It was the same with tiles. His concern for exactness of colour and finish within his cost limits is shown by the long correspondence with Minton's and then Maw's over the tiling of the new conservatory at Egerton Hall **[11]**.[43] At Blackmoor **[192]** he even used three different suppliers, Maw's, W. Godwin, and W. B. Simpson, in a single fireplace, presumably in order to secure precisely the combination of tints he wanted. However, after trying Minton's and Maw's, by the end of the 1860s he was commissioning almost all his tiles from William Godwin or Craven Dunnill & Co., for whom he even supplied one design for a geometric floor **[325]**.[44]

Yet Waterhouse was well aware of the limitations of even his favourite suppliers, using the Gloster Wagon Company for the large quantities of office furniture at the Prudential in Holborn (Pls. 130, 132, 178) and turning to Liberty's or Maple's for smaller items only.[45] For metalwork he first tried the well known firm of F. A. Skidmore in Coventry; but finding his deliveries unreliable, he turned to Robert Jones of Manchester and Hart, Son & Peard who supplied him for the rest of his career.[46] Heating systems he almost always got from Haden's of Trowbridge, or W. W. Phipson; and on the only occasion on which he turned to a completely new supplier, at the Natural History Museum, the result was a disaster.

The same preference for a tried and tested supplier, even if he was not in the forefront of fashion, applied to furniture. When he was aked to design this he frequently turned to Henry Capel, who, for instance, supplied the fine set of chairs for the Hall of Jesus College, Cambridge **[347]** (Pl. 175).[47] Always he sought to obtain good quality, and many of his regular suppliers' names, such as Guynan's for blinds or Gibbons of Wolverhampton for locks, are synonymous with quality.

Such close control is hardly surprising in an architect of Waterhouse's character, and the particular contractors he chose generally reflect his stylistic preferences. In stone carving, for instance, he employed Thomas Earp (later Earp, Hobbs and Miller) on a dozen or so occasions, notably at Harris's Bank, Leighton Buzzard **[204]** and at St Elisabeth, Reddish **[507]**. However, Earp's style was not entirely to Waterhouse's taste, and for preference he would turn to Farmer and Brindley.[48] This company, perhaps the largest firm of ornamental masons at that time, was involved in upwards of one hundred commissions for Waterhouse, ranging from a tombstone in Norwood cemetery **[181]** or a pulpit in Stanmore Church **[438]** to the extensive ornamental carving at Eaton Hall **[315]**. And it is surely significant that they were commissioned to produce models for almost all the Prudential buildings, which were then passed to the mouldmakers of the terracotta suppliers.

Selecting and supervising contractors and managing his highly productive office demanded

[43] OLB 97, 111, 112, 115, 116, 119, 122, 123, 135, 136, and 139 (1857). Private collection. Waterhouse had previously had an equally lengthy correspondence with Minton's over staining in their white glazed bricks, which he had used in alterations to the baths at Southport.

[44] Waterhouse's accounts show this work as done in 1869. It did not appear in Craven Dunnill's catalogue until 1878. Craven Dunnill & Co., *Catalogue of Plain and Encaustic Tiles* (1878), pl. 15. Ironbridge Gorge Museum Library.

[45] The office ledgers recall small payments to Maple's and Liberty's, but do not specify what for. Maple's supplied items for a number of other Prudential buildings, and some of the orders are identified as coal boxes: so their contribution to the Holborn offices may have been equally mundane.

[46] He regularly ordered materials from both firms for the same job. For instance, both had commissions at Eaton Hall, which indicates that Waterhouse was determinedly selecting precisely the item that he wanted for each location. A further attraction of Hart, Son Peard & Co. in the late 1880s was the extent to which they diversified into electric light fittings.

[47] Among Capel's other major works for Waterhouse was furniture for Blackmoor House, Eaton Hall, Balliol and Pembroke Colleges.

[48] An interesting comparison can be made between the work of Farmer & Brindley on Waterhouse's Manchester Town Hall and Earp's on Worthington's Manchester Magistrates' Court. The works are exactly contemporary and for the same client.

efficiency and acumen of a high order. In many ways Waterhouse was just as successful as a businessman as he was as an architect. He had fully mastered the commercial necessity for cost control in large or small buildings, and was clearly an efficient accountant and businessman. There is a further hint of this in his activities as a life Trustee of the Soane Museum. He was elected in 1878, but his attendance at meetings was distinctly patchy. He only attended on twenty-two occasions between 1879 and 1900, and not at all in 1884, 1886, 1887, 1890, 1895, and 1898. Yet his signature appears first on almost all the annual accounts; and the sale warrants of Consols (which provided some of the cash for running expenses) are invariably made out to 'Alfred Waterhouse and others'.[49] This business sense and efficiency are exactly the sort of virtues that would have appealed to clients, both private and corporate.

However, if one factor in his success was his ability to build within, or more or less within, cost limits, it seems in fact that he was not so very much more successful in this than other architects. None the less, he was ready to accept cost restrictions as a limiting factor in the design from the start, and to reduce the splendour of his buildings accordingly. He was not often lucky enough to build as extravagantly as at Eaton Hall; but a client as rich as the fourth Duke of Westminster would be a rarity in any architect's career. Instead he learnt the hard way by designing or altering houses for Lancashire merchants, and by dealing with groups of clients involved in putting up chapels or schools that were entirely dependent on public subscription.[50] This sharp schooling gave him a realistic understanding of financial constraints which fitted well with the frequent requirement to design larger schemes so that they could be built piecemeal.

Draughtsman, economist, and businessman, however, he still had another advantage that made him attractive as an architect, his charm. It was said that his smile was worth £10,000 a year to him; and the record of clients who returned again and again is certainly impressive. But above all it was his scrupulousness that distinguished him from architects in the mould of Phillip Bossiney, and made Waterhouse so especially attractive to larger organizations with substantial sums to spend, but who were responsible to shareholders or electors for the spending.

49 Minutes of the Trustees Meetings, Sir John Soane's Museum (1878–1900), *passim*. I am grateful to Dr Margaret Richardson for drawing my attention to this item. In the context of Waterhouse's own restoration work it is interesting that he did not attend any of the meetings at which James Wild, curator 1876–92, presented and carried his plans for building a new picture gallery in the place of Soane's old office. Waterhouse was not, therefore, implicated in the drastic suggestion of replacing the colonnade with a new 45 ft. clear gallery. Leighton expressed his approval, but the scheme mercifully foundered with Wild's death.

50 These small commissions, such as the Congregational Chapel at Ashton-on-Makerfield (costing less than £3,000 in 1865) **[191]** and the Prestwich National Schools **[219]**, cost the office a good deal of trouble. The latter in particular was still a source of complaints at the beginning of 1867, largely as a result of cheeseparing during construction (see OLB II 627, 903, 981. RIBA Library).

19

Waterhouse and the Profession

As early as the 1850s changes in the architectural profession and the pattern of patronage were already apparent. Waterhouse was himself well able to handle the new corporate or institutional clients, to manage the increasingly complex business of building, and to work with the new brand of large-scale contractors. But the changes brought inevitable pressures on the profession as a whole, and Waterhouse involved himself in them to the full. He was evidently a good committee man, and an active member of the Royal Institute of British Architects, acting as President for three years, rather than the usual two, from 1888 to 1891 at the time of the well-publicized row over registration. Yet his professionalism went far beyond handing over architecture to the 'care of a respectable body of practitioners, a close self-elected corporation'.[1] His concern with professional practice is reflected in his office production and in his career as a whole. His opposition to the 'artists' in the registration battle is understandable in the light of the shifting pattern of work and of his own experience.

His was not an office where creative designing was taught. That was the job of the architect. His pupils and staff were engaged in 'all the common routine of building'.[2] In this sense he was rather closer to the old style professional architect than to the newer 'art' architects. He would have felt that the creative side of architecture was best learnt from visits to the Continent, or at least to the South Kensington Museums, and would have regarded his role as the supervisor of pupils as being more concerned with the sort of practicalities set out in Robert Kerr's 'Seven Other Lamps of Architecture'.[3] This view is echoed by Lethaby's criticism of the education in architectural schools: 'It has been a wasteful system, too regardless of results, or too regardful of wrong results. It is absurd for instance that the writer should have been allowed to study cathedrals from Kirkwall to Rome and from Quimper to Constantinople; it would be far better to have an equivalent knowledge of steel and concrete construction.'[4]

It had been essential, in order to sustain the standards of his office work, that both pupils and assistants were reliably trained. This made him a natural supporter of the principle of competitive examinations for membership of the RIBA. However, it is too easy to polarize the registration debate, and to cast Waterhouse as the advocate of professionalism merely because he was not among

[1] T. G. Jackson, 'Essay XIII', in R. N. Shaw and T. G. Jackson (eds.), *Architecture: A Profession or an Art?* (London, 1892), 211.

[2] This was what Scott claimed to have learnt from his apprenticeship to Edmeston, begun in 1828. G. G. Scott, *Personal and Professional Recollections* (London, 1879).

[3] R. Kerr, 'Seven Other Lamps of Architecture: Letters from Architecturus to his Son', in *Builder*, 9 (1851), and 10 (1852).

[4] W. R. Lethaby, *Architecture* (London, 1911), 247.

the signatories of the opposition letter to *The Times* or the contributors to Norman Shaw's book of essays, *Architecture: A Profession or an Art?* He was in any case a friend of Shaw, and remained so, and retained the respect of some of the opponents such as T. G. Jackson.

None the less, Waterhouse has to be seen as a leading figure in the development of architecture into a profession, just as he was the driving force behind a thoroughly professional office. There are good reasons for this. He trained in a town where the scope for architecture and building of all standards was enormous. Yet the first Manchester Society of Architects, founded in 1837, died out in the 1840s. The need for professional standards was pressing, and it is clear that Waterhouse was closely involved in the refounding of the society in the early 1860s. He subscribed to the Manchester Architectural Association in 1861, although he had already joined the Institute of British Architects,[5] as soon as he had completed the required seven years in practice.[6] He was certainly committed to the idea of professional solidarity and status, and was even proposed as vice-president of the Architectural Exhibition Society in 1867.[7] In fact solidarity and integrity were key themes, to which he returned on more than one occasion when he had risen to the Presidency of the RIBA.

His Presidential Addresses reveal his concerns, and hardly support the view that he was opposed to the 'art' architects. It is even possible that his third term as President was the result of his even-handedness in the debate between them and the so-called 'professionals'. The problem was raised in the very first meeting of the Institute that he chaired as President *ad interim*.[8] His own position is set out in his first Presidential Address in terms close to those used by J. Macvicar Anderson (who had originally proposed him as President). They are the views of the Establishment but do not show the narrow-mindedness that Shaw's eloquence for the other side would lead us to expect:

> We have heard something lately of the conflicting terms 'professional man' and 'artist' as applied to the architect. Now in my opinion the true architect is both. The higher and more systematic education which we are hoping for and getting will train us in the efficient and easy practice of our profession—a profession which is open to all men of education, intelligence and industry, and one in which the greatest successes will attend those to whom a further artistic perception has been given and in whom it has been carefully nurtured.[9]

In professional terms Waterhouse was clear that education and the obligatory examination were the most important developments of the previous years. He had a firm belief in the essentiality of standardized training as the prerequisite for architectural efficiency. He claimed 'education is what we want, not registration';[10] though he believed that registration would follow naturally from the achievement of a profession consisting only of those who had succeeded in a single standardized examination. In effect, of course, this was to support the position attacked by Shaw and Jackson in which 'architecture would be handed over to a respectable body of practitioners'.[11]

Jackson continued his attack with a depressing prediction of the likely future and an attack on

[5] MS letter, AW to R. Knill Freeman (2 Apr. 1861). OLB 291. Private collection.

[6] He had been recommended for the national body by G. G. Scott. In his letter of application he cited James Bell, then Secretary of the RIBA, or William Burges as possible seconders. MS letter, AW to A. B. Thompson (20 July 1860). OLB 234. Private collection.

[7] MS letter, AW to Rowland Plumbe (26 Feb. 1867). OLB 739. RIBA Library. Waterhouse declined owing to pressure of business.

[8] Seventh ordinary meeting of RIBA (1888), reported in the *Builder* (25 Feb. 1888), 137 ff. His defence of the Council's opposition to the despised and unsuccessful Registration Bill makes it clear that he was not himself on the committee that was to report on the Bill.

[9] AW, Presidential Address for the session 1888–9, reported in the *Builder* (10 Nov. 1888), 336.

[10] Ibid.

[11] T. G. Jackson in *Architecture: A Profession or an Art?*, 211.

present practice. It was accepted that compulsory examination was only a step to certification; and certification, claimed Jackson,

> will guarantee a confiding public not only against ignorance of construction and bad drainage, but against bad design also, and *costly disappointments*. Membership of the Institute is to be taken as a warrant for '*the existence of a sufficient degree of designing*' power to prevent all such misfortunes. From this blessed prospect it is perhaps unkind to turn and ask who it is that perpetuates the artistic crimes which make our streets hideous at the present day.[12]

The whole debate was, of course, too courteous for names to be named; so it is not clear how many of Waterhouse's buildings would have been on Jackson's hit list. However, it is difficult to avoid the element of subjectivity in his view. He provides no criteria for hideousness; and the whole collection of essays never really faced the question of what was, or should be, involved in architecture. Waterhouse, on the other hand, was pressing for minimum standards to avoid what he called 'the blind and uncertain fashion in which I blundered on my way'.[13] He undoubtedly had a confident belief in the certainty of the path mapped out by books and selected examples of what were now accepted as great works of architecture. He also hoped that standardized qualifications would lead to more public recognition and better status for architects. He urged that the RIBA and the AA should consider amalgamation or alliance.[14]

Waterhouse's Presidency also saw the growth of alliances with previously independent local architects' societies.[15] After 1889 the RIBA was more than a metropolitan club, with nine presidents of local societies automatically nominated to the Council (always providing they were members of the Institute): and in his Presidential Address of 1890 Waterhouse was at pains to bemoan the fact that a number of leading architects still held aloof. From its foundation in 1837 the Institute had had the aim of securing 'uniformity and respectability of practice in the profession'. In the latest decades of the nineteenth century the need for this was, if possible, more pressing. The Local Government Act of 1888, which established bodies as powerful as the LCC, made no provision for a qualified architect among their officers. At the same time building regulations had multiplied to the point where there were manifest absurdities. Waterhouse devoted a large part of his 1890 Presidential Address to the need for reform of the Building Regulations,[16] and it is probably no accident that Thomas Blashill of the Local Government Board Office, in drafting the new Metropolitan Building Act of 1891, cited the Manchester by-laws as being the most modern and practical. Even within the profession there were difficulties arising from the increasing complexity of work; and another of his campaigns was to ensure fairness to builders. He was concerned that they should not be liable for errors in the bills of quantities, which were worked out by architects, or their surveyors, as the basis for estimating the builders' contracts.[17]

Waterhouse's greatest concern, however, was with the training of architects and with practical

[12] T. G. Jackson in *Architecture: A Profession or an Art?*, 217.

[13] AW, Presidential Address to RIBA students, reported in the *Builder* (2 Feb. 1889), 90.

[14] Waterhouse had been a member of the AA since 1866, but never held office in that body.

[15] The cause of the provincial societies was understandably close to Waterhouse's heart. He had publicly supported local societies that were in correspondence with the Institute at the RIBA Dinner of 1887, again linking their work with education. *Builder* (14 May 1887), 730.

[16] AW, Presidential Address for the session 1889–90, reported in the *Builder* (8 Nov. 1890), 360.

[17] AW, Presidential Address for the session 1888–9, reported in the *Builder* (10 Nov. 1889), 339. It is interesting to note that having established such an efficient copying shop of his own, Waterhouse is quick to link the provision of accurate drawings to his plan for fairer treatment in relation to quantities; and he presses for more use of the newly invented copying processes.

aspects of control and management of architectural works. In this he was extrapolating to the level of the profession solutions to problems of practice that he had already tried out in his own office. He had devised a working method that was admirably suited to the new building types, and which suited the wishes of his new corporate clients. It was really only in the private houses of the next generation that the businesslike efficiency of his system was found to be less palatable. As a thoroughgoing professional, Waterhouse takes on the appearance of a precursor of some of the great twentieth-century architects, such as Behrens or Mies van der Rohe. That he was himself instrumental in bringing about the growth of professionalism must be seen as one of his major contributions.

Professionalism was not, of course, unique to his office; but there can be little doubt that he had the leading reputation for this in the last quarter of the century. His reputation made him in demand not only as a designer in his own right, but also as a judge of his fellows. He had done well out of the competitive system, and now was well qualified to act as a competition assessor: and this was his real contribution in the field. It seems to have been his tact in dealing with the clients as well as his fairness to his brother architects that made him so successful. He acted as assessor in some sixty competitions,[18] far more than any of his contemporaries, and almost three times the number judged by Charles Barry II, who chaired the RIBA's committee on competitions. Waterhouse was in demand both for great competitions, such as the Imperial Institute or the Milan Cathedral façade, as well as humbler ones, such as the Emmanuel Congregational Church in Cambridge, a Baptist chapel in Sheffield, or the Leeds Exchange and Newsroom.

His decisions reveal a remarkable objectivity. With Waterhouse as judge, it seems no surprise that the award of the Birmingham Assize Courts competition went to a firm submitting under the pseudonym 'terra cotta'. Actually the architects, Aston Webb and Ingress Bell, used the material in a much more richly decorative, more plastic (and probably more expensive) manner than Waterhouse ever did. It is just as likely to have been their clear and logical planning that impressed him. In a later competition for the South Kensington Museum in 1891, when he again awarded the partnership first premium, there was little in the style or the materials that matched his own preferences, or his own neighbouring work. The position of assessor certainly gave Waterhouse the opportunity to favour his preferred Gothic style; and some such bias might be traceable in his choice of W. H. Lynn's design for Barrow Town Hall in preference to Collcutt's more sophisticated free Renaissance version. One or two names of architects well known to him also occur in the list of winners he selected. Brightwen Binyon benefited once from his former master's assessment. Generally, however, he seems to have judged very fairly on sound architectural grounds, even in spite of local pressure. It was he who selected E. W. Godwin's designs for Bristol Assize Courts. The first competition had only elicited one entry. Waterhouse judged the second competition; and his opinion was endorsed by Street in the notorious third competition, that was designed specifically to exclude Godwin. He could recognize a good architect, and it was as a result of his report on the Plymouth Guildhall competition that Godwin was appointed as consultant there. However, there was also the difficulty of convincing the commissioning committee. Waterhouse showed that he had learnt wisdom in this as assessor of the Birmingham Council House competition. His first choice was an exciting design by W. H. Lynn; but that was not popular with the city fathers, and his second award wisely went to the local competitor, R. Yeoville Thomason.[19]

[18] See App. 4.

[19] Thomason had been placed second in the first adjudication, and his premiated design was built. Part of the quality of Lynn's design was the provision of a dramatic bridge linking the

His genuine critical skill is perhaps best seen in the extent to which he was able, in later years, to subordinate his own stylistic preferences. His involvement in the competition for the completion of Milan Cathedral is simply an acknowledgement of his international reputation as a Gothic designer. But his choice of Aston Webb's rather etiolated Baroque design for the South Kensington Museum can hardly reflect his own personal taste.[20] It was a style he never indulged; and he would have been more in tune with E. W. Mountford's free Renaissance style that he selected for Sheffield Municipal Buildings. However, that design had clear planning merits, and was a practical work designed for easy extension. In selecting Brumwell Thomas's design for Belfast City Hall or E. A. Rickards's for Cardiff City Hall, Waterhouse again held back from his own preferences and chose two of the grandest Edwardian Baroque buildings that we have. In the latter case, especially, it is generally admitted that he picked a masterpiece.

In spite of his evident critical acumen and his popularity with the promoters (who obviously stood to gain by capturing the big name assessor), the final judgement should lie with the profession itself. It is remarkable how few complaints are recorded in competitions with which he was involved. The row at the Law Courts seems to have been as much that his rivals wanted the opportunity to produce similar volumes of photographs to his. At Manchester Town Hall it is not altogether surprising that Speakman and Charlesworth, who were placed second overall and first as a work of art, complained at his victory; but theirs was more or less a lone voice. As an assessor he seems to have escaped criticism altogether, which makes it all the more unfortunate that no drafts survive in the office archives of his competition reports. They would have been revealing.[21] The attitude of some of those he rejected is particularly impressive. T.G. Jackson, for instance, who had opposed Waterhouse in the profession or art debate, was placed second in the Imperial Institute competition. Recalling the decision later he wrote: 'the competition was won by Collcutt, . . . Waterhouse told me that when my design was unpacked he said, "Oh, that's the one!" but that my planning was inferior. I think he was right.'[22]

new buildings with Hansom's town hall next door. The competition brief had not envisaged such a solution, and so the bridge was a technical infringement of the building line. Lynn's design was therefore ruled unacceptable.

[20] Competition for the façade building of the South Kensington Museums (1891). First Aston Webb. The most dramatic of the designs was probably that of J. Belcher, placed second. Waterhouse admired it but ruled it out because it was less practical and more costly. For a detailed description of this competition see J. Physick, *The Victoria and Albert Museum* (London, 1982), 183 ff.

[21] Two of Waterhouse's Presidential Addresses to students discuss competitions and refer to the Milan Cathedral competition, *Builder* (2 Feb. 1889), and to the competitions for Sheffield Municipal Buildings and for Bristol Assize Courts, *Builder* (25 Jan. 1890).

[22] B. H. Jackson (ed.), *Recollections of T. G. Jackson* (Oxford, 1950), 210.

20

Planning

WHATEVER the fortunes of Waterhouse's reputation in the ebb and flow of taste, his ability as a planner has never been in doubt. Indeed his success may be said to lie not only in the winning of competitions, where so often it was the disposition of the required spaces which determined his prize, but in his ability to provide buildings which worked satisfactorily. The real demands on his skill as a planner came from the major public commissions, where he was faced with the task of combining large and small spaces and providing appropriate circulation, often of different kinds, within the same building. In this respect few commissions can have given him quite as much difficulty as the National Liberal Club **[514]** (Fig. 14), where the irregular shape and small size of the site greatly added to the problem. It is, however, a highly successful piece of planning, sophisticated and ingenious; and it can be taken as a demonstration of all his planning skills.

Waterhouse manipulated structure and variety of levels and turned the awkward site to advantage, producing a complex building that was none the less easy to use. The entrance, on the corner of Whitehall Avenue and Whitehall Place, leads into a long hall, which opens on one side into a reception lobby, where guests could wait. There were no major club rooms on this floor.[1] Beyond the staircase were sets of dressing rooms, bathrooms, and cloakrooms, all electrically lit, and at the back of the site a row of top-lit billiard rooms.[2] The principal rooms of the club were on the upper floors, so the key to this building was the vertical circulation provided by the spacious oval staircase in the centre of the site.

One flight led down to an immense double-height smoking-room.[3] The main stairs curved up to the two principal floors, each of which had two grand rooms, a dining-room and writing room, a smokers' reading room, and the Gladstone Library. These spaces filled the two long sides of each floor, and there were smaller, but still grand rooms, on the third open side of the site. The largest rooms are a stupendous size, almost 100 feet by 35 feet, which Waterhouse achieved by lining them with free-standing columns that diminish the span of the girders supporting the upper floors.[4] The

[1] The large conference room adjoining the reception lobby was designed to be used separately, and had its own entrance from the street. Significantly, both it and the reception lobby were 3 steps below the level of the club.

[2] They were fitted under the terrace at the rear of the building, and, being intended for use mainly at night, needed only to be top lit. They were redesigned as lavatories early in the club's history and the great smoking-room was converted to a billiard room.

[3] This was linked to the billiard rooms by the small balcony at one end, and was provided with its own bar.

[4] The columns are of steel in ash concrete, faced with faience. The floors are also of steel and concrete on Lindsay's patent system that Waterhouse favoured at this period. The mismatch between the large spans of the main rooms and the smaller members' bedrooms (each of which has of course its own fireplace and flue) is managed by having the fourth floor partly suspended by steel hangers.

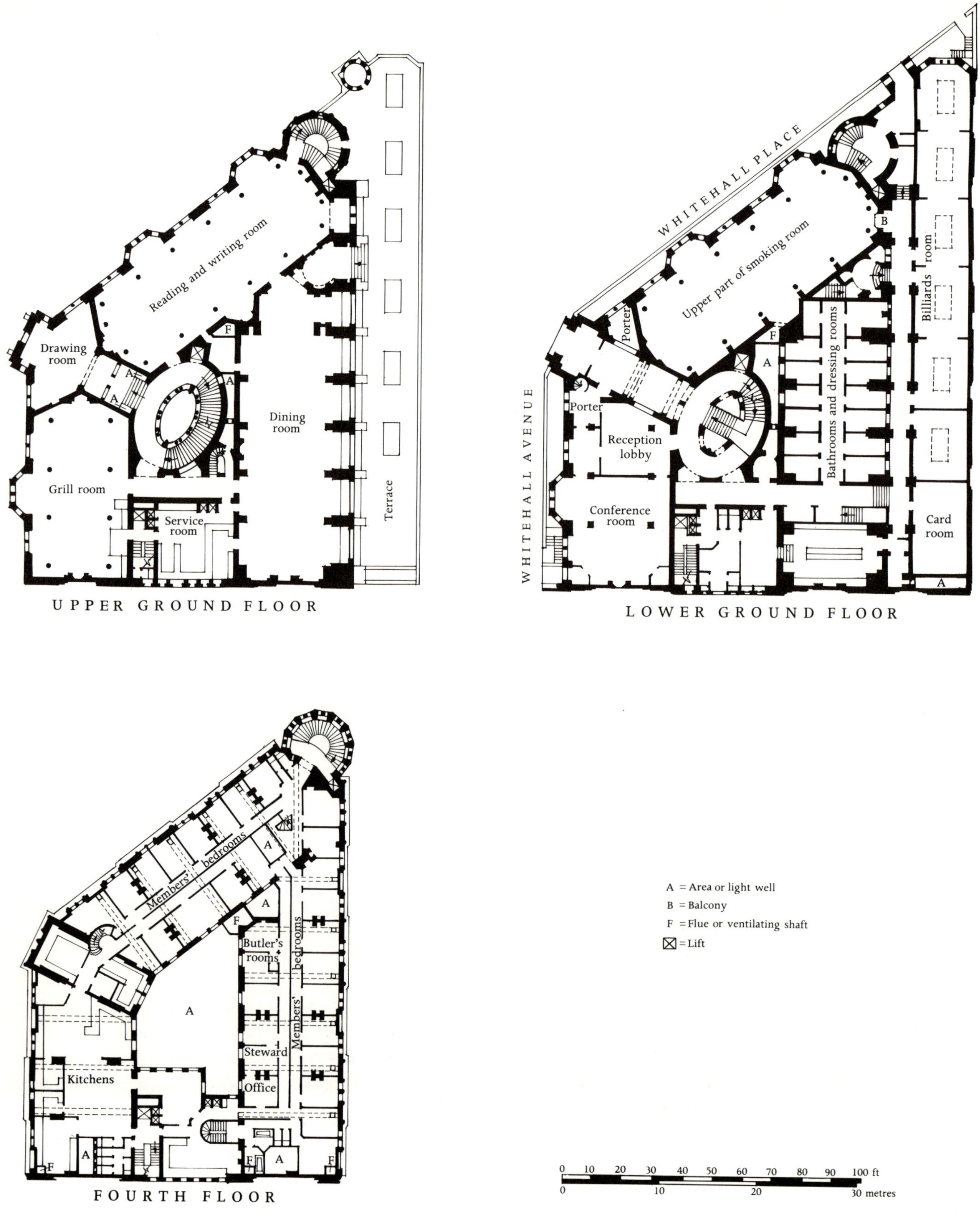

FIG. 14. London, National Liberal Club (1884) **[514]**: (*a*) plan of lower ground floor, (*b*) plan of upper ground floor, (*c*) plan of fourth floor

columns create bays along the sides of the rooms which are ideal for groups of friends meeting in the ambiguously public space of the club. The informal effect is further enhanced by the bay windows, which follow the logic of the façade and fall irregularly in the rooms. Waterhouse's mastery of planning is shown in the way he links these great rooms and the awkward angles of the site. The rooms themselves have canted ends that allow all the main spaces to be axially symmetrical but to interlock at the awkward angles; and the two main rooms on each floor are directly linked. Circulation space on each floor is kept to a minimum; and on the two principal floors there is only one short passage linking the front and rear rooms. This allows the spaces to be used by members in a thoroughly relaxed way; and anyone moving through the rooms is continually delighted by the gentle changes of direction. In combination with the monolithic splendour of the faience colonnades, the effect is one of affluence and relaxing informality. The only awkward spaces are the small lozenge-shaped drawing-room and the small smoking-room above; but even these work effectively as combinations of semi-private spaces.

Above the clubrooms were a further four floors of members' bedrooms, staffrooms, and, originally, the kitchen. This largely residential section was divorced from the main part of the club, and access for members was by way of the secondary staircase in the corner turret. The main staircase rose only to the first floor, where it was roofed; above that the stairwell became the light well for the four floors of bedrooms.[5] The secondary staircase rose the full height of the building, and afforded members a separate route to the street without passing through the club rooms. The service rooms are set on the blind side of the site, lit from the light well of Whitehall Court, and with three separate staircases serving the mezzanines. One of these rises the full height of the building, linking servants' rooms and the great kitchen on the fourth floor with the huge wine cellar and the menservants' rooms in the basement.

The whole complex arrangement of major spaces, vertical circulation, and service rooms is tightly organized, and makes a series of dramatic spaces on the awkward site. The ingenuity with which this is achieved is typical of Waterhouse at his best. The sheer quality of it is the way in which the ingenuity leads, not just to an acceptable solution, but is turned to creative advantage in unexpected ways. He had already played the game of modifying individual spaces at the New University Club **[208]** (Pl. 38) where he threw arches across all four corners of the morning room and put the fireplace in a bay to mirror the window, making a room full of intimate areas out of a potentially difficult square. This treatment reached its apogee in the National Liberal Club where there is ample accommodation for large numbers of members in spacious and elegant rooms,[6] whose awkward angles are completely disguised. The variety of bays, alcoves, and corners afford a range of semi-private spaces ideal for a club, and no space is wasted in unnecessary circulation. The end result is a complex and ingenious solution to a difficult brief.

Part of Waterhouse's skill was undoubtedly a refusal to be shackled by precedent. This is shown clearly in the planning of a couple of his early chapels. At both Rusholme and Ancoats, he was faced with problems that went beyond the mere adaptation of a traditional chapel form. At Rusholme **[164]**

[5] The light well is a substantial area, faced with glazed white bricks. However, members' bedrooms only face into it on the second and third floors. Above that, on the fourth floor above the main cornice, members' rooms face out on to Whitehall Avenue or over the river. A separate passage, parallel to the members' passage, gives access to the inward-facing servants' rooms. A fifth floor of attic bedrooms was kept for club servants, while the kitchen on the fourth floor was a double-height space occupying the whole of the side to Whitehall Place. It was set on top of the building in accordance with the current fashion for avoiding cooking smells.

[6] Membership of the club in 1884 was claimed to be over 4,300 (National Liberal Club Buildings Company Ltd., prospectus. University of Bristol Library, National Liberal Club Papers).

the landowner was unwilling to sell land to nonconformists,[7] so the site, though on a corner, was very small. At Ancoats **[105]** (Pl. 29) a triangular site tapered to a point at the junction of two streets. Both chapels required seating facing a central pulpit with schoolrooms and ancillary accommodation adjoining. Waterhouse gained the maximum available area by placing the tower at the corner of the site and arranging the seating to fan out round a pulpit, placed against the inside wall of the tower. It was a clear departure from the conventional arrangement of a chapel. At Rusholme the site was rectangular, yet he chose not to put the pulpit centrally on one of the sides; and the fan shape of the seating, in each case, afforded the best view of the pulpit for all members of the congregation. In both chapels the towers contained an entrance on the lowest level and vestries on two floors above, giving direct access for the minister to his high pulpit and providing a singers' gallery above.[8] In both cases his plan derives from the straightforward logic of the case, unfettered by precedent.

The same skill can be traced in almost any of his major schemes. The key seems to be an immediate acceptance of the constraints linked to a logical appreciation of the planning needs. Even where the basic layout was set, the thoroughness with which Waterhouse worked out the functions of each part of a building gives his planning an unrivalled smoothness. In each of his three largest schemes, the general form of the plan was to a great extent predetermined. In the Manchester Assize Courts **[67]** and the Law Courts **[278]** the competition conditions determined what was to be done, and in the Natural History Museum **[243]** the layout of the previous architect had the same effect. Yet even at the Assize Courts he demonstrated his ability in the sophistication with which he handled the basic scheme (Fig. 1). The great central hall, its long axis parallel to the main frontage, was the pivot, an arrangement which made the complex building easy to use. All the main circulation routes either passed through it or, on the upper floor, looked into it. An encircling corridor, lit by light wells at each corner, gave the most direct possible access to all parts of the building; and each court was grouped with its own offices. The combination of the double-height courtrooms and smaller offices was efficiently managed, and the whole building appears to have functioned very adequately. That so young an architect (Waterhouse was only 29) should manage so complex a scheme so effectively is a remarkable tribute to the thoroughness of his planning.

He was able to enlarge the idea and increase its complexity a decade later for the new Law Courts **[278]** (Figs. 2, 3), where his scheme was preferred by the bar committee, who would actually have to use it. In this case his logic was based on sound legal knowledge gained from his own court building and his brother's experience. The outer façade of his building contained offices, leaving the courts themselves segregated from noise beyond an internal street. With the top-lit courts at a higher level than the great hall there would have been a good deal of vertical circulation. Each court was provided with its own barristers' and public stairs, and the resulting unit was repeated as often as required without intervening space. However, the main system of interlocking staircases and carefully segregated broad straight corridors was masterly.

Practicalities aside there was always far more to Waterhouse's skill than merely arranging areas to relate usefully. Throughout the planning exercise he was aware of the enclosure of space three dimensionally, of the relationship of one space to another, and of movement from one space to another. In this context his Natural History Museum **[243]** and the Manchester Town Hall **[284]** are

[7] Benjamin Nightingale, *Lancashire Non-conformity* (Manchester, 1890–3), 162.

[8] The chapels both required schoolrooms. At Rusholme they were behind the chapel. A similar plan was tried at Ancoats, but, with insufficient land available, they were set beneath the church.

the best examples. The planning of the Natural History Museum seems a relatively simple proposition. Like Fowke, he had to base his plan on the original vision of Professor Owen, the instigator of the scheme. Unusually for Waterhouse the façade is rigidly symmetrical; but Owen's stipulation of a central court with rows of galleries to either side gave him his supreme spatial opportunity. There was no obligation here to provide complicated circulation around the building; the purpose of the central hall was primarily to provide an index section of the museum's contents. Waterhouse accordingly concentrated his effect on this tremendous space (Pl. 82). To take in the whole one has to enter the cathedral-like portal, to pass under the arch beyond, walk the length of the hall with the raised arcades on either side, then ascend the staircase and return behind one of the arcades finally climbing again up the flying staircase over the arch. At the end of that progress comes the view of the hall from this high vantage point.

The lack of public vertical circulation elsewhere in the building results from Owen's conception of completely distinct departments, and belongs to nineteenth-century concepts of the static museum. Besides one has to remember that the large public museum, like the major office, hospital, or town hall, was a new creation of the period. There were few precedents, and Waterhouse had to think out the solution virtually from scratch. At the Natural History Museum he was denied complete freedom, but the processional route he created from Owen's plan must rank as one of the grandest moments of high Victorian architecture.

Manchester Town Hall **[284]** (Fig. 4), however, combines the organizational complexity of the National Liberal Club **[514]** or the Law Courts **[278]** with the spatial qualities that Waterhouse achieved in the museum. Like the National Liberal Club there was the added imposition of the triangular site. Its central hall plays no part in the compositional massing so, unlike his competitors, he declined to force it on to the same axis as the tower and entrance. Instead he contrived the grand stairs and sculpture hall to shift the visitor from one axis to another. And by moving the principal stairs back to the further side of the corridor he not only achieved a greater circulation area but made it possible to use the great hall and the reception rooms either together or separately. The circular stairs at each point of the triangle 'present the advantage of a rapid ascent for those to whom this is an object'; and 'easier gradients for those for whom this is desirable'.[9] They also disguise the sharp angles and accommodate the varying floor levels. They become a principal feature of the whole concept and with the intermediate stairs on the long sides make for extremely simple and efficient circulation both vertically and horizontally.[10] In addition all the spaces, rooms, stairs, and corridors, have natural daylight without the use of light wells except for the one great well to the landing outside the main hall.[11]

What all these great buildings demonatrate is not just that Waterhouse was able to lay spaces out logically and conveniently. Logic and convenience had been the key to his efficient houses from the start, and they are characteristic also of his work at Girton **[346]**. There he broke with the traditional staircases of the Oxbridge tradition in favour of complete internal circulation. Butterfield was doing the same at Keble, setting his rooms on corridors, but still with frequent staircases. Waterhouse was more radical, linking all the buildings by one long corridor with stairs only at key junctions. What

[9] 'St. Valentine' (Waterhouse), written statement accompanying design submission. MCL Local History Collection.

[10] John Archer points out that there is no more than 18 m. from any office to a stairway.

[11] That is treated as a feature with the panels of its skylight inscribed with the names of the city's mayors.

was more he contrived his corridor in such a way that, with the exception of Orchard Wing, all the rooms face either south or west.

Of course the real value of his planning was in his ability to link logic and convenience with the creation of spatial effects. Even so one should not despise convenience and logic, especially as he had to devise new solutions for new types of buildings. The Refuge Offices in Manchester **[570]** (Fig. 12) provide the clearest demonstration of his approach. The structural system is allowed to dictate the whole layout with a complete grid of square or rectangular bays, only slightly modified for the stairs and corner entrance. The larger rectangular bays fill the two street façades and allow for larger private offices on the outer side of the building as well as the corridor. A central light well occupies six of the square bays with the surrounding space given over to an open office area on all floors. The main staircase occupies one bay beside the principal entrance; and a secondary staircase and strong rooms fill most of the southern end of the site. The main entrance,[12] on the commercially advantageous corner, leads through a lobby into the main public office with a screen of six columns along the counter separating the public from the clerical space. These columns are spaced at half the distance of the main grid, which nicely allows them to pick up the grid of the rectangular bays on each façade. By allowing the structure to dictate the planning, Waterhouse was able to provide a very flexible office layout using only partitions to divide the main clerical spaces. The adoption of deeper bays along the façades allowed space for a really grand public office, partly lit by the central light well, and for an even more sumptuous boardroom above.

At the Prudential in Nottingham **[584]** (Pl. 161) his planning skills were constrained by the site and the brief was very much more complicated. There, on a sloping triangular site, he provided a single main entrance up a flight of steps under the tower. At the head of the steps was a small lobby whose canted end with two doors shifted the approach parallel with each façade, and led into two separate offices. One was for the Prudential itself; the other, larger one, for lease, took in the centre of the site and was provided with a glazed top light. At the far end of the site were two further entrances. These gave separate access to the managers' offices, and led, one down to the restaurant in the basement, and one up a spiral staircase tower at the rear of the site to sets of lettable chambers on the upper floors. These consisted of from two to four linked rooms reached from a galleried corridor round the glazed roof of the main public office. What would have been a central light well was used for circulation and turned into an upper atrium with its own skylight at roof level.[13]

This last building demonstrates some of the ingenuity that Waterhouse could link to logical planning. It was a quality that was constantly called for in larger buildings, where the prime concept inevitably left awkward spaces or minor difficulties of circulation. What he so often managed to do was to turn ingenuity into elegance as in the way, at Caius College, Cambridge, he adjusts the axial entry from Kings Parade. A square lobby beneath the tower is linked to an octagonal one with the inner archway leading at an angle into the corner of the court. Such examples could be multiplied from most of his major buildings. They are typical of the man and fully justify his reputation as a planner.

To Waterhouse, this 'most ingenious and masterly of Victorian planners',[14] his gift was a moral

[12] With the addition of Number 2 building by Paul Waterhouse in 1912, the principal entrance was moved to a space under his tower, and the whole public office was reversed. The original lobby then fell into disuse.

[13] Sadly this atrium was floored over in the 1950s.

[14] John Summerson, *Victorian Architecture* (New York, 1970), 112.

necessity because of the position of trust between the architect and his client.[15] He summed up his ideas on planning in his addresses to 'the earnest students of practical architecture',[16] bringing together earlier ideas of the subservience of elevations to the plan, of economy and functional expression propounded by Scott,[17] into one all-embracing philosophy.

Your buildings must faithfully adapt themselves to those for whom you build: must embody their requirements, and give what they want, in the most direct way possible. Do not let the conventionalities of style interfere with this. First find out exactly what is wanted; never think about the elevation of your building, till you have ascertained this, and embodied it in your plans as fully and perfectly as you can. Afterwards clothe the building so planned in the most fitting dress you can devise. That dress may be in many cases extremely simple, in others ornate; it may have to be sometimes severe, sometimes of exquisite beauty, if you can make it so. As you have in the first instance been solicitous that your building should adapt itself in every way to the needs and conditions of the people who are to use it, so now strive that every detail of the dress in which you clothe it shall help to make its purpose clear.[18]

[15] Presidential address (1890), published in the *Builder* (8 Nov. 1890).
[16] Presidential address to students (1890), published in the *Builder* (25 Jan. 1890).
[17] G. G. Scott, *Remarks on Secular and Domestic Architecture* (London, 1858). Marked passage on p. 145 of Waterhouse's copy.
[18] Presidential address to students (1889), published in the *Builder* (2 Feb. 1889).

21

The Business of Building: Terracotta and Faience

WATERHOUSE'S earliest sketchbooks indicate his interest not only in the use of materials but in the materials themselves (Pls. 12, 13). There are notes of things seen at the architectural museum: fireproof floors, ornamented obscure glass, 'indurated stone—Bath and Caen stone preserved somehow and given a fine polish', 'Ransome's siliceous stone, looks cementy, and has every appearance of being cast'.[1] There are frequent detailed sketches of timber structures, dormers (Pl. 15) and porches, roofs and bridges with annotations as to their jointing. There are notes, too, on Edmund Sharpe's second all terracotta church at Platt Lane, Rusholme, of 1846. Waterhouse is so firmly linked with this material that his use of it deserves a special study.

A number of factors led to the rise in popularity of terracotta during the nineteenth century. Principal among these were the search for a material which would resist the effects of the increasing pollution in the coal-burning commercial and industrial centres and the desire for a material that was cheap but capable of ornamentation. Of those who promoted its use Waterhouse was the champion, though his initial approach was curiously hesitant. Asked to provide replacement finials for Rydal Church in 1857 **[35]**, he wrote to the client offering a choice of terracotta or Ancaster stone, 'a stone much used for church work as at Lincoln', adding 'I lean to the stone but do not like to give the order until I have your approval'.[2] It appears from later correspondence that the choice was for terracotta. However, his first effective use of it for new building was at Easneye **[258]** (Pl. 103) in 1866; almost simultaneously he followed this with his own house at Foxhill **[285]** (Pl. 117) and Reading Grammar School **[307]** (Pl. 205). Shortly afterwards he used it for the first time in a major public building as a fireproof interior surface at Manchester Town Hall **[284]** (Pl. 52). Then, at the Natural History Museum **[243]** (Pls. 77, 81, 164–7), he used terracotta on a grand scale both inside and out. South Kensington already had several buildings using the new material, which may have helped to make it acceptable there.

The supplier for the new museum was Gibbs and Canning of Tamworth who had made the material for the Albert Hall. Now they worked with Waterhouse to improve the design and manufacture of their product. And this was crucial. Not only did the forms have to be designed by

[1] SK 10 Sept. 1855–19 Feb. 1858.

[2] MS letter, AW to John Green (2 Mar. 1857). OLB 90. Private collection. The estimated cost of the two materials was the same.

the architect and models made by M. Dujardin of Farmer and Brindley; each block had to be drawn out and shrinkage scale details produced. The scale varied according to the size of the block, type of clay, and temperature at which it was fired; so this was usually a task for the manufacturers. Later it is clear that Waterhouse was so familiar with the process that he was able to produce shrinkage scale drawings in his own office.[3]

Not until the drawings were ready could manufacture begin; and manufacture involved first making a clay model from which plaster moulds could be taken. Then clay blocks were cast from the moulds; but, since the moulds begin to break down after about fifty castings, a scheme such as the Natural History Museum required huge numbers of repeat moulds. From the moulds the clay blocks had to be taken to drying sheds, where they could dry slowly enough to prevent warping. Finally the kiln turned the clay into an almost impervious material whose surface was supposed to be immune to the sooty atmosphere.

It was a complicated procedure, and Waterhouse's demands must have meant that Gibbs and Canning had to increase their staff, expand their drying sheds, and even build more kilns. Even so it took time to produce the finished blocks, and one failure could lead to weeks of delay.[4] Difficulties with supply abounded and at the Natural History Museum Waterhouse once had to defend a stoppage of the work because one 'whole course of terra cotta extending round the entire building has just come out of the kilns in a defective state and has had to be thrown aside'.[5] However, for the architect and builder even that was not the end of the story. The completed blocks had to be packed in straw and transported to the site without breakage or chipping. Then, since each shape was individually designed for one particular spot on the building,[6] they had to be sorted and laid out where they would not be in the way, but could be readily reached when needed. This demanded a whole new approach to the business of building for the potential problems were legion. Baker, the contractor at the Natural History Museum, was evidently defeated by the complexity and scale of it all and eventually went bankrupt. Significantly too, Waterhouse never again used terracotta for the complete cladding of a major building. However, he continued to use it along with brick and glazed tiles in most of his larger urban buildings. And he continued to experiment until the last years of his practice. When he determined on the deep plum-coloured terracotta for the Refuge building in Manchester **[570]**, he admitted that he was 'aiming at something quite out of the usual in the Terra Cotta . . . '; and he was clearly pleased to report that the makers 'have given us a quality of Terra Cotta which is quite exceptional and which it is very unlikely will be repeated in any other building'.[7] He was so pleased that, in a unique departure from his usual practice, he used the terracotta in place of granite for the wall base.[8]

His reputation with terracotta is such that he is often supposed to have used little else. However, his drawing back suggests he preferred to work well within the limits of a new material, and was aware of the demands on the trade imposed by very large commissions. It was not a problem that

[3] Shrinkage scale drawings with Waterhouse office lettering survive among the drawings for the Prudential building in Holborn from the 1890s (DC).

[4] The whole process from finished drawing to the final product took about 8 weeks.

[5] MS letter, AW to Office of Works (12 Nov. 1873). PRO.

[6] Unlike stone a terracotta block could not be cut or trimmed to make it fit anywhere but the place for which it was designed.

[7] MS letter, AW to W. Proctor, Refuge Assurance Co. (24 Sept. 1896). Private collection. Waterhouse explained that the new quality had carried a serious risk of delay to the building, because it had to be burned 'with extraordinary heat'. This put the contractors, Messrs Doulton, to 'enormous expense' in the way of kilns, which evidently had to be specially built. It also involved a considerable expense for the client.

[8] MS letter, AW to W. Proctor (15 Dec. 1890). Private collection.

could be solved by finding a reliable supplier. The firms in the 1870s simply were not big enough, which is probably why he turned to so many different suppliers. Gibbs and Canning, who supplied the terracotta for the Natural History Museum and for Manchester Town Hall, were clearly his preferred suppliers. Yet he also employed Doulton and J. C. Edwards of Ruabon and even the Watcombe Terra Cotta Company for his large development at Cook's Court, Carey Street **[370, 470]** (Pl. 46). Nor was this simply a matter of colour, although he certainly admired the variegated buff that Gibbs and Canning produced. When the directors of the Prudential preferred red terracotta he ordered the hard bright red from Ruabon; yet it was Gibbs and Canning who supplied the bright red for the chief office at Holborn Bars **[431, 602]** (Pl. 125).

The success of the new material lay in the way in which it imported colour to grimy city centres and in the economy that could be achieved by repetition of standard shapes. Waterhouse clearly enjoyed the former, and the latter suited his approach to relief decoration. However, by the time the Natural History Museum was complete terracotta was no longer regarded as a novelty. Although it had been proved technically, there were still aesthetic arguments based on its use as a substitute for stone. Waterhouse was adamant that it was a totally different material requiring its own particular treatment. The slight irregularities of shape could be accommodated by using a generous joint, while slight variations in colour he found attractive.[9] He liked the ease with which the blocks could be jointed, and in fact expressed this by allowing the joggle joints in flat arches to show on the face as, for instance, at St Paul's School, Hammersmith **[492]** and Girton College **[346]**.

It seems that he saw terracotta as a truly modern material, making use of the mechanical processes of production that were so much a part of his age. He was also attracted to its supposed washable qualities and to its real resistance to the corrosive atmosphere of the towns. Smoke damage was a considerable problem for architects, and there was a poetic justice in the new material 'made from a clay found in the same pit as the coal which did the mischief'. 'It seems', he continued, 'the only building material which can successfully withstand its corroding influence. In terra cotta the fire will at once give us those beautiful accidental tints of which we might avail ourselves if we chose boldly to use them.'[10] Waterhouse so enjoyed the rich palette that he continued to use terracotta even in country houses well away from the smoke and grime. In its use for city centre buildings he had merely led the way, and its appropriateness led to its widespread adoption in the closing decades of the century. In the country houses, however, he was really indulging his own personal preference, and he had very few imitators. Clearly he liked the material for itself, and it is that that gives him the title 'Mr Terracotta'.

However, when he began to use terracotta, there was no craft tradition in handling it, so Waterhouse had to match his designing closely to its technical potential. From the Natural History Museum he learnt lessons which he applied elsewhere. The Prudential Assurance Company liked the bright orange red that was felt to be a relief from the dull tones of stone and London stock brick. The red clays, however, were more liable to warp in the kiln; but he overcame these difficulties by regulating the size of the block. He also modified the very regular course height and block size to relieve the uniformity of colour which was a feature of the red clays. The repetitive nature of the material suited his desire for complete control over the texture of his buildings; and he was adept at producing simple shapes that could be effectively repeated. Clay was easily adapted for a variety of

[9] AW in a paper read to the Manchester Society of Architects, *Builder* (31 Aug. 1878).

[10] AW in a lecture to Birmingham art students, reported in *Building News* (1882), 245.

styles: Romanesque at the Natural History Museum **[243]** (Pls. 78, 167–8) and for the King's Weigh-house Chapel **[556]**, Early English Gothic at Girton **[346]** and Easneye **[258]**, or fifteenth-century Gothic for the window heads at St Paul's School **[492]** and the Prudential **[602]** (Pl. 124). The bolder, less undercut mouldings of the Romanesque were obviously easier to manufacture; but a great deal could be done by assembling mouldings from simpler shapes and the use of detached shafts, as at the Prudential. Waterhouse also developed an effective language of decoration in sunken rosettes and bold diapers, while the whole thing could be further enriched by the use of alternating blocks with quite slight differentiation.

All the same, his concern with supply led him to restrict terracotta to window dressings, string courses, and ornamental bands and the like (Pl. 177). His façades are for the most part of brick and terracotta combined, which allowed him to build in a variety of texture, as was almost essential in large structures. He preferred a mixture of red brick and buff terracotta, or some such colour contrast, and often enlivened plain coloured façades by the use of plain glazed tiles. In short, while Waterhouse certainly deserves his reputation as a master in terracotta, he was never in thrall to it. Instead he capitalized on its decorative potential as an adjunct to more traditional materials.

Terracotta is the exterior material with which he is associated; but it is really only half of the story. Waterhouse was equally keen on the use of glazed faience for interiors. This again was a new material of his age. Like terracotta glazed brick and tile had been available from the early years of the century; but it was the mid-century that saw the real growth in manufacture. True faience came even later when in 1879 James Holroyd went as manager to the Burmantofts works. His products were launched with the aid of designs by leading architects such as Maurice B. Adams and with publicity from the *Building News*.[11] Waterhouse, on his visits to the site of the new Yorkshire College, could have seen the new works, and was certainly known to Holroyd. He was enthusiastic about the new material which offered the same prospect for his interiors as terracotta had for the exteriors. It was easily cleaned, and it offered colour, above all permanent colour.

Waterhouse had used glazed tiles for dados in Manchester Town Hall; but the non-reflective terracotta was on the drab side for interiors. As soon as the new material was available, he began to put it in his buildings. The National Liberal Club **[514]** (Pls. 134, 135, 137) was one of the first major commissions on which he used this new material, and he used it extensively. He even devised a complete faience ceiling for the huge, 100 feet by 35 feet, smoking-room. He clearly revelled in the control and potential permanence it gave his colour schemes. But once again his use of faience was based on a thorough understanding of the processes of manufacture and the potential of the material. He made great use of simple moulded pieces which could be glazed in one colour. In the kiln the glaze ran into the depressions of the moulding, where its extra thickness gave a darker tone, a 'beautiful accidental tint' that enhanced the effect. As with terracotta he seems to have enjoyed the random variations caused by the way the glaze ran. He also applied assiduously the lessons learnt from the Natural History Museum, achieving rich effects by repetition of basic shapes and creating elaborate patterns by reversing or rotating simple designs. In almost all his early faience schemes, the individual pieces are of one colour, though they could be combined. Later he used pieces that were coloured in two tones, as on the directors' staircase of the Prudential in Holborn **[602]** (Col. Pl. 5; Pl.

[11] The full story of Burmantofts faience is told by Mark Anders and Alan Garlick in *Burmantofts Pottery*, exhibition catalogue published by Bradford Art Galleries and Museums and Leeds City Museums (1983).

169). But throughout he stuck to low key colour schemes, preferring plain drab, pearly greys, ivory, and honey gold, with only occasional touches of pale blue or brighter colours (Col. Pl. 4 (*c*), (*d*)).[12]

For the more important jobs, such as the National Liberal Club **[514]**, the public offices of the Prudential (Pls. 128, 129, 131), or Fosters' Bank in Cambridge **[565]** (Pl. 158), he used moulded and glazed faience for all the wall surfaces.[13] For less imposing spaces glazed faience could be alternated with glazed brick, which again the Burmantofts Company supplied in a range of colours. Waterhouse generally liked to use more than one colour, mixing French grey and white, cream and white, or honey gold and brown. These he regularly used to emphasize dados or to enhance window and door-heads.

There seem to have been far fewer technical problems with faience than terracotta, perhaps because the individual pieces were usually smaller than terracotta blocks. Also the material was not structural, and by the 1880s there was a considerable expertise in the trade in laying tiles, that was easily adapted to fitting faience. It is not clear how many of the faience pieces and tiles that Waterhouse used were to his own designs. He was certainly a big enough customer to have his designs made up, and for some commissions, such as the Prudential, his drawings for faience decorations survive. There, and in most of his commercial and institutional commissions from 1880 on, he used it for walls, ceilings, column casings, and fireplaces, creating in the last two decades of the century a series of unforgettable interiors. Sadly they were too unalterable, and, with the change of taste, became increasingly unpopular. Many were destroyed or plastered over. But the survivors are evocative displays of a late Victorian taste (Col. Pl. 5; Pls. 152, 158, 169, 170, 171) that became typical largely through Waterhouse's preference for this material as the counterpart to his external terracotta.

[12] See RIBA Presidential address to students (1891), published in the *Builder* (24 Jan. 1891).

[13] Even the problems of cladding columns complete with entasis in faience tiles do not seem to have deterred him, though each row of tiles had to be a slightly different size.

22

The Business of Building: Other Materials and Servicing

ALTHOUGH Waterhouse is known as the great user of terracotta, it should not be forgotten that he began practising in the great age of iron, and his career lasted into the era of structural steel. Cast and wrought iron were already well understood by 1850, and Rickman, a family friend, was well known for his two iron churches in Liverpool. However, Waterhouse seems to have shown no particular interest in these, probably because they were not acceptable within Puginian or Ruskinian canons. Terracotta, at least, was not specifically excluded. Yet an architect of the mid-century could hardly avoid the use of the modern material, and Waterhouse, though cautious, was never afraid to use iron. Indeed his Presidential Address to the Manchester Society of Architects in 1878 was largely a plea for the use of new materials.[1]

One early attempt at an iron structure, in Darlington, may have helped to reinforce his caution. He had been asked to design the market and public offices **[99]**, and probably hoped the commission would develop into a full-grown town hall (Pls. 200, 201).[2] However, the bulk of the site was to be covered with a simple market building. For this he designed a fairly conventional light structure of cast-iron columns supporting an iron roof. Because of the sloping site it was set up over a deep basement, whose vaults were themselves supported on cast iron brackets. The building was unadventurous both stylistically and structurally; but it was the scene of Waterhouse's only building failure. On opening day the market was hugely overcrowded, and a portion of the floor gave way, pitching two prize bulls and one bystander into the basement below.[3]

In the inquiry that followed the fault was traced to faulty casting. Waterhouse was cleared of all blame, and went on to receive several more commissions from his friends in the town. However, he never created a complete iron structure again; and his office correspondence shows enormous caution in the use of iron at the New University Club **[208]**. He had ordered Belgian wrought-iron girders, then considered to be the best available, to support the façade. When they arrived on site he was alarmed at the wide spacing of the rivets; and wanted to have each one individually tested. He even

[1] Presidential address to the Manchester Society of Architects (1878), published in the *Builder* (31 Aug. 1878).

[2] His sketch design shows a building much larger than was ever built, presumably intended as a complete façade for the market.

[3] The bystander unfortunately was killed, but the fate of the bulls is not known. I am grateful to Paul Jarvis for details of this tale from his unpublished B.Arch. diss., University of Newcastle, 1983.

sent his assistants to try to get a covert opinion on their soundness from another supplier.[4] There was no indication that the girders were in any way suspect, though he remained cautious, inviting suppliers in later years to tell him the required depth of girders rather than specifying what should be required from his own calculations. At the National Provincial Bank in Piccadilly **[576]** the load of numerous flues and chimney breasts had to be carried internally on steel beams surrounding the light well over the banking hall. Waterhouse evidently asked the suppliers how big a girder was needed.[5]

For steelwork he followed his usual practice of finding a reliable contractor, in fact two reliable contractors, and always turning to them. A. Handyside & Co. and J. S. Bergheim each supplied iron or steel for twenty-eight major commissions. With their help he used iron to support stairways at Manchester Town Hall **[284]**, the Natural History Museum **[243]** (Pl. 83), and Eaton Hall **[315]**. He inserted a splendid five-storey staircase with iron supports into the Lime Street hotel, and even built part of the roof of Eaton Hall with an iron frame. The Natural History Museum **[243]**, with ironwork from J. S. Bergheim, was his first major commission to use the material consistently. There the internal terracotta columns enclose iron stanchions, and iron beams support the floors throughout. By the 1870s this was a very standard structure, such as had been put into many of the Manchester warehouses. It was probably the only way Waterhouse could have achieved his interior spaces with their terracotta facing; and, of course, the regular grid of an iron structure is well suited to the repetitive bays of his terracotta façade. The museum also incorporates his only large iron roof, where the slim iron ribs seem to be designed for structural considerations only (Pl. 168). The great index hall was not a major span, so far as iron roofs went, and the ribs appear curiously light on top of his solid Romanesque shafts. Perhaps he was thinking of a more massive structure.

None the less, having proved the material for himself, Waterhouse went on to use it consistently in his larger commercial buildings. The first office for the Prudential **[431]** was built with cast-iron columns supporting an iron girder floor. In the securities block he used cast-iron columns with shallow brick arches. However, from the early 1880s he adopted Lindsay's patent system of corrugated steel decking for the floors, and used steel stanchions behind his faience pillars.[6] In the National Liberal Club **[514]** he even went as far as to combine the system with hanging floors in the upper part in order to maintain his great spaces below. A few years later the directors of the Metropole Hotel at Brighton **[546]** were determined that the floors of their entertainment rooms should be absolutely clear. Undaunted Waterhouse spanned 41 feet with a massive inverted queen post truss in steel, building the rooms above within the depth of the truss and threading the corridors through its members.[7] Still later, at University College Hospital **[588]**, when faced with the problem of wide spans in parts of the roofs over his cruciform wards, he simply switched from standard timber roofs to iron for the wider sections.

If the nineteenth century was the great age of iron and the dawn of the age of steel, it has to be admitted that Waterhouse was not one of its ardent supporters. He was prepared to use the material, and to leave it honestly exposed where status did not require something better. But he was never an

[4] A good deal of correspondence survives in the office letter books in the RIBA Library. The contractors also sublet the fixing of the girders which caused additional problems; but Waterhouse's chief concern seems to have been to discover whether the girders were likely to be sound. There is no evidence that they were rejected (13 Dec. 1866). OLB. RIBA Library.

[5] Among his drawings for the National Provincial Bank, Piccadilly, there is one with a note by the suppliers, A. Handyside & Co. informing him that the depth of the main girder must be 2 ft 4⅝ in. (DC).

[6] Waterhouse used this system of flooring at the Prudential and at the National Liberal Club. In the club it is combined with steel stanchions. At the Prudential the steel decking was supported on load-bearing brick cross walls.

[7] *Building News* (10 May 1889).

enthusiast for iron structures.[8] Instead his interest in iron was that of the craftsman, and his wrought-iron screens, balustrades, and sconces are some of the best of his decorative work. They, in contrast to his structural iron, have all the vitality and understanding of smithery that ever Pugin or Ruskin could have demanded. From his earliest years he had noted and sketched decorative ironwork along with other details; and he evidently enjoyed designing for it. The hinges he designed for a number of buildings show this particularly clearly (Fig. 15). His most consistent pattern appears like a bundle of rods with tight iron ties and ends coiled back pair by pair. The range of iron finials for buildings such as Manchester Town Hall **[284]** or Eaton Hall **[315]** also reveal how well he understood what was possible. There is a whole range of wrought-iron gates from his drawing board (those for the Natural History Museum **[243]** and Iwerne Minster **[445]** (Pl. 188) are particularly satisfying); and the window grilles he designed for Caius College **[268]** or Reading Grammar School **[307]** (Pl. 177), with their foliate heads revel in the possibilities of the material.

For structural work his preference was for traditional materials. Most of his buildings are of brick, which Waterhouse understood as well as any, and probably enjoyed every bit as much as his terracotta.[9] Where terracotta had become a standardized machine product, there was considerably

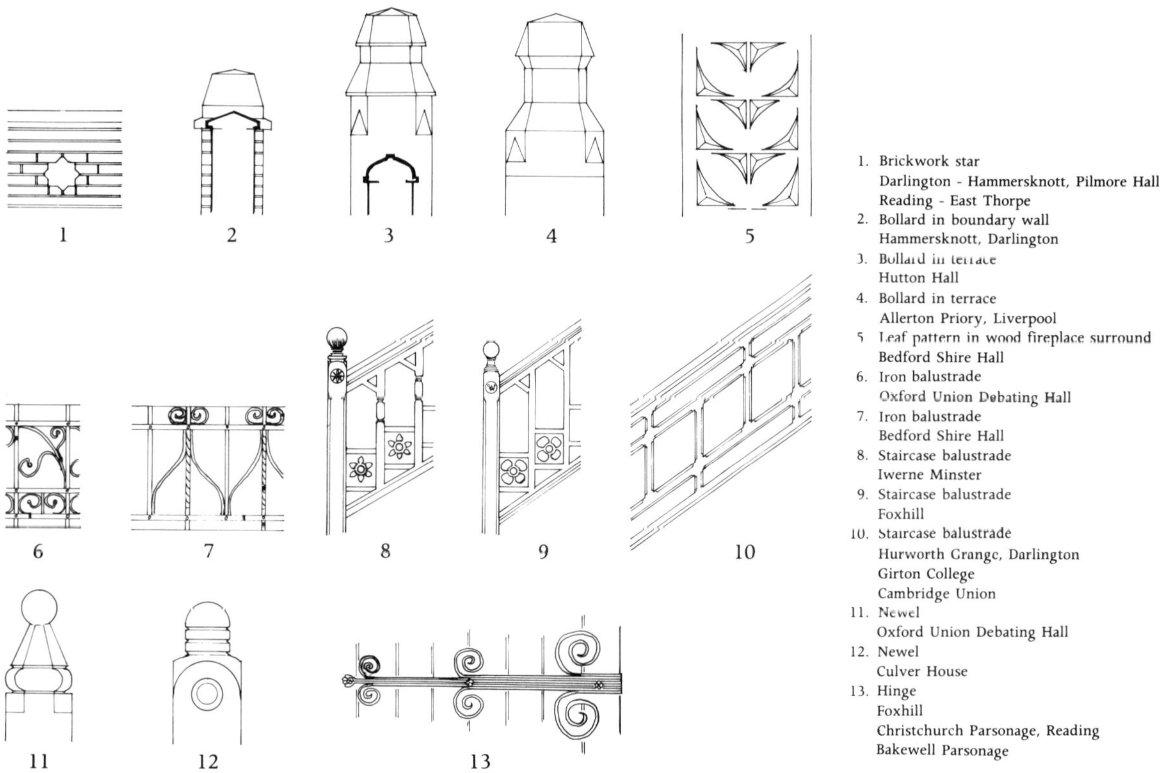

FIG. 15. Details of interior decoration, walls, and piers

[8] On small sites such as the corner of Piccadilly and Bond Street where lighting the interior was not a problem he preferred to build entirely in brick with piers and semicircular supporting arches (*Builder* (6 Nov. 1880)). It is interesting too that Waterhouse had seen and may possibly have restored John Carr's bold flying staircase at Arthington Hall **[426]** with its iron supports. Waterhouse refitted the staircase hall when he added a billiard-room wing. He cannot have been much impressed with the solidity of the staircases, which are alarmingly springy.

[9] Even Manchester Town Hall and the Natural History Museum are of brick, merely faced with stone or terracotta.

more variety in brick, since there was brick clay in many parts of the country. Also brick was usually the most economical material; yet Waterhouse was well able to achieve effects with very simple means.[10]

Brick was certainly the commonest material of his day, and, from the 1850s on, there was a huge increase in production. He used it for all types of buildings from humble chapels to the most sumptuous houses, combining it for decoration with stone, terracotta, or brick of a contrasting colour. Up until about 1870 much of his work, particularly in brick, was highly polychromatic. He used contrasting brick for horizontal banding, dressings of openings, and patterning of tympana and arches, for example at the Cambridge Union **[197]** (Pl. 176), and New Heys **[103]** and Mossley Park **[296]**, in Liverpool. Occasionally he opted for more subtle contrasts, selecting for the bank at Nantwich **[166]** a single brown-red shade from the local multi-colour bricks to match the sandstone which he used for dressings. In Reading, where brick is the common material, he used it extensively, for economy at the Caversham chapel **[225]** and decoratively on the town hall **[360]**, at St Bartholomew's **[450]**, and at Christchurch vicarage **[340]**. There, he achieved a masterly but simple elegance, choosing a grey brick for the main wall surface and a fresh orange-red brick for window openings and a deep band of chequer work at the sill level of the upper floor. The corbelled eaves, gable ends, and, unusually in his brickwork, quoining, were also in the red brick giving coherence to the whole.

There was more to his use of this traditional material than the mere choice of bricks. He was thoroughly well versed in the potential of inventive bricklaying. He had studied both old and modern examples carefully; and details of brickwork abound in his sketchbooks from Italy, Germany, and the Low Countries, as well as from the hands of Scott and others in this country.[11] This careful observation enabled him, for instance, to design the simple and solid buttresses of the Cambridge Union **[197]** with their attractive brick tumbling. That was a pattern of bricklaying that he was especially fond of, but he was also ready to experiment with other forms of bonding. At the bank at Nantwich **[166]**, for example, and in the house at Broxton **[127]**, the walls are built entirely in header bond; and tympana filled with herring-bone brickwork were another favourite motif. He achieved a range of effects with corbelled eaves courses, sometimes with stepped and corbelled gables, as at Crimplesham (Pl. 99). More often he used a deep bracketed cornice made up of three projecting courses on a model he had first seen on a modern building in Cologne in 1858 (pl. 195).[12] Corbelled brickwork seems to have delighted him, and was used at Crimplesham **[476]** and Hurworth **[381]** to transform octagonal ground-floor bays to square bays at first-floor level. At Reading Grammar School **[307]** he even managed bold arches to the undercroft with no less than thirteen courses of simply corbelled bricks. Another feature of his brickwork is the frequent use of coloured mortars. At Foxhill **[285]** (Pl. 117), Easneye **[258]**, Hurworth Grange **[381]**, and Girton College **[346]**, for example, black ash mortar sets off the bright bricks.[13] Finally he was able to create attractive ornament from inventive manipulation of simple bricks. One example that occurs frequently is an eight-pointed star made by manipulating four courses of a single brick type with one

[10] Waterhouse had a useful lesson at Strangeways, where, on a restricted budget, almost all the decorative work is achieved in brick. The extensive correspondence also suggests that there were problems in supplying the huge quantity of bricks and maintaining a consistent standard at the required price.

[11] Scott's Institute at Sandbach (Sketchbook 13 Aug. 1857–6, Oct. 1859, 11 Feb. 1859) seems to have been a particularly fruitful source (Pls. 9, 10).

[12] Sketchbook 13 Aug. 1857–6 Oct. 1859.

[13] The effect of this can be seen clearly by comparing the original work at Girton with the recent repointing in plain mortar.

corner chamfered. This was useful for pierced balustrades and garden walls, but could also be used for inset decoration on solid walls, as at Crimplesham **[476]** or East Thorpe in Reading **[482]** (Fig. 15).

Tiling, too, Waterhouse used with understanding, though he made less use of tile hanging than many of his contemporaries. He never used coloured tiles, although he had regularly used patterned slate roofs in his early years, and later at Manchester Town Hall **[284]**.[14] Instead he devised an economical means of achieving variety by laying bands of varying shapes, square, bullnosed, and fish scale, all in the same material. This he combined with a bold but simple pattern of eared ridge tiles that appears to be of his own designing. The roofs of Owens College **[321]** (Pl. 146), with their soft deep-red tiles, steep slopes, and dormers, were particularly good examples.[15] He used patterned tiling to enrich the roofs of Blackmoor **[192]**, Iwerne Minster **[445]**, and Girton **[346]**, for instance, adding complexity in the way that the alternating bands are of varying depth according to a simple numerical sequence. It is a demonstration both of his understanding of the qualities of the material, and of his ability to achieve effect with modest means.

Where the means allowed it was still common practice to build in stone, and Waterhouse acquired a good knowledge of the potential of the various stones available. His expertise is shown in a lecture on the stones of Westminster Abbey that he delivered as Visitor to the Royal Academy Schools.[16] For his own work he accepted the material gladly, often using local stone in the north of England, and readily adjusting his detailing and mouldings to the type he was using. However, with the amount of building being undertaken, stone was becoming increasingly expensive by mid-century. Stone was still Waterhouse's choice for most of his important buildings up to and including Manchester Town Hall. His Assize Courts **[67]** (Pl. 33) were a polychromatic success not least in the eyes of Ruskin; and in the sooty conditions of the time the contrasts of stone, Darley Dale and Forest of Dean, buff and greenish grey, with polished Dalbeattie granite for columns, were intended to retain some liveliness longer than a single colour. Even so the stonework was showing signs of trouble even before the building was complete.[17] He did use polychrome stone again for the Royal Insurance Building, Manchester **[108]** (Pl. 28), again choosing Forest of Dean stone for colour; but the main stone was the harder Huddersfield sandstone. By the time the selection had to be made for Manchester Town Hall **[284]** (Pls. 47, 60) a careful search suggested that Spinkwell stone from Bradford would be the most resistant to pollution, and he stuck to the single colour. This was not a sudden decision. He had used a single stone for the banks in Darlington **[175]** (Pl. 20) and Leighton Buzzard **[204]**, and for the massive Lime Street Hotel **[274]** (Pls. 203, 204).[18]

Stone was the conventional material of grandeur, and Waterhouse was conventional enough to prefer it whenever the budget allowed. His churches are mostly of stone, or they have stone dressings; though many of his chapels are of brick. His noblemen's houses, Eaton **[315]** (Pl. 87), Iwerne **[445]** (Pl. 104), and Blackmoor **[192]** (Pl. 106), are all of stone. So too are a number of smaller houses, such as Bakewell Parsonage in Derbyshire **[313]** or Adel Grange in Leeds **[177]**, where stone was more easily available than brick. The nature of each stone affects his style. His Lakeland houses,

[14] Soot has so dulled the pattern that it is now only visible on certain parts of the roof and in certain lights.

[15] Unfortunately, due to nail fatigue, they have recently been reroofed, and the replacement tiles are of an unsuitable hard harsh red.

[16] Address to Royal Academy students (1885), reported in the *Builder* (7 Mar. 1885).

[17] *Building News* (1862), 324. Pollution was having its effect and the authorities were to be asked if steps could be taken to abate the nuisance.

[18] His submission for the Law Courts also suggested Portland stone as the sole facing material.

for instance, conform to the local vernacular in their use of the rough-split slatey stone and the large roofing slates. Even the window surrounds at Fawe Park **[45]** are as rough as the rest: four narrow stones surround each opening, and only the sill is splayed to throw off the water. Yorkshire gritstone at Adel Grange **[177]** is handled with the dressings reduced to the simplest of sections; and the hard stone of Bakewell Parsonage **[313]** is relieved only by slender surrounds and strings in grainy sandstone of a similar tone, emphasizing the simple massing of the house. Polished granite was a favourite material for the plinths of commercial buildings, where it made an easily washable base. He also used it with alabaster and coloured marbles for the concourses of great buildings such as the Assize Courts **[67]** and the town hall at Manchester **[284]**; though he later claimed to prefer the less highly polished finish produced by emery which brought out the colour without a highly reflective surface.[19]

His choice of stone was related to a philosophy of colour, which he set out in his last Presidential Address to students. Portland stone he considered the only material that was improved by smoke; though harmonies in black and white were hardly great achievements in colour. He preferred other well-assorted unions; but soot was the great enemy, and a further difficulty lay in finding materials that would attract dirt in like degree. Thus glazed tiles in conjunction with stone he found quite inadmissible. He called for restraint in selecting mixed materials and urged students to avoid violent contrasts in colour and texture. He also claimed that his taste 'rebels against marble statues against a stone front'. His more basic argument for restraint was that form should not be sacrificed to colour except deliberately. 'It must be done from choice and not by accident: and when changes of colour are introduced into forms which are studied, and intended to produce a definite effect, the disturbing effect of the colour should always be taken into consideration'.[20]

None of this, however, marks him out as anything other than conventional. His only distinctive contribution was in the detailing, so that window sills and copings shed water efficiently, and in his system of laying stone facing in alternate deep and shallow courses, with the shallow course cut to brick depth and bonded in at every fourth course (Pl. 65). It is typical of Waterhouse that the watchword was efficiency within conventional constraints.

Much the same might be said of his timber work, which is characterized by generous sizing and great solidity. The construction of the roof of the Cooper Street tower at Manchester Town Hall **[284]** (Pl. 172) is enormously impressive and yet was never intended to be seen. Clearly the driving force was structural efficiency and strength; and several of his tower roofs, such as that at the Prudential in Holborn **[602]**, carry substantial water tanks. However, he saw no reason to stint himself, and seems to have preferred weightier sections whether they were strictly necessary or not. This can be seen in the striking timber umbrellas that cover the circular staircases in Manchester Town Hall, or in his open church roofs such as that at St Elisabeth, Reddish **[507]**. Even small features, such as his northern French style dormers and his porches, for all their charm, have a solidity given by the weight of the timbers, which are then gently articulated by chamfering.

Of course this approach to materials is closely paralleled by his concern for efficient planning, and is matched by the attention he gave to servicing. Heating and lighting in the vast new buildings of his age did cause problems, and new systems had to be invented. Beginning to practise in the second half

[19] AW, Presidential address to students (1891), published in the *Builder* (24 Jan. 1891).
[20] Ibid.

of the century, Waterhouse was fortunate that many of the early experiments had already been made for structures such as the Houses of Parliament; and he was able to benefit from the second generation of inventions. He was able to select and apply the most up-to-date systems, and, once a system was proved, could safely apply it elsewhere. The installation of Boyle's self-acting air-pump ventilators at Caius College **[268]**, for instance, was a novelty which was repeated in other Cambridge colleges.[21]

Waterhouse's written submission for the Law Courts competition **[278]** shows his thorough approach to services. He was particular about light and air. The width of the internal streets, at 50 feet or more, was to allow ample natural ventilation; and no room was to be without a window and some 30 square feet of glazing. Where corridors could not be lit by windows, borrowed lights were to be double glazed with obscure glass to protect the privacy of the rooms from sight and noise. He discussed the space about the building and how it could be improved to provide more light and air. The acoustics of the courts came under scrutiny. The floor area of each was compact and the ceilings as low as was 'consistent with the dignity of such an apartment'. The edges of the ceilings were to be coved, the wall surfaces broken up, and angles filled. The top light was to be flush with the ceiling rather than shaped as a lantern and the oak fittings would add resonance. Every last consideration was given to heating and ventilating, sanitation and drainage, movement of papers and coals, kitchens and refreshment rooms; and all the arguments were set out with reasons, especially where he departed from the conditions of the competition. Waterhouse described how he had relied on the researches of others,[22] and drew comparisons with the heating and ventilating of the Houses of Parliament, on which he hoped to improve. He intended to leave no doubt that the building would be convenient and efficient.

It was still common practice to provide open fires in most rooms, and Waterhouse's buildings did occasionally suffer from smoky chimneys. He usually turned the requirement for numerous substantial flues to good effect by making chimneys a feature of his picturesque skylines. However, to warm the really large buildings, something more than open fires (Pl. 58) was needed, and again he was concerned to combine the mundane need with architectural effect. His success here is best seen in the Manchester Town Hall **[284]**, where the central wells of the spiral staircases also act as rising shafts of warm air to help in the circulation. In the great hall, too, the dramatic ceiling is constructed with canted arches in its lower tier that conceal vents to draw off the vitiated air (Pl. 54). A few years earlier at Strangeways gaol **[113]** he had installed a magnificent plenum heating system that to this day provides three complete air changes every hour for both the men's and the women's prisons. Those systems were devised with the help of G. N. Haden of Trowbridge,[23] whose method Waterhouse had already 'selected out of many schemes' at the Assize Courts **[67]**.[24] Haden's rapidly became his favourites, and were involved in no less than ninety-two schemes for him.

His later buildings involved more modern technology, plenum ventilation, lifts, and electric light. He was not among the first to use electric light, and did not turn to it until about a decade after the

[21] *Builder* (6 Nov. 1880).

[22] Waterhouse's research was thorough and in a footnote he cites the case of Osmaston Manor, Derbyshire, as a satisfactory example of a similar installation: 'The roof line of the house is of that description of irregularity which generally ensures the most annoying complaints of smoky chimneys; but the most satisfactory draught is in each case secured by the system adopted.'

[23] G. N. Haden was apprenticed to James Watt, and subsequently invented the first true system of central heating and the first circulating pump. His first major system was installed in the Bank of England. His plenum system at Strangeways has been adopted for the recently refurbished prison almost without modification.

[24] AW paper to the RIBA on the Assize Courts, published in *Building News* (23 June 1865).

inventor Thomas Armstrong had installed it in his house at Cragside. However, by the early 1880s he was confident enough to light the National Liberal Club **[514]** with 850 Edison-Swan incandescent lamps. There were even automatic switches in the WCs and dressing-rooms. The *Builder* reckoned this a unique instance of a large installation in which every single lamp was controlled by a single fusible cut-out.[25] Waterhouse's brother Theodore may have had something to do with this, for he had formed a company to promote the carbon-filament lamp which was no doubt assisted by this project.

The National Liberal Club, in fact, boasted a long list of sophisticated equipment to ensure its luxury, and was probably as up to date as any building in the country. However, by the 1880s the primacy in inventions was already passing to America, and there is no evidence that Waterhouse ever looked across the Atlantic for new ideas. As with materials, his approach was to provide the best that was available having satisfied himself that each item was thoroughly proved. Being up to date was a fine thing, but being sure of what he offered was even better.

[25] *Builder* (21 May 1887).

23

The Complete Designer: Furniture and Interior Fittings

LIKE many architects of his day, Waterhouse was prepared to carry out the detailed design of all parts of a building right through to furniture and fittings. Unlike a number of his contemporaries, however, he does not seem to have designed furniture for any but his own buildings, so that his pieces have a consistency in their original location that does not always survive transplanting.[1] They are none the less an important element in his *œuvre*, and demonstrate a surprising range. The first comprehensive scheme was the Assize Courts **[67]** (Pl. 174) where he even designed the tableware for the Judges' Lodging. In the 1870s Manchester Town Hall **[284]** (Pls. 49, 180) provided the same opportunities on an even grander scale; while in the 1880s the church of St Elisabeth, Redddish **[507]**, allowed him to develop fine metal and fabric design in the altar plate and frontals (Pl. 140).

Waterhouse's earliest recorded design, a desk for his father **[5]**, is not particularly original and shows all the heaviness of the 1850s. However, he soon developed a more sophisticated system of designing, inspired at first by Pugin and soon seasoned with his own appreciation of the techniques of joinery. One example is the chair design used at the Assize Courts and later for the Cambridge Union **[197]**. All the sections are rectangular with some chamfering and the back panel, made from a single piece of wood, is punched through with four quatrefoils set in a circle.[2] A set of chairs and benches for the Prudential **[431]** (Pl. 178) develop the same approach of solid evident construction. As in his architecture, Waterhouse does not allow decoration and additive ornament in any way to distract from the function. In his later furniture he delighted in simple parallel grooves and lathe turning for uprights and for the spindles that he loved to use for the arms and backs of his chairs.

The early pieces show some variations on standard construction. One table for the Assize Courts, which owes a good deal to Pugin, has the stretcher turned down to form the feet. A similar treatment is shown in a sketch for a chair with hounds head ends to its arms and kick back legs.[3] Much later he designed a similar chair for the Prudential head offices with arms and legs formed in a continuous curve (Pl. 179). However, these idiosyncracies were generally abandoned in favour of more

[1] See Mark Girouard, 'Blackmoor House, Hampshire', *Country Life* (29 Aug. and 5 Sept. 1974).

[2] Examples of this Assize Courts chair survive in Manchester Town Hall, and are included in the catalogue of furniture at the Town Hall prepared for Manchester City Art Gallery by Sally MacDonald.

[3] Sketches for Assize Courts furniture. Private collection. The stretcher construction can be compared in the V. & A. with a chair by the architect William White. This chair of Waterhouse's shows, in its legs and stretchers, a move towards some turning between the square sections of the joints.

traditional styles. Painted furniture, of the sort produced by William Burges, had virtually no place.[4] Instead Waterhouse confined himself to plain timber and straightforward shapes. Much of his work is similar to that designed by Shaw, though with some of the solidity of Godwin's furniture. He may well have borrowed from Shaw, who was his friend; but he probably learnt the use of spindles from J. Bruce Talbert, who was working for the furniture makers Doveston, Bird, and Hull of Manchester not long before they were making pieces for the town hall. The resulting style has been rightly described as Old English—it has elements of Gothic, and even older styles, but its principal feature is its solidity.[5]

Waterhouse's recipe of basic forms allowed him to produce whole ranges of furniture that could be varied to suit different commissions or to reflect the hierarchy of different spaces. In addition to that for the grander rooms, he designed much of the ordinary furniture for Manchester Town Hall **[284]** (Pl. 181), clerks' desks for the Prudential head offices **[431, 602]** (Pls. 130, 131, 132), classroom desks for Reading Grammar School **[307]**, and fittings for his own assistants at New Cavendish Street **[180]**. Even some of the chairs for the National Liberal Club **[514]** (Pls. 134, 135, 137) and the Prudential offices are relatively simple and the chairs for the high table at Jesus College, Cambridge **[347]** (Pl. 175) are sturdy rather than grand.

Waterhouse acknowledged some difficulty in presenting the design of furniture, chairs in particular, on paper. They are 'peculiar things to design and cannot be judged in drawings so well as by their actual execution in wood';[6] but his grasp of design, construction, and the constraints of manufacture gave all his pieces a straightforward sturdiness. His preference for restraint in ornament was typical, as it was of his approach as an architect. He criticized some pieces in the Arts and Crafts Exhibition of 1890 which he considered overloaded with ornament. 'That shows . . . the advantage of the architect's direction in such work. He may not be able to design an arabesque panel better, or perhaps as well as an artist who devotes his life to such work: but he may know, and generally does know where to stop.'[7] His own designs, even some of the grandest such as the corner cupboard and the writing desk for Blackmoor House **[192]** (Pl. 186) show a strongly architectural character. The corner cupboard is almost a complete Gothic oriel in miniature topped by a crocketed gable with a peacock on its pinnacle.[8] The sideboard for the same room, in various woods with metal mounts, is the richest of all his cabinet work though the touch is equally light.[9]

Perhaps the most architectural of all his pieces is the buffet in the Banqueting Hall of Manchester Town Hall **[284]** (Pl. 180) which not only serves to house a radiator, but forms a part of the architectural expression of the room, sitting on the axis of the main façade between the two oriels of the Princess Street corner. Its upper half has his beloved turned shafts to support the shelves, and, in keeping with up-to-date taste of the 1870s, his drawing shows it sporting a large blue and white plate. In execution the robust floral motifs of the cove were replaced by aesthetic movement sunflowers and the busy bees of Manchester's coat of arms. Such a piece is typical of his concern to see that fitted

[4] On a few occasions Waterhouse did use restrained painting, for instance on the overmantels in the banqueting hall of Manchester Town Hall and in the coved panels of the buffet for the same room. He also accepted a large painted cradle by R. N. Shaw as a gift. (The cradle is now in the V. & A.)

[5] See Jill Lever, *Architects' Designs for Furniture* (London, 1982), 83.

[6] MS letter, AW to the Bursar of Jesus College. Jesus College Archives.

[7] AW, Presidential address to RIBA students (1891), published in the *Builder* (24 Jan. 1891).

[8] This piece is now in the V. & A. Museum.

[9] No drawings for the Blackmoor furniture survive, though the consistency of the pieces and their relation to other designs by Waterhouse make the attribution quite secure. The brackets of the writing desk, for instance, have the same leaf design as the committee room fireplace at Bedford Shire Hall.

furniture was designed both in conformity with the other furniture of the room and as an integral part of the overall scheme. He was most often involved with fitted furniture. The commercial offices, for instance, required quantities of counters, cupboards, desks, and so on. These were all designed by him, or his staff, complete with their fitted brass screens and lights. They were substantial pieces, though few, if any, survive; but early photographs reveal the same design theory and hierarchy of decoration.

For his private clients, however, Waterhouse was prepared to go even further. His involvement with them often went well beyond designing the house itself. In December 1865 he was advising a Mr Ashworth of Rawtenstall to visit Lamb's furniture warehouse to see a chimney-piece 'which would give him a valuable suggestion' and on another occasion telling him that his friend Tipping of Brasted had seen a Louis XIII chimney-piece at a dealers in Versailles (a bargain at £12) which could look good in Ashworth's low upstairs room **[167]**.[10] At Allerton Priory **[237]** he helped with the selection of furniture and with the purchase of pictures for the gallery. For the millionaire Julian Goldsmid he had chosen furniture and carpets and hung pictures in his house in Grosvenor Square **[299]**. At Hutton Hall **[176]**, besides designing some furniture and curtain fringes, he had helped with the selection of carpets and tablecloths, and had even set out to find an artist to work up a picture from photographs to record J. W. Pease's visit to Jerusalem.[11] That was probably exceptional, and Pease was in any case a relation by marriage; but it shows how far he was prepared to be involved.

Furniture, whether fitted or not was, of course, only part of the whole interior which Waterhouse was always keen to supervise (Pl. 182). Apart from the richly patterned vaults in the passages and the decorated ceilings of Manchester Town Hall **[284]** (Pls. 49, 52, 54), little survives of his schemes for painting.[12] He had employed Crace at the Assize Courts **[67]**, and expressed his admiration of his approach to colour in architectural decoration. So presumably Waterhouse was as happy as any of his contemporaries to design major decorative schemes or to work with designers on them. The lavish decoration of the Great Drawing-Room at Eaton **[315]** (Pl. 89) was apparently done under his direction, and was typical of the state rooms, though it is not clear how much of the decoration was actually designed by him. The office correspondence reveals that he took great pains with painted schemes such as the private museum for Edmund Christy at Bournemouth **[193]**. That, however, may be an exception among the smaller houses; though the contractor, Robert Pollitt, worked for him at seventeen other sites including Manchester Town Hall. Certainly stencilled friezes, and probably also stencilled ceilings, such as those in The Heath at Leighton Buzzard **[178]**, will have been quite common, and there is no reason to suppose that Waterhouse was reluctant to specify them. For the most part, however, it is likely that he allowed himself to be content with plain colour washes to interior walls or with the customer's choice of wallpaper. At Girton **[346]**, for instance, the rooms were decorated with wallpaper in the Japanese style.

Paint and paper were ephemeral materials, and Waterhouse preferred to concentrate on things that had rather more permanence. Floor tiling, for instance, occupied a good deal of his interest. He indulged in an extensive correspondence with Minton's and Maw's about the precise tints and patterns available when he was designing a conservatory for Edmund Ashworth at Egerton Hall

[10] MS letter, AW to Ashworth (23 Dec. 1864). OLB 210. RIBA Library.

[11] MS letter, AW to William Telbin (2 Feb. 1867). OLB 759–60. RIBA Library.

[12] A fragment of brightly coloured diapering preserved in a cupboard at Pierremont, Darlington, appears to date from the time of his alterations, and is a precious survivor. There are also painted ceilings at Hutton, which were probably done under his direction.

[11].[13] He generally preferred a strong diagonal pattern in his geometric floors (Col. Pl. 4 (*e*)), but in the main he was merely following the conventional styles of the time, using the buff, red, brown, and black that were readily available.[14] He was, of course, in the hands of the manufacturers, though he did provide at least one, Craven Dunnill, with designs of his own.

Along with tiled floors went fireplaces as a regular part of the architect's job, and Waterhouse designed a great many, developing an unmistakable style. He seems to have preferred them in timber, though Manchester Town Hall **[284]** (Pls. 57, 58), Eaton Hall **[315]**, and several of his early houses, such as Hutton **[176]** and Easneye **[258]**, have examples in stone. Later he produced a whole range of designs for faience fireplaces as well. His stone fireplaces generally have steep Gothic hoods instead of a mantelshelf, and were decorated with tiles and carved shafts. Only the richest had carved decoration on the hood. The earliest, in houses such as Rose Trees at Keswick **[28]**, are standard marble pieces presumably selected from a catalogue, but with added decoration in the form of carved initials or inset roundels of coloured marble. In the 1850s and 1860s he installed a number of equally conventional fireplaces, but showed a preference for shafts of dark marble. These were mostly supplied by the Lizard Serpentine Co. and there are good exaamples at Hutton Hall **[176]** and at Allerton Priory **[237]**. But the dark stone was a material he soon abandoned in favour of the paler buff Hopton Wood stone that took a pleasant polish but also retained the variety of its fossil contents. Hopton Wood stone met his preference for drab tones and he used it extensively in the 1870s and 1880s, and even clad one whole staircase in the material at the Prudential in the late 1890s **[602]**. For fireplaces he would use it as happily on its own or as a fire surround in conjunction with timber. But for fire surrounds he continued to work with a variety of other materials such as the fine coloured marbles of the directors' and managers' fireplaces in the Prudential.

His timber fireplaces and overmantels, however, are perhaps his most typical items, for he devised a simple system that could be enlarged or elaborated with ease to suit a variety of rooms (Pl. 184). The mantelshelf is supported on brackets that curve straight down into the side pieces, and the shelf can be of one or more levels, with turned spindles to support the upper shelf. The basic design could even be translated into other materials such as faience, but it was conceived for the timber fireplaces that grace most of his houses from the 1870s on.

As the key decorative element of most rooms, his fireplaces were also designed to reflect the hierarchy of each house or office. In the fireplaces of the Refuge building **[570]**, for instance, there is careful grading from the polished stone and timber of the Boardroom (Col. Pl. 6) and the faience of the main public office down through the managers' fireplaces to those of the humblest clerks'. The grandest fireplaces were adorned with sculpture, at the Prudential **[602]** (Pl. 183), or vases by William de Morgan, at the Refuge **[570]**, the latter presumably selected by Waterhouse. For private clients he would provide a range of overmantels which might be in the form of mirrors that would slide aside to reveal a window or conservatory, as at Blackmoor **[192]** or Crookhey **[401]**. Among his richest are the pair in the library at Eaton Hall **[315]** (Pl. 88) where the bracket supports are abandoned in favour of a miniature vault whose arcade was inlaid with floral patterns in mother of pearl and boxwood. Above these, sculptured panels supported a high mantelshelf into which was set,

[13] MS letters, OLB I: AW to Messrs Minton (20 Mar. 1857), 112; AW to Messrs Maw (23 Mar. 1857), 115; AW to E. Ashworth (25 Mar. 1857), 116; AW to Messrs Minton (13 Oct. 1857), 187; (22 Oct. 1857), 192; (2 Nov. 1857), 195–6. Private collection.

[14] Crimplesham Hall is a late work (1880) that has an unusually rich encaustic tile floor in red and buff. Presumably this was the particular choice of the client.

in one case a clock, and in the other an anemometer. Finally a further cove with more inlaid panels brought the whole piece up to the ceiling.

Fireplaces needed tiled surrounds and hearths, and these too he chose with care. He used a wide range of coloured tiles, and was evidently particular about the precise colours and patterns,[15] though he was restricted to the manufacturers' catalogues. Gothic Manchester Town Hall **[284]** contains fireplaces with tiles of a medieval pattern, blue and white tiles, tiles with Classical figures, and tiles of Turkish pattern. There is even one in the Mayor's apartments with Japanese *paterae*. Clearly fashion was an important factor, and he was as anxious to be up to date here as in his servicing. At Hutton Hall **[176]**, for instance, he retiled a fireplace and panelled a room in blue and white tiles in 1870, a mere two years after the opening of Poynter's grill room in the South Kensington Museums. Not surprisingly Waterhouse, or his clients, showed a marked preference in the 1870s and early 1880s for the paler tints and picture tiles of the Aesthetic Movement. These he usually set in a frieze beneath the mantelshelf. Most were supplied by W. B. Simpson and some, at Blackmoor **[192]** (Pl. 189), were made to the designs of, possibly even painted by, Lord Selborne's daughters. Elsewhere there was an attractive compendium of games in blue and white medieval dress for the billiard room at Hutton **[176]**, or extracts from Aesop's fables for the drawing-room in T. R. Hetherington's house in Manchester **[358]**.

Waterhouse was more able to be himself in other woodwork, doors, panelling, and balustrades, all of which he would regularly design. Some of his finest woodwork was for Eaton Hall **[315]** (Pl. 91), where he produced doors and panelling enriched with arabesque ribs, like the north door of Southwell Minster. At Heythrop **[350]** and at the Natural History Museum **[243]**,[16] more conventional panelling was enriched by black painted ribs set into the frames. As usual, however, he preferred his decoration to derive from simple cutting that reflected the construction. Many of his early front doors show this well in their inner rectangular frame that is decorated with segmental chamfering to make a pattern of squares and circles. He also enjoyed the richness of sunken roundels with geometric divisions (Pl. 187) that could be set into the top rail of panelling, as at Manchester Town Hall **[284]**, or used to decorate newel posts, as in his own house at Foxhill **[285]**, or even the strings of staircases.

The design of balustrades seems to have given him endless delight, whether in timber or iron (Pls. 114, 120, Fig. 15). A simple arrangement of widely spaced uprights could be linked by pieces set in opposition to the handrail to create an angular arcade, as at Iwerne **[445]** or Foxhill **[285]**. More complicated patterns could be devised to create a series of interlocking hexagons, as at the Hetherington house **[358]**; but both schemes were based on assemblages of simple chamfered pieces. The same simplicity ruled with iron balustrades where simple uprights could be almost endlessly enriched. Waterhouse particularly liked to create a running band of ornament which he would support between paired metal strips running parallel to the handrail and bolted to the uprights.

He could turn his hand to iron light fittings, designing some particularly splendid hanging sconces for Jesus College, Cambridge **[347]** (Pl. 173) and huge gasoliers for Manchester Town Hall **[284]**. Grilles and screens, such as those in his staircase at Balliol **[254]**, were a wrought iron artist's playground. He worked out a simple but rich system of designing iron staircase balustrades, and

[15] For one fireplace at Blackmoor he used tiles from 3 suppliers, presumably in order to get precisely the colour combination he wanted.

[16] Not all the woodwork of the Natural History Museum was supplied by Waterhouse. A good deal was inserted in later years by the Office of Works.

exterior ironwork, crestings, window grilles, and gates could be equally rich (Pls. 177, 188).[17] And always his designs expressed fitness for purpose as well as a clear understanding of the potential of the material.

The same was true of his ceilings. In his earlier years he fitted these with ceiling roses designed by J. W. Hindshaw, invariably selecting those with bold geometric designs. Later he became adept at patterning whole ceilings with simple ribs laid in a variety of interlocking geometrical shapes (Pls. 130, 131).[18] The patterns are simple but rich in their effect, and he generally preferred to avoid painted ceilings. Manchester Town Hall **[284]** and Eaton Hall **[315]** were obviously special cases, but the ceiling painting at Hutton Hall **[176]** seems to be a rare exception among his private works.[19]

The richest colours he generally reserved for the floors of halls and other public spaces. From his earliest travels he had admired the marble pavements of Italy, considering *Opus Alexandrinum* with its complex and severely geometric patterns the 'most satisfactory the world has ever seen'.[20] He was fortunate to have the opportunity to emulate them in the vestibule of Manchester Town Hall **[284]** (see also Col. Pl. 4 (*a*)) and in the sumptuous entrance hall at Eaton **[315]** (Pl. 191). He also managed to incorporate the lessons he had learnt in the borders of many public-office floors, most notably on the directors' staircase of the Prudential in Holborn **[602]** (Pl. 190). There the colour effects are less rich than at Eaton or Manchester Town Hall, but perhaps conform more to his general philosophy of achieving richness without extravagance.

The same approach governed his choice of stained glass. While most of his comments, in public at any rate, on work by his contemporaries were courteous, on the subject of stained glass he was at his most forthright.

> Of all the stained glass with which the churches of this country have been flooded within the last half century, there is not one bit in a hundred that could not in my opinion, be very easily spared. In the majority the drawing is bad, the sentiment is mawkish, and the colour, which is the real excuse for shutting out the cheerful light of day, is the worst thing about these stained glass windows; and yet they are there for all time, unless conflagrations, or another revolution conducted like our first by iconoclasts, intervene to rid us of them.[21]

He turned iconoclast himself at Yattendon **[494]** when carrying out alterations at the church and noted in his journal: 'The east window (or at least that part of it not hidden by the Reredos) I have doctored by stippling some umber over its most obtrusive parts. It was a very vile window by Hughes.'[22]

He preferred a very limited range of tones, usually grey and pale pink, and often used simple geometrical patterns of glazing enlivened only by rows of bullions (Col. Pl. 4 (*b*)). It is rare to find heraldic or narrative designs in his buildings; though Blackmoor **[192]** has a fine heraldic window and Eaton **[315]** had its four Arthurian scenes. Considering the attention that was given to themes in the decoration of the Great Hall of Manchester Town Hall, one might expect that its windows would have been particularly rich, but they contain only his pale grey and pink glass, since, as he said,

[17] The square iron rods of the frames often terminate in beasts' heads in Florentine manner. Both Gonville and Caius College and Reading Grammar School have window screens finished with wrought-iron flowers whose inspiration is Norman French.

[18] If the plaster ceilings and the panelling at Iwerne Minster are original it suggests that Waterhouse, as occasionally elsewhere, was not always successful in getting his own way.

[19] There is another painted ceiling in the extension he built to Arthington Hall; but its Adamesque patterns are so unlike him as to be almost certainly a later addition.

[20] AW, Presidential address to RIBA students (1891), published in the *Builder* (24 Jan. 1891).

[21] Ibid.

[22] AW Journal. There is a duplicate in St John, Hulme Street, Manchester.

'windows are made to admit light'.[23] Where he needed figured glass he relied on professional designers. Heaton Butler and Bayne designed for him a narrative window to put in a house for Joseph H. Fox at Wellington in Somerset **[186]**. They also did the mosaic panels for his alterations to Caius College Chapel **[268]**. Frederic Shields designed the glass for the apse of St Anne's, Manchester **[545]** and his small panels were brought from Houldsworth's chapel at Coodham **[369]** for the clerestory of St Elisabeth's, Reddish **[507]**. But Shields was a close friend who had also designed a series of panels for the chapel at Eaton Hall **[315]**. At St Anne's, artist and architect co-operated. The glass, in cool colours, was carefully designed 'in a style appropriate to the (Georgian) church'.[24]

Waterhouse took interior design just as seriously as any of the other aspects of architecture. It was obviously important to him to see that his buildings should look precisely as he wished when they were finally complete. Of course he recognized that interior decoration and fittings were liable to change; but his dismay at the effects of incautious refurbishment has been shared by many architects. It was one of the chief reasons why, in the later years, he liked to use faience for his interiors.

> It enables the architect to insure that his more important apartments shall remain as he designed them. Most of us have been occasionally disconcerted in discovering that some interior which depended greatly on its harmony of colour, and which we may have thought more or less a success when it left our hands, had been handed over when in need of repainting to the tender mercies of some decorator, who failing to appreciate the delicate scheme of colour upon which we had prided ourselves . . . had sown discord and vulgarity broadcast over our creation.[25]

As the creator of furniture and fittings Waterhouse was committed to solidity, simplicity, and constructional logic on the same lines as his architecture. That he was ready to be involved in such a wide range of details, and that he understood the materials so well, is typical of his thoroughness. The few Waterhouse interiors that survive more or less intact demonstrate the coherence he was able to give to a whole design when the opportunity allowed.

[23] AW, Presidential address to RIBA students (1891), published in the *Builder* (24 Jan. 1891).

[24] St Anne's Church, Manchester, 'Description of the restoration by A. Waterhouse.', May Day Bazaar Catalogue (1889).

[25] RIBA Presidential address (1890), published in the *Builder* (8 Nov. 1890).

24

Painting and the Picturesque

BEHIND all Waterhouse's understanding of materials, his efficient planning, and his constructional logic lies a strong sense of the picturesque. There is no doubt that he was very much aware of the visual massing of his buildings and much of his designing took strict account of perspective views. It is important to remember that he considered himself both artist and architect; and the two sides of the man are complementary. In 1877, at the peak of his architectural career he was elected an Associate of the Royal Academy, submitting as his Diploma work a large perspective of Manchester Town Hall (Col. Pl. 1).

The sheer quantity of his drawings would merit study, but their quality means they deserve special consideration. Disgruntled rivals were apt to complain that the quality of his perspectives was liable to sway competition juries. Attractive they may be, but they were intended to do more than influence clients. Their importance lies in the connection between his love of the picturesque and his concern as an architect to produce buildings that would look well on their site. If he thought of himself as a painter, it was only as an amateur. His works were not for sale, but were given away to friends; and his Royal Academy pictures were worked at on regular occasions by way of relaxation.

None the less, having originally wanted to be a professional artist, it must have been gratifying to have his landscape of Warkworth described as 'beyond question one of the most brilliant and facile watercolour drawings in an exhibition which is exceptionally strong in watercolours, even for the Academy'.[1] But that was in 1884, when he was already established as an associate. He was elected a full Academician the next year, and fully deserves his place among the great perspectivists. Yet the Warkworth picture was 'a landscape pure and simple without a trace of an architectural subject';[2] not quite what one might expect from a businesslike architect whose practice stood to gain by the public display of attractive perspectives. In fact his early Academy exhibits were all architectural. A *Cottage erecting in Cumberland* was hung as early as 1857 (Pl. 16);[3] but he showed nothing of the Assize Courts, and his next exhibit was Manchester Town Hall in 1868. Thereafter he exhibited every year, with the exception of 1873, until 1901; but until well after his election as an ARA his exhibits were all architectural perspectives. It was only in the closing decades of the century that he turned increasingly to landscapes (Col. Pl. 9); landscapes that were often carefully worked up from much earlier sketches. The architectural perspectives, however, are generally more attractive.

[1] *Building News* (1884), 817.
[2] Ibid.
[3] This was Fawe Parke **[45]**. See ch. 3.

The mechanical setting up was frequently carried out by a draughtsman in the office,[4] who may have added a good deal of the basic drawing of detail (Pls. 33, 142). Then they were finished by the architect himself, with a bravura treatment of skies, subtle colours, and, generally, with groups of busy street life. His diploma work of Manchester Town Hall sets the building under a sulphurous sunset, surrounded by the buildings and business of the town.[5]

Yet in spite of complaints from fellow competitors that his perspective skill gave him an advantage in competitions, it is unlikely that that skill helped to build his career. His reputation as a draughtsman and water colourist was deservedly high, but it would be wrong to assume that his clients were bowled over by attractive presentation. The one case where an enquiry is known to have followed one of his perspectives, that of Fawe Park, did not, apparently, lead to a commission. Indeed the accounts reveal that in several cases the perspective was not drawn until after the commission had been won and building had begun. The perspective of the Manchester Assize Courts (Pl. 33) records the final design after the original entry had been abandoned. His perspective of the National Liberal Club (Pl. 133) is dated 11 Nov. 1884, some months after he had been given the commission. At Ashton-in-Makerfield **[191]**, the perspective was not drawn until July 1867 when the chapel was almost completed;[6] and at St Matthew's, Blackmoor **[266]**, where Waterhouse charged 2.5 days of his own time on the perspective, it was not done until 1870 when the church was well under way. There are therefore good grounds for accepting that many of the fine perspectives were drawn more as aids to the appreciation of how an agreed design would appear than as part of the sales pitch. The great perspectives may have been attractive to clients, but as works of art. A number were hung in the boardrooms and committee rooms of the buildings themselves.

What is important is the way in which Waterhouse's draughtsmanship and eye for the picturesque related to his work as a designer. His working drawings are often breath-taking in their clarity and their detail. The clarity comes from the strict training under Richard Lane, in whose office he seems firmly to have put aside the idea of an artistic career. Yet his feeling for the picturesque was never suppressed, and comes across clearly from the sketchbooks. There, in delicate pencil line, he was able to set down the massing and grouping of buildings, silhouette, rhythm, light, and shadow, as well as to record accurate details of windows, gables, mouldings, and ornament. These sketchbooks provide the guide to his study of architecture, and to his own approach to design (Pls. 192, 194).

The tour volume of 1853–4 is the observation of a student freshly sent out into the field, noting eveything that strikes him irrespective of style or era. At the end of the book are a few pages of Swiss mountain scenes and a few of figures in the traditional costumes of southern France. These last reveal his artistic training based on conventional textbooks by Samuel Prout and J. D. Harding, who had been Ruskin's master. In fact Waterhouse was given one of Prout's books as a fourteenth birthday present, and copies of Harding's studies are pasted into his apprentice scrapbook.[7] From Prout he

[4] The perspective of the Assize Courts held in the Manchester Crown Courts is signed J. Battye. Registers contain entries for draughtsmen's time on perspectives: Murray for Ashton-in-Makerfield (July 1867); G. W. Nicolay (who was in the office 1870–5) for St Paul's School, Eaton Hall Chapel, Balliol Hall, and the Natural History Museum. The Register, 215, shows 3.5 days put in by Waterhouse, 7 by Steane, and 5 by Cooper for the perspective of Manchester Town Hall for the Royal Academy of 1868.

[5] This was hung in 1887.

[6] The Office Register No. 1, 60, records 3 days of a draughtsman's (Murray's) time charged for the perspective. It is almost the last item charged for design work.

[7] On his 14th birthday Waterhouse had been presented with a copy of *Hints on Light, Shadow and Composition etc. as applicable to Landscape for Juvenile Inquirers* by Samuel Prout FSA published by Ackerman and Co., 1838. It is inscribed: 'Alfred Waterhouse Jun. in remembrance of a short but very pleasant tour in Wales on his birthday 1844'. His student notebook fragments show he made use of J. D. Harding's *Principles and Practice of Art* (London, 1845).

learnt to compose the little groups of figures which, often highlighted in white, were so useful in giving life to architectural drawings (Pl. 11).

Drawing from life played no part in his studies,[8] and his few portrait sketches lack conviction. Only one, of his favourite brother Theodore at about 13, seems to have come off, and as late as 1883 his sketchbook contains unconvincing portraits, apparently done on request.[9] His forté was observation of places and scenes. The earliest surviving sketchbook records a picnic in Epping Forest, a view of the family home at Sneyd Park, Bristol, and a Frith-like thumbnail sketch of Queen Victoria's visit to the Sabbath Schools in Peel Park in October 1851, showing part of the huge crowd. A heavily worked pencil perspective of the drawing-room fireplace and panelling at Speke Hall in Lancashire, dated 1848, and a careful pencil and watercolour study of rocks near Weston-super-Mare of 1850 may be evidence of subsequent drawing lessons. But the chief impression is of the sheer quantity of the drawings through which he developed his skill in observation and recording.

The sketchbooks are for reference rather than for any purely artistic purpose, so that often a plan will be drawn of a building or of a part such as a doorway, window, or cornice (Pl. 8). Letters key the drawing to further enlarged details, small perspective views and sections of mouldings or decorative work, or to larger, more finished, pencil drawings or water colours, of which Waterhouse produced an astonishing number (Pl. 14). All the sketches from his first tour in 1853–4 are dated and numbered, the last number being 278, though only a small proportion have been traced. The subjects are almost entirely buildings but the scope is wider. Some are studies of individual buildings, others range from general perspective views, studying the massing of various elements, to buildings in their settings where the picturesque composition is more important than the architecture (Pls. 192, 193, 195). A few sketches show interior views, and a few are landscapes. Colour, in architectural subjects, is generally confined to almost monochrome washes over pencil but it is used more strongly to throw up polychromatic work and figures.

The earliest of the finished drawings from his 1853–4 tour is a delicate pencil and wash street scene in Dieppe (Pl. 6). In this, as in a number of such studies, the architecture is incidental to the effect of perspective and the composition of masses. Another shows the Temple of the Sybil at Tivoli. Around the roughly built base are groups of people chatting amongst the laundry which has been hung to dry at all convenient points over the venerable stones. One drawing, 13.25 in. by 10.25 in., shows a view of the Villa d'Este at Tivoli. The building is seen from below in full light, framed by the dark trees on either side of the steeply banked terraces. It is one of the relatively few large sketches finished entirely in pencil. The architectural detail is picked out in precise, sharp line. Dentils, balustrades, window surrounds, and cornices are all shown with the utmost economy; while the foreground trees contain grades of shadow down to the very blackest. His skill in perspective was well developed in colour too, and can be seen in views such as the aisle of Milan Cathedral looking towards the curve of the chevet.

It may have been one of the paintings from this tour that was exhibited in the first conversazione of the newly formed Bristol and Clifton Graphic Society in 1854. The exhibition included work by 'S. P. Jackson, Prout, Hardwick, and Franks also Mr. Waterhouse and other amateurs'.[10] In that

[8] A single academic life study survives among his office drawings. It is a perfectly competent student work, but it is not signed, and may not even be by him.

[9] A note on a page of ladies' heads in local costume in the sketchbook covering a visit to the Corniche in 1883 tells us: 'This sketching of ladies of the S. A. led to great disturbance. They were all anxious to be drawn!' SK 1879–85.

[10] Unidentified newspaper cutting (4 Mar. 1854), report of Bristol and Clifton Graphical Society exhibition in Waterhouse's scrapbook, 'Architectural Fragments'. This is in fact the first cutting in the book.

company he was firmly linked to the established picturesque taste; but he seems already to have turned his back on mere painting, and to have produced no further landscapes until shortly before his marriage in 1860. The sketchbooks return to architectural subjects with only a few picturesque views of hilltop castles from the family holiday in Switzerland and along the Rhine. Only in his written description of the trip does his enthusiasm for romantic landscape surface. The same is true of his visit to the Loire Valley in 1865, where a three week trip produced about forty small sketches, about 7 in. by 10 in., as well as all the other notes. However, this was clearly a study tour, and the drawings are noticeably rushed. But, in 1859, no doubt influenced by his bride to be, Elizabeth Hodgkin, herself a talented amateur painter, he produced a series of landscape sketches of the area around her home at Lewes.[11]

Once his career was established, Waterhouse allowed his landscape painting to develop and increase. The limited use of colour grew to a fuller range, which, while confined to natural, earthy colours, could express atmosphere and perspective in both land and sky. He enjoyed strong colour in nature. On a walk from Richmond to Staines in October 1863, he had noted: 'Wonderful colour in beds—Dark purple verbena, orange marigold, yellow calceolaria, light purple verbena',[12] but he had not attempted in his paintings to produce the reality of colours such as these. In fact he used watercolour quite conventionally; but his sketches have an intensity of observation and clarity of composition that raises them well above the merely amateur.

Without photography, sketching was still a highly popular pastime and it is no surprise that both Waterhouse and his wife continued it. The record of their travels reveals the allure of picturesque scenery, which was avidly recorded; though, increasingly, it was the sketchbook and not the portfolio that was filled. Their holidays for the first ten years of their marriage were to the popular areas of southern Germany and the Alps which provided the romantic and dramatic scenery which so obviously appealed to him. Among the watercolours of this period are some of the freshest and most accomplished that he produced.

In the spring of 1870 he started to take his family to Italy.[13] Two winter visits in 1873 and 1874 gave him further scope for recording views of Rome. In spite of his busy life they rarely missed an alpine holiday in August as well. They hardly ever stayed in Britain for holidays, though in 1873 he did some sketching in Scotland and in Cornwall.[14] In 1884 they ventured to Sicily and in August 1886 to Norway. Finally he visited Spain in 1890 (Pl. 198). Travel there was still uncomfortable but he was moved to go on hearing an address by Sir Frederick Leighton, then President of the Royal Academy. Afterwards he confirmed Leighton's opinion, warning that its architecture was unsuitable for the immature who might 'be led away by much that is as pernicious as it is alluring'.[15]

About this time his colouring becomes much more intense (Col. Pl. 9), which is possibly one of the reasons why his Warkworth landscape stood out so at the Royal Academy. However, for him the subject-matter was more important than the tints. The bold sweep of mountain dwarfing a village or reflected in a lake, the bridge flung across a deep gulf, the hilltop castle over a rugged gorge all gave him the opportunity to portray the romantic and the picturesque (Pls. 196, 199). His love of the

[11] It was at this point that he ceased to sign his work 'A. Waterhouse Jun.' and began to use the monogram of his own initials.

[12] SK 8 Oct. 1863.

[13] No doubt the Franco-Prussian war had some effect.

[14] On that occasion an anxious uncle Theodore wrote to his mother that he hoped the children would be safe on those narrow paths when Alfred had to leave them (*Theodore Waterhouse 1838–91. Notes on his Life and Extracts from his Letters and Papers*. Privately published, London, 1894).

[15] AW, Presidential address, RIBA (1890), published in the *Builder* (8 Nov. 1890).

juxtaposition of mountains and water shows up in the excitement of drawing views of a Norwegian fiord as the boat turns and turns again.[16]

In his later sketches he drew buildings more for scenic value than as architectural study. His sketches of Mont-Saint-Michel contain much that is typical of his art (Pl. 197). One in particular required all the skills of the perspective artist. He chose an eye level at the bottom edge of the picture in line with the lowest steps. The great buttressed supports of the towers flank the steps and dwarf a figure descending. The round towers break out from the bases in a series of mouldings marked by brush drawn shadows. A few courses of stone work are shown above to complete the overbearing close view, that is just relieved by the lightly touched-in view of sea and distant coast seen through the arch to one side.

The romantic and the picturesque in views like this are, however, firmly anchored in close study of the buildings; and that always mattered to Waterhouse. Drawing was a means of observation and of accumulating knowledge. However, the skill was not to be used merely for recording and repeating, as he made clear in his inaugural Presidential Address to the RIBA:

> While not neglecting to cultivate our powers of delineation, we might, perhaps, do more to make those powers of use to us than many of us do at present. If, for instance, instead of sketching ancient examples so much, we were content to do half as much in this way, and afterwards were to endeavour to reproduce our sketches and scale drawings from memory, we should accomplish two things,—marvellously sharpen our observing powers, and greatly increase our facility in design.[17]

Later he was more forceful:

> If the architect-student knew that it was not by the mere number and beauty of his sketches, and by the accuracy of his measured drawings of old edifices, that he could satisfy those who had sent him forth on his travels, but by the proofs he was able to afford that he had absorbed and digested what he had seen, . . . men would be more ready designers after having had such preparation. Having got the ancient examples into their heads, instead of merely into their sketchbooks, they would learn to lean not on illustrations and other extraneous help, but on what had become a part of themselves, and so prove not mere copyists, but self-reliant and original designers.[18]

All this is of obvious concern to an architect. It was what Waterhouse himself had set out to do from the start. He studied the setting of buildings, learning which parts were the most observed from different points and using these to the greatest effect. His observation of this sort of detail can be traced back to 1853–4, and certain subjects clearly had a strong appeal. Towers and spires obviously excited him. At Arles he recorded a spire decorated with crockets and set on an octagonal drum, in Vallais the Romanesque tower of Saint-Maurice, the Gothic clock tower of the Hôtel de Ville at Lausanne, another tower at Berne, and so on. Later he was fascinated by the tower of Les Halles at Bruges, whose octagonal drum was one source of inspiration for the Manchester Town Hall tower **[284]**. Then there are drawings of Italian campanili, in particular Vicenza with its bold double light opening and balconies in the upper storey, the tower of S. Pietro at Perugia, the spire of S. Croce in Florence, not then restored, as well as Giotto's campanile at the Duomo and another at Ravello. The leaning towers of Bologna make the subject of a dramatic watercolour.

[16] SK 1884–7.

[17] AW, Presidential address to RIBA, Nov 1888. RIBA Journal of Proceedings, published in the *Builder* (10 Nov 1888).

[18] AW, Presidential address, RIBA (1890), published in the *Builder* (8 Nov. 1890).

Such studies, though never providing a detailed prototype, offered him attractive solutions to many real architectural problems. He recorded a number of stone staircases on his tour, at Lausanne, at Berne, in the Palazzo della Loggia at Brescia, from the pulpit at Torcello, and in the Palazzo Barberini in Rome. Some of these later inspired picturesque or dramatic staircases in buildings such as Caius College, Cambridge **[268]**, or the National Liberal Club **[514]**. His penchant for dormer windows can be quite clearly traced to a series of studies of such details from Normandy. But in using his stock of information, Waterhouse was approaching his work from the stance of a man who has looked long and hard at the forms of existing buildings. Those that had stimulated his own love of the picturesque were a useful, though indirect, quarry for the design problems of the day.

For instance, his concentration on roofscape has a real purpose. The gloomy skies of nineteenth-century industrial Britain obscured the effect of fine detail. It was better to rely on a varied skyline using gables, dormers, high pitched roofs, and massive chimney stacks. Waterhouse manipulated the picturesque value of roofscape wherever possible to create effect. During the 1860s in particular, he used the steep pitches which in conjunction with other features such as gables and chimneys give vertical thrust to the whole, as at Easneye **[258]**, Foxhill **[285]**, or Eaton Hall **[315]**. The façade of his Lime Street Station Hotel **[274]**, with its tension between the vertically grouped windows and strong horizontals is finally given direction by what, at 80 degrees on the central tower and the end pavilions, must be the steepest pitches of any Waterhouse roof. Of course these steep pitches are uneconomical in relation to the width they span so that often only the outer pitch is exaggerated, the inner slope being flattened to gain floor space.[19] Above the roof line Waterhouse also strove for picturesque effect with wrought iron cresting and with chimneys. These he grouped boldly, giving them heavy squarish tops and extended height, particularly in domestic work. The lodges of Easneye **[258]** and Allerton Priory **[237]**, for instance, are each built to an appropriate cottage scale around huge single chimney stacks of the same size and pattern as the parent house.

Much of this compositional panache arises from Waterhouse's readiness to conceive the massing of a building in visual terms. His ability to design in perspective had a profound effect on the shaping of his buildings. He always advocated the making of perspective sketches during the design process, in which the building was blocked out in monotone so that its effect in silhouette could be studied.[20] He advised students to accustom themselves to designing in perspective, both inside a building and without, realizing the effect likely to be produced from different points.[21] And he applied these principles himself, visiting the sites of proposed buildings and sketching what was there. In this way he secured the dominance of the tower of Darlington market buildings **[99]** (Pl. 201) by a careful study of the awkwardly falling site and of the possible overbearance of the buildings in High Row above.[22] His tower, based on principles set out in Ruskin's *Lamp of Power*, allows the plainness of the greater part to disguise its height while its richly modelled upper storey and roof are clearly seen and form a focus for the heart of the town.

In the same way but from a much more advantageous point on the Shot Tower across the river, he studied the neighbourhood of the future Courts of Justice. The ensuing perspective has a drama that

[19] Later, and particularly when building to a budget, he tended to fall back on a 60° pitch, which, though the most economical to build, produces an uncompromising appearance to a gable end.

[20] AW, Address to students of Birmingham School of Art, *Building News* (Mar. 1883).

[21] AW, RIBA Presidential address to students (1889), published in the *Builder* (2 Feb. 1889).

[22] SK Nov. 1860.

belongs, however, not to the painting alone (Pl. 41).[23] Had he been commissioned to build the Law Courts, there is no doubt that his building would have displayed the benefits of his search for the picturesque solution as well as the planning advantages for which it was hailed. The same is true for all his major buildings, where he draws on past observation of picturesque motifs and structural solutions and then allows his artist's eye to bridge the gap between the sensible functionalism, for which he is famous, and the form of each structure.

[23] The perspective, submitted with his designs, was taken from the same point. It is now in the RIBA Drawings Collection.

25

A Delight in Firmness and Commodity: Waterhouse's Style and its Importance

It remains to summarize Waterhouse's achievement, to try to pin down just what it was that made him popular in his time and makes him important today. Critics will argue about the precise examples, but few would deny that his career produced a handful of real masterpieces. Not all his work was equally touched by that spark of genius that lifts some buildings into the realms of the sublime. He also produced an extraordinarily large number of buildings, most of which are substantial structures, but some of which are decidedly run of the mill. They have clouded the argument, and disapproval of some of their characteristics has turned into rejection of the whole mixture of qualities that is the essence of this practical and inventive architect.

His rise in a mere fifteen years from the obscure start of a provincial practice to a position where he was able simultaneously to handle three of the largest commissions of the decade, though meteoric, was by no means just a piece of good fortune. He was taken to task by at least one Quaker friend for overcommitting himself, and admitted with regret: 'I could have wished certainly that my 3 great works (Manchr. Town Hall Eaton Hall & the Nat: Hist: Musm.) had not come to me very nearly all at the same time.'[1] One can see why he showed little regret at losing the Law Courts competition with its condition embargoing the acceptance of other new work for three years. But there were good reasons why the Government, the wealthy Duke, and the hard-headed businessmen of Manchester should have turned with confidence to Waterhouse. By the end of the 1860s it was already clear that his office was run on what we would recognize as thoroughly professional lines, and that commissions entrusted to him would be efficiently handled. In this he was more like Scott, whose details one craftsman recalled were always ready and whose contractors never kept waiting, than Street, who was always behind.[2]

The confidence this inspired in his clients was evident from his early years. His work for William Tipping at Brasted is typical. He was commissioned in 1859, probably on the strength of Tipping's friendship with his father, to alter a house **[54]**. Waterhouse went on loyally to design alterations to the local National Schools **[76]** and to two farmhouses **[173, 234]**. This perseverance had its reward

[1] MS letter, AW to Jos. Boult (5 July 1872). MCL Archives. I am grateful to J. H. G. Archer for bringing this letter to my notice.

[2] Andrew Saint, *The Image of the Architect* (New Haven, Conn., 1983), 68.

when Tipping, as patron, secured him the commission to restore, or rather rebuild, St Martin's Church **[194]**. Then, as a director of the London and North Western Railway, Tipping was able to ensure that Waterhouse was among those invited to make designs for their prestige hotel at Lime Street **[274]** (Pls. 203, 204). His design was preferred, but there were doubts about the cost; so he suggested preparing an elaborate letter setting out the reasons for his estimates which Mr Tipping would then show to the Chairman.[3] It is typical of his career that Waterhouse could rely on such confidence from his clients; and the result on this occasion was one of his largest commercial buildings.[4]

But there is very much more to Waterhouse than the successful man of business. His designs were preferred over those of his colleagues and he had no shortage of commissions. It is easy to find an uncertain diversity of styles in his early work, which ranges from the Swiss chalet style in some of the Lakeland houses **[22, 42]** (Pl. 19) to the self-consciously Ruskinian Chester Street warehouse in Manchester **[154]** (Pl. 21); while the Gothic of the Royal Insurance building **[108]** (Pl. 28) in the same town seems to look back only as far as the composite style of Deane and Woodward. Before the Assize Courts competition **[67]** (Pl. 33) Waterhouse seems to have concentrated on the *application* of style *to* a building;[5] and after the completion of the Courts he still refers to particular sources for his ornament. It would be tempting to search the sketchbooks for specific sources throughout his career. However, as we have already seen, there is seldom a direct link between the buildings he sketches and his own works. As Goodhart-Rendel pointed out: 'it is very improbable that any architectural design, however original it may appear, can be other than a transmutation or an imitation of something in its author's experience, whether his memory of the model be conscious or subconscious. What that model may have been . . . is of no essential moment; what matters essentially is whether it has been transmuted or merely imitated.'[6] Waterhouse's attitude certainly is to grasp the nettle of nineteenth-century modernity until 'one forgets completely that [his] is a revived style. It has forgotten all about its medieval ancestry and belongs completely to the nineteenth century.'[7]

That is not to say, however, that his work was not rooted in tradition. We know, for instance, that his visit to Belgium in 1858 was intended to help him with ideas for the Manchester Assize Courts **[67]**. However, absorption and long consideration of ideas and forms is really the key to his historicism. Thus there is a gap of twenty years between his sketch of the arcading of the inside of the tower at Wimborne Minster,[8] and its echo in the chancel of St Elisabeth, Reddish **[507]**. There is much the same gap from his first sight of the south-western tower of the Abbaye aux Hommes at Caen,[9] and the tower of the King's Weigh-house Chapel in Mayfair **[556]**. From his first sight of the oval staircase at the Palazzo Barberini, in Rome,[10] to the design of the National Liberal Club **[514]** (Pl. 136) is almost thirty years; while the octagonal capitals of Milan Cathedral, first seen in 1853,[11] did not bear fruit until 1895 in the hall of the Holborn Prudential building **[602]** (Pl. 129).

[3] MS letter, AW to W. Tipping (26 Feb. 1867). OLB 753. RIBA Library.

[4] In this context it is interesting to note that Waterhouse himself, like many of the Victorian middle classes, was a railway shareholder, of the Midland Railway. He designed The Brand at Loughborough for Alfred Ellis, son of the Chairman of the Midland Railway.

[5] SK Oct. 1858. Trip to Belgium. 'Convinced of the application of the Gothic style for this purpose (Town Halls) as for others.'

[6] H. S. Goodhart-Rendel, *English Architecture Since the Regency* (London, 1953), 21.

[7] R. Furneaux Jordan, *Victorian Architecture* (Harmondsworth, 1966), 97.

[8] SK 1860–4.

[9] SK 1866.

[10] 'Scraps from France Switzerland and Italy', Waterhouse's bound volume of sketches made during his tour of 1853.

[11] Watercolour in private collection.

Equally Waterhouse did not ignore the stylistic debates of the time. In his early days his work was plainly derivative, as he acknowledged. 'When he was a boy the earnest character of the Medieval revival and the teaching of Pugin, Scott and above all Ruskin, carried away all enthusiasts, himself among the number . . . When commencing practice, his buildings . . . were affected by the influence which the Roman and Pointed architecture of Italy could not fail to exercize on those who studied it.'[12] His early admiration for Ruskin soon became tempered by reality, but, following the Lille Cathedral competition,[13] his appetite was fed by Nesfield's published sketches of secular buildings and simple details, issues of Johnston's *France*, and by his purchase in 1859 of Verdier and Cattois's *Architecture civile et domestique du Moyen-age à la Renaissance.*[14] Two years later he bought the first complete edition of Viollet-le-Duc's *Dictionnaire raisonnée de l'architecture française* and his *Mobilier générale*. All these helped to undermine his belief in Ruskin and to develop his own modern Gothic.

Style and the choice of styles bulked large in the architectural debates of the mid-century; but style was often seen as an added feature. It might afford the sense of delight that was the third essential characteristic of good architecture laid down by Vitruvius. But Waterhouse, as he developed his own style, was looking to the two other virtues of firmness and commodity first. Style was important to him, but only as the end result of a logical process of designing and constructing. Whether he was in the mainstream of the high Victorian movement in his philosophy, he was in tune with Fergusson's view of architecture that it should be 'practised only for the sake of supplying the greatest amount of convenience attainable, combined with the most appropriate elegance'.[15]

In admitting his indebtedness 'to Mr. Scott for a hint or two as to the treatment' of the upper-floor windows of the Assize Courts **[67]**,[16] Waterhouse reveals a specific influence. His copy of Scott's *Remarks on Secular and Domestic Architecture*, bought in the year of publication, has markings, notes, and even illustrations which indicate admiration and discipleship. He has marked passages which prove Scott's recognition of Puginesque principles and extend them to a method of building for the times. Scott's summary of what was needed is marked and marked again.[17] Passages on the planning of houses in the country and on the design of frontages of urban and public buildings are all underlined. Scott also touches on a subject not directly the concern of the architect, that is, the moral duty of a 'landed proprietor' to provide suitable accommodation for those to whom he is responsible. Waterhouse put these precepts into practice in his own work.[18] There are no marks in the chapter on restoration, for Waterhouse deplored the methods of Scott's time. He was committed to the new philosophy of retaining everything possible of an ancient building, promulgated by Morris and promoted by the Society for the Protection of Ancient Buildings, of which he was a founder member.[19] Methods of window design described in Scott's text are illustrated by Waterhouse in the margin; passages on the use of banding in wall surfaces both decoratively and as a means of economy

[12] *Building News* (2 Mar. 1883). A report of an address by Waterhouse read at the annual meeting of the Birmingham School of Art of which he was at that time President.

[13] One of the winning designs was by Waterhouse's friend William Burges working with Henry Clutton.

[14] These titles are included in the list of books purchased in the first office ledger. Private collection.

[15] J. Fergusson, *History of the Modern Styles of Architecture* (2nd edn., London, 1862), 329.

[16] AW on the Assize Courts, address to an ordinary meeting of the RIBA, published in *Building News* (23 June 1865).

[17] G. G. Scott, *Remarks on Secular and Domestic Architecture* (London, 1857), 270. The more detailed discussions on the method of achieving his ideal by learning from historic examples without copying them have strong echoes in Waterhouse's own later statements in published addresses.

[18] Ibid. 122, 135, 141.

[19] Waterhouse wrote to the incumbent at Brasted: 'I . . . very much regret that I do not see my way to [yield] in the matter of new Church. As a mere matter of business, it would of course be more to my interest that an entirely new building shld be put up: but wld it not be a matter of regret that so interesting a piece of antiquity, almost unique in one point at any rate shld be suffered to fall into decay?' AW to Revd C. J. Astley (7 Mar. 1865) OLB 672. RIBA Library.

are stressed as are the pages on the careful use of brick and brickwork ornament,[20] and Waterhouse adds references to Ruskin and Street and to French sources for further examples.[21] However, he went on to work Scott's ideas and the application of sources out for himself in his own buildings, and in so doing created his own style.

Street once referred to his 'medieval architecture', and Waterhouse corrected him: 'I hope that when he sees Manchester Town Hall completed, he will find the building essentially of the nineteenth century, and adapted to the wants of the present day.'[22] Waterhouse, for instance, consistently avoided the contrived lack of symmetry favoured by Pugin as a medievalizing touch. Instead he allowed his asymmetry to evolve as an expression of function. A footnote in Scott's *Remarks* begins 'it is a favourite notion, that to be Gothic a building must be irregular'.[23] The pencil marking in Waterhouse's copy indicates that he agreed with Scott's excuse for symmetry where it occurs naturally as the expression of the ordered use of the building.

Though he admired Classical work,[24] Gothic was the obvious, even the conventional starting-point; and for Waterhouse this had to be Gothic in its broadest sense embracing everything from the twelfth to the sixteenth centuries, from Italy to northern France as well as England. He allowed the round and the pointed arch the same status as the shouldered and the segmental. The choice was one of logic and practicality; and he described his Assize Courts **[67]** as 'gothic of the thirteenth century, albeit with departures'.[25] When asked in what style he intended to build Strangeways prison **[113]** (Pl. 37) he replied that 'all the details will be Gothic of the same period as the [assize] courts but the round arch used instead of the pointed as being equally Gothic and better adapted for the requirements of the case'.[26]

Yet the revived twelfth-century style that he preferred is hardly typical of the high Victorian movement; and more was involved than merely pandering to currently fashionable taste. Indeed the range of eclectic styles available at the end of the 1860s was so bewilderingly wide that clients in any case needed firm guidance. Besides the very size and scope of the buildings that flowed from the Waterhouse office is a reminder that other factors were more important in a successful architect than the ability to manipulate a style or styles. Not only were the costs of building rising in the 1860s and 1870s, but the increase in urban population was leading to the need for larger and larger buildings, schools, hospitals, colleges, fewer and fewer of which were the result of the patronage of a single person. Waterhouse was very much an urban architect and not only well connected but well able to cope with disparate groups of municipal and commercial clients. His abilities approximate more closely to Robert Kerr's 'Seven Other Lamps of Architecture' than to Ruskin's mystical and aesthetic *mélange* with its implied rejection of current values.[27] His reputation was built on factors other than mere aesthetics.

[20] Scott, *Remarks on Secular and Domestic Architecture*, 96 ff.

[21] Scott had recommended these to him 'plus a fortnight in France'. Waterhouse's sketchbook for 1857–9 includes notes on sash windows, and 2 pages of details from Scott's Institute at Sandbach.

[22] AW paper on Manchester Town Hall, *RIBA Sessional Papers* (1876–7), 117 ff.

[23] Scott, *Remarks on Secular and Domestic Architecture*, 119.

[24] AW, Presidential address to students (1889), published in the *Builder* (2 Feb. 1889). 'I suppose the most perfect buildings the world has ever seen were the Greek temples and Propylaea.'

[25] AW on the Assize Courts, address to an ordinary meeting of the RIBA, published in *Building News* (23 June 1865).

[26] MS letter, AW to R. B. Dockray (23 Nov. 1864). Friends' House Library. The letter head is printed Royal Insurance Buildings, 24, New Street, Spring Gardens, SW, that is, from Waterhouse's first London office.

[27] Robert Kerr, 'Seven Other Lamps of Architecture' in a series of 7 articles in the *Builder*, 9 (1851), 700–2, 731–3, 764; 10 (1852), 3–4, 50–2, 148–9, 178–9. Kerr's recipe was: the lamp of Art, the lamp of Delineation, the lamp of Science, the lamp of Building, the lamp of Learning, the lamp of Teaching, and the lamp of Making a Living. The list is more prosaic than Ruskin's but by no means entirely satirical. Kerr's Lamp of Art, however, states unequivocally that 'the architect is an artist' and 'is in short the builder gifted with artistic skill', but he must be builder first.

First, he built solidly, and took few risks. Then he could select and translate Gothic features that were appropriate, and which could make a building picturesque. To be in fashion was certainly not his aim. He regretted the failure of the medieval revival to grow into a modern style, instead becoming the forerunner of a continuing change from style to style.[28] But he was himself adept at creating dramatic shapes, particularly in city centres, by the use of towers with tourelles, as at Darlington market **[99]** (Pl. 201), or steep roofs, as at the Seamen's Orphans' Institution in Liverpool **[337]** (Pl. 142). Oriel windows set across a corner were another favourite motif, which he used at Caius **[268]**, Jesus **[290]**, and Girton **[414]** Colleges in Cambridge. His love of dormers has already been described; as has his delight in manipulating staircases, especially when they were linked to other spaces. When talking to students he commended a design in the Sheffield Municipal Offices competition which showed a principal and a business staircase separated by an open arcade to good effect. In his own work he combined a spiral service staircase with the grand staircase at Manchester Town Hall **[284]** (Pl. 53), and later repeated the trick at Owens College **[515]**.

Always, however, there was a sound practical reason for this sort of indulgence. Waterhouse soon rejected the purely visual effects of Ruskinian decoration. As early as the Assize Courts **[67]** he was having doubts about the O'Shea's sculptural decoration: 'The fact that much of their carving, in common with other people's, looked much better when half worked, inclines me to believe that if our allies the stone carvers, would put more thought and less work into their foliage, architecture would be greatly the better for it.'[29] His approach to ornament was really much closer to that of Owen Jones. It is for instance the complex interlocking of simple geometric shapes that makes much of the success of the Natural History Museum **[243]** (Pl. 83) He did enjoy constructional polychromy; but that was very much part of his understanding of materials, and he translated it into the logical, supposedly washable terracotta. He also developed his own theory of architectural colour and texture: that single materials should be carved and patterned but where several colours of one material were used the surfaces should be kept flat.

Colour, texture, Gothic and other forms were, however, all seen as part of a building, and subservient to its prime purpose. It is this that makes his houses so consistent. The concentration on convenience makes them appear earnest and solid rather than fanciful and fun; but neither their owners nor their architect saw architecture as mere fun. Serious-minded solid work with carefully selected eclectic details is really the hallmark of Waterhouse's style. His rejection of any flamboyance was an expression of good mannered English reticence.

Many of his buildings are Gothic in essence, but they are also essentially Waterhouse. The same character comes out in his neo-Romanesque work and in wider eclecticism. The circular parrot house at Eaton Hall **[315]** (Pl. 90), with its Ionic colonnade and terracotta caryatids within, demonstrates his ability to handle the Classical orders. The lodges and coachmen's cottages also show his readiness to adopt new styles as fashion directed (Pl. 92). But all are handled with the same consideration for ducal splendour that characterized the mansion itself. By contrast the Turner Memorial Home **[497]** (Pl. 202), a charity for retired seamen, is a rather gaunt pile on the banks of the Mersey. Yet it, too, is given interest by a range of eclectic features: a late Gothic chapel, an early French turret, and a half-timbered porch with a richly decorated clock gable above. The seaward front is fitted out with a two-

[28] *Building News* (2 Mar. 1883). A report of an address by Waterhouse read at the annual meeting of the Birmingham School of Art of which he was at that time President.

[29] AW on the Assize Courts, address to an ordinary meeting of the RIBA, published in *Building News* (23 June 1865).

storey verandah with workaday cast-iron pillars, but also has a single pierced buttress that doubles as a garden gate. There is enough display to remind us of the princely patronage,[30] and enough variety to enliven a substantial pile of building. Like his schools and colleges and other substantial buildings, it is both creative and economical.

Here one other strand of Waterhouse's characteristic style is involved. His attention to cost served as a useful brake on unnecessary decoration and wasted space. This waste not want not attitude and opposition to frivolity was deeply rooted in him from his Quaker upbringing, and reinforced by his northern industrial training. Early on he was brought up against the problems of commercial as opposed to religious or private patronage. The pressure for economy is a positive part of the commercial client's approach almost by definition. It cannot always be turned to creative effect. Thus his design for the Lime Street Hotel **[274]**, won in competition with E. M. Barry and others, lacks the panache of Scott's contemporary Midland Grand Hotel. True it has a more restricted site, but it was also built to a much tighter budget; so the only available effect was in the repetition of units along the immense façade. No attempt is made to conceal the great bulk of the building and the two central towers hardly interrupt the façade. His Bryce Smith warehouse in Manchester **[372]** shows a similar simple approach. There is only minimal ornament, but the building gains great effect from the two small gables that flank the roof, and incidentally give extra light to offices at the end of the building.

There is a directness and an honesty in this approach that reflects his whole background. But the size and solidity of his urban buildings also makes them intensely Victorian, monuments of their own age. His buildings were in no sense extreme, neither flamboyantly modern nor excessively old fashioned. They fulfilled a need for a confident conservatism of taste while expressing the solid financial basis of his clients in the carefully graded richness of their decoration. His own move from Quakerism to the Church of England reflects his acceptance of the Establishment. He was in many ways a man of the middle way. The implied moderation is evident in his designs, which are far from being those of a dedicated goth like Burges, Street, or Scott. Indeed, while it is possible to see his love of the dramatic skyline and picturesque asymmetry as derived from Gothic architecture, it is probably a fairer assessment of his mature work to see it as that of a creative eclectic working happily in a variety of styles so far as surface decoration was concerned. What his buildings have in common is a great sense of solidity that matches the confidence of the times; and in this context it is less surprising that the style he made his own was the bold decorated neo-Romanesque of the Natural History Museum **[243]** (Pl. 78).

This generation still finds uncomfortable his almost total disregard of the mellow qualities favoured by his successors. He preferred, however, to accept the mechanical means of workmanship that were the norm of his time, working with existing patterns rather than hankering after past values or pushing for new techniques. But if his stylistic conservatism is not to everyone's taste, it is still important to consider the reasons for his huge success and to rediscover the basis for his popularity with his clients and contemporaries. One thing is certain, his buildings do not represent any unthinking factory-style production. Each is a carefully thought out solution to an individual problem. The plan came first from a proper understanding of the client's needs; so commodity was the prime concern. His cautious and meticulous approach to structure and materials made firmness

[30] The memorial was originally intended to be a simple clock tower on the sea front. Evidently a great deal more money was forthcoming than would have been needed for that.

the second consideration. Delight, in Waterhouse's eyes, should be the product of a happy marriage of the two.

His success and the wide range of his buildings make him important, simply because he is so typical of his age. His buildings are a reminder that stylistic games were not the only thing in architecture so far as many Victorian clients were concerned. But his other characteristics make him important in the wider history of architecture. In his attention to materials, though not in his rejection of hand craftsmanship, he is close to the thinking of the Arts and Crafts Movement. In his ready acceptance of the technology of the day and in his concern for planning his position is parallel to that of the later Modern Movement. But above all his creative eclecticism and the traditional roots of his style make him triumphantly and typically Victorian.

Epilogue

26

The Dynasty

WATERHOUSE'S story does not simply end with his own career as a successful and energetic designer and an efficient businessman. Of course his efforts needed the support of many assistants, men such as Cooper, Evans, and the indomitable Willey, his chief clerk. But he was also the successful paterfamilias. So, when his eldest son Paul followed his father into the office, Waterhouse gained a new assistant whom he could train in all his thorough ways. He was also launching the succession that was to make Waterhouse, Waterhouse and Son and finally Waterhouse and Ripley one of the longest lasting dynasties in architecture.[1]

The eldest of the family's five children, Paul, was educated at Eton and Balliol, a very different upbringing from his father. It developed a literary bent inherited from his mother and his love of the classics, making him an essayist and speaker of some repute.[2] He left Eton early in order to begin in architecture before going up to Oxford. How much he was involved in designing at that stage is not clear, though he must have spent almost two years in the office. He then studied at Balliol, taking a second class degree in Classics, and enjoying all the university life that his father had never had. He coxed the Balliol eight for four years, became captain of the College Boat Club, and was even a candidate for the University boat. Then in 1884 he joined his father for good.

He now had to study for exams to qualify as an architect, another thing his father had not had to do. On the way he collected the RIBA Silver Medal for essays, and almost won the Ashpitel Prize. He, like his father, took the study of architecture very seriously, and, although a classicist, produced some excellent scholarly essays on Pugin and wrote a score of entries for the *Dictionary of National Biography*. Recognition as a scholar came with his election, in 1918, as a Fellow of the Society of Antiquaries.

His interests were wide like those of his father, for he was a member of both the Foreign Architectural Book Society (William Burges's high-spirited band, which still lingered on in a second generation) and the Surveyors' Club. He was also a regular member of the Bach Choir and a talented madrigal singer. In 1924 he was elected a Fellow of the Royal Society of Literature.[3] Like his father, he was an accomplished committee man, serving the RIBA on one committee or other from 1890 onwards. His particular concern was architectural education, a concern he must have inherited from

[1] Waterhouse opened his practice in 1854. David Waterhouse, his great-grandson, retired from the practice in 1989. A dynasty of 135 years.

[2] *Paul Waterhouse: A Collection of Certain of his Papers and Addresses* was published by Oxford University Press and Humphrey Milford, London, in 1930 after his death.

[3] His brother-in-law was the poet laureate, Robert Bridges, which presumably explains this link.

his father; and in due course he became Chairman of the Board of Architectural Education.[4] In 1924 he became a founder member of the Franco-British Union of Architects, a continental involvement his father had not had, though his buildings were widely admired there. The final achievement of Paul's career was his election in 1921 as President of the RIBA, as his father had been before. There he was memorable for his use of gentle wit to take the heat out of any argument.

However, an architect needs to be judged by his buildings, and Paul carried on the traditions of his father's office. He will have worked on most of the schemes that passed through the office from the late 1880s. Waterhouse took him into partnership in 1890, and, after a stroke in 1901, passed the whole control of the firm to him. Thus Paul took over a number of major commissions such as University College Hospital **[588]** and the Prudential headquarters **[602]**. He continued building at Leeds and Manchester **[627]** Universities,[5] and at Girton College added the Chapel, Hall, and Woodlands Court. At the Refuge in Manchester he built the block containing the tower and a new entrance. The link with the Prudential Assurance Company also continued resulting in the offices at Stockton-on-Tees, Darlington, Leicester, Dunfermline, Aberdeen, Middlesbrough, and Lewisham. Churches, offices, and a large house at Mount Melville near St Andrews all formed part of his practice, in which Waterhouse's tradition of eclecticism was tempered with stronger doses of the prevailing Classical taste. Paul was never one of the innovators of the kind that formed the Modern Movement. He belonged to a slightly older generation, and his work can be seen as the continuation of the line of sound traditional building that Waterhouse had begun.

Paul died in 1924, and his eldest son, Michael, followed him. One of his best-known buildings is the Classical Wellcome Institute in Euston Road, London, in a traditional style that is probably more conservative than much of his grandfather's work. He also continued the tradition of deep involvement in the life of the RIBA. Like his father and grandfather, he also became its President, leading it at a time of difficulty and adjustment soon after the end of the Second World War. The line has continued with a fourth generation of architects, one of whom, David, was Chairman of the Architects' Registration Council from 1974 to 1977. Four generations have headed the family business that Waterhouse created, and the firm of Waterhouse and Ripley is still in business.

[4] This was an interest PW shared with another of Waterhouse's assistants, L. Sylvester Sullivan, who was selected as the RIBA's representative in the discussions that finally allowed Canadian architects their own system of qualification.

[5] At Leeds he built the Textile Extension, the Electrical and the Agricultural blocks and at Manchester the Museum extension, additions to the Library, and the little octagon lodge near the south-west exit from the large court.

27

The Critical Heritage

NOT long after the Victorian Society was founded there appeared in *Punch* a set of rhymes poking gentle fun at the architectural masters of that age. Waterhouse was characterized with the verse:

> That Prudential man Alfred Waterhouse
> used to tell all his clients he thought a house
> should be beefy and thick,
> built of Ruabon brick,
> just the colour of underdone Porterhouse.[1]

A more scurrilous version had him suggest to clients that a house should be:

> partly chateau
> and partly gateau
> but chiefly municipal slaughterhouse.

The two versions may sum up his penchant for glazed faience interiors, for terracotta, and for muscular detail, but they also reflect the attitude which condemned his works for so long to shocked disapproval or at best amused contempt.

For years his buildings were largely ignored, and even those who noticed them scarcely studied them; with the result that Pevsner could dismiss the Leeds Prudential (Col. Pl. 7) with its gay Renaissance detailing as 'needless to say: by Waterhouse, and needless to say: Gothic'.[2] The reaction was partly the taste of the times, and tastes change. But a proper study of architecture needs to take account of more than just external form. Only a rounded study of all the aspects, in which style and detail are assessed along with planning, structure, massing, and economics can give Waterhouse's works their true value.

From the time of the Assize Courts competition, appreciation of his buildings varied; though success made his work noteworthy. But his reputation was such that even during his last illness, although he had been retired for several years, his progress was reported in the *Building News*.[3] In the obituaries later the same month it is quite clear that his buildings had lately been receiving considerable adverse criticism, being considered by the younger school of architects without favour

[1] Peter E. Clarke in *Punch* (25 Dec. 1957).
[2] N. Pevsner, *The Buildings of England: Yorkshire, West Riding* (2nd edn. rev. E. Radcliffe, Harmondsworth 1967). 318.
[3] *Building News* (18 Aug. 1905).

or sympathy.[4] Only eighteen years earlier the *Building News* had invited its readers to name the leading British architect and Waterhouse had come out top with 90 per cent of the votes. By the time of his death, although he was recognized as an able planner and a successful architect, it was the man himself who was being remembered rather than his buildings. Obituaries recalled his good health, cheerfulness, and good nature, his quiet dignity and charm of manner that had contributed to RIBA council meetings.[5] 'Even those who did not like his architecture liked the man.'[6]

The Assize Courts **[67]**, partly through Ruskin's approval,[7] had given him national acclaim for his involvement in the stylistic arguments of the day. But praise had been by no means universal. William Morris described them as a 'dreary pretentious heap'.[8] Critics of his work had varying points of view, some relying almost entirely on artistic criteria, others, as the judges in the Manchester Town Hall competition **[284]** had done, viewing each building also as a functional whole. It was said of Owens College **[321]**: 'The style is Gothic, somewhat freely treated, being like so many of Mr. Waterhouses's works, an adaptation of mediaeval features to the requirements of the present age; the building, however, is remarkable more for the excellence and convenience of its internal arrangements, than architectural pretensions externally.'[9] Late in his career the patrons of Mansfield College, Oxford **[525]**, had considered commissioning him, but his design 'a tired piece of French Renaissance' was rejected in favour of Basil Champney's Gothic,[10] probably largely on grounds of style. And mere style was never what attracted Waterhouse.

Beginning his practice in the 1850s it was almost inevitable that he would be caught up in the tide of the Gothic style. Only recently has it been recognized that 'mediaeval architecture required a special treatment to adapt it to nineteenth century requirements, the sort of treatment it has received at the hands of Waterhouse, whose high reputation is owing to his skilful adaptation of its spirit and beauty to the conditions of the time in which we live'.[11] It was this special treatment that led Lethaby at the turn of the century to categorize him as one of the 'hard' goths, those who thought Gothic and built according.

Waterhouse's free, or even cavalier, attitude to style was recognized by early critics. Eastlake, in stressing the importance of considerations other than the solely artistic at the Assize Courts **[67]**, echoes the assessors' decision on the town hall, which at the time he wrote was still under construction.[12] At Balliol **[254]**, Eastlake conceded there was 'a marked improvement . . . in the breadth and vigour'. He disagreed with the critics of the buildings, but preferred to leave the final judgement to 'the hand of time, whose artistic touch has exercized perhaps a more potent influence than we suppose on the opinion of modern amateurs'.[13] Others, though, were more hopeful: Manchester Town Hall **[284]** was described as being 'in that manner of Gothic treatment which its author seems to have finally made up his mind to, and which he has to some extent invented for himself; a manner which is not highly picturesque in detail, but has a sterling common-sense look

[4] *Builders' Journal and Architectural Record* (30 Aug. 1905), quoting *The Times*.

[5] *Building News* (25 Aug. 1905).

[6] *The Times* obituary of the late Alfred Waterhouse, quoted in *Architectural Record* (30 Aug. 1905).

[7] Kenneth Clark states that this is the only favourable comment by Ruskin he can find after 1860 on a completed Gothic revival building: Kenneth Clark, *The Gothic Revival* (London, 1950), 289. Ruskin made the remark in his lecture 'Traffic', delivered at Bradford (21 Apr. 1864).

[8] Philip Henderson, *Letters of William Morris* (London, 1950).

[9] Alfred Brothers, FRAS, with introduction by James Croston, FSA, *Manchester As It Was: A Series of Views* (Manchester, 1878).

[10] Clyde Binfield, *So Down to Prayers: Studies in English Nonconformity 1780–1920* (London, 1977).

[11] Alfred Darbyshire, *An Architect's Experiences: Professional, Artistic, Theatrical* (Manchester, 1897).

[12] C. Eastlake, *The History of the Gothic Revival* (London, 1872), 314.

[13] Ibid. 361–2.

about it which will probably always be recognized as satisfactory and appropriate in its way, whatever changes of taste we may go through'.[14]

'Always satisfactory' was, however, poor praise at a time when Ruskin's notions still held a good deal of weight and when the Queen Anne style was ushering in a new taste for elegance. Waterhouse's preference remained for sturdiness and solidity. In later years he admitted a 'singular fascination' for the work of the American architect H. H. Richardson, whom he regarded as having an 'original, creative mind free from prejudice and unfettered by precedent', producing buildings which were 'unmistakably his own creation'.[15] Richardson's predilections for high Victorian Gothic would have been an initial point of contact but Waterhouse's admiration for him embraced qualities which they both share: 'the discarding of the orders, the Romanesque feeling, admirable planning, appearance of strength and solidity, the value put upon mere wall surface whenever attainable, the treatment of this surface by the varied coursing of the stonework, and the contrast between tooled and rough surfaces; the use of coloured materials; battered bases, sparing use of string courses, the introduction of deep voussoirs. . . ';[16] but all this does not preclude some adverse criticism of Richardson's sense of proportion. Richardson's debt to Waterhouse in some of his buildings, in particular the Glessner house in Chicago, is a mark of approval of the same qualities in the older architect. And that approval comes more often from architects than from professional critics.

Waterhouse's search for modernity was recognized in the early 1870s with some misgivings:

> in one sense it would seem that [he] has effected the union between Medieval and modern feeling in architecture to an extent that has not been achieved elsewhere . . . there is a level standard of neatness and formal finish which can hardly be found fault with and yet which impresses us as cold and deficient in feeling . . . Mr. Waterhouse appears, however to have settled his path and we have little doubt that the buildings he will leave in this style will always command approval and commendation, even when they fail to rouse enthusiasm.[17]

Opinions only hardened in the last years of his life, until his buildings were described as 'wholly devoid of charm',[18] though one writer thought that the prejudice against his architecture had been carried, in what might be called 'aesthetic' circles, to an absurd degree.[19] For almost half a century after his death, his reputation remained unchanged. Architectural historians and commentators excluded him or actively disliked his work. Lethaby, in his clipped anecdotal biographies of nineteenth-century architects, while admitting his greater sense of reason, gently puts his methods in another stream from the then current 'softer effects of stylism'.[20] Cecil Stewart put it far more strongly as marching 'firmly in the wrong direction'.[21] Sir Charles Reilly in Liverpool invented the tag 'pastrycooks' architecture' for his work;[22] and even Hitchcock dubbed Manchester Town Hall an inferior work.[23]

A few critics did see his qualities, and Kenneth Clark, while not discussing any of his buildings individually, credited him with a sense of construction which makes his buildings more real than Scott's.[24] These very qualities, together with a belief in function as the prime concern of the

[14] *Builder* (17 Oct. 1874).
[15] AW, Presidential address to RIBA (Nov. 1888), published in the *Builder* (10 Nov. 1888). [16] Ibid.
[17] *Builder* (16 May 1874). Architecture at the Academy.
[18] *British Architect* (30 Oct. 1896).
[19] *Builders' Journal and Architectural Record* (30 Aug. 1905).
[20] W. R. Lethaby, *Philip Webb and his Work* (London, 1935) 74.
[21] Cecil Stewart, *The Stones of Manchester* (London, 1956), 71.
[22] Quoted in T. Kelly, *For the Advancement of Learning: The University of Liverpool 1881–1981* (Liverpool, 1981).
[23] Henry-Russell Hitchcock, *Architecture: Nineteenth and Twentieth Centuries* (Harmondsworth, 1968), 186.
[24] Clark, *The Gothic Revival*, 263.

architect, were surely the core of his design philosophy. Function and construction first was the leitmotif of Pevsner's classic *Pioneers of the Modern Movement*. However, in that work Waterhouse does not even rate a mention. Nor is he referred to in Siegfried Giedion's *Space Time and Architecture*, that other classic of the interwar period. It is an indication of how deep the reaction against his stylistic effects had gone.

It was left to political, economic, and social historians to appreciate his worth. Sir Robert Ensor, writing at the same time as Pevsner and Giedion, considered that while T. G. Jackson's Oxford Examination Schools of 1882, 'discarding Gothic for English Renaissance, afforded one of the first large examples of the return to styles more congruous with modern needs and capacities, the typical architect of the sixteen years 1870–86 was Alfred Waterhouse, whose Manchester Town Hall was still an orthodox, though very skilful essay in Gothic, but who subsequently worked his way to modified styles, showing a much greater sense of reality'.[25]

After the war H. S. Goodhart-Rendel, witty champion of Victorian architects and their buildings, still felt bound to put Waterhouse's stock low, suggesting that, while he himself found the buildings 'definite, logical and brave', Oxford was still waiting for Eastlake's expectations on Balliol to be fulfilled.[26] Goodhart-Rendel recognized his supreme skill as a planner;[27] and it has since been accepted wisdom that his works are probably very practical but undeniably ugly.

Eventually Robert Furneaux Jordan tried to 'evoke just a little of the vigour, the self-assurance, the sheer artistry of . . . Waterhouse's High Victorian architecture in Manchester'.[28] Even so the turn of the tide of taste which has directed attention to the architecture of the nineteenth century has been slow to embrace his work. The cult of ugliness which made Butterfield suddenly attractive was not extended to Waterhouse, who was still felt to be dour and heavy handed. The cleaning of the Natural History Museum provided a salutary jolt to this way of thinking. A more recent popularity poll has placed him sixth among leading architects of the last 150 years and his Natural History Museum fourteenth among buildings of the same period.[29] Then the Museum was selected for inclusion in the first volume of *Timeless Architecture*,[30] along with Soane's Dulwich Picture Gallery, Barry's Reform Club, E. S. Prior's Church of St Andrew Roker, and Victor Horta's Tassel House in Brussels. Waterhouse could hardly want for better company since, as Mark Girouard wrote in his introduction, those buildings were all 'works of architects who could be described as creative professionals. They knew their materials, they knew how to construct and build buildings that would weather well, and they were prepared to use the latest technology. Those who had offices ran them

[25] *Sir Robert Ensor, England 1870–1914* (Oxford, 1936), 156.

[26] H. S. Goodhart-Rendel, *English Architecture Since the Regency* (London, 1953), 156. In his introduction to *Seven Victorian Architects* (London, 1976) Nikolaus Pevsner records his experience at his inaugural lecture, as Slade Professor in Cambridge in 1949, of being ridiculed by the students for attempting to analyse the buildings of Matthew Digby Wyatt, 1820–77. One of the authors was among the audience. Such was the attitude to 19th-cent. architecture in general that it was with great diffidence that I made it known to my fellow architectural students that the subject of my thesis was to be Alfred Waterhouse. Some tolerance was shown on account of his being my great-grandfather. P. W.

[27] Goodhart-Rendel, *English Architecture since the Regency*, 158.

[28] R. Furneaux Jordan, *Victorian Architecture* (Harmondsworth, 1966), 18.

[29] In 1984 *Building Design* conducted a survey of British architecture of the previous 150 years. Waterhouse came 6th among the architects following Macintosh, Lutyens, Stirling, Foster, and Voysey. The Natural History Museum came 14th among buildings. In a programme on BBC TV's *Saturday Review* series in conjunction with the 'New Architecture' Exhibition at the Royal Academy in Oct. 1986, James Stirling, Richard Rogers, and Norman Foster were interviewed. Manchester Town Hall was described by Foster as a great building, almost the only completed building mentioned by the 3 architects during the programme.

[30] Don Cruikshank (ed.), *Timeless Architecture*: i (London, 1985). This was a collection of essays first printed in the *Architects Journal*.

efficiently. They kept close, or close enough, to their schedules and their estimates. They found out what would suit their clients and set out to provide it; none of them indulged their own foibles or played elaborate games against their client's interest but at their expense. But the end products were more than convenient buildings at a price which the clients could afford. They were also creative works of art which caught the imagination of contemporaries or posterity, or were admired and imitated around the world.'[31] What better critical epitaph could an architect have.

[31] M. Girouard's Introduction, ibid.

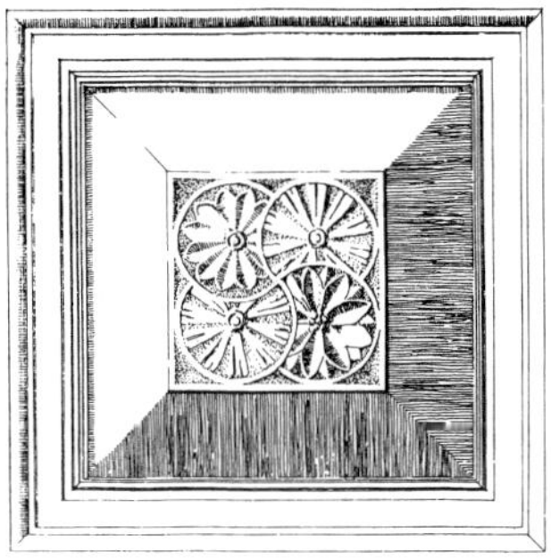

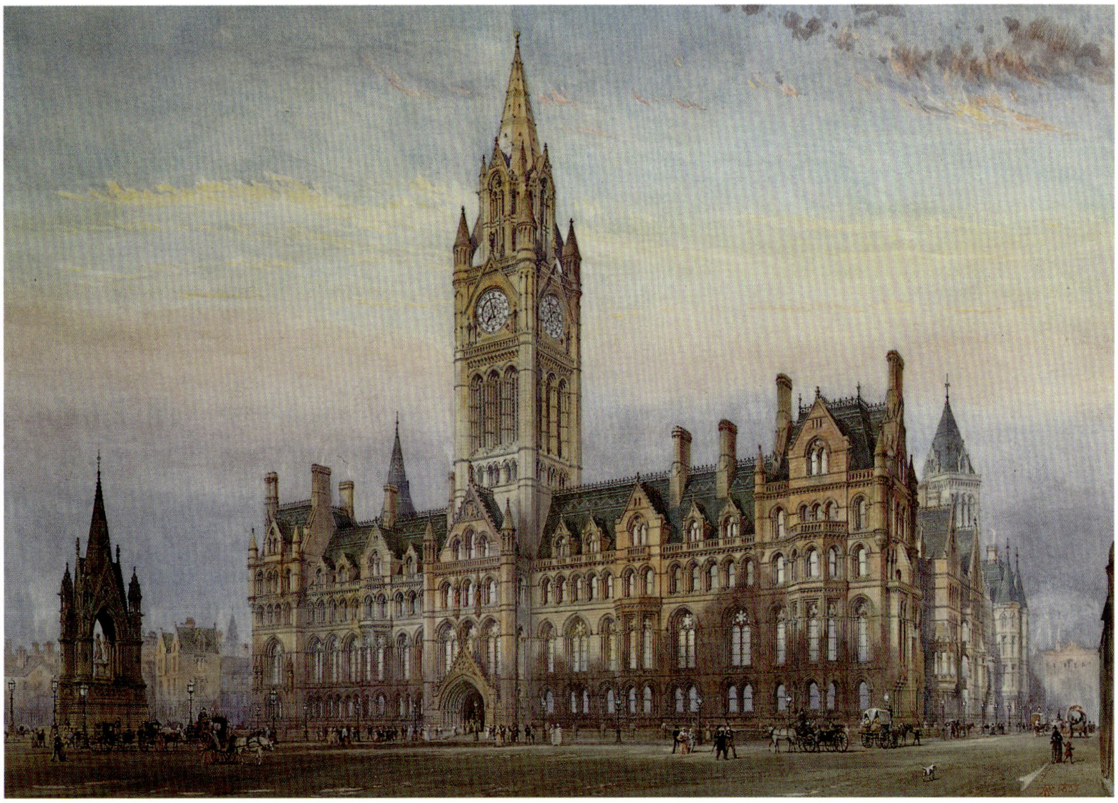

1. *Manchester Town Hall* **[284],** perspective 1887, exh. Royal Academy (1887), Diploma Work

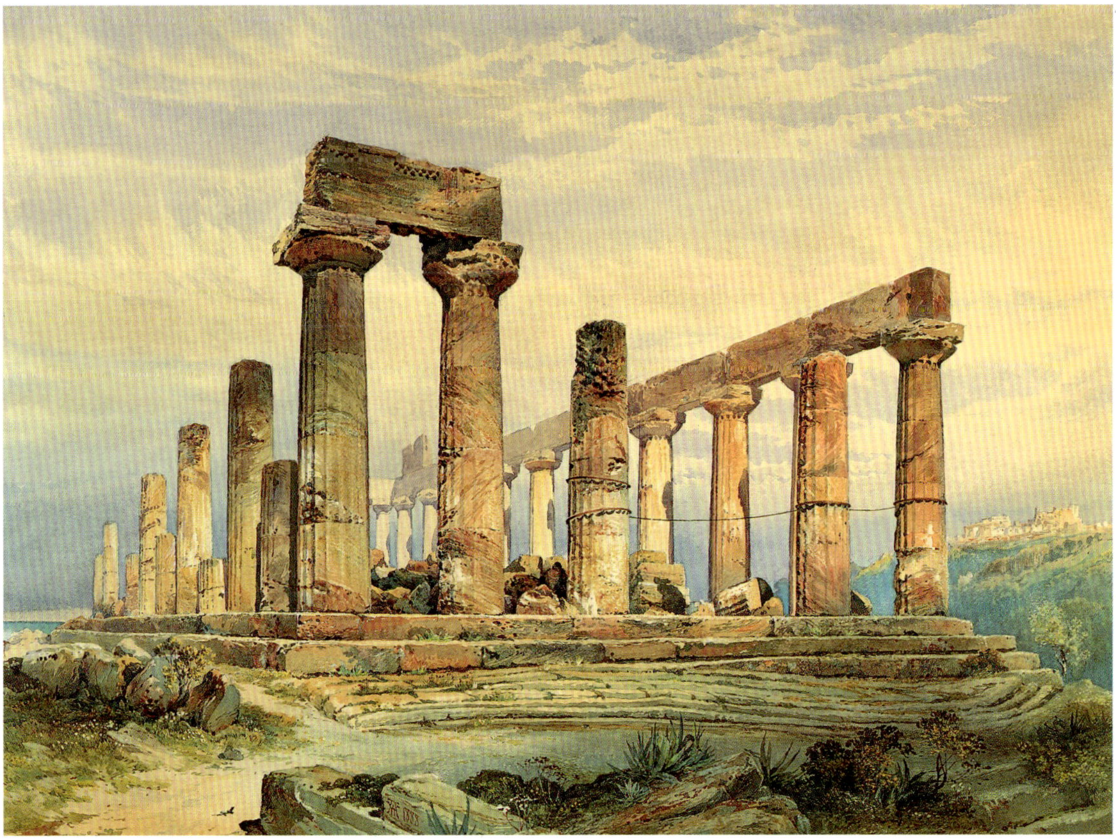

2. *Temple Ruins at Agrigentum*, 1888, exh. Royal Academy (1888)

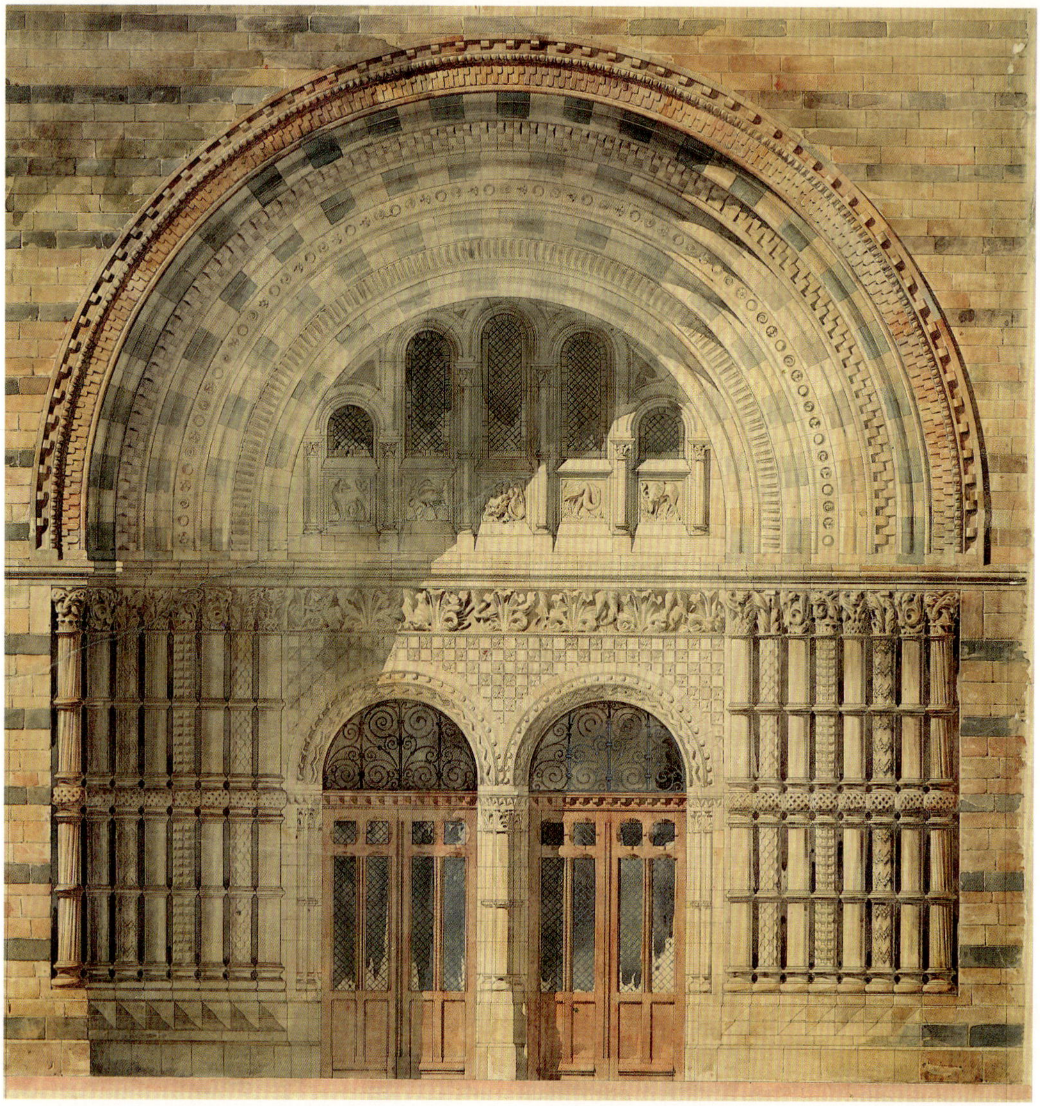

3. Natural History Museum, London **[243]**, elevation of main entrance, *c.* 1876

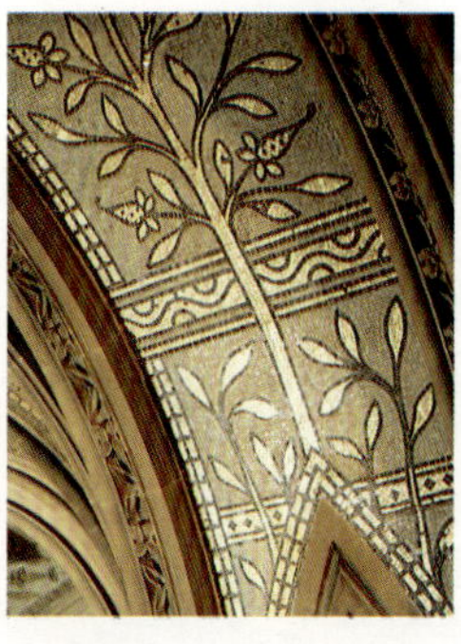

(*a*)

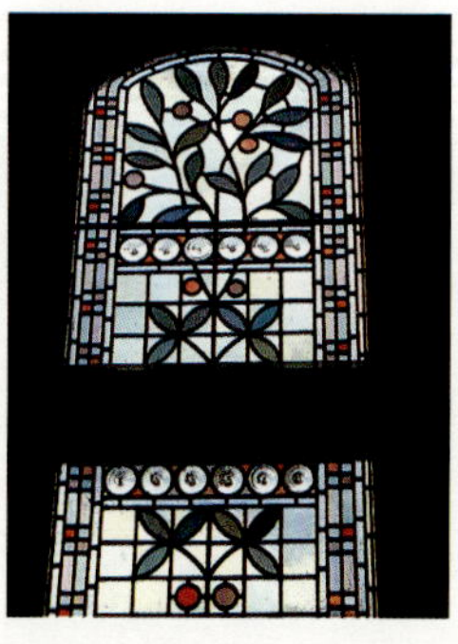

(*b*)

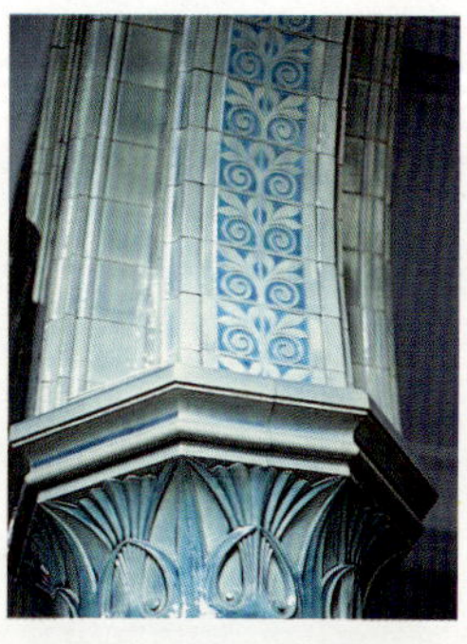

(*c*)

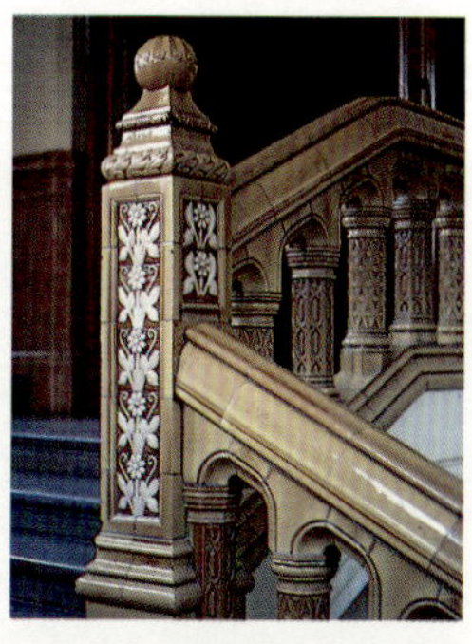

(*d*)

4. Details of Coloured Materials: (*a*) glass mosaic, Manchester Town Hall **[284]**; (*b*) stained glass, Silwood Park **[439]**; (*c*) faience, Liverpool Royal Infirmary **[532]**; (*d*) faience, Leeds University, Great Hall **[574]**; (*e*) geometric tile floor, Oxford Union Society **[464]**; (*f*) stained glass, Cambridge Union Society **[197]**

(*e*)

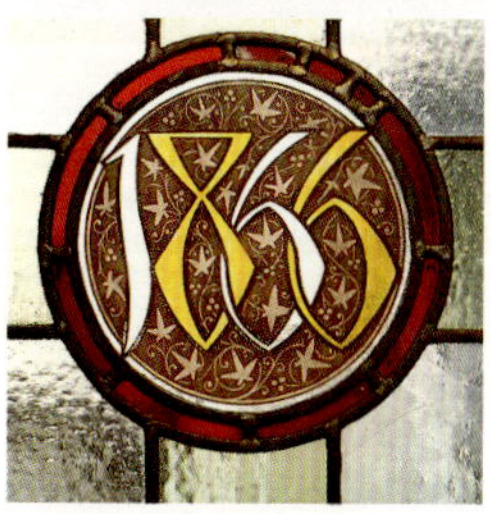

(*f*)

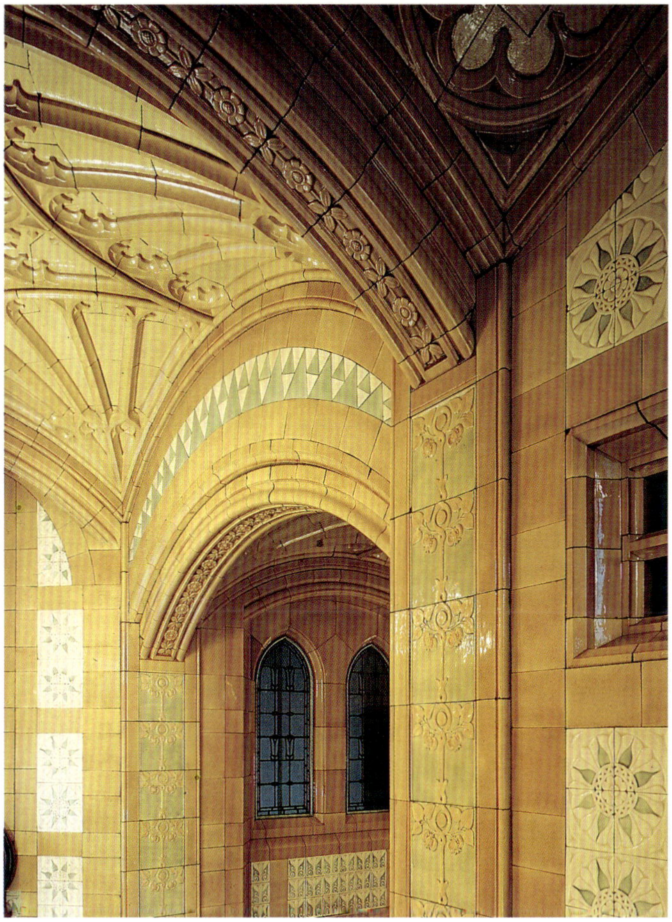

5. Prudential Assurance, Chief Office, Holborn, London, 1895–1905, principal staircase [602]

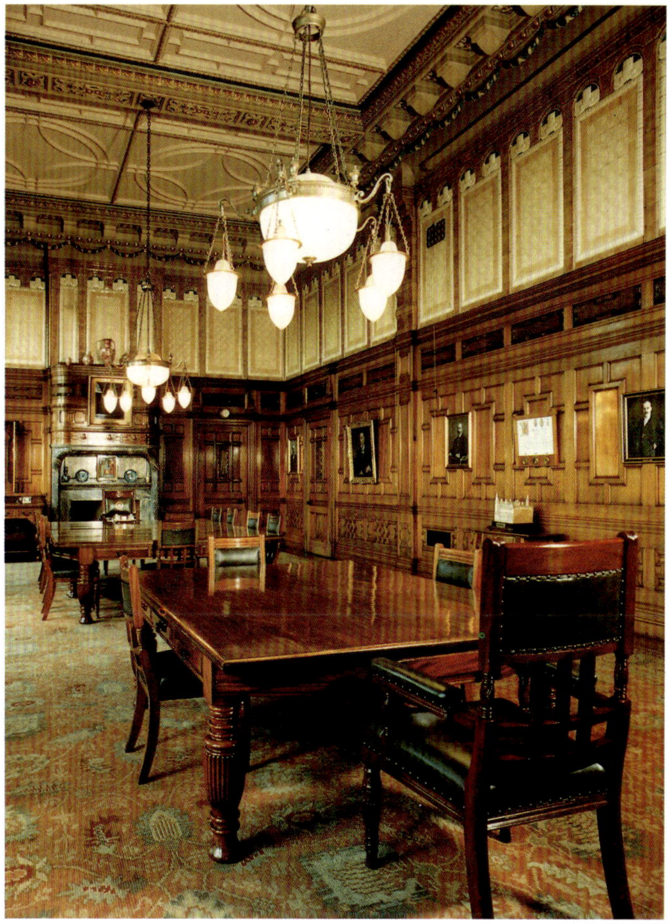

6. Refuge Assurance, Manchester, 1891–6, boardroom [570]

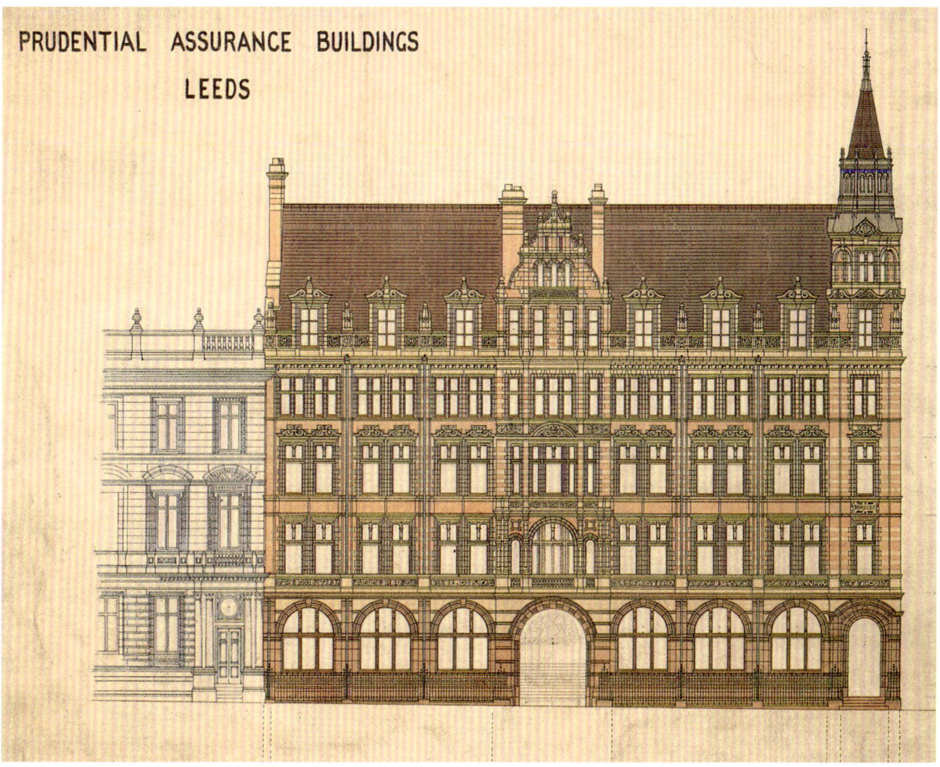

7. Prudential Assurance Building, Leeds, 1891, elevation [559]

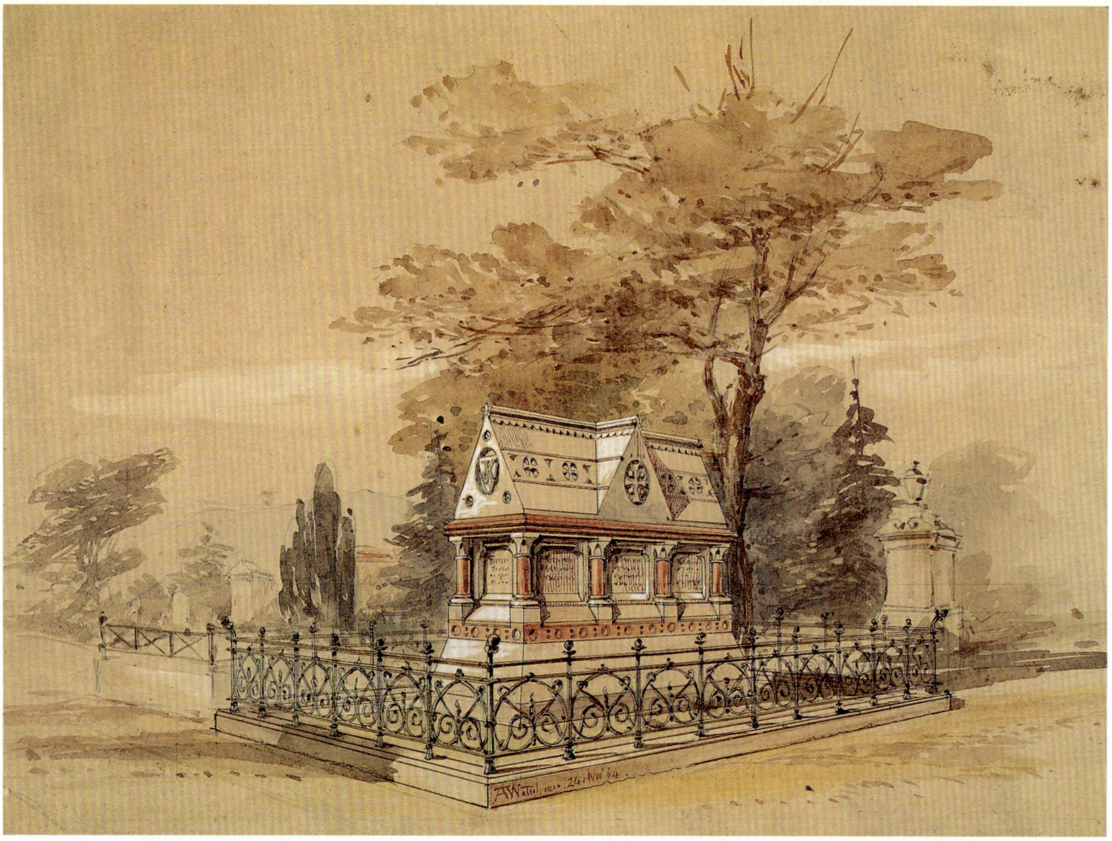

8. Design for a monument to James Kershaw, MP, Norwood Cemetery, London, 1864 **[181]**

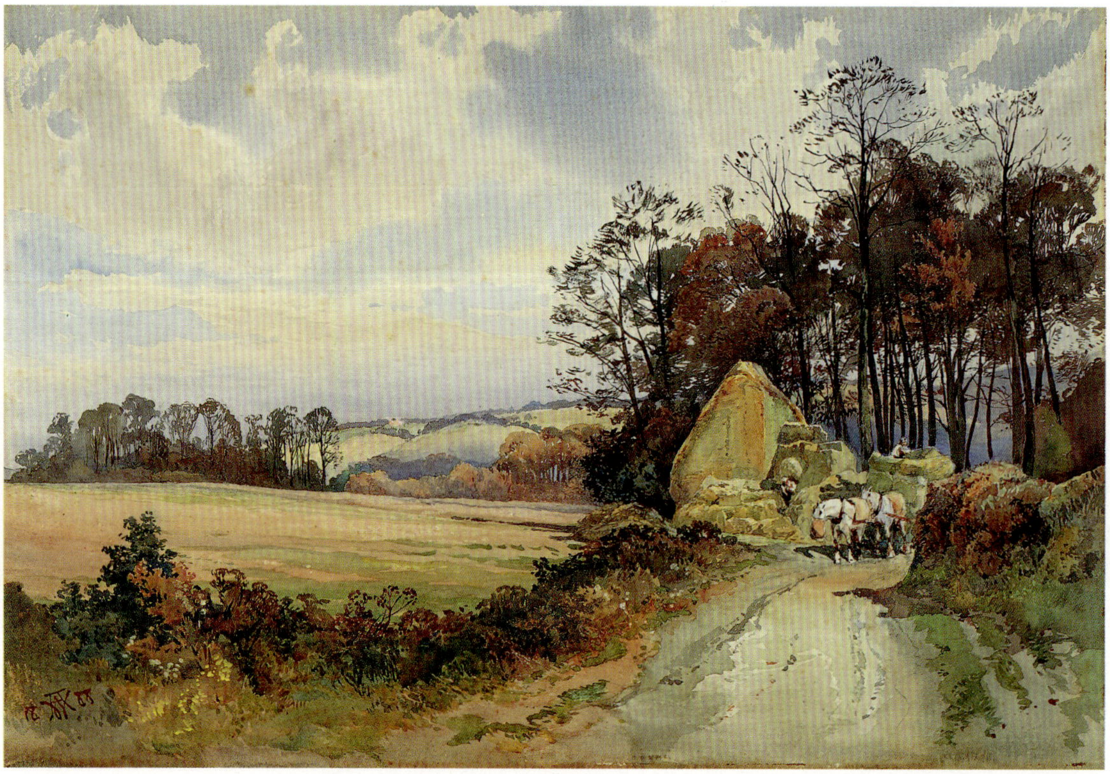

9. *A Berkshire Lane at Easter*, 1889, exh. Royal Academy (1889)

Appendices

Appendix 1

LIST OF WORKS

This list has been compiled from the various office archives, letter books, and from the surviving drawings and buildings themselves. Buildings are listed according to the date of the commission, which in some cases precedes the beginning of construction by more than a year. Dates of completion are also shown. Separate entries for later extensions by AW are included when those were the subject of a separate commission. Major alterations by AW are dated and described where they are significant. Alterations by other architects are noted only where they substantially affect the original design.

The list includes some details of contractors and cost, and the name of the clerk of works, where these are available. The names of the main contractor and of certain key subcontractors are included, as they provide an invaluable insight into the pattern of AW's work. Precise figures of cost are difficult to establish and in some cases even the architect's own records give different totals. Wherever possible we have based our figure on the certificates issued (to the nearest £5); the costs therefore do not include the architect's fees, usually 5 per cent, or the site purchase.

This list is presented in chronological order, with entries alphabetically for each year. The entries are numbered, and the numbers are used for cross reference to the text and to the geographical concordance, Appendix 2. The following abbreviations are used to indicate the documentation for these commissions:

c = information in office correspondence or archives in private collection;
C = information in office correspondence (2 vols.) in RIBA Library;
D = drawings in the RIBA Drawings Collection;
d = drawings and/or other archival material in other collections, or in possession of the building owner;
Dem = building known to be demolished at the time of writing;
M = drawings in Manchester University School of Architecture;
O = information from the office archives (private collection);
P = drawings and other material in the Public Record Office;
S = cited in S. A. Smith's thesis (see Bibliography) but not otherwise corroborated;
T = information or drawings in Manchester Central Library archives, principally the Thompson letters;
V = drawings or other items in the Victoria & Albert Museum.

[1] 1852
MANCHESTER: 23 Ardwick Green [S Dem]
Alterations to house: enlargement of dining-room
Client: H. Waterhouse
Drawings bear note 'Richard Lane Architect'

[2] 1854
AMBLESIDE, Cumbria: Rothay Holme [c d O]
House of stone with a slate roof, elaborate barge boards, and timber verandah of two storeys. Repairs 1860. Alterations 1867, 1871, and 1870 by J. Fleming
Client: Elizabeth Head
Contractors: general, E. Wilson & Sons; drains etc., Miles Clark; grates, W. Wilson; ironmongery, Hart & Son
Cost: professional charge £20; 1860 £220

[3] 1854
BRISTOL, Avon: Sneyd Park [d]
Stables
Client: A. Waterhouse Sen.

[4] 1855
AMBLESIDE, Cumbria: Brathay Hall [c O]
Late eighteenth-century mansion. Alterations to entrance hall, kitchen, and service wing, redecoration of mansion. Repairs 1862
Client: R. R. Redmayne
Contractors: general, Ed. Wilson; furniture, sketch for desk by AW
Cost: *c.*£1,500

[5] 1855
BRISTOL, Avon: Sneyd Park [d]
Parlour desk (for his parents' home)
Client: A. Waterhouse Sen.

[6] 1855
HINDERTON, Cheshire [O]
Alterations to cottage (this may be Hinderton Hall Farm)
Client: ? Christopher Bushell
Cost: estimate £2,000

[7] 1855
MANCHESTER, Chester Street: Binyon & Fryer Warehouse [c d O Dem]
Warehouse and sugar refinery with two-storey façade of brick with stone dressings, Italianate detail (Pl. 22)
Alterations 1863 **[154]** q. v.
Client: Thomas Binyon
Cost: £12,000

[8] 1855
TAUNTON, Somerset: Assize Courts and Judges' Lodging [c]
Designs for competition entry

[9] 1855–6
LOCATION UNKNOWN: Queenwood [c O]
Unspecified alterations to house
Client: G. W. Edmundson
Cost: £960

[10] 1855
LOCATION UNKNOWN: [c O]
Chimney piece and flue
Client: P. Clare

[11] 1856–7
BOLTON, Lancashire: Egerton Hall [c Dem]
Addition of conservatory to mansion
Alterations see 1860 **[74]** Egerton Hall
Client: Edmund Ashworth
Contractors: masonry, Radcliffe, Smith; tiles, Maw & Co.
Cost: £1,900

[12] 1856–7
?BOWDON, Cheshire [c]
Minor works including desks, rails, stoves, cupboards
Client: William Nield
Contractors: Warburton
Cost: £65

[13] 1856–7
HINDERTON, Cheshire: Hinderton Hall [c C d O]
Mansion of red sandstone with a slate roof. Stables, gardener's cottage and boundary walls (Pl. 26)
Alterations 1868, 1875
Client: Christopher Bushell
Contractors: general, Haigh & Co; tiles, Minton; T. Oakden; chimney-pieces, Bonehill; Patterson; W. Bennett; iron work, Hart & Son; F. A. Skidmore
Sir Thomas Deane advised on alterations to porch

[14] 1856–7
INCE-IN-MAKERFIELD, Lancashire: Cemetery [c]
Two chapel buildings (one Anglican; one double, Roman Catholic and Nonconformist), lodge and gates for cemetery. All of stone with slate roofs. Anglican chapel in Gothic, other in Norman style (Pl. 25)
Contractors: general, James Fairclough; carving, J. Bonehill; iron work, Hart & Son

[15] 1856
KENDAL, Cumbria: Unidentified House [c]
Client: William Wilson

[16] 1856–7
KESWICK, Cumbria: Pettitt's Art Gallery [c D O]
Two-room gallery of local stone with slate roof and wooden verandah
Client: A. and G. W. Pettitt
Contractors: general, Edward Wilson; iron fittings, W. Wilson

[17] 1856
LYMM, Cheshire: Stables [c]
Client: Thomas Morris
Contractors: general, John Ardens; locks, Hart & Son

[18] 1856
MANCHESTER: location unknown [c]
Refurbishment of shop and offices
Client: J. Barratt

[19] 1856
MANCHESTER, Higher Lane: ? Whitefield [c]
Front boundary wall for new house
Client: G. C. Dewhurst

[20] 1856
MANCHESTER, Prestwich: Oak Hill [c]
New conservatory for house
Client: Samuel Eveleigh

[21] 1856
SOUTHPORT, Lancashire: Victoria Baths [c]
Refurbishment of public baths, additions and alterations
Contractors: general, Stanley; tiles, Minton; T. Oakden
Cost: £1,150

[22] 1856–61
WINDERMERE, Cumbria: Elleray Bank [c O]
House of rubble walling in local stone with ashlar dressings and a shallow pitched slate roof, including garden layout. see 1866 **[261]** Elleray Bank, stables (proposed enlargement 1869; not built)
Client: G. G. Cunningham
Contractors: general, Abraham Pattinson; chimney-pieces, J. & H. Patteson

[23] 1856
LOCATION UNKNOWN [c]
Alterations to house, including drawing-room and stables
Client: Mr Norris
Contractors: Joseph Ardens

[24] 1857
BRADFORD, Darley Street: Bradford Old Bank (now Midland Bank) [C O]
Stone faced bank building in round arched/Italianate style. Greatly enlarged in matching style 1866; additional storey 1881
Client: Bradford Banking Co. (A. Harris)
Contractors 1865–8: general, J. & W. Beauland (including fittings); structural steelwork, W. H. Lindsay; fireproof construction, A. & W. Dennett; stone carving, Farmer & Brindley; tiles, Maw & Co.; chimney-pieces, W. H. Burke & Co.; gas fittings and umbrella stands, J. Hart & Son; iron work, Robert Jones; plaster centre flowers, J. W. Hindshaw
Contractors 1882: general, Wm. Brunten
Clerk of works: J. S. Wilson
Cost: 1865–8 £13,2450; 1881–2 £7,170

[25] 1857
EASENHALL, Warwickshire [c]
Survey of estate and first proposal for house; unexecuted. see 1871 **[348]**
Client: Edward Cropper

[26] 1857
GRASMERE, Cumbria [c]
Proposed house
Client: J. Harwood

[27] 1857
KESWICK, Cumbria [c]
House and lodge; possibly design only
Client: Robert Wilson

[28] 1857–61
KESWICK, Cumbria: Rose Trees [c O]
Small house of local stone with a low pitched slate roof, also advice on garden layout
Client: Gilbert Henderson

Contractors: foundation, Watson; mason, Bromley
Clerk of works: S. Harrison
Cost: *c.*£1,500 (house and grounds on sale for £7,000, 1861)

[29] 1857
LIVERPOOL: 39 Catherine Street [c C d O]
Terraced house of brick in two colours with stone dressings (Pl. 18)
Client: Sebastian Waterhouse
Cost: £2,700

[30] 1857
LIVERPOOL: Proposed new Friends' Meeting House [c O]
Quaker meeting house to seat 1,000; not built
Client: Henry Crosfield and Rogers Waterhouse
Cost: estimated £5–6,000

[31] 1857
LYMM, Cheshire [c]
Alterations to roof of cottage
Client: Mr Brocklebank
Contractors: Joseph Ardens

[32] 1857
MANCHESTER, Ardwick Green: Ragged and Industrial School [c O Dem]
Three-storey building of brick for school, workshop, and dormitory, added to exisitng school
Client: Commissioners of the Ardwick Green Ragged School (per Jos. Adshead)
Cost: £4,000

[33] 1857
MANCHESTER, Fairfield (unidentified) [c O]
House, possibly only alterations. See 1866 **[252]**
Client: Gaylord Hadwen

[34] 1857
MANCHESTER: Greenfield House [c]
Designs for building
Client: Revd B. Benson

[35] 1857
RYDAL, Cumbria: St Mary's Church [c]
Restoration of church tower, with replacement pinnacles
Client: John Green

[36] 1857
WELLINGTON, Somerset: Swallowfield [c]
Alterations to house
Client: Henry Fox

[37] 1857
WIGAN, Lancashire: The Minorca Inn [c]
New canopy over doorway
Client: John L. Hunter

[38] 1857
LOCATION UNKNOWN: Bellamont [c]
Survey of house and advice on potential alterations
Client: Edward Cropper

[39] 1857
LOCATION UNKNOWN (Cumbria) [c]
Design for cottage
Client: D. Bailey Jones

[40] 1857
LOCATION UNKNOWN [c]
House
Client: Matthew Kennedy
Cost: estimate £1,800-£1,900

[41] 1857
LOCATION UNKNOWN [c]
Designs for cottages
Client: Mr Millar

[42] 1858
AMBLESIDE, Cumbria: Brathay Hall [c O]
Gate lodge of stone with a slate roof, timber verandah (Pl. 19)
Client: R. R. Redmayne
Contractors: general, Miles Clark
Cost: £350

[43] 1858
DARLINGTON, Co. Durham: Proposed porter's lodge [c O]

Unexecuted
Client: Dr Nathan Plews (per Compton Birchall)
Cost: bad debt

[44] 1858–60
DARLINGTON, Co. Durham: Blackwell House [c O]
Alterations to house
Client: E. Backhouse
Contractors: general, Thomas Robson; Richardson & Ross
Cost: £1,115

[45] 1858
KESWICK, Cumbria: Fawe Park [c d O]
Small house of local stone with slate roof and half-timbered gallery, also lodge and design for garden (Pl. 16). Much altered and extended by others
Client: James Bell
Contractors: mason, W. Bromley; R. Knubley; joiner, J. Crosthwaite; chimney-pieces, Hopton Wood Stone Co.
Cost: estimate £1,200

[46] 1858–60
LEEDS, W. Yorkshire, Headingley [O]
Unspecified alterations to house
Client: W. W. Hewitson
Cost: £250

[47] 1858
LIVERPOOL, Aigburth: Mossley Bank [O]
Alterations and additions to house
Client: Rogers Waterhouse

[48] 1858–9
LYMM, Cheshire: Eagle Brow, High Bank and Low Bank [c O]
Pair of semi-detached houses of brick with stone dressings
Client: G. C. Dewhurst
Cost: £1,400

[49] 1858–60
MANCHESTER, Concert Hall Street off Fountaine Street: Concert Hall [c C O Dem]
Reconstruction of auditorium and enlargement of gallery with other internal alterations. Alterations 1865
Client: Manchester Concert Hall Committee
Contractors: general, Robert Neill; heating and ventilating, W. Wilson; gas fittings, W. Wilson; decoration, Brandt; upholstery, Hulme; organ case, Kirtland & Jardine; ornamental plaster, G. Jackson
Cost: £2,145; alterations £510

[50] 1858–9
MANCHESTER, Droylesden: Institute [O Dem]
Reading Room etc. of red brick with blue brick patterning and a slate roof (Pl. 23)
Cost: £1,255

[51] 1858
?MANCHESTER: Greengate [c]
Alterations and refurbishment of three cottages
Client: J. B. Eveleigh

[52] 1858
MANCHESTER, Victoria Park: Barcombe Cottage [c d O Dem]
House of stone with a low pitched slate roof. Alterations 1863 (Pl. 17)
Client: A. Waterhouse Jun. (for himself and his wife)
Contractors: Stone carving, T. Woolner
Cost: alterations £440

[53] 1858
LOCATION UNKNOWN (possibly London) [c]
Survey of cement render of house
Client: Dr Phillips

[54] 1859, 1864, and 1870–2
BRASTED, Kent: Brasted Park [C O]
Alterations and enlargement of house. Further alterations 1887–9, 1896
Client: W. Tipping
Contractors (1870–2): general, Dartnell & Wells; steelwork, J. S. Bergheim; terracotta, Gibbs & Canning
Cost: 1864–72 £4,735; 1887 £1,350; 1896 £30

[55] 1859–61
BELFAST, Lisburn [O]
Alterations to house and new verandah. Alterations 1863–4
Client: R. Barbour
Cost: £610; alterations £420

[56] 1859
CARTMEL, Cumbria: Friends' Meeting House [O]
Unspecified alterations
Client: per Robert Jackson
Cost: £400

[57] 1859
COALBROOKDALE, Shropshire: Ness Grange or Nesscliff [O]
Unidentified house
Client: Henry Dickinson
Cost: £1,220

[58] 1859
KENDAL, Cumbria: Derwent Grange [O]
Unidentified house of Lakeland stone
Client: Robert Wilson
Cost: £3,010

[59] 1859–60
KESWICK, Cumbria: Thornthwaite Parsonage [c O]
Design for new house, or for alterations; not carried out
Client: Revd H. Martin Short

[60] 1859
LEEDS: Friends' Institute [O]
Unspecified work
Cost: £360

[61] 1859
LYMM, Cheshire [C O]
Semi-detached house for Dr Simpson
Client: G. C. Dewhurst
Cost: £4,000

[62] 1859
MANCHESTER or BELFAST [S]
Unspecified work
Client: R. Barbour

[63] 1859
MANCHESTER: Audenshaw [O]
Unspecified work
Client: William Hadwen
Cost: £70

[64] 1859–62
MANCHESTER: ?Fairfield [O]
New mill tower and gallery, grate and gate piers
Client: Richard Christy
Cost: £340

[65] 1859
MANCHESTER: Fallowfield, Victoria Park [O]
Design for a house; not built
Client: John Barratt

[66] 1859
MANCHESTER: Fallowfield, Victoria Park [c O]
Design for house; not built
Client: E. F. Saul

[67] 1859–65
MANCHESTER, Great Ducie Street: Assize Courts [c C d D O V Dem]
Assize Courts building of polychrome stone with a patterned slate roof, consisting of hall, two court-rooms, and other offices and tower, all symmetrically planned with a judges' lodging to the west. Won in competition (Pls. 33, 34, 35, 36, 174, 187; Fig. 1)
Client: Manchester Assize Courts Committee
Contractors: foundation, H. Southern & Co.; superstructure and finishing, Samuel Bramall; heating and ventilating, G. N. Haden; stone carving, T. Woolner; Farmer & Brindley; O'Shea & Whelan; Banks*; Lane* (*samples only); tiles, Thomas Oakden; stained glass, R. B. Edmundson; Lavers & Barraud; Geo. Shaw; Heaton Butler & Bayne; decoration, J. G. Crace; furniture/fittings, Doveston, Bird & Hull; Jas. Lamb; Kendal & Co.; J. Beaumont; Minton & Co.; Marsh & Jones Co.; iron work, F. A. Skidmore; R. Jones; chimney-pieces, J. Bonehill; W. Wilson; J. & H. Patteson; plaster ceiling roses, J. W. Hindshaw
Clerk of works: John Shaw; G. O. Roberts; Henry Littler
Cost: £118,050

[68] 1859
MANCHESTER: Grosvenor Street [O]
Design for a seat and board
Client: Aspinall and Monkhouse

[69] 1859
MANCHESTER, Moss Side: Britannia Cottage [c C O]
Unidentified house. Alterations 1863; 1865; 1869; 1871
Client: Robert Corkling
Contractors: general, Samuel Bramall; ironwork, R. Jones; Hart & Son
Cost: £1,210; 1863 £410

[70] 1859
MANCHESTER: Smedley Lane, Cheetham Hill [c O]
Minor alterations to house; possibly not carried out
Client: E. L. Glover
Cost: Bad debt

[71] 1859
READING, Berkshire, Whiteknights Park: Erlegh Park [c O Dem]
House, stables, and furniture
Client: Thomas Porter
Contractors: general, J. Wheeler & Sons; heating and ventilating, D. O. Boyd; chimney-pieces, D. O. Boyd; W. Wilson; furniture/fittings, Holland & Sons; iron work, F. A. Skidmore
Clerk of works: W. R. Turnbull
Cost: £10,085

[72] 1859–60
STANDISH, Lancashire [O]
Designs for church restoration, windows, bench ends, etc.; not carried out

[73] 1860
BOLTON, Lancashire: Eagley Works [O Dem]
Addition of verandah
Client: Arthur Greg
Cost: £115

[74] 1860
BOLTON, Lancashire: Egerton Hall [c Dem]
Alterations to porch, library, etc. of existing mansion. Alterations 1867, 1870
Client: Edmund Ashworth

[75] 1860
BOLTON, Lancashire: Egerton Reading Room [c O]
Sketch design only; not built
Client: Edmund Ashworth

[76] 1860
BRASTED, Kent: National Schools [c O]
Alterations to school building
Client: W. Tipping
Contractors: general, H. Singer
Cost: £1,000

[77] 1860
?DARLINGTON [O]
Unspecified work
Client: J. C. Backhouse & Son

[78] 1860
GREAT AYTON, Cleveland: Cleveland Lodge [c O]
Additions and alterations to existing mansion. Alterations 1883–6
Client: John Hodgkin Sen.
Contractors 1883–6: general, Braithwaite; heating and ventilating, D. O. Boyd; tiles, Craven Dunnill; chimney-pieces, W. H. Burke; Hopton Wood Stone Co.
Cost: 1883–6 £790

[79] 1860
HALIFAX, Yorkshire: South Wood End [O]
Unspecified work to house
Client: Joseph Thorp
Cost: £160

[80] 1860
HAMSTERLEY, Co. Durham: Shull [O Dem]
Unspecified alterations to house
Client: J. C. Backhouse
Cost: £460

[81] 1860
HINDERTON, Cheshire [O]
Alterations to cottage
Client: Christopher Bushell
Cost: £100

[82] 1860
LIVERPOOL [O]
Unspecified work
Client: Isaac Hadwen
Cost: £1,900

[83] 1860
?MANCHESTER or BELFAST [S]
Unspecified work
Client: R. Barbour & Co.

[84] 1860
MANCHESTER, Buxton Street: Infants School [O Dem]
Alterations to school building
Cost: £130

[85] 1860
MANCHESTER: Cheetham Hill [O]
Partial preparation of plans for pair of semi-detached houses
Client: Messrs Sheldon and Beaumont

[86] 1860
MANCHESTER: Chepstow Street [O]
Sketches for extension; not built
Client: Watkin & King

[87] 1860–2
MANCHESTER, Droylesden: St Mary's Parsonage [c C O]
Parsonage house and design for garden layout
Client: Revd Philip Thompson; Gaylord Hadwen
Contractors: general, J. Bates & Co.; tiles, Thos. Oakden
Cost: £2,065

[88] 1860–6
MANCHESTER, Fallowfield, Victoria Park: ?Hillside [c C O]
House including garden layout, boundary walls and gates, and bookcases by G. T. Redmayne
Client: Mr Hewett

[89] 1860
MANCHESTER, Quay Street: Owens College [O]
Minor alterations and fitting out of building for Owens College
Cost: £70

[90] 1860–6
MANCHESTER, Swinton: Swinton Park [c O Dem]
House or possibly extensive alterations to existing mansion
Client: Thomas Binyon, then James Radcliffe from 1862
Contractors: general, S. Bramall; stained glass, Hayworth; chimney-pieces, W. Wilson; plaster ceiling roses, J. W. Hindshaw; plaster ornaments, G. Jackson
Cost: £9,755

[91] 1860
REETH, N. Yorkshire [c O]
Design for school buildings selected in competition
Contractors: Joseph Cleasby

[92] 1861
ALLONBY, Cumbria: Reading Room and Teacher's House [c O]
Reading room over an open ground floor and attached house, of brick, slate roof. Clock turret now demolished
Client: J. W. Pease
Contractors: brickwork, Eaglesfield; joinery, Longcake; plumbing, McKnight
Cost: £610

[93] 1861
BELFAST [c O]
Design for offices and warehouse, not built under AW's supervision
Client: W. Barbour and Son
Cost: Professional charges £12. 12*s*.

[94] 1861
?BELFAST
New conservatory and porch to house [c]
Client: R. Barbour

[95] 1861
BOWDON, Cheshire: High Lawn [c O]
Alterations to mansion
Client: William Nield
Cost: £1,880

[96] 1861
CHEADLE, Cheshire: The Arbour [c O]
Alterations to house
Client: Thomas Ashworth
Contractors: general, R. Niell; S. Bramall; glass, R. B. Edmundson; iron work, Best
Clerk of works: P. Wilds
Cost: £1,960

[97] 1861
CLAPHAM, N. Yorkshire [O]
Unspecified work
Client: Revd John Marrinet
Cost: £220

[98] 1861
DARLINGTON, Co. Durham: Brinkburn [c]
Examination and sketches of alterations to designs by Richardson & Ross; house completed to their designs 1862
Client: Henry Fell Pease

[99] 1861–3
DARLINGTON, Co. Durham: Market and Public Offices [c C D O]
Market building of iron with glass roof. Two blocks of offices and tower of Pease bricks with matching stone dressings. Repairs to collapsed floor and extension of market buildings 1865–6 (Pls. 200, 201)
Client: Darlington Local Board of Health (per J. W. Pease)
Contractors: general, R. Stapp; chimney-piece, Joseph Bonehill; iron work, F. A. Skidmore; J. W. Russell & Son
Clerk of works: S. Harrison
Cost: £9,851; 1865–6 £2,615
Note: The tower was the gift of J. W. Pease

[100] 1861–4
DARLINGTON, Co. Durham, Hurworth: Pilmore Hall [C O]
New coachmen's cottages, alterations and additions to house amounting to rebuilding, new stables, extensions to hall. Further alterations and extensions 1873, 1875 (proposed alterations and tower set aside 1877–9) (Fig. 15)
Contractors for both campaigns: general, W. Richardson & Co.; Thomas Robson (1861–4); J. Parnell & Sons (1876–9); heating and ventilating, G. N. Haden; D. O. Boyd; stone carving, Farmer & Brindley; tiles, T. Oakden; Taylor; W. B. Simpson & Son; stained glass, Lavers & Barraud; R. B. Edmundson; Heaton Butler & Bayne; chimney-pieces, W. Wilson; J. & H. Patteson; Joseph Bonehill; Hopton Wood Stone Co.; W. H. Burke; decoration, Best & Lea; furniture and fittings, Henry Capel; iron work, F. A. Skidmore; R. Jones; W. MacFarlane; Hart Son Peard & Co.; bells, Warner & Sons
Clerk of works: G. G. Hoskins
Cost: £14,335; 1876–9 £15,000

[101] 1861
DARLINGTON, Co. Durham: Woodlands [c O]
New chimney-piece and other alterations to mansion (*c.*1829, enlarged by Richardson & Ross 1860)
Client: J. W. Pease
Contractors: chimney-pieces, Leclerc; tiles, Minton; iron work, F. A. Skidmore
Cost: £200

[102] 1861
LEWES, Sussex: Shelleys [O]
Layout of terrace
Client: J. Hodgkin (father-in-law of AW)
Cost: £55

[103] 1861–5
LIVERPOOL, Allerton: New Heys [c d O]
House of stone with a slate roof; also stables, conservatory, garden layout, and furniture. Alterations 1866, 1868, 1869–70 (Pl. 98)

Client: W. G. Bateson
Contractors: general, Thomas Gore; stained glass, Lavers & Barraud; chimney-pieces, W. Wilson; furniture and fittings, Bradley; garden, with advice from Kemp
Clerk of works: W. Coulthurst (J. Battye)
Cost: £6,700

[104] 1861
LIVERPOOL [c]
Report on stone for Wellington Memorial
Client: Liverpool City Council

[105] 1861–5
MANCHESTER, Ancoats: Congregational Chapel [c C D M O Dem]
Chapel and schools of brick on triangular site with pyramid roofed tower at apex (Pl. 29)
Client: Joseph Thompson
Contractors: general, Joseph Robinson Jun.; heating and ventilating, G. N. Haden; plaster ceiling roses, J. W. Hindshaw
Cost: £5,670

[106] 1861
MANCHESTER, Cross Street: Cross Street Chambers [O]
Minor alterations to office
Client: R. Fisher
Cost: £30

[107] 1861
MANCHESTER, Fairfield: ?Summer Hill [O]
Unspecified work. Alterations 1866–7
Client: Joseph Hadwen

[108] 1861
MANCHESTER, King Street: Royal Insurance Building [c D O Dem]
Bank building with offices of three arcaded storeys, basement, and dormered atttics, of Huddersfield stone with Forest of Dean voussoirs (Pl. 28)
Client: Royal Insurance Company
Contractors: general, Joseph Thompson; stone carving, Ibberson; iron work, F. A. Skidmore
Clerk of works: Latham
Cost: £7,025
Note: AW's involvement with the commission included help with finding tenants for the offices. He was also, in 1862, invited to design the company's offices in Bristol, but declined

[109] 1861
MANCHESTER: The Lancastrian School [O]
Minor alterations
Cost: £120

[110] 1861–4
MANCHESTER, Mount Street: Friends' Meeting House and Institute [c C O Dem]
Alterations to meeting hall and to institute, including conversion of basement office let to AW. Alterations 1865
Client: Manchester Friends Meeting
Contractors: general, Kay; S. Bramall; tiles, Minton; T. Oakden; glass, R. B. Edmundson; office fittings, W. Wilson; grates, D. O. Boyd; F. A. Skidmore; iron work, Hart Son Peard & Co.
Cost: £420; alterations £195

[111] 1861
MANCHESTER, Piccadilly: Royal Infirmary [c O]
Renovation and valuation, reventilation and design for memorial tablet to J. C. Harter (see 1865 **[218]**)
Client: House Stewards of the Royal Infirmary (per J. Butterworth)

[112] 1861
MANCHESTER, Princess Street: Royal Institution [S]
Unspecified alterations

[113] 1861–9
MANCHESTER, Salford: Strangeways Gaol [c C d O]
Gaol of radiating blocks of red brick with stone dressings with a slate roof, also boiler chimney; governor's house, gate house, and boundary walls (Pl. 37)
Client: Salford County Gaol Committee
Contractors: general, Mrs Bramall; heating and ventilating, G. N. Haden; tablet, Farmer & Brindley; tiles,

J. Grundy; Woolscroft; glass, R. B. Edmundson; chimney-pieces, W. Wilson; iron work, R. Jones; F. A. Skidmore; gas fittings, Hart Son Peard & Co.
Clerk of works: Henry Littler
Cost: £126,940

[114] 1861
MANCHESTER: York House [c O]
Alterations and new laundry
Client: S. J. Steene
Cost: estimate £600

[115] 1861
NEWTON-IN-CARTMEL, Cumbria: Parsonage [c O]
Redrawing of plans by joiner for alterations to parsonage
Client: Revd I. M. Ransome
Cost: professional charge £21

[116] 1861–2
READING, Berkshire, Whiteknights Park: Whiteknights [c C O]
Alterations to house and stables. Further alterations 1872–3
Client: A. Waterhouse Sen.
Cost: £780

[117] 1861
SALTBURN-ON-SEA, Cleveland [O]
Proposed hotel; not built
Client: Stockton and Darlington Railway Co.

[118] 1861
TAMWORTH, Staffordshire [O]
Survey of defective weaving and sizing shed
Client: J. & R. Pope

[119] 1861
LOCATION UNKNOWN: Moses Gate [c]
Survey and valuation
Client: J. & R. Pope

[120] 1861
LOCATION UNKNOWN [O]
Unspecified work
Client: John A. Simpson
Cost: £80

[121] 1861–2
LOCATION UNKNOWN [c O]
Design for working people's cottages
Cost: professional charge £8

[122] 1862
ALDERLEY EDGE, Cheshire [c C O]
House offices and stable
Client: W. W. Hoyland
Contractors: general, J. Haywood; ironwork, W. Young; tiling, T. Oakden
Cost: £1,940

[123] 1862
ALLONBY, Cumbria: School [c]
Alterations to school building
Client: ?J. W. Pease
Contractors: general, W. Dodgson; bell, Naylor Vickers & Co.

[124] 1862
AMBLESIDE, Cumbria: Low Wood [c]
Houses on Little Lowther estate. AW was commissioned to produce plans for the layout of houses. It is unclear whether any were built to his designs
Client: John Brooks

[125] 1862–5
BINGLEY, W. Yorkshire: Mechanics' Institute [C O V]
Hall, reading room, etc., in the Gothic style, of stone with a slate roof (Pl. 24)
Client: Alfred Sharp (per Alfred Harris)
Contractors: heating and ventilating, G. N. Haden
Cost: £2,100

[126] 1862
BOWDON, Cheshire [O]
Design for house, postponed
Client: Alfred Fryer

[127] 1862
BROXTON, Cheshire: Bankhead [c O]
House, stables, and farm buildings of brick (facings of house all in headers) with sandstone dressings and tiled roof. Alterations 1873
Client: Isaac Hadwen
Contractors: general, R. Beckett; stained glass, Heaton Butler & Bayne; chimney-pieces, W. Wilson; wallpapers, J. Robertson; G. F. Graham
Clerk of works: R. Wood
Cost: £5,205

[128] 1862–4
DARLINGTON, Co. Durham: Uplands [c C D O Dem]
Mansion, stables, and lodge (drive shared with Hummersknott) of brick and stone with a patterned slate roof. Alterations 1868
Client: J. W. and A. Pease for Rachel Leatham (*née* Pease)
Contractors: foundation, J. Windle; superstructure and finishing, Thos. Robson; heating and ventilating, G. N. Haden & Sons; stone carving, W. Farmer; tiles, Minton; D. Conway; Whetstone; Garrett Bros.; stained glass, R. B. Edmundson & Son; chimney-pieces, W. Wilson; Thomas Oakden; iron work, F. A. Skidmore; R. Jones
Clerk of works: R. Robinson
Cost: £7,180

[129] 1862
GOTTESBURG, Sweden [O]
Designs for lodge; not built under AW's supervision
Client: Robert Troost

[130] 1862–3
KESWICK, Cumbria: Underscar [c O]
Unspecified alterations to house
Client: W. Otley
Cost: £520

[131] 1862
LIVERPOOL [c]
Proposed fountain; not built
Client: G. F. Lyster

[132] 1862–4
LIVERPOOL, Aigburth: Prince Park (originally Gisburne House) [c C O]
House of banded brick with stone dressings and a slate roof
Client: G. F. Lyster
Contractors: Holme & Nicoll
Cost: £4,500
Certificates for only £3,000 are entered and there is no record of subcontractors. The certificates clearly do not represent the total cost of the building, and it is possible that it was completed under different supervision. The figure of £4,500 is that on which AW charged commission in 1866

[133] 1862
LIVERPOOL: New Exchange Buildings [O]
AW sent for particulars of this competition but it is not clear whether he entered a design

[134] 1862–3
MANCHESTER, Fulshaw [c O]
House including observatory. Alterations 1866
Client: Alfred Fryer
Contractors: P. Farrell
Cost: £3,660

[135] 1862
MANCHESTER, Greenheys: The Orchard [c C O]
Additions to conservatory. Alterations 1865
Sir W. L. Bardsley, MP
Cost: £365

[136] 1862–5
MANCHESTER, Rusholme [c O Dem]
Wesleyan Chapel
Client: Rusholme Wesleyan Chapel Committee

[137] 1862
ROSTHERNE, Cheshire [c]
Monument
Client: Joseph Simpson
Contractors: Ibberson

[138] 1862
TOWYN, Denbigh: Towyn Estate [c C]

Design for layout of estate
Cost: professional charge £25

[139] 1862
WINDERMERE, Cumbria [C O]
House (unidentified)
Client: B. A. Irving
Contractors: general, Abraham Pattinson; plaster ceiling roses, J. W. Hindshaw
Cost: £1,210

[140] 1862–4
LOCATION UNKNOWN, Monmouthshire: Cefn Ila [c C O]
House (unidentified), stables, bailiff's house, and alterations to lodge
Client: E. Lister
Cost: tender £3,940

[141] 1862
LOCATION UNKNOWN
Alterations to chimneys, etc. Alterations 1872 [c O]
Client: John Eden Wilson
Contractors: chimney-pieces, W. H. Burke; W. Wilson
Cost: Alterations £1,500

[142] 1862
LOCATION UNKNOWN [c]
Proposed house
Client: D. A —— (Illegible in correspondence)
Cost: estimate £1,000

[143] 1862
LOCATION UNKNOWN [O]
Unspecified work
Client: M. Martindale
Cost: £10

[144] 1863–6
ABERDOVEY, Merionethshire: Parsonage [C O]
Alterations to house to make new parsonage house
Client: Revd J. Williams
Contractor: general, Evans
Cost: £850

A troublesome job in which the contractor broke the contract, possibly through bankruptcy.

[145] 1863
BURTINSHAW, Lancashire [O]
Minor alterations to unidentified house
Client: Thomas Ashworth

[146] 1863
CHEADLE, Cheshire: Royal Lunatic Asylum [C D O]
Additional villas to lunatic asylum by Richard Lane, 1848–9. Alterations 1868 **[293]** q. v.
Contractors: general, R. Hirst
Cost: £2,620

[147] 1863–4
DARLINGTON, Co. Durham: Hummersknott [C O]
Large house of brick and stone with low pitched slate roof; also stables, lodge, and boundary walls (Pl. 109; Fig. 15) Now a school
Client: A. Pease
Contractors: general, Thos. Robson; heating and ventilating, W. Wilson; stained glass, R. B. Edmundson; grates, D. O. Boyd; iron work, F. A. Skidmore
Cost: *c.*£6,000

[148] 1863
ECCLES, Lancashire [C]
Unspecified alterations to house
Client: Charles Tysoe

[149] 1863
KENDAL, Cumbria: Parsonage [C O]
Design for house and gardens; not built
Client: W. H. Wakefield

[150] 1863–5
LIVERPOOL, Aigburth: Beech Mount [C O]
Alterations and extensions, including new billiard room
Client: Edward Lawrence, MP
Contractors: general, Samuel Brookfield; stained glass, R. B. Edmundson; grates, W. Wilson
Cost: £2,910

[151] 1863
MANCHESTER, Ashton Park [C O Dem]
House and stables, layout of gardens and fernery. Alterations 1868
Client: W. H. Ormerod
This property was sold to G. Lord 1868
Contractors: general, Joseph Thompson; tiles, T. Oakden; chimney-pieces, W. Wilson; J. Bonehill; painting, J. Robertson; iron work, F. A. Skidmore; R. Jones; plaster ceiling roses, J. W. Hindshaw
Clerk of works: J. Handford
Cost: £7,475

[152] 1863
MANCHESTER, Brooklands: St John's Church [C d D O]
Church of Hollington stone with aisles, transepts, and flèche, patterned brickwork internally (Pl. 27)
Client: Samuel Cunliffe Brooks
Contractors: foundation, Humphrey Davies; superstructure, J. Robinson Jun.; heating and ventilating, G. N. Haden; stone carving, Thos. Earp; tiles, W. Godwin; stained glass, Lavers & Barraud; iron work, R. Jones; reredos, Heaton Butler & Bayne; communion plate, John Hall & Co.; lectern, altar cross, etc., F. A. Skidmore; upholstery and robes, J. Brown
Cost: £7,945 excluding boundary walls, organ, and other fittings

[153] 1863
MANCHESTER: Brown Street [O]
Alterations to warehouse
Client: McKean Bull
Cost: £300

[154] 1863
MANCHESTER, Chester Street: Binyon & Fryer Warehouse [C O V Dem]
Upper storeys of warehouse in diapered brickwork. See 1855 **[7]** (Pl. 21)
Client: Messrs Fryer, Benson and Forster
Contractors: general, Mark Foggett; gas fittings, Hart Son Peard & Co.
Cost: £2,570

[155] 1863–8
MANCHESTER: Corporation Street [d]
Competitive design for temperance hotel with pseudonym 'Iris'; not built

[156] 1863
MANCHESTER: Cross Street [O]
Alterations to building
Client: R. Gladstone
Contractors: J. Thompson
Cost: £320

[157] 1863
MANCHESTER, 20 Cross Street: India Building [c O]
Alterations to offices. Further alterations 1865
Client: Thomas Hilton
Contractors: Farrell
Cost: £3,295

[158] 1863
MANCHESTER, Fallowfield, Victoria Park: Egerton Lodge [c O]
Alterations to house; not carried out
Client: Mr Partington

[159] 1863
MANCHESTER, Fallowfield, Victoria Park: The Lodge [O]
New billiard room
Client: J. Slagg
Cost: £320

[160] 1863
MANCHESTER: King Street [O]
Design for note heading
Client: S. Brooks

[161] 1863
MANCHESTER: King Street [O]
Design for sign board
Client: S. Brooks

[162] 1863
MANCHESTER: Pendlebury [d O]
Tomb

Client: Thomas Cooke
Cost: £150

[163] 1863
MANCHESTER, Prestwich: Besses o' th' Barn Congregational Chapel [C O]
Chapel and school of red brick with yellow and blue brick bands and stone dressings with a slate roof. Part of the school seems to be a later addition (Pls. 31, 32)
Client: I. Richardson and the Congregation of Besses o' th' Barn
Contractors: general, Mark Foggett; heating and ventilating, G. N. Haden; iron work, R. Jones
Clerk of works: J. Bedford
Cost: £4,165

[164] 1863
MANCHESTER, Rusholme: Congregational Chapel and Schools [C D M O Dem]
Church and schoolroom, designed possibly with G. T. Redmayne, of red brick and stone dressings with a corner tower and round headed openings (Pl. 30)
Client: Rusholme Congregational Church Committee
Contractors: general, S. Bramall; stained glass, R. B. Edmundson; iron work, R. Jones
Cost: £4,588

[165] 1863
MANCHESTER: next to Birch Villa [c O]
Alterations and additions to house and design for garden
Client: Charles J. Samuels
Contractors: T. Peak

[166] 1863–6
NANTWICH, Churchside: Manchester and Liverpool District Bank [C O]
Bank building including manager's house of red brick, all in headers, with black brick bands and sandstone dressings and a patterned tile roof.
Client: Mr Hobson
Contractors: general, Mr Beckett; tiles, T. & R. Boote; stained glass, R. B. Edmundson; chimney-pieces, W. Wilson; plaster ceiling roses, J. W. Hindshaw
Clerk-of-works: R. Wood
Cost: £4,480

[167] 1863–4
RAWTENSTALL, Lancashire: ?Ashlands [C O]
Alterations and additions to house including the insertion of 17th-cent. fireplace from Versailles; also new conservatory and lamps to entry
Client: James Ashworth of Hall Carr
Contractors: finishing, Rushton; decoration, J. G. Crace; furniture/fittings, Lamb; iron work, F. A. Skidmore; grates, W. Wilson

[168] 1863–6
WARRINGTON, Cheshire: Grappenhall [C O]
House and offices
Client: W. Long
Contractors: Gibson & Sons
Cost: £3,060

[169] 1863–6
WINDERMERE, Cumbria [O]
House (unidentified) offices and stables
Client: W. Long
Contractors: general, Gibson & Sons; tiles, W. Taylor; chimney-pieces, J. Bonehill
Cost: £2,846

[170] 1863
LOCATION UNKNOWN: Broughton Hall [O]
Unspecified alterations
Client: J. C. Harter
Cost: £600

[171] 1863
LOCATION UNKNOWN [O]
Tomb
Client: J. Slagg

[172] 1864
BANBURY, Oxfordshire: Wood Green [C O]
Suburban mansion, lodge, and stables of brick with a slate roof
Client: Charles Gillett
Contractors: general, C. Claridge; tiles, Thomas Oakden; stained glass, R. B. Edmundson & Son; chimney-pieces and grates, W. Wilson; plaster ceiling roses, J. W. Hindshaw; wallpapers, Mitchell Hammond & Co.
Cost: £3,600

[173] 1864–6
BRASTED, Kent: Vinesgate House [C O]
Alterations to house
Client: William Tipping

[174] 1864
CHEADLE, Cheshire: Belmont (now a school) [C O]
House of red brick with stone dressings and coloured brick bands with a slate roof, exaggerated projecting gable to garden front
Client: James Milne for his daughter Mrs Smith
Contractors: general, Thomas Clay; chimney-pieces, grates, W. Wilson
Cost: £2,125

[175] 1864–7
DARLINGTON, Co. Durham: Backhouse Bank (now Barclays Bank) [C d O]
Bank building of stone with a slate roof; three storeys and dormered attic. Manager's house etc. to rear by G. G. Hoskins 1867 (Pl. 20)
Client: J. Backhouse & Co.
Contractors: general, W. Richardson; heating and ventilating, G. N. Haden; stone carving, W. Farmer; marble flooring, T. H. Hartley; chimney-pieces, W. Wilson; plaster ceiling roses, J. W. Hindshaw; iron work, F. A. Skidmore & Co.; Hart Son Peard & Co.
Clerk of works: G. G. Hoskins
Cost: £12,185

[176] 1864–71
GUISBOROUGH, Cleveland: Hutton Hall [C D O]
Large mansion of red brick with Dunhouse stone dressings with a slate roof. Also game larder, bridges, lodge, walled garden, gardeners' cottages, conservatories, vineries, etc., plan of garden, advice on furniture, selection of carpets and tablecloths. Alterations: 1871–4 billiard room and new conservatory; 1875 cottages; 1876 alterations and stables (Pls. 108, 112; Figs. 7, 15)
Client: J. W. Pease, MP
Contractors: general, W. Richardson; structural steelwork, J. S. Bergheim; heating and ventilating, G. N. Haden; stone carving, Farmer & Brindley; terracotta and pots, J. M. Blashfield; tiles, Thos. Peake; W. Godwin; Minton; Cullis; J. M. Allen; stained glass, etc., Lavers & Barraud; Heaton Butler & Bayne; Morris & Co.; R. B. Edmundson; F. T. Odell; chimney-pieces, W. H. Burke & Co.; Hopton Wood Stone Co.; W. Wilson; grates, D. O. Boyd; plaster ceiling roses, J. W. Hindshaw; furniture/fittings, Farmer & Brindley; Hart Son Peard & Co.; Gillow; iron work, Hart Son Peard & Co.; R. Jones; F. A. Skidmore; Lester & Hodkinson; clock, Edward White; bells, Warner & Son; fountain, Pulham
Clerk of works: R. Robinson, followed by Ed. Robson
Cost: £10,036; 1871–4 £14,067; stables £5,673; lodge £770
Though there were separate stages to this commission the main contractors are listed from all stages.

[177] 1864–6
LEEDS, Adel: Adel Grange [C O]
Suburban mansion of stone with a slate roof. Also lodge, stables and clock turret, fruit wall, garden walls, and bridge to carriage drive. Terrace 1872
Client: R. L. Ford
Contractors: general, John Wood; stone carving, J. W. Seale; tiles, W. Godwin; T. Oakden; stained glass, Heaton, Butler & Bayne; chimney-pieces, Lizard Serpentine Co.; W. Wilson; cast iron newels, Kitchen, Brown & Co.; gas fittings, Hart & Son; plaster ceiling roses, J. W. Hindshaw
Clerk of works: R. Cairns
Cost: £7,270

[178] 1864–7
LEIGHTON BUZZARD, Bedfordshire: The Heath [C O Dem]
Large house and stables of brick with stone banding and dressings, conservatory and stables (Pls. 114, 116)
Client: Frances Bassett
Contractors: general, Albert Kimberley; stone carving, W. Farmer; tiles, W. Godwin; W. Whetstone; T. Oakden; chimney-pieces, W. Wilson; plaster ceiling roses, J. W. Hindshaw; wallpaper, Robertson
Clerk of works: W. Barker
Cost: £5,295

[179] 1864–7
LONDON, Lombard Street: Alexander's and Cunliffe's Bank [C O Dem]

Bank building, stone faced of three storeys with dormered attic (Pl. 39)
Client: Alexander, Cunliffe & Co.
Contractors: foundation, J. & G. Coleman; superstructure and finishing, Jackson & Shaw; heating and ventilating, D. O. Boyd; stone carving, J. W. Seale; tiles, W. Godwin; stained glass, Heaton Butler & Bayne; furniture/fittings, Holland & Sons; W. Wilson; Hart Son Peard & Co.; iron work, F. A. Skidmore
Clerk of works: E. Harrison
Cost: £17,485

[180] 1864–5 LONDON: 8 New Cavendish Street (later No. 20, now No. 61) [C O]
Alterations to existing house to create drawing office and London home. Alterations 1867, 1873, 1896–1903, 1905
Client: Alfred Waterhouse
Contractors: heating and ventilating, G. N. Haden; D. O. Boyd; stained glass, Heaton Butler & Bayne; Lavers & Barraud; R. B. Edmundson; decoration, Robert Pollitt; furniture/fittings, Heal; Lamb; W. Wilson; iron work, R. Jones; F. A. Skidmore; plaster ceiling roses, J. W. Hindshaw; advice on garden, R. Marnock
Cost: £3,250

[181] 1864
LONDON: Norwood Cemetery [C d O]
Monument for James Kershaw (Col. Pl. 8)
Client: Benjamin Armitage
Contractors: stone carving, Macdonald Field & Co.; iron work, F. A. Skidmore
Cost: £590

[182] 1864
NANTWICH, Cheshire [C]
Designs for alterations to unidentified premises
Cost: professional charge £6. 6*s*.

[183] 1864
NANTWICH, Cheshire [C]
House (alterations)
Client: Mr Hobson
Contractor: chimney-piece, W. Wilson

[184] 1864
NEWCASTLE, Northumberland: Benwelldene [C O]
Stone faced house with octagonal hall. Alterations and extensions 1872, 1877, see also **[591]** 1894
Client: Thomas Hodgkin
Contractors: general, George Graden; heating, D. O. Boyd; stone carving, W. Farmer; Farmer & Brindley; tiles, Hawes Denny & Hargreaves; W. Godwin; stained glass, Heaton Butler & Bayne; R. B. Edmundson; F. T. Odell; iron work, Hart Son Peard & Co.; R. Jones; chimney-pieces, Lizard Serpentine Co.; W. Wilson; Hopton Wood Stone Co.; decoration, R. Pollitt; furniture, H. Capel; plaster ceiling roses, J. W. Hindshaw
Clerk of works: Hartley
Cost: £5,295; 1872–4 £2,865

[185] 1864
REIGATE, Surrey: Parkside [C O]
Stables for house. Alterations 1868
Client: W. C. Alexander
Contractors: H. Thornton
Cost: £1,070

[186] 1864–7
WELLINGTON, Somerset: The Cleve [C O]
House and stables of rubble stone with ashlar dressings and low pitched slate roof, including plan for garden. Alterations 1868–9; 1878
Client: Joseph H. Fox
Contractors: general, W. Brock; tiles, W. Godwin; stained glass, Heaton Butler & Bayne; chimney-pieces, W. Wilson; Lizard Serpentine Co.; iron work, Hart Son Peard & Co.
Cost: £4,860

[187] 1864
LOCATION UNKNOWN: Neptune Hall [C]
Sketch designs for proposed house; probably not built
Client: Thomas Jones

[188] 1864
LOCATION UNKNOWN (?Saltburn): Zetland Hotel [C]
Unspecified work measured by M. Robinson
Cost: £7,000

[189] 1864–5
Design for a dressoir [C O]
Client: J. E. Hodgkin
Manufacturer, F. A. Skidmore
Cost: estimate £60

[190] 1865
ACKWORTH, W. Yorkshire [C O]
New Master's House for Ackworth School
Client: The Flounders Institute
Cost: £650

[191] 1865–7
ASHTON-IN-MAKERFIELD, Lancashire: Congregational Chapel [C O Dem]
Chapel and school
Client: Joseph Evans
Contractors: general, George Harris; decoration, J. Robertson; gas fittings, J. W. Dovey; heating and ventilating, G. N. Haden
Cost: £3,335

[192] 1865, 1866–7, 1868–73
BLACKMOOR, Hampshire: Blackmoor House [C D O V]
Large mansion of stone with a tile roof, offices, stables, lodges, gardens, and furniture designed by AW. The commission began with alterations to Blackmoor farmhouse. Then designs for a new house were drawn and set aside. Finally the mansion was created around the core of the old house. Additions 1882–3 (Pls. 106, 186, 189)
Client: Roundell Palmer, MP , later Lord Selborne
Contractors: general, 1865 Gibson Bros.; 1868–73, Dyer; stone carving, Farmer & Brindley; tiles, W. Godwin; Heaton Butler & Bayne; W. B. Simpson; stained glass, Heaton Butler & Bayne; 1882, F. T. Odell; R. B. Edmundson; chimney-pieces, Hopton Wood Stone Co.; grates, D. O. Boyd; decoration, Robert Pollitt; furniture/fittings, Henry Capel; iron work, Robert Jones; Hart Son & Peard & Co.; wallpapers, Scott Cuthbertson & Co.; clock, J. Moore & Son
Cost: 1865 £2,445; 1868–73 £16,675; Alterations 1882–3 £1,490

[193] 1865–7
BOURNEMOUTH, Hampshire: Eastwood [C O Dem]
New wing with museum and conservatory added to mansion
Client: Edmund Christy
Contractors: general, J. D. Cowland; heating and ventilating, G. N. Haden; decoration, J. Robertson (with advice from Robert Pollitt); furniture, Wright and Mansfield; gas fittings, F. A. Skidmore & Co.
Cost: £2,920

[194] 1865–6
BRASTED, Kent: St Martin's Church [C D O]
Parish church of stone with tile roof: restoration of the ruinous church amounting to the rebuilding of all but the tower, and the addition of the north aisle and the south window of the tower (altered after the fire 1990)
Client: Revd J. C. Astley and Edward Udal Thompson for the Brasted Church Committee
Contractors: general, Thomas Blake; heating and ventilating, G. N. Haden; tiles, W. Godwin; stained glass, F. T. Odell; east window, O'Connor; reredos, Heaton Butler & Bayne; wrought iron, Robert Jones
Clerk of works: W. Matthews
Cost: £3,250 (AW subscribed £20) AW took advice from G. E. Street about potential builders

[195] 1865
BRASTED, Kent: Chapel-of-ease [C]
Proposed small new church in addition to the restoration of St Martin. The accounts are not separated from the restoration of the parish church and this chapel may not have been built.

[196] 1865–8
BRISTOL, Avon: Goldney House (now part of Bristol University) [C O]
Refacing in stone and additions to early 18th-cent. house, including new staircases, entrance cloister, and belvedere tower, also new kitchen wing in brick and design for sideboard
Client: Lewis Fry
Contractors: general, J. & J. Foster; tiles, W. Godwin; Heaton Butler & Bayne; stained glass, Heaton Butler & Bayne; panelling in drawing-room,

Howard & Son; hall and staircase chimney-pieces, Benham & Co.; W. H. Burke & Co.; oak chimney-piece, W. Farmer; decoration, J. Hankins; iron work, F. A. Skidmore & Co.; Hart Son Peard & Co.; plaster ornament, J. W. Hindshaw
Clerk of works: Alexander Gray
Cost: £12,850

[197] 1865–7
CAMBRIDGE, Cambridgeshire: Union Society [C D O]
Debating hall and rooms of coloured brick and stone with a tiled roof, including furniture and sketch for garden. Alterations 1882; 1884–6 new wing; 1891; 1898–1901; extensively altered 1930s (Col. Pl. 4(*f*); Pls. 62, 69, 176, 182; Fig. 15)
Client: Committee of the Cambridge Union Society (per Henry Peto)
Contractors: general, Jackson & Shaw; 1884–6, J. Parnell & Sons; heating and ventilating, G. N. Haden; stone carving, W. Farmer; tiles, W. Godwin; chimney-pieces, Ed. Pearce; Yates; Harwood & Drabble; stained glass, Lavers & Barraud; decoration, Green & King; furniture/fittings, W. Wilson; Jackson & Shaw; 1884–6 Maple & Co.; Rattee & Kett; iron work, F. A. Skidmore & Co.; Hart Son Peard & Co.; plaster ceiling roses, J. W. Hindshaw
Clerk of works: J. Dickson
Cost: 1882 £645; 1884–6 £8,335; 1891 £190; 1898–1901 £460

[198] 1865
CHESTER, Cheshire: Grosvenor Hotel [C O]
Signboard on iron bracket for new hotel by T. M. Penson (1863–6)
Client: The Rt. Hon. Earl Grosvenor, MP, 2nd Marquis of Westminster
Contractors: iron work, F. A. Skidmore; Robert Jones

[199] 1865
CHESTER, Cheshire: East Gate [O]
Proposed rebuilding of east gate of city walls as memorial to 1st Marquis of Westminster
Client: The Rt. Hon. Earl Grosvenor, MP, 2nd Marquis of Westminster

[200] 1865
DARLINGTON, Co. Durham: Woodlands [O]
South-east lodge
Client: Joseph Whitwell Pease
Cost: £650

[201] 1865
DARLINGTON, Co. Durham [C]
Alterations to chimneys of house
Client: Gurney Pease

[202] 1865–8
KINGSTON UPON THAMES, Surrey: Woodbines [C O]
House and stables, also lodge and garden layout. Alterations 1869, 1870
Client: The Misses Christy
Contractors: general, Jackson & Shaw; heating and ventilating, G. N. Haden; D. O. Boyd; tiles, W. B. Simpson; stained glass, Lavers & Barraud; chimney-pieces, W. H. Burke; Lizard Serpentine Co.; fittings, W. Wilson; iron work, Hart & Son; plaster ceiling roses, J. W. Hindshaw
Clerk of works: J. White
Cost: £7,000; 1870 £980

[203] 1865
KNUTSFORD, Cheshire [C O]
Extensions to gaol. Further extensions 1869
Client: Cheshire County Magistrates per Lee P. Townshend

[204] 1865–7
LEIGHTON BUZZARD, Bedfordshire: Bassett and Harris Bank (now Barclays Bank) [C O]
Bank building of stone, two floors and dormered attic with slate roof, also manager's house. Openings on ground floor altered
Client: Bassett Son and Harris
Contractors: general, Albert Kimberley; heating and ventilating, G. N. Haden; D. O. Boyd; stone carving, Thomas Earp; tiles, W. Godwin; chimney-pieces, W. Wilson; furniture/fittings, Wm. Smee & Sons; iron work, F. A. Skidmore; Hart Son Peard & Co.; wallpaper, Scott Cuthbertson
Clerk of works: W. Barker
Cost: £7,150

[205] 1865–6
LEWES, Sussex [O]
Lodge and viaduct for unidentified house
Client: Burwood Godlee
Cost: £1,520

[206] 1865
LIVERPOOL [C O]
Report on sample roof, similar to roof of new Hay Market
Client: Liverpool Corporation

[207] 1865–6
LONDON: New Bond Street [C O]
Alterations to dining-room
Client: G. Redmayne & Co.
Cost: £200

[208] 1865–9
LONDON, St James's Street: New University Club [C d D O T Dem]
Club house of brick and steel with stone facing built as two blocks on an irregular site running back to Arlington Street. Alterations 1878 (Pls. 38, 40)
Client: Committee of the New University Club
Contractors: general, Wm. Brass; structural steelwork, J. S. Bergheim; heating and ventilating, G. N. Haden; D. O. Boyd; stone carving, Farmer & Brindley; stained glass, Heaton Butler & Bayne; chimney-pieces, D. O. Boyd; Lizard Serpentine Co.; decoration, Heaton Butler & Bayne; furniture/fittings, H. Capel; iron work, R. Jones; Hart & Son; plaster ceiling roses, J. W. Hindshaw
Cost: £25,000. Alterations £875

[209] 1865
LONDON: Highgate [O]
Attention to drains etc. of existing house
Client: Edward Fry

[210] 1865
MACCLESFIELD, Cheshire: Infirmary [C O]
Competitive designs for Infirmary building. Withdrew from competition with compensation

[211] 1865
MANCHESTER [O]
Plan of the chamber floor of Mr Frith's house
Client: E. Salomons

[212] 1865
MANCHESTER: Cheetham Hill [C O]
Survey and valuation of property

[213] 1865–6
MANCHESTER: Chester Street [C O]
Hot water fountain
Client: probably Fryer, Benson and Forster
Contractors: general, Mrs S. Bramall; tiles, D. Conway
Cost: £80

[214] 1865
MANCHESTER: King Street [O]
Minor alterations to bank building
Client: Messrs Brooks & Co.

[215] 1865
MANCHESTER: King Street/Spring Gardens [C O]
Designs for proposed new Exchange; not built (Pl. 45)
Client: Committee of the Manchester Exchange

[216] 1865
MANCHESTER: Market Street [O]
Alterations to premises
Client: Petschlers Photographic Co.

[217] 1865–70
MANCHESTER, Pendlebury: High Field [C O Dem]
Alterations and additions to house, new stables, coach house, conservatory, and vineries
Client: Peter Maclaren
Contractors: general, W. Southern; heating and ventilating, G. N. Haden; tiles, Oppenheimer; Hargreaves & Craven; chimney-pieces, W. Wilson; iron work, Robert Jones; plaster ceiling roses, J. W. Hindshaw
Cost: £10,045

[218] 1865
MANCHESTER, Piccadilly: Royal Infirmary [C O Dem]

New stables, etc. Alterations 1867–8 including decoration of chapel to designs of G. T. Redmayne. See 1861 **[111]**
Client: House Stewards of the Royal Infirmary (per Mr Roy)
Cost: £340

[219] 1865–7
MANCHESTER, Prestwich: National Schools [O]
Alterations 1867
Client: Revd H. M. Birch for the Prestwich National Schools Committee
Contractors: general, J. Robinson Jr.; iron work, R. Jones
Cost: £2,860

[220] 1865–6
MANCHESTER, Rusholme Road: Free Library [C O Dem]
Conversion of school to library
Client: Manchester Free Library Committee
Contractors: finishing, H. F. Ledger; heating and ventilating, G. N. Haden; tiles, Hargreaves & Craven; chimney-pieces, W. Wilson; iron work, R. Jones
Clerk of works: John Handford—dismissed for inattention
Cost: £3,140

[221] 1865
MANCHESTER, Whalley Range: Whalley Church [O]
Sketches for new window
Client: W. C. Brooks
Cost: Professional charge £1. 1*s*.

[222] 1865
?MANCHESTER [C O]
Detail of plaster cornice
Client: Thomas Hibbert

[223] 1865
NANTWICH, Cheshire [O]
Report regarding alterations and measurement of unidentified premises

[224] 1865–9
PENMAENMAWR, Caernarvon: St Seiriol's Church [C O]
Church of stone with free-standing bell tower linked by porch to main building; nave with high arched trusses, chancel, and south aisle in white brick with red bands. Organ and lectern designed by Ernest Geldart
Client: Penmaenmawr Church Committee (M. Gladstone and H. Kneeshaw)
Contractors: general, H. Atkinson; heating and ventilating, G. N. Haden; stained glass, F. T. Odell; iron work, R. Jones; fittings, Mark Foggett; organ, Bevington; font, T. L. Carter
Cost: £3,610

[225] 1865–6
READING, Berkshire, Caversham: Free Church [C O Dem]
Chapel building, see 1872 **[376]**; 1876–7 **[437]**
Client: Ebenezer West
Contractors: general, A. Barnicoat; heating and ventilating, G. N. Haden; iron work, Hart Son Peard & Co.
Cost: £1,535

[226] 1865–6
SUNNINGHILL, Berkshire: Sunningdale Estate [C O]
Designs for house; not built
Client: John Everitt

[227] 1865–6
WELLINGTON, Somerset [C]
Proposed house; not built
Client: F. T. Elworthy

[228] 1865
LOCATION UNKNOWN [C]
Competitive design for a memorial hall
Client: Revd Thomas Jones

[229] 1865
LOCATION UNKNOWN [C]
Unspecified alterations
Client: Mr East

[230] 1865
Design for monogram [O]
Client: St Leger Glynn

[231] 1866
ALTRINCHAM, Cheshire: Broadheath Congregational Church [O Dem]
Schoolroom and chapel of brick with slate roof
Client: Joseph Thompson
Cost: £200

[232] 1866–8
BLACKMOOR, Hampshire: Cottages and Schoolroom [O]
Row of three and row of four cottages with school/reading room attached, of brick with tile roofs; cottages have alternate porches and pentice roofs
Client: Roundell Palmer, MP, later Lord Selborne
Contractors: general, Thomas Kemp; ironmongery, Ginger; clock, E. White
Cost: £845
The costs for the work on the Blackmoor Estate are not clearly separated. They include payments for house, offices, lodge, stables, church, vicarage, cottages and schoolroom, and new school

[233] 1866
BOWDON, Cheshire: Springbank [D O]
Alterations to house; not completed as designed
Client: J. B. Forster
Contractors: general, J. Douglas; chimney-pieces, W. Wilson
Cost: £655

[234] 1866–8
BRASTED, Kent: Farmhouse [C O]
Conversion of cottage to farmhouse, and new stables
Client: William Tipping
Clerk of works: W. Matthews

[235] 1866
CHIPPING SODBURY, Avon: Ridge House [C O]
Alterations to house, new stables, and lodge
Client: Edward Burges
Contractors: heating and ventilating, W. Wilson
Clerk of works: H. Hartley
Cost: £1,500

[236] 1866–9
KIRKCASSOCK, Co. Down: Kirkcassock House [C O Dem]
House and offices, lodge and vinery. Alterations 1894 for W. Christie Miller
Client: J. J. Richardson
Contractors: general, Harvey & McLaughlin; heating and ventilating, G. N. Haden; D. O. Boyd; tiles, Hargreaves & Craven; chimney-pieces, Sibthorpe & Sons; furniture/fittings, W. Wilson; iron work, R. Jones; Hart Son Peard & Co.; plaster ceiling roses, J. W. Hindshaw
Clerk of works: McMillan
Cost: £12,170; 1894 £190

[237] 1866–75
LIVERPOOL: Allerton Priory [C O]
Large house of brick with red sandstone dressings and slate roof and tower over entry; also lodge, garden layout, and some furniture. 1866 design accepted; construction 1868–71; 1872–6 designs for picture gallery (Pl. 115; Figs. 7, 15)
Client: John Grant Morris
Contractors: general, Holme and Nichol; heating and ventilating, G. N. Haden; stone carving, Farmer & Brindley; tiles, W. Godwin; L. Oppenheimer; stained glass, Heaton Butler & Bayne; F. T. Odell; R. B. Edmundson; chimney-pieces, Lizard Serpentine Co.; W. H. Burke; W. Wilson; grates, Hart Son Peard & Co.; D. O. Boyd; furniture/fittings, Gillow; iron work, R. Jones; F. A. Skidmore; Lester & Hodkinson; plaster ceiling roses, J. W. Hindshaw
Clerk of works: J. Dickson
Cost: £16,590; 1872–6 estimate £5,000

[238] 1866–72
LONDON [O]
Site plan and company seal and advice *re* Private Bill
Client: Metropolitan Estates Co. (see Serle Street and Cooks Court Improvement Co. **[370]**)

[239] 1866–8
LONDON: 92 Lancaster Gate [C O]

Negotiating purchase of house, valuing, specification for completion, supervision of accounts
Client: C. P. Stewart
Contractors: grates, W. Wilson
Cost: professional charge £75

[240] 1866–8
LONDON: 107 Lancaster Gate [C O]
Negotiating purchase of house, valuing, specification for completion and alteration, supervision of accounts. Alterations 1875 decoration
Client: Dr C. D. F. Phillips
Cost: professional charge £75. Alterations £250

[241] 1866
LONDON: Ludgate Hill [O]
Proposed building; not built
Client: H. Bewlay and T. B. Smithies

[242] 1866–8
LONDON, Stoke Newington: Kennaway Hill, The Willows [C O Dem]
Proposed alterations to existing house (1865) set aside and new house built with lodge, conservatory, etc.
Client: G. W. Alexander
Contractors: general, Jackson & Shaw; heating and ventilating, D. O. Boyd; stone carving, Thos. Earp; stained glass, Heaton Butler & Bayne; decoration, Hawkins; iron work, J. Hart & Sons; plaster ceiling roses, J. W. Hindshaw
Clerk of works: S. Harrison
Cost: £13,085

[243] 1866 and 1870–80
LONDON, South Kensington: Natural History Museum [C d D O P V]
Museum with central hall and galleries of London stock brick with dressings and façade walls of load-bearing blue and buff terracotta and a patterned slate roof, internally a grid of iron stanchions cased in terracotta. Alterations 1882 spirit room; 1884 lodge (Col. Pl. 3; Pls. 77–83, 164–8)
Client: HM Government
Contractors: general, G. Baker; 1882 G. Shaw; 1884 Mowlem; structural steelwork, J. S. Bergheim; heating and ventilating, Stevens & Son; terracotta, Gibbs & Canning; modelling, M. Dujardin; tiles, W. B. Simpson; mosaic flooring, W. H. Burke; stained glass, F. T. Odell; iron work, Hart Son Peard & Co.; window sill marbles, Farmer & Brindley
Cost: including fittings £549,045. Alterations 1882 £7,200; 1884 £2,500

[244] 1866
LONDON, Southwark [O]
Survey plan of existing offices
Client: Theodore Waterhouse

[245] 1866–7
LONDON: 32 Wimpole Street [C O]
General renovation of house and addition of conservatory and balcony
Client: R. P. Amphlett
Contractors: general, G. F. Graham; iron work, Hart Son Peard & Co.; R. Jones; fountain, W. H. Burke & Co.
Cost: £685

[246] 1866
MANCHESTER [O]
Alterations to house
Client: Sir James Watts

[247] 1866–8
MANCHESTER, Ancoats: Free Library [O]
Library building of brick with stone dressings
Client: Manchester Free Libraries Committee
Contractors: general, James Terras; heating and ventilating, G. N. Haden
Cost: £3,140

[248] 1866
MANCHESTER: Ardwick Cemetery [O]
Tomb
Client: Joseph Thompson
Contractors: stone carving, Joseph Bonehill; iron work, R. Jones
Cost: £60

[249] 1866–8
MANCHESTER: 56 Brown Street [C O]
New offices

Client: Messrs Cunliffe & Leaf
Contractors: general, R. Neill & Sons; chimney-pieces, W. Wilson
Cost: £2,625

[250] 1866–8
MANCHESTER, Cotham Street: Stipendiary Magistrates Court [C D O Dem]
Court-house of brick with stone dressings and a slate roof. Asymmetrical plan with bow façade to court room
Contractors: general, Mrs S. Bramall; heating and ventilating, G. N. Haden; stained glass, F. T. Odell; chimney-pieces, W. Wilson; iron work, R. Jones
Clerk of works: Henry Littler
Cost: £5,400

[251] 1866
MANCHESTER: Every Street Free Library [O Dem]
Library building of brick and stone
Client: Manchester Free Libraries Committee
Cost: £2,850

[252] 1866–8
MANCHESTER: Fairfield [O]
Alterations to house built for Gaylord Hadwen. See 1857 **[33]**
Client: Wm. Hadwen
Contractor: Thos. Clay
Cost: £1,690

[253] 1866
MANCHESTER [O]
Unspecified work in connection with statue of Cobden
Cost: no charge for design work

[254] 1866–71
OXFORD, Oxfordshire: Balliol College [C d D O]
New Master's Lodge and range of buildings to Broad Street, stone faced with tile roof. See 1873 **[392]**, 1886 **[537]** (Pls. 63, 64)
Client: Master and Fellows of Balliol College (and Miss Brackenbury)
Contractors: general, Wm. Brass; heating and ventilating, G. N. Haden; D. O. Boyd; stone carving, T. L. Carter; Farmer & Brindley; tiles, Maw & Co.; stained glass, G. G. Hodson; Heaton Butler & Bayne; F. T. Odell; fittings, W. Wilson; iron work, R. Jones; Hart & Son; plaster ceiling roses, J. W. Hindshaw
Cost: £25,615

[255] 1866
READING, Berkshire, Caversham: Amersham Hall Cottages [C O]
Two cottages (one demolished) and boundary walls
Client: Ebenezer West
Contractors: general, A. Barnicoat
Cost: £510

[256] 1866
REIGATE, Surrey [O]
Proposed reading room, probably not built
Client: W. C. Alexander

[257] 1866
SKELWITH, ?Cumbria [C]
Advice on proposed house
Client: H. E. Robson

[258] 1866
STANSTEAD ABBOTS, Hertfordshire: Easneye Park (now All Nations Christian College) [C d O]
Large mansion, stables, and lodge of patterned brick and terracotta with stepped gables and tile roof. Also south lodge. Other lodges, cottages, and farm buildings may be by another hand. Advice on layout of gardens by Robert Marnock (Pl. 103; Fig. 7)
Client: Thomas Fowell Buxton
Contractors: general, Wm. Brass; heating and ventilating, G. N. Haden; modelling and carving, Farmer & Brindley; tiles, W. B. Simpson; stained glass, Heaton Butler & Bayne; F. T. Odell; chimney-pieces, Lizard Serpentine Co.; W. H. Burke; fittings, W. Wilson; iron work, Hart Son Peard & Co.; R. Jones; Lester & Hodkinson; plaster ceiling roses, J. W. Hindshaw; advice on gardens, R. Marnock
Clerk of works: Burton
Cost: £32,800

[259] 1866–8
STONY STRATFORD, Buckinghamshire: Hanslope House [C O Dem]

House and stables
Client: Reginald Walpole
Contractors: general, John Watkin; heating and ventilating, D. O. Boyd; chimney-pieces, D. O. Boyd
Cost: £3,300

[260] 1866
WARWICK, Warwickshire: The Priory [C O]
Alterations to house, originally of the 17th-cent., as consulting architect (the house was moved to Virginia in 1927)
Client: T. Lloyd
Cost: professional charge £31. 10*s*.

[261] 1866–7
WINDERMERE, Cumbria: Elleray Bank [C O]
Stables for house, of red brick with Welsh slate roof. See 1856 **[22]**
Client: G. G. Cunningham

[262] 1866
LOCATION UNKNOWN: Bunkers Hill [O]
Valuation of old works
Client: Executors of the late Thos. Drinkwater

[263] 1866
LOCATION UNKNOWN [C]
Advice on cost of house
Client: Wm. or Thos. Baker
Cost: estimate £3,500–£4,000

[264] 1866
LOCATION UNKNOWN [O]
New oriel window for house
Client: T. Merrick

[265] 1866
LOCATION UNKNOWN [O]
Unspecified work
Client: T. Rowland

[266] 1867–70
BLACKMOOR, Hampshire: St Matthew's Church [C d O]
Church with nave, chancel, and west tower, of banded stone with tile roof
Client: Roundell Palmer, MP, later Lord Selborne
Contractors: general, Thomas Kemp; heating and ventilating, G. N. Haden; stone carving, Farmer & Brindley; tiles, W. Godwin; stained glass, Heaton Butler & Bayne; F. T. Odell; furniture/fittings, Christopher Pratt; Heal & Son; iron work, Lester & Hodkinson; R. Jones; grate, D. O. Boyd; organ, Bevington; clock, E. White; bells, J. Warner & Son
Cost: £8,635

[267] 1867
BLACKMOOR, Hampshire: Vicarage [O]
House and stables of stone with tiled roof
Client: Roundell Palmer, MP, later Lord Selborne
Contractors: Thomas Kemp
Cost: £1,280

[268] 1867–76
CAMBRIDGE, Cambridgeshire: Gonville and Caius College, Tree Court [C d D O V]
Sixty sets of college rooms on four floors, stone faced with leaded and slated roof around three sides of court. Also refacing, alteration, and extension of other buildings in court; refurbishment of chapel with addition of apse, and alterations to Master's Lodge. Plans for planting of Fellows' garden (Pls. 68, 70, 71, 162). Alterations 1878–9. See also 1883 **[502]**
Client: Master and Fellows of Gonville and Caius College
Contractors: general, George Trollope & Sons; heating and ventilating, G. N. Haden; D. O. Boyd; stone carving, Farmer & Brindley; tiles, W. B. Simpson & Sons; stained glass, Heaton Butler & Bayne (also mosaics in chapel); glazing, F. T. Odell; chimney-pieces, W. H. Burke; fittings, minor work, Rattee & Kett; iron work, Robert Jones; iron fittings, Hart Son Peard & Co.; Lester & Hodkinson
Clerk of works: Mr Tift
Cost: £34,425; decoration of Chapel £1,500 (excluding glass and mosaics); 1878–9 £523; 1881 £130

[269] 1867
CAMBRIDGE, Cambridgeshire: Trinity College [O]
Proposed rebuilding or reroofing of Chapel; not proceeded with

Client: Warden and Fellows of Trinity College (probably through Henry Peto. See Cambridge Union Society **[197]**)

[270] 1867
EAST GRINSTEAD, Sussex [C O]
Turkish bath for unidentified house
Client: Captain Hamond

[271] 1867
GORMELLICK, Cornwall [c C]
New porch
Client: Alfred Waterhouse

[272] 1867
GUISBOROUGH, Cleveland: Hutton Railway Station [D O]
Station building of 'Pease' brick with slate roof
Client: J. W. Pease, MP.

[273] 1867
?LEIGHTON BUZZARD, Bedfordshire [C]
New gate or alterations to entrance
Client: Mr Sharman

[274] 1867
LIVERPOOL, Lime Street: London and North Western Hotel [C O]
Railway hotel, stone faced with lead and slate roofs. Cast-iron door canopy, iron and stone grand staircase under a stained glass skylight, faience decoration to main rooms (Pls. 203, 204)
Client: London and North Western Railway Co.
Contractors: general, Haigh & Co.; heating and ventilating, G. N. Haden; D. O. Boyd; stone carving, Farmer & Brindley; tiles, Hargreaves & Craven; stained glass, Heaton Butler & Bayne; chimney-pieces, W. H. Burke; iron work, Lester & Hodkinson; R. Jones; plaster ceiling roses, J. W. Hindshaw
Cost: £80,268

[275] 1867
LONDON: Bryanston Square [O]
Proposed additional storey to house; not built
Client: John Murray

[276] 1867–71
LONDON: Holborn Viaduct [O Dem]
Wine and spirit warehouse and offices; destroyed 1941
Client: H. B. Fearon & Co.
Contractors: foundation, Jackson & Shaw; tiles, W. B. Simpson & Sons; chimney-pieces, W. H. Burke & Co.; D. O. Boyd; cast-iron work, W. Macfarlane & Co.; wrought-iron-work, Lester & Hodkinson; plaster ornaments, G. Jackson
Cost: £13,100

[277] 1867
LONDON: St James's Place [O]
Proposed rebuilding of house
Client: J. Haynes

[278] 1867
LONDON, The Strand: Royal Courts of Justice [C d D O P]
Competition entry. Unsuccessful (Pls. 41, 42, 43, 44; Figs. 2, 3)
Cost: estimate £1,419,845

[279] 1867–8
MANCHESTER: Abington Street [O]
Additional floor to warehouse
Client: Samuel Schuster
Contractor: Joseph Thompson
Cost: £3,830

[280] 1867–8
MANCHESTER: Bankruptcy Court [O]
Consultations and possible sketch designs for proposed transfer of court from Athenaeum to Assize Courts

[281] 1867–9
MANCHESTER, Heaton Mersey: West Bank House [C O Dem]
House of red brick with stone dressings and a shallow pitched slate roof (Pl. 101; Fig. 7)
Client: James Buckley
Contractors: general, Thos. Clay; heating and ventilating, G. N. Haden; tiles, Dale & Sons; stained glass, Heaton Butler & Bayne; chimney-pieces, D. O. Boyd; W. Wilson; decoration, Robertson; iron work, R.

Jones; gas fittings, Hart Son Peard & Co.; plaster ceiling roses, J. W. Hindshaw
Cost: £8,925

[282] 1867–71
MANCHESTER: Northenden [C O]
House and stables (unidentified)
Client: John King Jnr.
Contractors: general, Joseph Thompson; stone carving, Birchenough; tiles, D. Conway; stained glass, F. T. Odell; chimney-pieces, W. Wilson; decoration, W. H. Lomas; plaster ceiling roses, J. W. Hindshaw; iron work, Lester & Hodkinson; fingerplates, Hart Son Peard & Co; garden, John Kellett
Cost: £5,140

[283] 1867
MANCHESTER, Salford: Sacred Trinity Church [C O]
Proposed restoration. AW proposed demolition and rebuilding. Not carried out
Client: Trustees of Booths Charities, Salford: N. Shelmerdine

[284] 1867 1st competition, 1868 2nd competition 1868–77 *et seq.*
MANCHESTER: Town Hall [d D M O T V]
Town Hall of yellow brick with Spinkwell stone facing with a patterned slate roof. Interior lined with terracotta and tiles (Col. Pls. 1, 4(*a*); Pls. 47–61, 172, 180, 181, Fig. 4)
Client: Manchester City Corporation
Contractors: foundation, Thos. Clay; superstructure, Geo. Smith; structural steelwork, J. S. Bergheim; fireproof construction, Dennett & Co.; heating and ventilating, G. N. Haden; finishing, Thos. Clay; stone carving, Farmer & Brindley; terracotta, Gibbs & Canning; G. & J. Fenning; tiles, Craven Dunnill & Co.; W. Godwin; W. B. Simpson; marble flooring, W. H. Burke & Co.; mosaic flooring, J. Rust; stained glass, F. T. Odell; R. B. Edmundson; chimney-pieces, Hopton Wood Stone Co.; Farmer & Brindley; decoration, Heaton Butler & Bayne; R. Pollitt; Best & Lea; furniture/fittings, Doveston Bird & Hull; H. Capel; G. Goodall; iron work, F. A. Skidmore; Hart Son Peard & Co.; R. Jones; clock, Gillett & Bland; bells, Taylor & Co.; murals, Ford Madox Brown; organ, Cavaillé Coll; organ case, Farmer & Brindley; curtains, designed by R. E. Holding, made by Royal School of Art Needlework, S. Kensington
Clerk of works: K. J. Osborne
Cost: £859,000

[285] 1867
READING, Berkshire, Whiteknights Park: Foxhill (now a part of Reading University) [d O]
House and stables of red brick with blue diaper work and a tile roof. Alterations 1875 (Pl. 117; Figs. 8, 15)
Client: Alfred Waterhouse

[286] 1867
?CHELMSFORD, Essex: Brooklands House [O]
Sketch designs for alterations
Client: Mr Occleston

[287] 1867 or earlier [c]
Octagonal table in walnut and mahogany with ebony edging, crossed stretchers. Now in the V. A. Museum
Client: Alfred Waterhouse

[288] 1868–9
BOWDON, Cheshire, St Margaret's Road: Devisdale [O Dem]
House and stables, including sketches for garden
Client: Arthur Ransome
Contractors: general, Joseph Thompson; tiles, Thomas Dale & Son; chimney-pieces, W. Wilson & Co.; iron-work, R. Jones; fingerplates, Hart Son Peard & Co.
Cost: £3,065

[289] 1868–9
CAMBRIDGE, Cambridgeshire: Christ's College [O]
Report on existing buildings and proposed new works
Client: Master and Fellows of Christ's College

[290] 1868–70
CAMBRIDGE, Cambridgeshire: Jesus College [D O]
New range of undergraduate rooms of three storeys in red brick with stone dressings, low pitched slated roof. Central tower with turrets. Windows follow the

pattern of existing Tudor work in the College. Also restoration of Master's Lodge and new gateway. Includes laying out grounds. See also 1871 **[347]**
Client: Master and Fellows of Jesus College, Cambridge
Contractors: general, John Loveday; heating and ventilating, D. O. Boyd; stone carving, Rattee & Kett (Master's Lodge); Farmer & Brindley (new range); chimney-pieces, W. H. Burke; iron work, R. Jones; plaster ceiling roses, J. W. Hindshaw
Clerk of works: Mr Tift
Cost: £8,430

[291] 1868
CAMBRIDGE, Cambridgeshire: Trinity College
Memorial brass for the Revd W. J. Beaumont
Contractor: Hardman
Cost: £83

[292] 1868–71
CAMBRIDGE, Cambridgeshire: YMCA [D O Dem]
Lecture room and offices of brick
Client: Cambridge YMCA
Contractors: general, J. Loveday; heating and ventilating, G. N. Haden; glass, F. T. Odell; grates, D. O. Boyd; fittings, W. Wilson; gas fittings, Hart Son Peard & Co.; plaster ceiling roses, J. W. Hindshaw
Cost: £3,080

[293] 1868–9
CHEADLE, Cheshire: Royal Lunatic Asylum, Chapel [O]
Design work by E. Geldart and G. T. Redmayne

[294] 1868–9
GRASMERE, Cumbria [O]
Design for house; not built
Client: J. F. Green

[295] 1868–73
KIRKBY LONSDALE, Cumbria: Lunefield [O Dem]
Large house of stone with tower over entrance, also terraces
Client: Alfred Harris
Contractors: general, J. & W. Beauland; heating and ventilating, D. O. Boyd; stone carving, Farmer & Brindley; tiles, Hargreaves & Craven; F. T. Odell; stained glass, Heaton Butler & Bayne; furniture, Gillow; iron work, R. Jones; Hart Son Peard & Co.; plaster ceiling roses, J. W. Hindshaw
Cost: £14,900

[296] 1868–71
LIVERPOOL, Aigburth: Mossley Park [O]
House of brick in coloured bands with stone dressings and tile roof; also garden terraces
Client: Lloyd Rayner
Contractors: general, Holme and Nichol; heating and ventilating, D. O. Boyd; stone carving, Farmer & Brindley; tiles, W. Godwin; flooring, L. Oppenheimer; stained glass, G. E. Cook; R. B. Edmundson; F. T. Odell, chimney-pieces, H. Sibthorpe; Farmer & Brindley; T. Harrison & Sons; W. H. Burke & Co.; iron work, Hart Son Peard & Co.; R. Jones; plaster ceiling roses, J. W. Hindshaw; garden, W. Henderson
Cost: £13,000

[297] 1868–70
LONDON: Cornhill: Commercial Union Assurance Building [O]
Alterations and installation of heating in office building
Contractors: general, W. Cubitt & Co.; heating and ventilating, G. N. Haden
Cost: £1,185

[298] 1868
LONDON: 20 Elvaston Place [O]
Inspection, valuation, and advice as to alterations
Client: Revd Canon McGrath
Cost: professional charge £15

[299] 1868–71
LONDON: 49 Grosvenor Square [O]
Alterations to house, advice on furniture and carpets. Alterations 1870–71 additional storey
Client: Julian Goldsmid
Contractors: heating and ventilating, D. O. Boyd; chimney-pieces, W. H. Burke; gas fittings, Hart Son Peard & Co.
Cost: £2,655. Alterations £1,215

[300] 1868–9
LONDON, Hampstead Heath: The Pryors [O]
New billiard room for house. Alterations 1870
Client: Wm. Strickland Cookson
Contractors: general, H. Gray; heating and ventilating, D. O. Boyd
Cost: £1,275. Alterations £320

[301] 1868–70
LONDON: 36 Queens Gate [O]
Superintending completion of house including survey and selection of carpets and decoration
Client: James Houldsworth
Contractors: heating and ventilating, D. O. Boyd; stone carving, Farmer & Brindley; stained glass, R. B. Edmundson; chimney-pieces, Lizard Serpentine Co.; W. H. Burke; furniture/fittings, H. Capel; iron work, Hart Son Peard & Co.; plaster ornament, G. Jackson; clock, E. White
Cost: £6,000 (total cost of house £16,900. AW's estimate of value of work done at purchase £10,900)

[302] 1868–72
LONDON, Stoke Newington, Abney Park Cemetery: Monument to J. H. Spreat [O]
Vertical monument of stone with iron railing. Additional burial inscription 1884
Client: Mrs Spreat
Contractors: stone carving, Farmer & Brindley; wrought-iron work, Lester & Hodkinson
Cost: £86

[303] 1868
MANCHESTER: Greenfield [O]
New offices; possibly not built
Client: Mr Bottomley

[304] 1868–9
MANCHESTER: Moss Side [O]
Alterations and additions to house. This commission was handed over to G. T. Redmayne
Client: Peter Spence

[305] 1868–72
MANCHESTER: Portland Street [C O Dem]
Block of five warehouses, symmetrically designed block of four storeys and attic with central pavilion, rusticated ground floor with round arched openings, load-bearing stone faced external walls, and iron beams and stanchions
Client: Frederick W. Grafton
Contractors: general, Thos. Clay; stone carving, Birchenough; chimney-pieces, W. Wilson; iron work, Lester & Hodkinson
Cost: £33,405

[306] 1868–70
MANCHESTER, Whalley Range: Birkfield House [O]
Restoration of house. This commission partly in the hands of G. T. Redmayne
Client: Frederick W. Grafton
Contractors: general, J. Robertson; chimney-pieces, W. Wilson
Cost: £900

[307] 1868–72
READING, Berkshire, Erleigh Road: Grammar School [d D O]
Schoolrooms, hall, and masters' houses, of red brick and terracotta with a tile roof, symmetrically planned but with an asymmetrical turret. See 1873 **[394]** (Pls. 177, 205)
Client: Trustees of Reading Grammar School
Contractors: general, J. Parnell & Sons; heating and ventilating, D. O. Boyd; stone carving, Farmer & Brindley; tiles, R. M. Taylor; stained glass, F. T. Odell; chimney-pieces, Thos. Harrison; furniture/fittings, H. Capel; iron work, R. Jones; plaster ceiling roses, J. W. Hindshaw; bells, J. Warner & Co.
Cost: £19,709

[308] 1868
SEVENOAKS, Kent [O]
Proposed house, not built
Client: Mrs Janson

[309] 1868–9
STANSTEAD ABBOTS, Hertfordshire: Church School [C O]
School building of red brick with tiled roof
Client: T. F. Buxton

Contractors: general, Green & Son; heating and ventilating, D. O. Boyd; iron fittings, Lester & Hodkinson
Cost: £1,495

[310] 1868
SWANSEA, Glamorgan [O]
Tracing plan of property
Client: T. Waterhouse

[311] 1868
SWITZERLAND [O]
Proposed house; probably not built
Client: J. Blumenthal

[312] 1869–70
ASPLEY GUISE, Bedfordshire: Rectory [O Dem]
Rectory house of brick with a slate roof. Original design for a different site
Client: Revd J. Chadwick
Contractors: general, Winn & Foster; heating and ventilating, D. O. Boyd; tiles, Hargreaves and Craven; glass, F. T. Odell; plaster ceiling roses, J. W. Hindshaw
Cost: £2,350

[313] 1869–73
BAKEWELL, Derbyshire: Parsonage [O]
Parsonage house of stone with a slate roof. First design set aside (Figs. 7, 15)
Client: Dr Thomas Balston
Contractors: general, Thomas Clay; chimney-pieces, G. Redfern; grates, D. O. Boyd; glass, F. T. Odell; plaster ceiling roses, J. W. Hindshaw; wallpapers, Scott Cuthbertson & Co.; garden laid out with advice from H. E. Milner
Cost: £5,570

[314] 1869–71
BOWDON, Cheshire: Merle Bank (now Merlewood) [O]
House of yellow brick with red brick bandwork, stone surrounds, and a slate roof. Some fitted furniture to AW's designs (Pl. 102, Fig. 7)
Client: G. H. Fryer
Contractors: general, J. P. Cardwell; tiles, R. M. Taylor; glass, F. T. Odell; chimney-pieces and grates, W. Wilson & Co.; chimney-piece, Farmer & Brindley; iron work, R. Jones; decoration, Robert Pollitt; plaster ceiling roses, J. M. Hindshaw; garden, J. Kellett
Cost: £3,240

[315] 1869–83
EATON, Cheshire: Eaton Hall [C d D O V Dem]
Reconstruction of large mansion (by Porden and Gummow, refaced and extended by Burn) with new library, billiard room, kitchen, and private wings, stone faced with slate roof. Chapel, stables, kennels, coachmen's cottages, lodges, gas works, and parrot house of various materials; also garden layout. Mansion demolished. Alterations 1884–91; 1893; 1898–9 (Pls. 84–97, 184, 185, 191; Figs. 5, 6)
Client: Hugh Lupus Grosvenor, 3rd Marquis (later 1st Duke) of Westminster
Contractors: general, George Smith; some buildings Thomas Hughes; heating and ventilating, G. N. Haden; structural steelwork, J. S. Bergheim; A. Handyside; stone carving, Farmer & Brindley; terracotta, J. C. Edwards; tiles, Milner & Allen; W. Godwin; W. de Morgan; Minton; R. M. Taylor; Leeds Fireclay Co.; W. B. Simpson; mosaic flooring, J. Rust; W. H. Burke; L. Oppenheimer; stained glass, Heaton Butler & Bayne; R. B. Edmundson; F. T. Odell; chimney-pieces, Hopton Wood Stone Co.; Farmer & Brindley; decoration, Best & Lea; furniture/fittings, H. Capel; murals, F. J. Shields; carton pierre, G. Jackson; iron work, F. A. Skidmore; R. Jones; Hart Son Peard & Co.; clock, Gillett & Bland; bells (advice from Dr Stainer) Severin van Aerschodt; statue (Psyche), Stonier & Co.; anemometer, Beck; mother of pearl inlay, R. A. Brangan; heraldry, Metcalfe
Cost: £740,550; 1884–91 £3,725; 1898–9 £1,120

[316] 1869–70
HITCHIN, Hertfordshire: College for Women [O]
Report on sites and alterations to existing building, and plans for a temporary iron building. Not carried out. See Girton College **[346]**
Client: Miss Emily Davies

[317] 1869
LIVERPOOL, Aigburth: Parish Church [O]
Unspecified work; probably not carried out

[318] 1869–73
LONDON: 1 Ennismore Gardens [O Dem]
House. Alterations 1877–9, 1887
Client: Sir Thomas Edwards Moss
Contractors: general, Thorn & Co.; heating and ventilating, D. O. Boyd; stone carving, Farmer & Brindley
Cost: £16,620

[319] 1869
LONDON: 43 Weymouth Street [O]
Minor alterations to house
Client: Theodore Waterhouse
Cost: £105

[320] 1869
MANCHESTER [O]
Proposed warehouse; not built
Client: J. C. Harter

[321] 1869
MANCHESTER, Oxford Road: Owens College [D O]
Symmetrically planned college buildings and medical school, stone faced with tiled roofs, including sketch layout for whole site. Alterations 1882 **[499]**, 1884 **[515]**, 1886 **[536]**, 1891 **[569]**, 1894 **[590]**, 1895 **[605]**, 1898 **[627]** (Pls. 145, 146, 147)
Client: Building Committee of Owens College
Contractors: general, Thos. Clay; structural steel-work, J. S. Bergheim; fireproof construction, Dennett & Co.; stone carving, Farmer & Brindley; heating and ventilating, G. N. Haden; D. O. Boyd; tiles, Craven Dunnill & Co.; mosaic flooring, Jesse Rust; stained glass, R. B. Edmundson; F. T. Odell; chimney-pieces, Hopton Wood Stone Co.; furniture/fittings, Lamb; H. Capel; iron work, Hart Son Peard & Co.; R. Jones
Cost: £105,000

[322] 1869
PORTSLADE, ?Sussex [O]
Decoration and carving in chapel
Cost: £210

[323] 1869–73 and 1875–82
STEEP, Hampshire: Coldhayes [d D O]
Large house and stables of brick faced in polygonal local stone with ashlar dressings and tiled roof, built in two stages. Includes sketch for laying out garden (Pl. 111; Fig. 7)
Client: Revd G. Horsley Palmer
Contractors: general, 1871–2, Dyer; W. & H. Lewis; 1876–82, Stephens & Bastow; stone carving, 1882, Farmer & Brindley; tiles, W. B. Simpson & Sons; grates, D. O. Boyd; furniture/fittings, Henry Capel; iron work, Hart Son Peard & Co.; columns in hall, Lizard Serpentine Co.
Cost: 1876–82 £1,555

[324] 1869
LOCATION UNKNOWN [O]
Design for organ case
Client: W. H. Houldsworth

[325] 1869
Design for tiles [O]
Client: Messrs Craven Dunnill & Co.

[326] 1870
ABINGER, Surrey: Abinger Hall [O Dem]
Large country house of brick and terracotta with a tile roof. Also cottages, terrace walls, layout of garden, and selection of carpets (Pl. 113)
Client: Thomas Henry Farrer
Contractors: general, G. W. Booth; heating and ventilating, G. N. Haden; D. O. Boyd; modelling, Farmer and Brindley; terracotta, Gibbs and Canning; tiles, W. Godwin; J. M. Allen; parquet flooring, L. Oppenheimer; stained glass, R. B. Edmundson; F. T. Odell; decoration, R. Pollitt; fittings, Wedgwood; iron work, Hart Son Peard & Co.; R. Jones; plaster ornaments, G. Jackson
Cost: £14,895

[327] 1870
ALDFORD, Cheshire: Church of St John the Baptist

(by J. Douglas, 1866) [O]
New reredos
Client: 2nd Marquis of Westminster
Contractors: mosaic, Salviati

[328] 1870–1
BLACKMOOR, Hampshire: South Lodge [O]
Lodge house of stone with tiled roof. Porch with turned baluster shafts
Client: Roundell Palmer, MP, later Lord Selborne
Contractors: general, Thomas Kemp
Cost: £550

[329] 1870
BLACKMOOR, Hampshire: School [O]
Schoolroom of stone with a tiled roof, with teacher's house attached
Client: Roundell Palmer, MP, later Lord Selborne
Contractors: general, Thomas Kemp; school grate, D. O. Boyd; clock, E. White
Cost: £1,700

[330] 1870–71
BRISTOL, Avon [O]
Child's monument
Client: J. F. Norris
Contractors: stone carving, Farmer & Brindley; iron work, Robert Jones
Cost: £26

[331] 1870–3
BUCKHURST HILL, Essex: Knighton House [D O]
Addition of domestic wing, dining-room, and bedrooms round central hall to double the size of an existing house by W. E. Williams, all in brick. Alterations 1881–2
Client: E. N. Buxton
Contractors: general, J. Egan; tiles, Gibbs & Moore; stained glass, Heaton Butler & Bayne; grates, D. O. Boyd; iron work, R. Jones
Cost: £6,240

[332] 1870–3
CAMBRIDGE, Cambridgeshire: Trinity Hall [O]
East range of South Court, sets of rooms on four storeys, ashlar faced with tile roof and stepped gables (Pl. 65)
Client: Master and Fellows of Trinity Hall, Cambridge
Contractors: general, P. Horsman; heating and ventilating, D. O. Boyd; stone carving, Farmer & Brindley; glass, F. T. Odell; chimney-pieces, Hopton Wood Stone Co.; furniture/fittings, H. Capel; iron work, R. Jones
Clerk of works: Mr Tift
Cost: £6,400

[333] 1870–3
CAMBRIDGE, Cambridgeshire: Pembroke College [d D O]
New Master's Lodge and south range of college rooms. Separate buildings of brick with stone dressings and slate roofs. Master's Lodge distinguished by rich use of stone mouldings and banding. South range with steep pitched French roof. Includes proposal for complete rebuilding of college and layout of garden. See also **[415]** 1875 (Pl. 72)
Client: Master and Fellows of Pembroke College, Cambridge
Contractors: general, P. Horsman; constructional steelwork, J. S. Bergheim; heating and ventilating, D. O. Boyd; stone carving, Farmer & Brindley; tiles, W. Godwin; W. B. Simpson; J. M. Allen; stained glass, R. B. Edmundson; F. T. Odell; decoration, Morris & Co.; H. Capel; W. Woollam; furniture/fittings, H. Capel; woodwork, Rattee & Kett; iron work, R. Jones; plaster ceiling roses, J. W. Hindshaw; garden, W. Mudd
Clerk of works: Mr Tift
Cost: £25,000

[334] 1870–1
HEYSHAM, Lancashire: Heysham Hall [O]
Alterations to existing mansion and to cottages. Alterations 1883–4
Client: F. W. Grafton
Contractors: heating and ventilating, D. O. Boyd; tiles, Hargreaves & Craven; W. B. Simpson; iron work, R. Jones
Cost: £155; 1883–4 £770
The relation of cost to designer's time charged indicates that only a small portion of the original work was carried out

[335] 1870
HITCHIN, Hertfordshire: College for Women [O]
Alterations to house and plans for temporary iron building
Client: Miss Emily Davies

[336] 1870–2
LEEDS: Royal Insurance Building [O Dem]
Insurance office and chambers of stone with a slate roof
Client: Royal Insurance Co. (per Percy M. Dove)
Contractors: general, Hall & Co.; structural steelwork, Sauvée & Bergheim; heating and ventilating, D. O. Boyd; stone carving, Thomas Earp; chimney-pieces, H. Welsh; iron work, R. Jones
Cost: £9,340

[337] 1870–5
LIVERPOOL: Seamen's Orphans' Institution (now Park Hospital) [O]
Orphan institution and chapel of stone with a slate roof. L-plan with hall and tower separating boys' and girls' wings (Pl. 142)
Contractors: general, Haigh & Co.; structural steelwork, J. S. Bergheim; heating and ventilating, G. N. Haden; D. O. Boyd; stone carving, Farmer & Brindley; tiles, W. Godwin; glass, F. T. Odell; chimney-pieces, Hopton Wood Stone Co.; iron work, R. Jones; Hart Son Peard & Co.; bell, J. Warner & Sons
Cost: orphanage £26,925; chapel £6,550

[338] 1870–4
LONDON, Lincolns Inn Fields: Equity and Law Life Assurance Co. [O]
Office building, stone faced with Gothic detailing to façade and dormer windows breaking through cornice. Previous work and rebuilding 1867–8
Client: Equity and Law Life Assurance Co.
Contractors: general, Newman & Mann; structural steelwork, J. S. Bergheim; heating and ventilating, D. O. Boyd; stone carving, Farmer & Brindley; tiles, W. Godwin; stained glass, R. B. Edmundson; iron work, R. Jones; Hart Son Peard & Co.; plaster ornament, G. Jackson; clock, E. White
Cost: £11,515

[339] 1870
MANCHESTER: Halliwell Lane [O]
Formation of archway
Client: W. H. Houldsworth
Contractors: Thos. Clay
Cost: £20

[340] 1870–2
READING, Berkshire, Whiteley: Christchurch Parsonage [D O]
Parsonage house and stables of grey brick with red patterning and tiled roof. Alterations 1878–80 (Fig. 15)
Client: Revd Matthew Dawley
Contractors: general, J. Parnell & Son; glass, F. T. Odell
Cost: £2,170; 1878–80 £475

[341] 1870–1
ROCHESTER, Kent: The Deanery [O]
Client: Revd Dr Scott (formerly Master of Balliol)
Alterations to house
Contractors: general, Ford & Son; heating and ventilating, D. O. Boyd; chimney-pieces, W. H. Burke & Co.; H. Capel; overmantel, H. Capel; iron work, R. Jones
Cost: £1,350

[342] 1870
WINTERTON, ?Lincolnshire: Parish Church [O]
Unspecified work and sketches for restoration
Client: F. W. Hume

[343] 1870
LOCATION UNKNOWN [O]
Memorial tablet
Client: Mrs St Leger Glynn

[344] 1871
BROMBOROUGH, Cheshire: MacIvor Memorial Fountain [O]
Public drinking fountain in granite
Client: MacIvor Fountain Committee
Stone carving, Farmer & Brindley
Cost: £65

[345] 1871–2
CAMBRIDGE, Cambridgeshire: Emmanuel College [O]
Proposed alteration to Master's Lodge, sketches for new lodge, examination of Hall roof and library, survey of college buildings; not carried out
Client: Master and Fellows of Emmanuel College, Cambridge

[346] 1871–5
CAMBRIDGE, Cambridgeshire: Girton College [d D O]
New college for women of red brick and terracotta with a tile roof, also layout of grounds. The college was built in a succession of separate campaigns. First section: 18 sets of undergraduate rooms, Fellows' rooms and small hall. See also 1875 **[414]**, 1878 **[459]**, 1880 **[475]**, 1883 **[501]**, 1886 **[531]**, 1898 **[621]** (Pl. 75; Fig. 15)
Later additions by Paul (chapel) and Michael Waterhouse (library wing)
Client: The College for Women (per Miss Emily Davies)
Contractors: general, J. Loveday; heating and ventilating, G. N. Haden; tiles, W. Godwin; glass, F. T. Odell; chimney-pieces, Hopton Wood Stone Co.; grates, D. O. Boyd; iron fittings, Hart Son Peard & Co.; Robert Jones; clock, J. Moore & Son; bell, J. Warner & Sons
Cost: £11,370

[347] 1871–7
CAMBRIDGE, Cambridgeshire: Jesus College [D O]
Restoration of hall oriel and screens, and alterations to service wing and staircases (Pls. 173, 175). See also 1868 **[290]**
Client: Master and Fellows of Jesus College, Cambridge
Contractors: general, Rattee & Kett; heating and ventilating, D. O. Boyd; stone carving, Farmer & Brindley; bay window (Master's Lodge), Rattee & Kett; stained glass, Heaton Butler & Bayne; F. T. Odell; chimney-pieces, Hopton Wood Stone Co.; furniture/fittings, Henry Capel; iron work, R. Jones
Clerk of works: Mr Tift
Cost: £6,190

[348] 1871–6
EASENHALL, RUGBY, Warwickshire: Easenhall Hall, now known as Town Thorns [d O]
House, stables, lodges, etc. First design set aside, new plan 1873 in the Classical style of red brick with stone dressings. Alterations 1874 (q.v.), 1885 **[403]** (Pl. 118)
Contractors: general, P. Horsman; heating and ventilating, G. N. Haden; tiles, Minton Hollins & Co.; Craven Dunnill & Co; J. M. Allen; stained glass, R. B. Edmundson; chimney-pieces, W. H. Burke; iron work, Hart Son Peard & Co.; plaster ceiling roses, J. W. Hindshaw
Client: Washington Jackson
Cost: £7,000 (this figure is almost certainly a part only of the total cost, no further certificates are recorded); 1885 £430

[349] 1871–2
HAMSTERLEY, Co. Durham: Dryderdale
Mansion, lodge, and stables of stone with a slate roof (possibly including design of fencing and garden bridges) (Pl. 107)
Client: Alfred Backhouse

[350] 1871–7
HEYTHROP, Oxfordshire: Heythrop Hall (now Management Training Centre, National Westminster Bank) [O]
Reconstruction of ruined mansion, attributed to T. Archer, stables, walled garden, and lodge, also cottages
Client: Albert Brassey
Contractors: general, William Cubitt & Co.; heating and ventilating, D. O. Boyd; stone carving, Farmer & Brindley; tiles, J. M. Allen; W. Godwin; W. de Morgan; stained glass, R. B. Edmundson; F. T. Odell; chimney-pieces, W. H. Burke & Co.; iron work, Hart Son Peard & Co.; R. Jones; plaster ornaments, G. Jackson; clock, J. Moore; E. White; terracotta fountain, Pulham
Cost: £153,000

[351] 1871–5
KESWICK, Cumbria: Lingholm [O]
Large house and lodge of stone with a slate roof; much

altered by other hands
Client: Lt.-Col. Greenall
Contractors: general, W. Haigh & Co.; heating and ventilating, D. O. Boyd; tiles, W. Godwin; Minton; stained glass, R. B. Edmundson; F. T. Odell; decoration, R. Pollitt; furniture/fittings, H. Capel; iron work, Hart Son Peard & Co.; R. Jones; bell, J. Warner & Sons
Cost: £15,700

[352] 1871–2
KNUTSFORD, Cheshire: Market Hall [O]
Market hall with assembly hall over, of red and blue brick with tiled roof
Client: Lord Egerton
Contractors: general, J. Parnell & Sons; heating and ventilating, G. N. Haden; stone carving, Farmer & Brindley; tiles, W. Godwin; stained glass, R. B. Edmundson; F. T. Odell; iron work, R. Jones; plaster ceiling roses, J. W. Hindshaw
Cost: £6,740

[353] 1871
LIVERPOOL [O]
Examination and report as to defects in gables of house

[354] 1871–2
LONDON, Marlborough Mews: 15 Blenheim Street [O Dem]
New warehouse
Client: D. O. Boyd
Contractors: G. W. Booth
Cost: £4,200

[355] 1871–2
LONDON, Gordon Square: University Hall [O]
Alterations and new windows
Client: University College London
Contractors: general, G. F. Graham; stained glass, Clayton & Bell
Cost: £445

[356] 1871–2
LONDON, Norwood Cemetery: Monument for C. J. Elworthy [O]
Client: Executors of Mr and Mrs Charles Elworthy
Contractors: stone carving, Farmer & Brindley; iron work, Hodkinson
Cost: £190

[357] 1871–3
LONDON: 15 Upper Phillimore Gardens [O]
Enlargement of house. Alterations 1891, 1897–8
Client: William Shaen
Contractors: general, Newman & Mann; structural steelwork, J. S. Bergheim; heating and ventilating, D. O. Boyd; stone carving, Farmer & Brindley; stained glass, Heaton Butler & Bayne; F. T. Odell; iron work, R. Jones
Cost: £1,130; alterations £90

[358] 1871–4
MANCHESTER: Victoria Park (now part of Xaverian College) [O]
House of brick with stone dressings and slate roof (Pls. 110, 120; Fig. 7)
Client: T. R. Hetherington
Contractors: general, T. Clay & Son; structural steelwork, J. S. Bergheim; heating and ventilating, D. O. Boyd; tiles, J. M. Allen; Heaton Butler & Bayne; D. Conway; W. Godwin; stained glass, J. M. Allen; Heaton Butler & Bayne; F. T. Odell; chimney-pieces, J. & H. Patteson; decoration, R. Pollitt; plaster ceiling roses, J. W. Hindshaw; garden, Kellett
Cost: £9,300

[359] 1871
MANCHESTER: William's Estate [O]
Inspection and report on estate
Client: Waterhouse and Winterbotham

[360] 1871–6
READING, Berkshire: Municipal Buildings [D O]
Alternative designs for new town hall set aside; design of new Council Chamber and offices as extension to Town Hall of red and blue brick. Alterations 1881 include clock and carillon
Client: Reading Borough Council
Contractors: general, J. Parnell & Sons; structural steelwork, J. S. Bergheim; carving, Farmer & Brindley; tiles, Craven & Co.; stained glass, F. T. Odell;

decoration, R. Pollitt; furniture/fittings, H. Capel; iron work, R. Jones; Hart Son Peard & Co.; clock and bells, Gillett & Bland (1881)
Cost: £8,650

[361] 1871
TEMPLECOMBE, Dorset [O]
Survey and report on estate
Client: W. R. Sandbach

[362] 1872
BASINGSTOKE, Hampshire: Basingstoke School [O]
Advice on proposed enlargements; not built to AW's design

[363] 1872–3
BRACKLEY, Northamptonshire: Evenley Hall [O]
Alterations and additions to late 18th-cent. stone mansion, including new porch, billiard room, and reconstruction of kitchen
Client: Lt.-Col. George Campbell
Contractors: general, G. Smith & Co.; grates, D. O. Boyd
Cost: £1,890

[364] 1872
CAMBRIDGE, Cambridgeshire [O]
Inspection and proposed enlargement of house
Client: John Linton

[365] 1872–5
FALMOUTH, Cornwall: The Crag (now hotel) [O]
House of brick with slate roof. Alterations 1880. Largely destroyed by fire and rebuilt
Client: Misses Sterling
Contractor: Roberts
Cost: £2,980

[366] 1872–5
GLASGOW, Thornliebank: Rouken House [O Dem]
House and stables, lodge and cottages, possibly by enlargement of existing house. Alterations 1877
Client: Alexander Crum
Contractors: heating and ventilating, D. O. Boyd; glass, F. T. Odell; R. B. Edmundson; chimney-pieces, W. H. Burke; iron work, Hart Son Peard & Co.; R. Jones; plaster ceiling roses, J. W. Hindshaw; plaster ornaments, G. Jackson
Cost: £10,000

[367] 1872–3
GUISBOROUGH, Cleveland: Hutton Hospital [D O]
Dispensary building of brick
Client: J. W. Pease
Contractors: general, Windale; Radge; terracotta, Gibbs & Canning; tiles, Cullis; W. Godwin; Clay & Son; chimney-pieces, Beauland; grates, D. O. Boyd
Cost: not distinguished from Mission Room (q.v.)

[368] 1872–3
GUISBOROUGH, Cleveland: Hutton Mission Room [D O]
Reading room and cottage of brick with timber and lead turret (Pl. 119)
Client: J. W. Pease, MP
Contractors: general, Windale; Radge; terracotta, G. Gunton; tiles, Cullis; W. Godwin; Clay & Son; glass, F. T. Odell; iron work, R. Jones
Cost of Hospital (q.v.) and Mission Room £3,535

[369] 1872–9
KILMARNOCK, Ayrshire: Coodham [O]
Additions to large house, with chapel, music room, and conservatory; also lodge, cottages, farm buildings, and design for garden
Client: William Houldsworth
Contractors: general, William Scott; heating and ventilating, G. N. Haden; D. O. Boyd; stone carving and reredos, Farmer & Brindley; M. Muir; tiles, W. Godwin; stained glass, Heaton Butler & Bayne; F. T. Odell; chimney-pieces, W. H. Burke & Co.; furniture/fittings, H. Capel; decoration, Best & Lea; iron work, Hart Son Peard & Co.; R. Jones; plaster ornaments, G. Jackson; murals, F. J. Shields; chapel reredos, Farmer & Brindley
Cost: £13,365

[370] 1872
LONDON, Carey Street: New Court [d D O Dem]
Speculative development of chambers in brick and

terracotta, first block. Alterations see 1879 **[470]** (Pl. 46)
Client: Serle Street and Cooks Court Improvement Society
Contractors: general, Perry; structural steelwork, J. S. Bergheim; heating and ventilating, D. O. Boyd; terracotta, Watcombe Terracotta Co.; Gibbs & Canning; Architectural Terracotta Co.; modelling, Farmer & Brindley; tiles, Craven Dunnill & Co.; chimney-pieces, Hopton Wood Stone Co.; furniture/fittings, Henry Capel; iron work, Hart Son Peard & Co.
Cost: £66,425

[371] 1872 and 1874–7
LONGDOWN, Devon: Culver House [O]
Alterations and additions to Tudor-style house of 1836 and new stables. (Extensions only partially executed.) Lodge in AW style. Interior decoration included new staircase, organ gallery, and stained glass window. (Organ and case now in parish church.) (Fig. 15)
Client: E. Byrom
Contractors: tiles, W. Godwin; decoration, Robert Pollitt; fountain, W. H. Burke & Co.

[372] 1872–5
MANCHESTER, Nicholas Street: Bryce Smith Warehouse [O]
Warehouse building of brick with stone dressing, three storeys and dormered attic with slate roof
Client: Bryce Smith
Contractors: general, Herd; heating and ventilating, D. O. Boyd; stone carving, Farmer & Brindley
Cost: £8,625

[373] 1872–6
MANCHESTER, Reddish: St Elisabeth's School [O]
School buildings of red brick with stone dressings and a slate roof
Client: W. H. Houldsworth
Contractors: general, Thos. Clay; heating and ventilating, G. N. Haden; tiles, Craven Dunnill & Co.; chimney-pieces, W. Wilson; iron work, R. Jones; Hart Son Peard & Co.; bell, J. Warner & Sons
Cost: £11,880. This figure probably includes the cost of the Working Men's Club, 1873 **[388]**, q.v.: the certificates are not clearly distinguished

[374] 1872–3
MARLBOROUGH, Wiltshire: Marlborough College Cricket Pavilion [D O]
Pavilion and store of red brick with wooden verandah and balcony with corrugated iron roof
Client: Marlborough College
Contractors: general, Samuel Clarke; heating and ventilating, D. O. Boyd
Cost: £1,200

[375] 1872
MENTON, Italy: La Mertola [O]
Unspecified alterations to existing house, and garden buildings
Client: Thomas Hanbury

[376] 1872 and 1874–7
READING, Berkshire, Caversham: Free Church School [O]
Schoolroom for Caversham Free Church 1876 **[437]** (q.v.). See also 1865 **[225]**
Client: Ebenezer West
Contractors: A. Barnicoat
Cost: £410

[377] 1872–3
READING, Berkshire: Sonning [O]
Minor alterations to house
Client: J. B. Lee

[378] 1872–5
TWYFORD, Berkshire, Bowsey Hill: Upcroft [O]
Farm buildings. Additions 1879–80, 1881–2. See also 1875 **[425]**, 1878 **[466]**
Client: Theodore Waterhouse
Contractors: 1872–5 Parnell & Son; 1879–80 R. Silver; 1881–2 Hussey
Cost: 1872–5 £840; 1879–82 £660

[379] 1872
LOCATION UNKNOWN [O]
Design for house
Client: Herr Troëltch, per Robert Corkling

[380] 1873–6
DARLINGTON, Co. Durham: Pierrmont [D O]
Gatehouse, clock tower, and conservatory, additional wing to existing mansion and other alterations including decoration of hall
Client: Henry Pease
Contractors: general, J. W. Beauland; heating and ventilating, D. O. Boyd; stone carving, Farmer & Brindley; tiles, W. Godwin; Minton; stained glass, R. B. Edmundson; iron work, Hart Son Peard & Co.; clock, J. Moore; fountain, Fenning
Cost: £9,110

[381] 1873–5
DARLINGTON, Co. Durham, Hurworth: Hurworth Grange [O]
Large house of brick with stone dressings and patterned tile roof and stables. Additional wing with billiard room 1886–7 (Pl. 100; Fig. 15)
Client: J. E. Backhouse
Contractors: general, Parnell & Sons; heating and ventilating, D. O. Boyd; stone carving, Farmer & Brindley; tiles, W. Godwin; J. M. Allen; Craven Dunnill & Co. (1886); stained glass, R. B. Edmundson; F. T. Odell; furniture/fittings, Henry Capel; iron work, Hart Son Peard & Co.; R. Jones; plaster ornaments, G. Jackson; bell, J. Warner & Sons
Cost: £15,000; 1886–7 £1,500

[382] 1873
?DARLINGTON, Co. Durham [O]
Unspecified work
Client: Chas. Pease

[383] 1873–4
GLASGOW: Thornliebank Schools [O]
Schoolroom. Not built
Client: Alexander Crum

[384] 1873–5
LONDON: 45 (later 36) Grosvenor Place [O]
Survey of three houses. Renovation and alterations to existing house, advice on furniture and carpets. Alterations 1878 rebuilding stables, alterations to kitchens, etc., and furniture
Client: J. G. Morris
Contractors: general, Rhodes & Roberts; Holland & Hannen; grates, D. O. Boyd; furniture/fittings, Gillow; iron work, Hart Son Peard & Co.
Cost: £6,325. Alterations £1,870

[385] 1873
LONDON: 18 New Cavendish Street [O]
Minor alterations to house
Client: T. H. Farrer

[386] 1873
LONDON: 18 Princes Gardens [O]
Minor works to existing house
Client: J. W. Pease

[387] 1873–8
MANCHESTER, Long Millgate: Chethams Hospital [D O]
New schoolroom (built 1876) and alterations to old buildings, together with plans (1873) for extensive rebuilding of old school; red sandstone and slate roof to match old work
Client: Feoffees of Chethams Hospital
Contractors: general, Geo. Clay; heating and ventilating, D. O. Boyd; stone carving, T. Earp & Co.; stained glass, F. T. Odell; chimney-pieces, Hopton Wood Stone Co.; iron work, Hart Son Peard & Co.
Cost: £4,890

[388] 1873
MANCHESTER, Reddish: Working Men's Club
Meeting rooms of red brick with stone dressings and a slate roof. See 1872 **[373]**
Client: W. H. Houldsworth

[389] 1873–7
MIDDLESBROUGH, Cleveland: Grammar School [O]
College buildings of red brick with stone dressings of two storeys with dormered attic and central tower with slate roofs. See also 1884 **[516]**, 1885 **[524]**, 1888 **[553]**
Client: J. W. Pease, MP
Contractors: general, J. Johnson; heating and ventilating, D. O. Boyd; carving, Farmer & Brindley; tiles, W. Godwin; mosaic flooring, J. Rust; chimney-

pieces, Hopton Wood Stone Co.; furniture/fittings, G. Goodall; iron work, R. Jones; Hart Son Peard & Co.; clock, Gillett & Bland
Cost: £10,295

[390] 1873–5
MIDDLESBROUGH, Cleveland: Town Hall [O]
Design for municipal offices and town hall; not built
Client: Middlesbrough Corporation

[391] 1873
OTTAWA, Canada [O]
Advice on terrace walls
Client: Marshall Wood, sculptor

[392] 1873–8
OXFORD, Oxfordshire: Balliol College [d D O]
New hall and sets of rooms of stone with a slate roof; also alterations to other buildings. See also 1866 **[254]**, 1886 **[537]**
Client: Master and Fellows of Balliol College
Contractors: general, P. Horsman; structural steelwork, J. S. Bergheim; fireproof construction, Dennett & Co.; heating and ventilating, D. O. Boyd; stone carving, Farmer & Brindley; terracotta, Gibbs & Canning; tiles, W. Godwin; stained glass, Clayton & Bell; F. T. Odell; Metcalfe; decoration, R. Pollitt; furniture/fittings, Henry Capel; iron work, R. Jones; Hart Son Peard & Co.
Cost: £36,490
In 1881 after completing this commission, and perhaps charged with it, Waterhouse designed the Balliol College barge, which was built by Messrs Salters in 1882

[393] 1873–4
READING, Berkshire, Whiteknights Park: The Wilderness [O]
House, lodge, stables of patterned brick with stone dressings and a slate roof, also garden
Client: Julian Goldsmid, MP
Contractors: general, Claridge; heating and ventilating, D. O. Boyd; stone carving, Farmer & Brindley; tiles, W. Godwin; J. M. Allen; stained glass, R. B. Edmundson; F. T. Odell; decoration, R. Pollitt; iron work, R. Jones; bells, J. Warner & Sons; garden, Hayes
Cost: £7,120

[394] 1873–4
READING, Berkshire, Earley Road: Grammar School Chapel [D O]
Chapel of red brick and terracotta with a tiled roof. See also 1868 **[307]**
Client: Revd Dr Stokoe
Contractors: general, Wright and Goodchild; heating and ventilating, G. N. Haden; terracotta, Watcombe Terra Cotta Co.; tiles, W. Godwin; stained glass, F. T. Odell; furniture/fittings, H. Capel; iron work, R. Jones
Cost: £2,760

[395] 1873–5
SLOUGH, Buckinghamshire [O]
Enlargement of chapel
Client: Revd J. W. Hawtrey
Cost: £1,775

[396] 1873
STROUD, Gloucestershire: Temperance Hall [O]
Unspecified work

[397] 1873–8
WIGAN, Lancashire: Free Library [O]
Library building of brick with stone dressings and library fittings
Client: Thomas Taylor, Esq.
Contractors: general, Hughes (Liverpool); heating and ventilating, G. N. Haden; D. O. Boyd; stone carving, Earp & Co.; tiles, Craven Dunnill & Co.; stained glass, F. T. Odell; chimney-pieces, Hopton Wood Stone Co.; decoration, R. Pollitt; furniture/fittings, G. Goodall & Co.; iron work, R. Jones; Hart Son Peard & Co.
Cost: £9,955

[398] 1873–8
WIGAN, Lancashire: Grammar School [O Dem]
School building of brick with stone dressings and classroom fittings
Client: Wigan Grammar School
Contractors: heating and ventilating, G. N. Haden;

tiles, W. Godwin; stained glass, J. Jones
Clerk of works: J. Varty
Cost: Tenders £17,995–£21,860

[399] 1874–5
CAMBRIDGE, Cambridgeshire: St Edmund's Church [O]
New vestry and organ chamber in rubble and ashlar to match existing structure
Client: Dr Geldart (Trinity Hall)
Contractors: Rattee & Kett
Cost: £600

[400] 1874–7
CANNES, France: Villa Florentina [O]
House and stables
Client: Sir Julian Goldsmid
Cost: £7,000+

[401] 1874–80
COCKERHAM, Lancashire: Crookhey Hall [O]
Mansion with lodge, stables, and kennels of stone with slate roof; *porte cochère* and tower with steep French roof; top-lit double-height hall
Client: Capt. C. H. Bird
Contractors: foundation, Murgatroyd; superstructure and finishing, Haigh & Co.; heating and ventilating, G. N. Haden; D. O. Boyd; stone carving, Farmer & Brindley; tiles, W. B. Simpson; stained glass, R. B. Edmundson; glass, R. B. Edmundson; F. T. Odell; chimney-pieces, W. H. Burke; Hopton Wood Stone Co.; decoration, Robert Pollitt; furniture/fittings, Henry Capel; iron work, Hart Son Peard & Co.; R. Jones; clock, J. B. Bryce
Cost: £42,115

[402] 1874
DURGAN, ?Co. Down [O]
Proposed house; not built
Client: J. W. Pease

[403] 1874
EASENHALL, RUGBY, Warwickshire: Town Thorns [O]
Addition of new kitchen and conservatory to house of 1871 **[348]** (q.v.)
Client: Washington Jackson
Contractors: general, P. Horsman; heating & ventilating, G. N. Haden; tiles, Minton Hollis & Co.; Craven Dunnill & Co.; J. M. Allen; iron work, Hart Son Peard & Co.; plaster ceiling flowers, J. W. Hindshaw

[404] 1874
LEWES, Sussex: Redlands Knoll [C O]
Report and proposed alterations; not carried out
Client: J. Hodgkin

[405] 1874–5
LIVERPOOL, Aigburth: Oakwood [O]
New porch and additions to existing house
Client: Benson Rathbone
Contractors: Haigh
Cost: £500

[406] 1874
LIVERPOOL, Knotty Ash: Parish Church [O]
Proposed new chancel; not built
Client: S. J. Arnold

[407] 1874–7
MANCHESTER, Reddish: St Elisabeth's Parsonage [O]
Parsonage house of red brick with stone dressings and a slate roof
Client: W. H. Houldsworth
Contractors: Webster
Cost: £3,025

[408] 1874–6
SALTBURN-ON-SEA, Cleveland: Overdene [O]
House and stables of red brick with stone dressings and a tile roof, with walled garden and glasshouses. The house plan is a mirror image of its neighbour (Woodlands 1874 **[409]**, q.v.)
Client: William Whitwell
Contractors: general, Shafto & Barry; heating and ventilating, D. O. Boyd; stone carving, Farmer & Brindley; tiles, W. Simpson; Craven Dunnill & Co.; chimney-pieces, Hopton Wood Stone Co.
Cost: £9,000

[409] 1874–6
SALTBURN-ON-SEA, Cleveland: Woodlands [O]
House of red brick with stone dressings and a tile roof. The house plan is a mirror image of its neighbour, (Overdene, 1874 **[408]** q.v.)
Client: Francis Fox
Contractors: general, Shafto & Barry; heating and ventilating, D. O. Boyd; stone carving, Farmer & Brindley; stained glass, F. T. Odell; chimney-pieces, Hopton Wood Stone Co.; iron work, Hart Son Peard & Co.; plaster ceiling roses, J. W. Hindshaw
Cost: £6,050

[410] 1874
SANDY, Bedfordshire: Sandy Lodge [O Dem]
Cure of dry rot and new porch to house (by Clutton, 1869)
Client: Arthur W. Peel, MP
Contractors: general, Waller & Son; glass, F. T. Odell
Cost: £1,920

[411] 1874
SHREWSBURY, Shropshire: Shrewsbury School [O]
Unspecified work
Client: Master and Governors of Shrewsbury School

[412] 1874
STANMORE, Middlesex: Stanmore Park [O]
Repairs and underpinning to mansion. Extensive subsequent decoration was later carried out under Brightwen Binyon, using the firm of Morris & Co.
Client: Dowager Lady Wolverton
Contractors: Cubitt
Cost: £765

[413] 1874
Design for a bookcase [O]
Client: Edwin Waterhouse

[414] 1875–7
CAMBRIDGE, Cambridgeshire: Girton College [D O]
Extension to college buildings in brick and terracotta with tile roof. Additional sets of rooms, classroom, laboratory, and gymnasium. See also 1871 **[346]**, 1878 **[459]**, 1880 **[475]**, 1883 **[501]**, 1886 **[531]**, 1898 **[621]**
Client: Mistress and Fellows of Girton College
Contractors: general, William Bell; tiles, Craven Dunnill & Co.; glazing, F. T. Odell; grates, D. O. Boyd
Cost: £3,020

[415] 1875–9
CAMBRIDGE, Cambridgeshire: Pembroke College [D O]
New hall on site of old of brick with stone dressings in the Gothic style under a slate roof. Also Fellows' sets and new library (Pls. 73, 74). Additional floors over hall 1925, by Maurice Webb. See also 1870 **[333]**
Client: Master and Fellows of Pembroke College, Cambridge
Contractors: general, Rattee & Kett; structural steelwork, J. S. Bergheim; heating and ventilating, G. N. Haden; D. O. Boyd; tiles, W. B. Simpson; Craven Dunnill; chimney-piece, Farmer & Brindley; fittings, Rattee & Kett; mosaic for clock dial, Jesse Rust; glass, F. T. Odell; decoration, Best & Lea; iron work, Hart Son Peard & Co.; R. Jones; clock, Gillett & Bland
Clerk of works: Mr Tift
Cost: £29,675

[416] 1875
ETON, Berkshire: Mr Daman's House [O]
Decoration of house
Client: Dr C. D. F. Phillips

[417] 1875
FALMOUTH, Cornwall [D O]
Design for house; not built
Client: J. W. Pease

[418] 1875
GUISBOROUGH, Cleveland [O]
Public ornamental fountain
Client: J. W. Pease
Contractors: Stone carving, Farmer & Brindley
Cost: £100

[419] 1875–7
LONDON: 105 Piccadilly [O]
Renovation of existing house and stables. Alterations

1877 **[448]** q.v.
Client: Julian Goldsmid, MP
Contractors: general, Cordwell; heating and ventilating, D. O. Boyd; iron work, Hart Son Peard & Co.
Cost: £2,920

[420] 1875–6
LONDON: 30 Portland Place [O]
Redecoration of existing house
Client: Roundell Palmer, Lord Selborne
Contractors: general, Cordwell; heating and ventilating, D. O. Boyd; tiles, W. Godwin; decoration, Robert Pollitt; furniture/fittings, H. Capel
Cost: £1,360

[421] 1875–8
LOUGHBOROUGH, Leicestershire, Woodhouse Eaves: The Brand [O]
Large house of locally quarried slate with red sandstone dressings and a local slate roof. Alterations 1891 (Fig. 7)
Client: Alfred Ellis
Contractors: general, Lindley; heating and ventilating, D. O. Boyd; carving, Farmer & Brindley; tiles, Craven & Co.; J. M. Allen; stained glass, F. T. Odell; R. B. Edmundson; iron work, R. Jones
Cost: £6,105

[422] 1875
NORBURY BOOTHS, ?Lancashire: Church [O]
Design for organ
Client: J. Houldsworth

[423] 1875
PINCHINTHORPE, ?Cleveland [O]
Elevations for cottages
Client: J. W. Pease

[424] 1875–7
READING, Berkshire: Whiteley [O]
House with plans for stables and cottages
Client: W. S. Darter
Contractors: general, Tunewell; iron work, R. Jones
Cost: £2,900

[425] 1875
TWYFORD, Berkshire, Bowsey Hill: Upcroft [O]
Lodge and stables. See also 1872 **[378]**, 1878 **[466]**
Client: Theodore Waterhouse
Contractors: general, J. Silver; stone carving, Farmer & Brindley; tiles, J. M. Allen; grates, D. O. Boyd
Cost: £1,675

[426] 1876–8
ARTHINGTON, W. Yorkshire: Arthington Hall [O]
Additions and alterations to late 18th-cent. mansion, including new library and billiard room wing, new conservatory at entrance, refurbishment of service wing, and reconstruction of staircase; redecorations. Library shelving designed by AW
Client: Revd Thomas Sheepshanks
Contractors: general, J. Wood; steelwork, J. S. Bergheim; heating, G. N. Haden; carving, Farmer & Brindley; tiles, Craven Dunnill; W. B. Simpson; mosaic flooring, W. H. Burke & Co.; fittings, Hart Son Peard & Co.; fireplaces, Hopton Wood Stone Co.
Cost: £13,400

[427] 1876
CAMBRIDGE, Cambridgeshire: Trinity Hall [O]
Decoration of Chapel
Client: Master and Fellows of Trinity Hall, Cambridge
Cost: £800

[428] 1876
GREAT AYTON, Cleveland: Cleveland Lodge Cottages [O]
Pair of large cottages of brick with ornamental half-timbering to one wing, with slate roof
Client: J. B. Hodgkin
Contractors: general, Woodward; Grierson; tiles, Craven Dunnill & Co.; iron work, Hart Son Peard & Co.
Cost: £2,910 (includes cost of shop **[429]** q.v.)

[429] 1876
GREAT AYTON, Cleveland: Cleveland Lodge Shop [O]
Cottage and shop of brick with a slate roof, with ornamental half-timbering to upper part of gable
Client: J. B. Hodgkin

Contractors: See Cleveland Lodge Cottages **[428]**
Cost: See Cleveland Lodge Cottages

[430] 1876
LIVERPOOL [O]
Funeral monument to Lloyd Rayner
Client: Mrs Rayner
Cost: estimate £200

[431] 1876–9
LONDON, Holborn Bars: Prudential Assurance Company Offices [D O]
Corporate headquarters building of red brick and terracotta of four storeys and attic with a slate roof. Load-bearing outer walls and shallow brick vaults on iron stanchions. Fireproof valuation department. Alterations 1885 **[522]** **[523]**, 1895 **[602]** q.v. (Pls. 123, 125, 178, 179)
Client: Prudential Assurance Company
Contractors: general, Holland & Hannen; structural steelwork, J. S. Bergheim; W. H. Lindsay; heating and ventilating, W. W. Phipson; D. O. Boyd; terracotta, Gibbs & Canning; modelling and polished granite, Farmer & Brindley; tiles, W. B. Simpson; faience, ? Burmantofts; chimney-pieces, W. H. Burke; decoration, L. Liberty & Co.; Decorators Association; furniture/fittings, H. Capel; Gloucester Wagon Co.; Maple & Co.; Marchant; iron work, Hart Son Peard & Co.
Clerk of works: Dowdall (dismissed 1877)
Cost: £144,940

[432] 1876
LONDON: Kensington [O]
Sketch designs for proposed houses
Client: J. G. Morris
Cost: professional charge £40

[433] 1876–8
LONDON: 38 Sheffield Terrace [O]
House of three storeys with basement and attic of stock brick with slate roof, set in two-storey stucco terrace
Client: Edward C. Sterling
Contractors: Lascelles
Cost: £5,690

[434] 1876–8
MANCHESTER, Cross Street: Haworths Buildings [O]
Stone faced office building. Alterations 1880, 1890
Client: Abraham Haworth
Cost: Alterations 1880 £1,810; 1890 £915

[435] 1876–80
MANCHESTER, Whalley Range: Lancashire Independent College [O]
New master's house, new hall library and classrooms, conversion of entry, and alterations to wings in brick with stone dressings or stone facing to match existing
Client: Lancashire Independent College
Contractors: general, Thomas Clay; structural steelwork, J. S. Bergheim; heating and ventilating, G. N. Haden; D. O. Boyd; carving, T. Earp; tiles, Craven Dunnill; glass, F. T. Odell; chimney-pieces, Hopton Wood Stone Co.; decoration, R. Pollitt; iron work, Hart Son Peard & Co.
Cost: £20,315

[436] 1876
MISERDEN, Gloucestershire: Miserden [O]
Alterations to existing mansion and new domestic wing
Client: E. A. Leatham, MP
Contractors: general, Eastcourt; structural steelwork, J. S. Bergheim; heating and ventilating, Skinner; D. O. Boyd; stone carving, Farmer & Brindley; tiles, W. Godwin; W. B. Simpson; stained glass, Metcalfe; F. T. Odell; chimney-pieces, Hopton Wood Stone Co.; iron work, Hart Son Peard & Co.
Cost: £15,270

[437] 1876–7
READING, Berkshire, Caversham: Free Church [O]
Chapel of red brick with coloured brick bands and tile roof. Low tower at south-west corner. See Caversham Free Church School 1865–6 **[225]** and 1872 **[376]**
Client: Ebenezer West
Contractors: general, Searle; heating and ventilating, G. N. Haden; D. O. Boyd; tiles, Craven Dunnill & Co.; glass, F. T. Odell; chimney-pieces, Hopton Wood Stone Co.; iron work, R. Jones; Hart Son Peard & Co.
Cost: £3,870

[438] 1876
STANMORE, Middlesex: St John's Church [O]
Pulpit of stone with four carved figures
Client: Dowager Lady Wolverton
Contractors: Stone carving, Farmer & Brindley
Cost: £1,300

[439] 1876–9
SUNNINGHILL, Berkshire: Silwood Park. Now part of Imperial College [O]
Large mansion with central double-height hall of red brick with stone dressings and a tile roof; also stables and two lodges. Earlier plans for modifying old house for £22,000 set aside (Col. Pl. 4(*b*))
Client: Charles P. Stewart
Contractors: general, W. Hearne; structural steelwork, J. S. Bergheim; heating and ventilating, G. N. Haden; D. O. Boyd; stone carving, Farmer & Brindley; tiles, W. B. Simpson; stained glass, R. B. Edmundson; F. T. Odell; furniture/fittings, H. Capel; iron work, R. Jones; Hart Son Peard & Co.; clock, Gillett & Bland
Cost: £37,655

[440] 1876
TWYFORD, Hampshire: St Mary's Church [d D O]
New church, on site of old, of stone with flintwork; nave with aisles and west tower; spire not built. Aisle columns and capitals and clerestory windows were incorporated from the old church. Alterations 1887
Client: Committee of St Mary's Church. Sir T. Fairbairn, donor for at least part of the church
Contractors: general, Dyer; heating and ventilating, G. N. Haden; tiles, W. Godwin; stained glass, Heaton Butler & Bayne; F. T. Odell; iron work, Hart Son Peard & Co.
Cost: £7,655. Alterations £200

[441] 1876
WORCESTER, Worcestershire: Guildhall [d O]
Report as to reconstruction and sketch designs for rebuilding with new façade; not built
Client: Worcester City Corporation
Cost: professional charge £81

[442] 1877
CAMBRIDGE, Cambridgeshire: King's College [d]
Proposed new sets of rooms to replace the Wilkins screen. Limited competition for four architects: Scott, Street, Burges, and AW, by invitation. AW declined to submit. Competition won by G. E. Street
Client: Master and Fellows of King's College, Cambridge

[443] 1877–9
GUISBOROUGH, Cleveland: Overbeck [O]
House of pink brick with red brick dressings with a Westmorland slate roof. Alterations 1889
Client: W. C. Trevor
Contractors: general, Johnson; tiles, Craven Dunnill; grates, D. O. Boyd; G. N. Haden; furniture/fittings, Henry Capel; iron work, R. Jones; Hart Son Peard & Co.; plaster ceiling roses, J. W. Hindshaw
Cost: £4,730; 1889 £610

[444] 1877–80
HEYSHAM, Lancashire: Hope Hall [O]
Alteration and extension of existing mansion
Client: F. W. Grafton
Contractors: general, Wm. Southern; heating and ventilating, G. N. Haden; D. O. Boyd; tiles, Craven Dunnill; W. B. Simpson; decoration, Robert Pollitt; iron work, Hart Son Peard & Co.
Cost: £11,520

[445] 1877–82
IWERNE, Dorset: Iwerne Minster. Now Clayesmore School [O]
Large mansion of Ham Hill and Portland stone with a slate roof, also gates, lodge, and stables of brick and tile. Alterations 1883 arcade, 1885 south porch (Pls. 104, 188; Fig. 15)
Client: Lord Wolverton
Contractors: general, G. Smith; heating and ventilating, G. N. Haden; D. O. Boyd; stone carving, Farmer & Brindley; terracotta chimney pots, J. C. Edwards; tiles, W. B. Simpson; glass, F. T. Odell; chimney-pieces, Hopton Wood Stone Co.; Farmer & Brindley; decoration, Best & Lea; furniture/fittings, H. Capel; marble mosaic, W. H. Burke; J. Rust; iron work, Hart Son Peard & Co.; R. Jones; clock, Gillett

& Bland; gardens, E. Milner
Cost: £80,240; 1883 £120; 1885 £1,440

[446] 1877
?LIVERPOOL: Monument for W. R. Callender [O]
Stone tomb
Client: J. S. Mayson
Contractors: Stone carving, Farmer & Brindley
Cost: £125

[447] 1877
LONDON, Gower Street: University College Hospital [O]
Sketch plans for rebuilding; not undertaken. See 1894 **[588]**
Cost: Professional charges £80

[448] 1877–9
LONDON: 105 Piccadilly [O]
New ballroom, etc. see **[419]** 1875
Client: Julian Goldsmid, MP
Contractors: general, W. H. Lascelles; structural steelwork, J. S. Bergheim; heating and ventilating, D. O. Boyd; tiles, W. B. Simpson; iron work, Hart Son Peard & Co.
Cost: £5,840

[449] 1877
READING, Berkshire: Coffee Shop [O]
Temperance café of brick with tiled roof and half-timbering
Client: Mrs Waterhouse
Contractors: general, Barnicoat; heating and ventilating, D. O. Boyd; tiles, Craven; furniture/fittings, Cowie; iron work, Hart Son Peard & Co.; R. Jones
Cost: £2,165

[450] 1877–80
READING, Berkshire, Earley: St Bartholomew's Church [O]
Church of red and blue brick and stone dressings with a tiled roof. Porch and chancel by Bodley
Client: Committee of St Bartholomew's Church
Contractors: general, Stephens & Bastow; heating and ventilating, G. N. Haden; tiles, W. Godwin; iron work, R. Jones; Hart Son Peard & Co.; granite shafts, G. & J. Fenning; hangings, Watts & Co.
Cost: £3,435

[451] 1877–9
REIGATE, Surrey: The Wilderness [O Dem]
House and stables, etc., including furniture and carpeting
Client: Henry Heap
Contractors: general, Lascelles; heating and ventilating, D. O. Boyd; stone carving, Farmer & Brindley; tiles, Craven Dunnill & Co.; stained glass, F. T. Odell; chimney-pieces, Hopton Wood Stone Co.; furniture/fittings, H. Capel; iron work, R. Jones; Hart Son Peard & Co.; plaster ornaments, G. Jackson
Cost: £14,145

[452] 1877
YATTENDON, Berkshire: Reading Room [c]
Village reading room with caretaker's cottage attached, brick with tile roof
Client: Alfred Waterhouse
Cost: £570

[453] 1877
YATTENDON, Berkshire: Yattendon Court [c D Dem]
House of red brick with terracotta dressings and a tile roof, laundry, stables, lodge, and layout of garden with advice from E. Milner. House rebuilt 1926 (Pl. 121, Fig. 9)
Client: Alfred Waterhouse
Contractors: Terracotta, J. C. Edwards
Cost: £11,865

[454] 1877
LOCATION UNKNOWN: King's Hill [O]
Report on cost of repairs
Client: Mrs Lucas

[455] 1877
Use of cartoons for tile designs [O]
Client: W. B. Simpson
Cost: professional charge £4

[456] 1877
Cartoons for tile designs [O]
Client: M. Brydon
Cost: charge for use of cartoons £3

[457] 1878–81
BEDFORD, Bedfordshire: Assize Courts and Shire Hall [D O]
Court house of brick and terracotta with slate roof built in two stages. Alterations 1881–3 (Fig. 15)
Contractors: general, John Wood; heating and ventilating, G. N. Haden; D. O. Boyd; stone carving, Farmer & Brindley; tiles, Craven Dunnill & Co.; W. Godwin; glass, F. T. Odell; chimney-pieces, Hopton Wood Stone Co.; furniture, Wells & Co. (Bedford); iron work, Hart Son & Peard; R. Jones; ironmongery, J. Gibbons
Cost: 1878–81 £14,495; 1881–3 £10,345

[458] 1878–80
BLAGDON, Somerset [O]
Alterations to house
Client: T. R. Hetherington
Contractors: general, L. C. Webster; iron work, R. Jones
Cost: £,2070

[459] 1878
CAMBRIDGE, Cambridgeshire: Girton College [D O]
Additional college rooms. See also 1871 **[346]**, 1875 **[414]**, 1880 **[475]**, 1883 **[501]**, 1886 **[531]**, 1898 **[621]**
Client: Mistress and Fellows of Girton College
Contractors: general, Claridge; heating and ventilating, Hayward Tyler & Co.; wrought iron, Robert Jones; Hart Son Peard & Co.; grates, D. O. Boyd
Cost: £6,470

[460] 1878–81
LEEDS, W. Yorkshire: Yorkshire College (University) [D O]
Textile Industries and Dyeing Department of red brick with stone dressings and a slate roof. Alterations 1883 **[504]**, 1884 **[510]**, 1885 **[520]**, 1892 **[574]**, 1898 **[622, 623, 624]** q.v. (Col. Pl. 4 (*d*); Pl. 150, Fig. 11)
Client: Clothworkers Company through the Yorkshire College
Contractors: general, Woolley & Son; structural steelwork, J. S. Bergheim; Fairbairn, Kennedy & Naylor; heating and ventilating, G. N. Haden; D. O. Boyd; stone carving, Farmer & Brindley; tiles, W. Godwin; iron work, Hart Son Peard & Co.; R. Jones; chimney-pieces, Hopton Wood Stone Co.
Cost: £9,785

[461] 1878
LONDON: Kensal Green Cemetery [O]
Monument for Mrs Bevan
Client: Executors of Mrs Bevan
Contractors: stone carving, Farmer & Brindley; iron work, Hart Son Peard & Co.

[462] 1878
OXFORD, Oxfordshire: Oriel College [d O]
Design for new sets of rooms next to college buildings; not built
Client: Master and Fellows of Oriel College, Oxford

[463] 1878
OXFORD, Oxfordshire: Radcliffe Camera [O]
Installation of heating. Alterations 1888, new staircase
Contractors: general, Symm & Co.; structural steelwork, J. S. Bergheim; heating and ventilating, Rosser and Russell; tiles, W. B. Simpson; iron work, Hart Son Peard & Co.
Cost: £700; 1888 £1,725

[464] 1878–80
OXFORD, Oxfordshire: Union Society [D O]
New debating hall of red brick and buff terracotta with coved cornice and tiled roof. Also designs for new coffee room, reading and writing rooms, not built (Col. Pl. 4 (*e*); Pls. 66, 67; Fig. 15)
Client: Oxford Union Society
Contractors: general, J. Parnell & Sons; heating and ventilating, G. N. Haden; terracotta, Gibbs & Canning; modelling, Farmer & Brindley; tiles, W. Godwin; iron work, R. Jones; Hart Son Peard & Co.
Cost: £6,030

[465] 1878–82
SWANMORE, Hampshire [O]
House, offices, and stables
Client: Charles Myers
Contractors: general, Stephens and Bastow; heating and ventilating, Rosser & Russell; D. O. Boyd; stone carving, Farmer & Brindley; tiles, W. B. Simpson; W. Godwin; stained glass, R. B. Edmundson; F. T. Odell; chimney-pieces, Blackmore & Nixson; decoration, Best & Lea; furniture/fittings, H. Capel; iron work, R. Jones; Hart Son Peard & Co.; plaster ornaments, G. Jackson
Cost: £36,520

[466] 1878
TWYFORD, Berkshire: Bowsey Hill [O]
Pair of cottages. See also 1872 **[378]**, 1875 **[425]**
Client: Theodore Waterhouse
Contractors: Stephens & Bastow
Cost: £760

[467] 1878-*c.*1900
YATTENDON, Berkshire [c]
Various cottages, singly, in pairs, and in threes, in brick with tile roofs; also village shop and farm buildings
Client: Alfred Waterhouse

[468] 1878
LOCATION UNKNOWN [O]
Conservatory for unidentified house
Client: Rudolf Lehmann
Contractors: Lascelles
Cost: £86

[469] 1879–80
HELMSLEY, N. Yorkshire: Duncombe Park [O]
Unspecified work (probably conversion of stables to dwelling after a fire in the mansion)
Client: Earl of Faversham
Contractors: Barry
Cost: £1,345

[470] 1879–85
LONDON, Carey Street: New Court [D O Dem]
Speculative development of chambers in brick and terracotta, second block (see 1872 **[370]**). Alterations 1889 to accommodate Ye Mecca restaurant; 1892; 1896; 1898–9, minor alterations 1901–4 (Pl. 46)
Client: Serle Street and Cooks Court Improvement Society
Contractors: general, Stephens and Bastow; structural steelwork, J. S. Bergheim; W. H. Lindsay; heating and ventilating, G. N. Haden; D. O. Boyd; terracotta, possibly Gibbs and Canning; modelling, Doulton; faience (restaurant), Burmantofts; stained glass, R. B. Edmundson; F. T. Odell; chimney-pieces, Hopton Wood Stone Co.; iron work, Hart Son Peard & Co.; R. Jones
Cost: £87,560. Alterations £3,030

[471] 1879–83
LONDON: 24 Kensington Palace Gardens [D O]
Repairs and alterations to existing house and stables
Client: J. W. Pease, MP
Contractors: general, Higgs & Hill; heating and ventilating, D. O. Boyd
Cost: £1,960

[472] 1879–80
READING, Berkshire, Earley: St Bartholomew's Vicarage [O]
Parsonage house of patterned brick with a tiled roof; now flats
Contractors: Stephens and Bastow
Clerk of works: F. W. Barr
Cost: £1,870

[473] 1879–82
ROCKCLIFFE, Kircudbrightshire: Baron's Craig [O]
Granite faced house in rubble and dressed stone with battlemented tower; now hotel
Client: Christopher Morris
Contractors: general, Crackston & Sons; heating and ventilating, D. O. Boyd; stone carving, Farmer & Brindley; terracotta, J. C. Edwards; tiles, W. Godwin; W. B. Simpson; chimney-pieces, J. & H. Patteson; furniture/fittings, H. Capel; iron work, R. Jones; Hart Son Peard & Co.
Cost: £9,550

[474]) 1879
YATTENDON, Berkshire: Well House [c]
Well house in village square, brick and open timber work with a tile roof, now bus shelter
Client: Alfred Waterhouse

[475] 1880–2
CAMBRIDGE, Cambridgeshire: Girton College, Gate Lodge [O]
Lodge cottage of brick under a tiled roof, with timber porch. See also 1871 **[346]**
Client: Mistress and Fellows of Girton College
Contractors: William Bell
Cost: £360

[476] 1880–2
CRIMPLESHAM, Norfolk: Crimplesham Hall [O]
House and stables of yellow Cambridgeshire brick with stone string courses and low pitched slate roof. Alterations 1885 (Pl. 99)
Client: John Grant Morris for his daughter Mrs A. T. Bagge
Contractors: foundation, R. Gage; superstructure and finishing, Stephens & Bastow; heating and ventilating, G. N. Haden; stone carving, Farmer & Brindley; tiles, W. Godwin; flooring, L. Oppenheimer; chimney-pieces, Hopton Wood Stone Co.; iron work, R. Jones; Hart Son Peard & Co.
Cost: £8,535; 1885 £35

[477] 1880–1
OXFORD, Oxfordshire, Headington: Headington Hall. Now part of Rye St. Anthony School [O]
Medium-sized house of brick with tile roof, unusual verandah with stone arcade, and separate stable block
Client: Ferdinand Willert
Contractors: general, C. Claridge; heating and ventilating, D. O. Boyd; stone carving, Farmer & Brindley; tiles, Craven Dunnill; furniture/fittings, Henry Capel; iron work, R. Jones; Hart Son Peard & Co.
Cost: £4,880

[478] 1880–3
HOVE, Sussex: Town Hall [D O Dem]
Town hall, municipal offices, and fire station of brick and terracotta with a slate roof
Client: Hove Borough Council
Contractors: general, John T. Chappell; heating and ventilating, D. O. Boyd; stone carving/modelling, Farmer & Brindley; tiles, W. Godwin; granite columns, G. & J. Fenning; stained glass, F. T. Odell; chimney-pieces, Hopton Wood Stone Co.; decoration, Salviati; iron work, Hart Son Peard & Co.; R. Jones; clock, Gillett & Bland; organ, Willis; plaster ceiling roses, J. W. Hindshaw
Cost: £39,920

[479] 1880
LONDON: 32 Belgrave Square [O]
Unspecified alterations to house
Client: Earl of ClanWilliam
Contractors: general, Bywaters; heating and ventilating, D. O. Boyd; iron work, Hart Son Peard & Co.
Cost: £2,220

[480] 1880–2
LONDON: 30–9 Brooke Street [D O Dem]
Shops, offices, and chambers of red brick and terracotta
Client: Prudential Assurance Company
Contractors: general, B. E. Nightingale; iron work, Hart Son Peard & Co.
Cost: £17,325

[481] 1880–1
LONDON: 1, 1a Old Bond Street [O]
Client: Wakefield Christy
Contractors: general, Holland & Hannen; heating and ventilating, D. O. Boyd; terracotta, Gibbs & Canning; modelling, Farmer & Brindley; glass, F. T. Odell; chimney-pieces, W. H. Brooke & Co.; iron work, Hart Son Peard & Co.
Cost: £11,310

[482] 1880–2
READING, Berkshire, Redlands Road: East Thorpe; now St Andrew's Hall of Reading University [D O]
House and stables of brick and terracotta. Alterations 1888 (Fig. 15)
Client: Alfred Palmer

Contractors: general, J. Parnell & Sons; heating and ventilating, D. O. Boyd; terracotta, J. C. Edwards; modelling, Farmer & Brindley; iron work, R. Jones
Cost: £10,990; 1888 £105

[483] 1880–2
STEVENTON, Hampshire: Steventon Manor [O]
Extension to stables of Old Manor, new house and offices
Client: Henry Harris
Contractors: general, Henry Lovatt; heating and ventilating, G. N. Haden; D. O. Boyd; stone carving, Farmer & Brindley; mosaic flooring, J. F. Ebner; stained glass, R. B. Edmundson; chimney-pieces, Hopton Wood Stone Co.; W. H. Burke & Co.; iron work, R. Jones; plaster ornaments, G. Jackson
Cost: £15,585

[484] 1880
SWANMORE, Hampshire: Church [O]
Memorial pulpit and tablet
Client: Executors of C. Myers
Contractors: Stone carving, Farmer & Brindley
Cost: £150

[485] 1880
LOCATION UNKNOWN: Southington Lodge [O]
Probably alterations only. Alterations 1885–6, 1889–90
Client: W. W. Portal
Contractors: general, Sanders; stained glass, F. T. Odell; chimney-pieces, Hopton Wood Stone Co.
Cost: £1,125; 1885–6 £315; 1889–90 £245

[486]*c.* 1880
PIANO [V]
Grand piano in mahogany with inlaid mother of pearl decoration. Now in V. & A. Museum
Client: Alfred Waterhouse

[487] 1881–5
DARLINGTON, Co. Durham: Bushell Hill (now Mowden Hall) [D O]
Large house and stables of patterned red brick and terracotta with tiled roof built as an extension to house by Jno. Ross, 1862. Layout of gardens 1886–8 with advice from E. Milner
Client: J. B. Pease
Contractors: general, Wm. Southern; heating and ventilating, G. N. Haden; D. O. Boyd; modelling, Farmer & Brindley; tiles, Craven Dunnill; mosaic flooring, J. F. Ebner; chimney-pieces, Hopton Wood Stone Co.; iron work, Hart Son Peard & Co.; R. Jones
Cost: £13,195; 1886–8 £670

[488] 1881–3
LIVERPOOL: University College [O]
Alterations to old asylum building to form Liverpool University. See also 1884 **[512]**, 1887 **[542]**, 1888 **[549]**, 1895 **[598]**, 1899 **[633]**
Client: Liverpool University College
Contractors: general, Wm. Tomkinson & Son; heating and ventilating, G. N. Haden; D. O. Boyd, tiles; D. Conway; chimney-pieces, Blackmore & Nixson; furniture/fittings, North of England Furnishing Co.; iron work, Hart Son Peard & Co.
Cost: £4,540

[489] 1881–3
LOFTUS-IN-CLEVELAND, Cleveland: Grinkle Park [D O]
Mansion of stone with a slate roof, tower, and conservatory. The stables were designed by another hand (Pl. 105)
Client: C. M. Palmer, MP
Contractors: general, T. Dickinson; heating and ventilating, D. O. Boyd; stone carving, Farmer & Brindley; terracotta, J. C. Edwards; tiles, W. Godwin; chimney-pieces, Hopton Wood Stone Co.; Blackmore & Nixson; iron work, R. Jones; Hart Son Peard & Co.
Cost: £8,500

[490] 1881–6
LONDON, Exhibition Road: City and Guilds of London Institute [d D O Dem]
College of red brick and terracotta with two symmetrical wings and central entrance (Pl. 144)
Contractors: foundation, George Munday; superstructure, Henry Lovatt; structural steelwork, W. H. Lindsay; heating and ventilating, G. N. Haden; D. O.

Boyd; terracotta modelling, James Gamble; Farmer & Brindley; granite, G. & J. Fenning; tiles, Craven Dunnill & Co.; chimney-pieces, Hopton Wood Stone Co.; W. H. Burke & Co.; furniture/fittings, Stephens & Bastow; Maple & Co.; iron work, Hart Son Peard & Co.; R. Jones; plaster ornaments, G. Jackson; clock, Gillett & Co.
Cost: £88,120

[491] 1881
LONDON: 68 Grosvenor Street [O]
Minor alterations to house
Client: George Palmer
Contractors: James Clark
Cost: £215

[492] 1881–7
LONDON, Hammersmith Road: St Paul's School [d D O Dem]
Symmetrically planned Gothic school building in dark brick and red terracotta with a slate roof. Matching master's house, caretaker's lodge, and gazebo. Playing field walls in red brick and terracotta with ornamental railings (Pl. 143)
Client: Mercers Company
Contractors: general, J. Parnell & Sons; structural steelwork, W. H. Lindsay & Co.; J. S. Bergheim; heating and ventilating, G. N. Haden; D. O. Boyd; modelling, Farmer & Brindley; granite, G. & J. Fenning; tiles, Craven Dunnill & Co.; glass, F. T. Odell; chimney-pieces, W. H. Burke; Hopton Wood Stone Co.; furniture/fittings, Maple & Co.; North of England School Furniture Co.; iron work, Hart Son Peard & Co.; R. Jones; clock, Gillett & Bland; garden Osborn
Clerk of works: T. M. Rickman
Cost: £105,735

[493] 1881–3
MANCHESTER: 66 Spring Gardens/Fountain Street [d O]
Warehouse of three storeys with basement and attics; ground floor in stone and upper floors in brick with stone dressings
Client: J. H. Gartside & Co.
Contractors: general, R. Neill; fireproof construction, Dennett & Ingle; heating and ventilating, G. N. Haden; D. O. Boyd; stone carving, Earp & Hobbs; tiles, D. Conway; stained glass, R. B. Edmundson; chimney-pieces, J. & H. Patteson; Hopton Wood Stone Co.; iron work, R. Jones; Hart Son Peard & Co.
Cost: £22,965

[494] 1881
YATTENDON, Berkshire: Parish Church of SS Peter & Paul [O]
Brick and timber porch with tile roof; vestry in stone with tile roof over heating chamber, new altar, etc. Alterations 1889 new organ; 1896 rebuilding of lower part of tower
Client: Alfred Waterhouse
Contractors: general, J. Jones (vestry); heating and ventilating, G. N. Haden; tiles, Craven Dunnill; vestry screen, Rattee and Kett; altar, Farmer & Brindley; hangings, Watts & Co.; lamp standards, Hart Son Peard & Co.
Cost: 1896 £370

[495] 1882–3
CANNES, France: Villa Allerton [O]
Villa of stone with steep pitched roof. Alterations 1888–9
Client: John Grant Morris (See Allerton Priory 1866 **[237]**)
Contractors: general, Casson *aîné*; chimney-pieces, W. H. Burke & Co.; iron work, R. Jones; Hart Son Peard & Co.
Cost: £8,750; 1888–9 15,000 fr.

[496] 1882–5
EMSWORTH, Hampshire: Watergate House [O]
Mansion. Alterations 1886, 1887, 1888, 1892
Client: Richard Christy
Contractors: general, Foster & Dicksee; heating and ventilating, G. N. Haden; D. O. Boyd; stone carving, Farmer & Brindley; tiles, W. Godwin; stained glass, R. B. Edmundson; iron work, Hart Son Peard & Co.; R. Jones; plaster ornaments, G. Jackson
Cost: £13,580; 1886 £150; 1887 £115; 1888 £34; 1897 £25

[497] 1882–5
LIVERPOOL: Turner Memorial Home [d O]
Home and chapel for seamen, of stone with a tiled roof and half-timbered porch. Alterations 1887–9 (Pl. 202)
Client: Mrs Turner
Contractors: general, Holme & Nichol; heating and ventilating, G. N. Haden; D. O. Boyd; stone carving, Earp Son & Hobbs; granite columns, G. & J. Fenning; tiles, Craven Dunnill & Co.; stained glass, Heaton Butler & Bayne; R. B. Edmundson; chimney-pieces, W. H. Burke; Blackmore & Nixson; Hopton Wood Stone Co.; decoration, Watts & Co.; iron work, R. Jones; Hart Son Peard & Co.; clock, Gillett & Co.; organ, Gray & Davison
Cost: £32,170

[498] 1882
LONDON: 4 Chester Place [O]
Alterations to house
Contractors: tiles, W. de Morgan; grates, D. O. Boyd; furniture/fittings, H. Capel; iron work, Hart Son Peard & Co.
Cost: £840

[499] 1882–3
MANCHESTER, Oxford Road: University [D O]
Extension to University Medical School. See also 1869 **[321]**, 1884 **[515]**, 1886 **[536]**, 1891 **[569]**, 1894 **[590]**, 1895 **[605]**, 1898 **[627]**
Client: Building Committee of Owens College
Contractors: general, Wm. Southern; fireproof construction, Dennett & Ingle; heating and ventilating, G. N. Haden; D. O. Boyd; stone carving, Earp & Hobbs; iron work, Hart Son Peard & Co.; decorative stonework, Hopton Wood Stone Co.
Cost: £12,980

[500] 1882–5
WREXHAM, Shropshire: Bryn-y-Pys Hall [O]
Internal alterations and new stables and tower
Client: Edmund Peel
Contractors: general, R. Beckett; heating and ventilating, D. O. Boyd; iron work, R. Jones; plaster ornaments, G. Jackson
Cost: £7,795

[501] 1883–5
CAMBRIDGE, Cambridgeshire: Girton College [D O]
Additional rooms and service rooms, extension of Hall, Stanley Library with Mistress's flat. See also 1871 **[346]**, 1875 **[414]**, 1878 **[459]**, 1880 **[475]**, 1886 **[531]**, 1898 **[621]** (Pl. 76)
Client: Mistress and Fellows of Girton College
Contractors: general, George Grimwood; tiles, Craven Dunnill & Co.; glazing, F. T. Odell; chimney-pieces, Hopton Wood Stone Co.; grates, D. O. Boyd; furniture/fittings, Maple & Co.; minor fittings, Rattee & Kett; wrought iron fittings, Hart Son Peard & Co.; Robert Jones
Cost: £13,315

[502] 1883
CAMBRIDGE, Cambridgeshire: Gonville and Cauis College [O]
Two-storey lecture hall of brick with stone dressings with a slate roof; now converted to a garage. See also 1867 **[268]**
Client: Master and Fellows of Gonville and Caius College, Cambridge
Cost: £2,735

[503] 1883
METHLICK, Grampian: Haddo House, Aberdeenshire [O]
Wall and archway to private graveyard
Client: Earl of Aberdeen
Contractors: William Kieth; iron work, Robert Jones
Cost: £940

[504] 1883–5
LEEDS, W. Yorkshire: Yorkshire College (University) [D O]
Classrooms and laboratories and Baines memorial wing of red brick and stone with a slate roof. Alterations see 1878 **[460]**, 1884 **[510]**, 1885 **[520]**, 1892 **[574]**, 1898 **[622, 623, 624]**
Client: Committee of Yorkshire College
Contractors: general, J. Wood & Sons; chimney-pieces, Hopton Wood Stone Co.; iron work, Hart Son Peard & Co.; R. Jones
Clerk of works: J. Battye
Cost: £34,135

[505] 1883–7
LONDON, Hampstead: Lyndhurst Road Congregational Church [D O]
Hexagonal church of dark brick and red terracotta with matching schoolrooms. Alterations 1896–9 (Pl. 141; Fig. 10)
Client: Trustees of Hampstead Congregational Church
Contractors: general, J. Parnell & Son; structural steelwork, W. H. Lindsay; heating and ventilating, D. O. Boyd; modelling, Farmer & Brindley; tiles, Craven Dunnill & Co.; iron work, Hart Son Peard & Co.; R. Jones
Cost: £15,970; 1896–9 £550

[506] 1883
LONDON: 41 Wimpole Street [O]
Minor alterations to house
Client: Henry Head
Contractors: B. Cordwell
Cost: £75

[507] 1883–5
MANCHESTER, Reddish: St Elisabeth's Church [D O]
Church of red brick and stone with apsidal east end and south chapel and south-east tower. Clerestory windows to designs by F. J. Shields. These were originally ordered for Coodham, see 1872 **[369]**. Alterations 1890, pulpit (Pls. 138, 139, 140)
Client: W. H. Houldsworth
Contractors: foundation, C. H. Normanton; heating and ventilating, G. N. Haden; superstructure and finishing, Wm. Southern; stone carving, Earp & Hobbs; ?granite, G. & J. Fenning; marble work, W. H. Burke; J. & H. Patteson; tiles, W. Godwin; stained glass, F. T. Odell; Heaton Butler & Bayne; decoration, Watts & Co.; iron work, R. Jones; Hart Son Peard & Co.; grounds, J. Kellett
Cost: £19,425; 1890 £125

[508] 1884
ALLOA, Cambus, Stirlingshire: Arnsbrae [O]
Country house of local sandstone with a slate roof. Also lodge, stables, and offices. Alterations 1896–8, 1900, and later
Client: James Younger
Contractors: general, Robert Macfarlane; G. & R. Cousins; R. Cairns; heating and ventilating, D. O. Boyd; stone carving, Farmer & Brindley; chimneypiece, Hopton Wood Stone Co.; iron work and gas fittings, Robert Jones; Hart Son Peard & Co.
Cost: £7,720; 1896–8 £2,792

[509] 1884–5
CHART SUTTON, Sussex: Chart Place [O]
Alterations to manor house
Client: Thomas Balston
Contractors: Wallis
Cost: £655

[510] 1884–6
LEEDS, W. Yorkshire: Yorkshire College (University) [D O]
Extension of Textile Industries Department of red brick with stone dressings and a slate roof. Alterations see 1878 **[460]**, 1883 **[504]**, 1885 **[520]**, 1892 **[574]**, 1898 **[622, 623, 624]**
Client: Committee of Yorkshire College
Contractors: general, J. Wood & Son; stone carving, Farmer & Brindley; iron work, R. Jones; Hart Son Peard & Co.
Cost: £8,830

[511] 1884
LEEK, Staffordshire: Town Hall
Competition entry. Placed first but not commissioned

[512] 1884–7
LIVERPOOL: University, Chemical Laboratories [D O]
Laboratories and teaching rooms of brick with red terracotta dressings and slate roofs. See also 1881 **[488]**, 1887 **[542]**, 1888 **[549]**, 1895 **[598]**, 1899 **[633]**
Client: Committee of Liverpool University College
Contractors: general, Jones & Sons; heating and ventilating, G. N. Haden; iron work, R. Jones; Hart Son Peard & Co.; decorative stonework, Hopton Wood Stone Co.
Cost: £15,805

[513] 1884–6
LONDON: Park Lane [O]
Chambers and stables
Client: Sir Julian Goldsmid
Contractors: general, Foster & Dicksee; heating and ventilating, D. O. Boyd; iron work, Hart Son Peard & Co.; stone carving, Farmer & Brindley; stonework, Hopton Wood Stone Co.
Cost: £17,025

[514] 1884–7
LONDON, Whitehall Place: National Liberal Club [d D O]
Clubhouse of steel and brick with Portland stone facing and Westmorland slate roof (Pls. 133–7; Fig. 14)
Client: National Liberal Club Buildings Co. Ltd.
Contractors: foundation, Henry Lovatt; superstructure and finishing, George Shaw; structural steelwork, W. H. Lindsay; fireproof construction, Dennett & Ingle (tower); heating and ventilating, W. W. Phipson; stone carving, Farmer & Brindley; C. Smith; tiles, Carter Johnson & Co.; faience, Wilcock & Co.; mosaic flooring, J. F. Ebner & Son; chimney-pieces, Hopton Wood Stone Co.; grates, D. O. Boyd; furniture/fittings, Morris & Norton; W. James & Co.; Maple & Co. (carpets); iron work, Hart Son Peard & Co.; R. Jones; plaster ornaments, G. Jackson; electric lighting, Edison & Swan
Clerk of works: Thomas Warburton
Cost: £165,950

[515] 1884–8
MANCHESTER, Oxford Road: University [D O]
Extension of college buildings to include museum, council room, and tower. Alterations 1889–90. See also 1869 **[321]**, 1882 **[499]**, 1886 **[536]**, 1891 **[569]**, 1894 **[590]**, 1895 **[605]**, 1898 **[627]** (Pl. 149)
Client: Building Committee of Owens College
Contractors: general, Wm. Southern & Son; fireproof construction, Dennett & Ingle; heating and ventilating; G. N. Haden; D. O. Boyd; stone carving, Farmer & Brindley; Earp Son & Hobbs; tiles, Craven Dunnill & Co.; faience, Burmantofts; chimney-pieces, Hopton Wood Stone Co.; J. & H. Patteson; iron work, Hart Son Peard & Co.; R. Jones
Cost: £75,725

[516] 1884–5
MIDDLESBROUGH, Cleveland: Grammar School [O]
New wing to school. See also 1873 **[389]**, 1885 **[524]**, 1888 **[553]**
Client: Trustees of Middlesbrough High School
Contractors: general, J. Johnson; tiles, Craven Dunnill & Co.; iron work, Hart Son Peard & Co.
Cost: £3,820

[517] 1884–8
PANGBOURNE, Berkshire: Buckhold [O]
Large house of red brick and buff terracotta with half-timbered entrance porch, tile hanging, and tile roof
Client: Dr H. Watney
Contractors: general, Wheeler Bros.; heating and ventilating, D. O. Boyd; modelling, Farmer & Brindley; tiles, Doulton; iron work, R. Jones; Hart Son Peard & Co.
Cost: £34,600

[518] 1884–5
SALTBURN-ON-SEA, Cleveland: Concert Hall [O]
Hall and service rooms of yellow brick and terracotta with iron framed roof
Contractors: general, T. D. Ridley; heating and ventilating, G. N. Haden; D. O. Boyd; grates, D. O. Boyd; furniture/fittings, Maple & Co.; iron work, Hart Son Peard & Co.; R. Jones
Cost: £4,165

[519] 1884
LOCATION UNKNOWN: Chateau St Michel [O]
Alterations
Client: Mrs Elder
Contractors: Geo. Russell
Cost: £490

[520] 1885
LEEDS, W. Yorkshire: Yorkshire College (University) [D O]
Engineering block of red brick with stone dressings with a slate roof. Alterations see also 1878 **[460]**, 1883 **[504]**, 1884 **[510]**, 1892 **[574]**, 1898 **[622, 623, 624]**
Client: Committee of Yorkshire College

Contractors: general, J. Wood & Son; heating and ventilating, D. O. Boyd; chimney-pieces, Hopton Wood Stone Co.; iron work, Hart Son Peard & Co.; R. Jones
Cost: £6,420

[521] 1885–8
LIVERPOOL: Prudential Assurance Company Buildings [D O]
Office building, terracotta faced with a slate roof, faience decoration to public offices and stairways. Alterations 1890–2, 1899, and 1904–6 by Paul Waterhouse
Client: Prudential Assurance Company
Contractors: general, Holme & Green; heating and ventilating, D. O. Boyd; terracotta, J. C. Edwards; tiles, Craven Dunnill & Co.; faience, Wilcock & Co.; Burmantofts; W. W. Brown & Co.; stained glass, R. B. Edmundson; chimney-pieces, Hopton Wood Stone Co.; iron work, Hart Son Peard & Co.; Blackmore & Co.
Cost: £15,360

[522] 1885–8
LONDON, Brooke Street: Prudential Assurance Company Offices [D O]
Extension of offices of red brick and terracotta with floors of steel decking on load-bearing walls. Glazed brick interior decoration. See also 1876 **[431]**, 1895 **[602]**
Client: Prudential Assurance Company
Contractors: general, Holland & Hannen; structural steelwork, W. H. Lindsay & Co.; terracotta, Gibbs & Canning; mosaic flooring, J. F. Ebner
Cost: £33,770

[523] 1885–8
LONDON, 17–19 Greville Street: Prudential Assurance Company Offices [D O Dem]
Extension of offices of red brick and terracotta with floors of steel decking on load-bearing walls. Glazed brick interior decoration. Alterations 1892, 1893, 1894, 1898. See also 1876 **[431]**, 1895 **[602]**
Client: Prudential Assurance Company
Contractors: general, Holland & Hannen; structural steelwork, W. H. Lindsay & Co.; heating and ventilating, W. W. Phipson; terracotta, Gibbs & Canning; mosaic flooring, W. H. Burke; furniture/fittings, Gillow & Co.; iron work, Hart Son Peard & Co.
Cost: £20,455; alterations £6,765

[524] 1885–6
MIDDLESBROUGH, Cleveland: Grammar School [O]
Chemical laboratories. See also 1873 **[389]**, 1884 **[516]**, 1888 **[553]**
Client: Trustees of Middlesbrough High School
Contractors: general, J. Johnson; iron work, Hart Son Peard & Co.; R. Jones
Cost: £1,385

[525] 1885
OXFORD, Oxfordshire: Mansfield College [d]
Design for college buildings, in competition with Basil Champneys; not selected
Client: Building Committee of Mansfield College (per Abraham Haworth and Joseph Thompson)

[526] 1885
READING, Berkshire, Silver Street: The Rising Sun [O]
Minor alterations
Contractors: John Bottrill

[527] 1885–8
ROCHDALE, Lancashire: Town Hall [D O]
New tower to replace original designed by W. H. Crossland, destroyed by fire
Client: Rochdale Borough Council
Contractors: general, W. A. Peters & Sons; stone carving, Earp Son & Hobbs; iron work, Hart Son Peard & Co.
Cost: £11,900

[528] 1885–7
WINDLESHAM, Surrey: Elm Ridge [O]
House
Client: Mrs Mereweather
Contractors: general, J. Bottrill; heating and ventilating, G. N. Haden; chimney-pieces, Hopton Wood Stone Co.; iron work, R. Jones
Cost: £3,350

[529] 1885–6
YATTENDON, Berkshire: Village School [c]
New school building and teacher's house in brick with Ham Hill dressings. Alterations 1891
Client: Alfred Waterhouse; 1891 Executors of T. Waterhouse
Cost: £2,077

[530] 1886–9
ALLOA, Stirlingshire: Town Hall and Library (Thompson Hall) [D O]
Public hall of Polmaise stone with slate roof
Client: J. Thompson Paton
Contractors: general, G. & R. Cousin; heating and ventilating, W. W. Phipson; D. O. Boyd; faience, Burmantofts; mosaic flooring, W. H. Burke & Co.; stained glass, R. B. Edmundson & Sons; decoration, Reed & Downie (Edinburgh); furniture supplied by Taylor & Sons; Whitlock & Reed (Edinburgh); enamelled and painted ceiling lights, Edmeston of Manchester; gas fittings, Hart Son Peard & Co.; organ, Foster and Andrews
Cost: £18,008

[531] 1886–9
CAMBRIDGE, Cambridgeshire: Girton College [D O]
Gate tower and new ranges of college rooms, garden layout, tables for reading room. Minor alterations 1890. See also 1871 **[346]**, 1875 **[414]**, 1878 **[459]**, 1880 **[475]**, 1883 **[501]**, 1898 **[621]**
Client: Mistress and Fellows of Girton College
Contractors: general, J. Shillitoe; structural steelwork, W. H. Lindsay & Co.; heating and ventilating, W. W. Phipson; terracotta, Gibbs & Canning; tiles, Craven Dunnill & Co.; fittings, minor work, Rattee & Kett; Maple & Co.; ironwork, gas light fittings and lift, Hart Son Peard & Co.; bell, J. Warner & Sons; memorial tablet, Hart Son Peard & Co.
Cost: £16,295

[532] 1886–92
LIVERPOOL: Royal Infirmary [d D O]
Hospital, nurses' home, and medical school of red brick and terracotta with a slate roof. Glazed brick and faience decoration to interior. A mixture of circular and Nightingale wards. Alterations 1898. See 1895 **[599]** (Col. Pl. 4(*c*); Pls. 155, 156, 171)
Client: Governors of the Royal Infirmary
Contractors: general, Holme & Green; structural steelwork, W. H. Lindsay & Co.; heating and ventilating, W. W. Phipson; terracotta, J. C. Edwards; modelling, Farmer & Brindley; tiles, Carter Johnson & Co.; Craven Dunnill & Co.; faience, Burmantofts; chimney-pieces, J. & H. Patteson; Blackmore & Co.; Hopton Wood Stone Co.; iron work, W. H. Peake & Co.; Hart Son Peard & Co.
Cost: £123,500

[533] 1886
LONDON: 29 Berkeley Square [O]
Unspecified minor alterations to house
Client: Albert Brassey
Contractors: Wm. Cubitt & Sons
Cost: £165

[534] 1886
LONDON, Lancaster Gate [O]
Alterations to house
Client: Lord Wolverton
Contractors: G. Howe
Cost: £355

[535] 1886–9
MANCHESTER, King Street: Prudential Assurance Company Building [D O]
Office building of red brick and terracotta with central tower and slate roof. Alterations see 1895 **[606]**
Client: Prudential Assurance Company
Contractors: general, Wm. Southern & Co.; structural steelwork, W. H. Lindsay; heating and ventilating, D. O. Boyd; terracotta, J. C. Edwards; tiles, W. Godwin; faience, Burmantofts; stained glass, Heaton Butler & Bayne; R. B. Edmundson; chimney-pieces, J. & H. Patteson; furniture/fittings, Maple & Co.; iron work, Hart Son Peard & Co.; clock, Self Winding Clock Co.
Cost: £8,230

[536] 1886–8
MANCHESTER, Oxford Road: Owens College [O]
Engineering Laboratories. Extensions 1896. See also

1869 **[321]**, 1882 **[499]**, 1884 **[515]**, 1891 **[569]**, 1894 **[590]**, 1895 **[605]**, 1898 **[627]**
Client: Building Committee of Owens College
Contractors: W. W. Harrison
Cost: £2,230; extensions £905

[537] 1886–7
OXFORD, Oxfordshire: Balliol College [O]
New organ gallery in chapel. See 1866 **[254]**, 1873 **[392]**
Client: Master and Fellows of Balliol College, Oxford
Contractors: Symm & Co.
Cost: £110

[538] 1886–93
PORTSMOUTH, Hampshire, Commercial Road: Prudential Assurance Company Building [D O]
Office and sets of offices in two parallel blocks with façade of red brick and terracotta with stepped gables and slate roof. Alterations 1896
Client: Prudential Assurance Company
Contractors: general, Stephens & Bastow; structural steelwork, A. Handyside; heating and ventilating, D. O. Boyd; terracotta, Gibbs & Canning; modelling, Farmer & Brindley; tiles, W. Godwin; mosaic flooring, J. F. Ebner; faience, Burmantofts (1896); furniture/fittings, Maple & Co.; iron work, Hart Son Peard & Co.; clock, Self Winding Clock Co.
Cost: £20,356

[539] 1886
LOCATION UNKNOWN [O]
Monument
Client: Ralph Brocklebank Jun.
Contractors: stone carving, Farmer & Brindley
Cost: £155

[540] 1887
CAMBRIDGE, Cambridgeshire: St John's College [O]
Unspecified work
Client: Master and Fellows of St John's College, Cambridge
Contractors: T. Wontner Smith & Son
Cost: £1,545

[541] 1887–9
GUISBOROUGH, Cleveland: Guisborough Grammar School; now Prior Pursglove Sixth Form College [O]
Master's house and school; ground floor of grey stone with red brick and yellow terracotta above, with a slate roof. Alterations 1897
Client: Chairman and Governors of Pursglove Charity (per J. W. Pease, MP)
Contractors: general, Adam White; chimney-pieces, Hopton Wood Stone Co.; iron work, Hart Son Peard & Co.; faience, Burmantofts
Cost: £4,210

[542] 1887–91
LIVERPOOL: Liverpool University, Walker Engineering Block [D O]
Teaching rooms of brick with red terracotta dressings and a slate roof. See also 1881 **[488]**, 1884 **[512]**, 1888 **[549]**, 1895 **[598]**, 1899 **[633]**
Client: Committee of Liverpool University College
Contractors: foundation, Wm. Tomkinson & Sons; superstructure, Brown & Backhouse; structural steelwork, W. H. Lindsay & Co.; heating and ventilating, G. N. Haden; terracotta, J. C. Edwards; modelling, Farmer & Brindley; tiles, Craven Dunnill & Co.; faience, Burmantofts (?one fireplace); chimney-pieces, J. & H. Patteson; iron work, Hart Son Peard & Co.
Cost: £13,845

[543] 1887
LONDON, High Holborn: Staple Inn [O]
Removal of 18th-cent. stucco front and windows, repair, restoration, and reconstruction of half-timbered buildings to street and restoration of hall
Client: Prudential Assurance Company
Contractors: general, Holland & Hannen; heating and ventilating, W. W. Phipson; D. O. Boyd; iron work, Hart Son Peard & Co.
Cost: £9,900

[544] 1887–90
LONDON: Putney Hill House [O Dem]
House of brick and terracotta
Client: Alexander Miller

Contractors: general, J. Parnell & Sons; structural steelwork, W. H. Lindsay; terracotta, Gibbs & Canning; modelling, Farmer & Brindley; tiles, Craven Dunnill & Co.; W. de Morgan; stained glass, R. B. Edmundson; chimney-pieces, W. H. Burke; Hopton Wood Stone Co.; iron work, Hart Son Peard & Co.
Cost: £11,310

[545] 1887–9
MANCHESTER: St Anne's Church [D O]
Restoration of church, new floor, windows, stained glass, and choir seating. Alterations 1896, 1901, 1904–7
Client: Vicar and Churchwardens of St Anne's Church
Contractors: general, Wm. Southern; marble work, J. & H. Patteson; furniture/fittings, Doveston, Davey, Hull & Co.; iron work, Hart Son Peard & Co.; cartoons for stained glass, F. J. Shields
Cost: £3,100; 1896 £445; 1901 £205; 1904–7 £1,815

[546] 1888–9
BRIGHTON, Sussex: Hotel Metropole [D O]
Sea front hotel, six storeys and attic of red Rowlands Castle brick and terracotta, Ruabon tile and zinc roofs; attached ballroom; garden court
Client: Frederick Gordon & Co.
Contractors: foundations, J. T. Chappel; structural steelwork, A. Handyside & Co.; terracotta, Burmantofts; Joseph Cliff & Sons; Gibbs & Canning; clock, Gillett & Co.; lifts, Waygood
Clerk of works: T. Holloway
Cost: £14,720

[547] 1888
CAMBRIDGE, Cambridgeshire: Girton College [D]
Design for prize bookplate
Client: Mistress and Fellows of Girton College, Cambridge

[548] 1888–93
GLASGOW: Prudential Assurance Buildings [D O]
Insurance offices and chambers of red brick and pink sandstone with a slate roof
Client: Prudential Assurance Company
Contractors: general, Guthrie & Co.; terracotta chimney pots, J. C. Edwards; faience, Burmantofts; chimney-pieces, W. H. Burke & Co.; iron work, Hart Son Peard & Co.; clock, J. Edward & Sons; clock case, Wylie and Lockhead
Cost: £30,040

[549] 1888–93
LIVERPOOL: Liverpool University, Victoria Building [D O]
Hall, library, and lecture rooms with Jubilee Tower, all of brick with red terracotta dressings with slate roofs and faience decoration to the interior (Pls. 151, 152, 153). See also 1881 **[488]**, 1884 **[512]**, 1887 **[542]**, 1895 **[598]**, 1899 **[633]**
Client: Committee of Liverpool University College
Contractors: foundation, Isaac Dilworth; superstructure, Brown & Backhouse; structural steelwork, A. Handyside & Co.; modelling, Farmer & Brindley; Earp & Hobbs; faience, Burmantofts; mosaic flooring, J. Rust; chimney-pieces, Hopton Wood Stone Co.; J. & H. Patteson; iron work, Hart Son Peard & Co.
Cost: £47,580

[550] 1888
LONDON, Trafalgar Square: Gordon Memorial Pedestal [O]
Granite pedestal for bronze memorial statue by Hamo Thorneycroft
Client: Hamo Thorneycroft
Construction: stone carving, Farmer & Brindley
Cost: £680

[551] 1888–93
MANCHESTER, Spring Gardens/York Street: National Provincial Bank [D O]
Bank and offices on a quadrant site, stone faced with Renaissance detailing
Client: National Provincial Bank Ltd.
Contractors: general, R. Neill & Sons; structural steelwork, A. Handyside & Co.; heating and ventilating, W. W. Phipson; stone carving, Earp & Hobbs; tiles, Craven Dunnill & Co.; faience, Burmantofts; chimney-pieces, W. H. Burke; Hopton Wood Stone

Co.; W. Wilson; iron work, Hart Son Peard & Co.
Cost: £36,495

[552] 1888
MELBOURNE, Australia: Gordon Memorial [O]
Design work for pedestal of the replica of the London Gordon Memorial
Client: Hamo Thorneycroft
Contractors: Stone carving, Farmer & Brindley
Cost: £650

[553] 1888–90
MIDDLESBROUGH, Cleveland: Grammar School [O]
Addition of wing to school. See also 1873 **[389]**, 1884 **[516]**, 1885 **[524]**
Client: Trustees of Middlesbrough High School
Contractors: general, J. Johnson; tiles, Craven Dunnill & Co.; iron work, Hart Son Peard & Co.
Cost: £2,575

[554] 1889–92
BIRMINGHAM: Prudential Assurance Company Offices [D O]
Insurance offices and chambers with restaurant in red brick and terracotta with a slate roof. Alterations 1895–6
Client: Prudential Assurance Company
Contractors: general, J. Barnsley & Sons; structural steelwork, A. Handyside & Co.; modelling, Farmer & Brindley; terracotta, Burmantofts; tiles, Craven Dunnill & Co.; faience, Burmantofts; mosaic and parquet flooring, J. F. Ebner; L. Oppenheimer; chimney-pieces and grates, Hopton Wood Stone Co.; W. Wilson & Sons; fittings, Hart Son & Peard; furniture, Norton & Co.; iron work, Hart Son Peard & Co.; clock, J. Moore & Sons
Cost: £16,390; 1895–6 £625

[555] 1889
BOLTON, Lancashire: Prudential Assurance Buildings
Offices and chambers of red brick and terracotta, polished granite plinth and slate roof. The buildings were designed by Mr Smith, architect of Bolton. AW may have acted as consultant, but no records survive for this building

[556] 1889–93
LONDON, Duke Street, Mayfair: King's Weigh-house Chapel [d D O]
Chapel of red brick and orange terracotta with tiled roof, with oval nave and south-west corner tower. Alterations 1901–3 redecorations and new pulpit
Client: Trustees of Old King's Weigh-house Chapel
Contractors: general, John Shillitoe & Son; structural steelwork, A. Handyside & Co.; heating and ventilating, G. N. Haden; tiles, Craven Dunnill & Co.; faience, Burmantofts; iron work, Hart Son Peard & Co.
Cost: £26,495

[557] 1889–92
MALSHANGER, ?Kent [O]
House
Client: W. S. Portal
Contractors: general, Wheeler Bros.; heating and ventilating, G. N. Haden; carving, Farmer & Brindley; tiles, Craven Dunnill & Co.; mosaic flooring, J. F. Ebner; chimney-pieces, W. H. Burke & Co.; decoration, Heaton Butler & Bayne (painted shields); iron work, Hart Son Peard & Co.; plaster ornaments, G. Jackson
Cost: £9,955

[558] 1889
RUGBY, Warwickshire: Hillmorton Paddox [O]
Additional wing to late Georgian mansion in red brick with stone dressings
Client: Col. R. Cooper
Contractors: J. Parnell & Sons
Cost: £1,415

[559] 1890–4
LEEDS, W. Yorkshire: Prudential Assurance Building [D O]
Insurance office and chambers of five storeys in pink brick and buff terracotta with asymmetrically placed turret under a tiled roof (Col. Pl. 7). Alterations 1895

Client: Prudential Assurance Company
Contractors: general, Charles Myers; structural steelwork, A. Handyside & Co.; terracotta, Burmantofts; tiles, Craven Dunnill & Co.; faience, Burmantofts; chimney-pieces, Hopton Wood Stone Co.; furniture/fittings, Maple & Co.; iron work, Hart Son Peard & Co.; clock, J. Moore
Cost: £44,785; 1895 £790

[560] 1890
LONDON, Stoke Newington: Abney Park Cemetery [O]
Monument
Client: Executors of J. H. Spreat
Contractors: stone carving, Farmer & Brindley; iron work, Hart Son Peard & Co.
Cost: £40

[561] 1890–3
MAIDSTONE, Kent: Springfield [O]
Large house of brick and pink terracotta. Alterations 1897–8
Client: R. J. Balston
Contractors: general, J. Parnell & Son; structural steelwork, A. Handyside & Co.; heating and ventilating, G. N. Haden; terracotta, Gibbs & Canning; J. C. Edwards; ?modelling, Farmer & Brindley; faience, Leeds Fireclay Co.; chimney-pieces, Hopton Wood Stone Co.; furniture/fittings, Maple & Co.; iron work, Hart Son Peard & Co.; clock, J. Moore
Cost: £32,190. Alterations £810

[562] 1890
PANGBOURNE, Berkshire: Coffee House [O]
Coffee house/Sunday school building of red brick and tiled roof
Client: Dr H. Watney
Contractors: general, Holland & Hannen; faience, Burmantofts; iron work, Hart Son Peard & Co.
Cost: £3,545

[563] 1890–1
READING, Berkshire: Leighton Park School [O]
Alterations and additions, boys dormitories, classrooms, and extension of dining-room. See 1892 **[579]**
Client: Friends' Public School Company
Contractors: general, Wheeler Bros.; tiles, Craven Dunnill & Co.; chimney-pieces, Hopton Wood Stone Co.; iron work, Hart Son Peard & Co.; bells, J. Warner & Sons
Cost: £12,635

[564] 1891
BURNHAM, Buckinghamshire: Britwell Court [D O]
New floors (from marble in client's possession); alterations and additions to house. Pigeon house and stables 1895; entrance lodge and alterations 1896; central heating and minor repairs 1902
Client: W. Christie-Miller
Contractors: general, Holland and Hannen; stone carving, Farmer & Brindley; mosaic flooring; J. F. Ebner; L. Oppenheimer; stained glass, Heaton Butler & Bayne; iron work, Hart Son Peard & Co.; clock, J. Moore & Son
Cost: 1894–8 £24,755; 1902 £45

[565] 1891–4
CAMBRIDGE, Cambridgeshire: Fosters' Bank (now Lloyds Bank) [d O]
Bank of brick and stone with tiled roof; large octagonal banking hall. Additions by Munro Cautley, 1935 (Pl. 158, Fig. 13)
Client: E. B., J. E., and C. F. Foster
Contractors: general, William Sindall; structural steelwork, A. Handyside and Co.; stone carving, Farmer & Brindley; F. W. Pomeroy; tiles, Craven Dunnill & Co.; faience, Burmantofts; mosaic flooring, J. F. Ebner; J. Rust; iron work and light fittings, Hart Son Peard & Co.; lighting, Belshaw & Co.; clock, John Moore & Co.; plaster ornaments, G. Jackson

[566] 1891–4
CARDIFF, Glamorgan: Prudential Assurance Company Building [D O]
Insurance offices and chambers. Alterations 1897
Client: Prudential Assurance Company
Contractors: general, T. Collins; structural steelwork, A. Handyside & Co.; modelling, F. W. Pomeroy; terracotta, Leeds Fireclay Co.; J. C. Edwards;

faience, Burmantofts; mosaic flooring, J. F. Ebner; chimney-pieces, Hopton Wood Stone Co.; furniture/fittings, Maple & Co.; iron work, Hart Son Peard & Co.; clock, J. Moore & Sons
Cost: £8,375; 1897 £100

[567] 1891–3
LONDON: 21 St James's Place [O]
Alterations to house
Client: W. Christie-Miller
Contractors: general, G. House; Holland & Hannen; heating and ventilating, G. N. Haden; stone carving, Farmer & Brindley; tiles, W. de Morgan; iron work, Hart Son Peard & Co.
Cost: £3,165

[568] 1891–2
LONDON, Northumberland Avenue: SPCK Offices [O]
Two additional storeys to building by John Gibson
Client: Society for the Promotion of Christian Knowledge
Contractors: general, J. Shillitoe & Son; structural steelwork, A. Handyside & Co.; iron work, Hart Son Peard & Co.
Cost: £3,250

[569] 1891–6
MANCHESTER, Oxford Road: Owens College [O]
Medical School extension of yellow brick and stone dressings with a tiled roof. See 1869 **[321]**, 1882 **[499]**, 1884, **[515]** 1886 **[536]**, 1894 **[590]**, 1895 **[605]**, 1898 **[627]**
Client: Building Committee of Owens College
Contractors: general, R. Neill; structural steelwork, A. Handyside & Co.; heating and ventilating, G. N. Haden; tiles, D. Conway; furniture/fittings, Mack Hamilton & Co.; iron work, Hart Son Peard & Co.; decorative stone, Hopton Wood Stone Co.
Cost: £40,830

[570] 1891–6
MANCHESTER, Oxford Street: Refuge Assurance Company Building [d O]
Corporate headquarters building of light red brick and plum coloured terracotta with corner entrance and turret with slate roof (Col. Pl. 6, Pls. 157, 170; Fig. 12)
Client: Refuge Assurance Company
Contractors: foundation, C. H. Normanton; superstructure and finishing, Wm. Southern & Sons; structural steelwork, A. Handyside & Co.; heating and ventilating, G. N. Haden; terracotta, Doulton & Co.; modelling, Earp & Hobbs; tiles, W. de Morgan; D. Conway; faience, Burmantofts; mosaic flooring, J. F. Ebner; chimney-pieces, J. & H. Patteson; W. Wilson; decoration, Heighway & Son; furniture/fittings, G. Goodall & Co.; iron work, Hart Son Peard & Co.; R. Jones
Cost: £86,525

[571] 1891 construction 1899–1901 *et seq.*
MANCHESTER: St Mary's Hospital [D O Dem]
Maternity hospital of brick and terracotta
Client: The Trustees of St Mary's Hospital
Contractors: foundation, C. H. Normanton; superstructure and finishing, H. Vickers; structural steelwork, A. Handyside & Co.; heating and ventilating, James Keith; terracotta, Gibbs & Canning; modelling, Earp & Hobbs; granite (doorway), G. &. J. Fenning; tiles, D. Conway; faience, Leeds Fireclay Co.; marble work, J. & H. Patteson; iron work, Hart Son Peard & Co.
Cost: £65,140

[572] 1891–7
NEWCASTLE, Northumberland: Prudential Assurance Company Building [D O]
Office building on corner site of red brick with red sandstone dressings. Alterations 1898–1900
Client: Prudential Assurance Company
Contractors: general, Walter Scott; structural steelwork, A. Handyside & Co.; stone carving, Farmer & Brindley; tiles, Craven Dunnill & Co.; D. Conway (1898); faience, Burmantofts; mosaic flooring, J. F. Ebner; iron work, Hart Son Peard & Co.; clock, J. Moore & Sons
Cost: £37,155; 1898–1900 £1,090

[573] 1891–2
STANMORE, Middlesex: Herondale [O]
House of brick with tile roof. Alterations 1894–5 stables

Client: Frederick Gordon
Contractors: general, J. Kirby (house); W. Bailey (stables); tiles, Doulton; iron work, Hart Son Peard & Co.
Cost: £3,000. Alterations £1,320

[574] 1892–5
LEEDS, W. Yorkshire: Yorkshire College (University) [d D O]
Hall, library, and lecture rooms of red brick with stone dressings and a slate roof. Alterations see 1878, **[460]** 1883 **[504]**, 1884 **[510]**, 1885 **[520]**, 1898 **[622, 623, 624]** (Col. Pl. 4(*d*); Pl. 148)
Client: Committee of Yorkshire College
Contractors: general, J. Wood & Son; structural steelwork, A. Handyside & Co.; tiles, Craven Dunnill & Co.; faience, Burmantofts; mosaic flooring, J. F. Ebner; chimney-pieces, Hopton Wood Stone Co.; furniture/fittings (library), Carr Bros.; iron work, Hart Son Peard & Co.
Cost: £21,025

[575] 1892–6
LEICESTER, Leicestershire: Prudential Assurance Buildings [D O]
Insurance office and chambers of red brick and buff terracotta with a slate roof
Client: Prudential Assurance Company
Contractors: general, T. & H. Herbert; terracotta, Burmantofts; mosaic flooring, J. F. Ebner; iron work, Hart Son Peard & Co.; clock, J. Moore & Sons

[576] 1892–6
LONDON, Piccadilly: National Provincial Bank [D O]
Bank and offices of steel and brick with stone facing
Client: General Manager, National Provincial Bank
Contractors: general, Wm. Brass & Son; structural steelwork, A. Handyside & Co.; stone carving, Farmer & Brindley; F. W. Pomeroy; tiles, Campbell Tile Co.; Craven Dunnill & Co.; mosaic flooring, J. F. Ebner; furniture/fittings, Maple & Co. (small contract); iron work, Hart Son Peard & Co.; plaster ornaments, G. Jackson
Cost: £31,975

[577] 1892
LONDON, Great Portland Street: St Paul's Church [O]
Minor alterations, probably new heating
Client: Revd C. J. Williamson
Contractors: heating and ventilating, Shuffrey & Co.; iron work, Hart Son Peard & Co.
Cost: £190

[578] 1892
LONDON: Cleveland Yard [O]
Stables
Contractors: general, Charles Jackson; faience, Leeds Fireclay Co.; iron work, Hart Son Peard & Co.
Cost: £4,020

[579] 1892–5
READING, Berkshire: Leighton Park School [D O]
Extension of school buildings and new master's house. See 1890 **[563]**
Client: Grove House School Trustees
Contractors: general, J. H. Margetts; mosaic flooring, J. F. Ebner; iron work, Hart Son Peard & Co.
Cost: £8,090

[580] 1892–4
READING, Berkshire: Leighton Park School [D O]
New sanatorium
Client: Friends' Public School Co.
Contractors: general, Chas. Jackson; iron work, Hart Son Peard & Co.
Cost: £1,625

[581] 1893–6
BRADFORD, W. Yorkshire: Prudential Assurance Company Building [D O]
Insurance office, chambers, and Turkish bath in pink brick and terracotta with polished granite plinth and slate roof. Alterations 1899; 1901
Client: Prudential Assurance Company
Contractors: general, J. Moulson & Co.; terracotta, J. C. Edwards; modelling, Farmer & Brindley; F. W. Pomeroy; tiles, Craven Dunnill & Co.; faience, Burmantofts; mosaic flooring, J. F. Ebner; furniture/fittings, J. Moulson & Son; iron work, light fittings,

Hart Son Peard & Co.; chimney-pieces, Hopton Wood Stone Co.; Turkish bath apparatus, J. Constantine & Son; clock, W. Potts & Son; plaster ornaments, G. Jackson
Cost: £19,070

[582] 1893
BURNHAM, Buckinghamshire: St Peter's Church [O]
New window
Client: W. Christie-Miller
Contractor: stained glass, Heaton Butler & Bayne
Cost: £310

[583] 1893–4
LONDON, 92 Cheapside: Atlas Insurance Offices [O]
Alterations and rear extension to office building. Alterations 1903
Client: Atlas Insurance Co. Ltd.
Contractors: general, Henry Lovatt; stone carving, Farmer & Brindley; tiles, Craven Dunnill & Co.; faience, Doulton; mosaic flooring, J. F. Ebner; iron work, Hart Son Peard & Co.
Cost: £18,130

[584] 1893–8
NOTTINGHAM, Nottinghamshire: Prudential Assurance Company Building [D O]
Office building of red brick and terracotta on triangular site with tower at apex. Alterations 1901, 1903 (Pl. 161)
Client: Prudential Assurance Company
Contractors: foundation, H. Vickers; superstructure and finishing, J. Hutchinson; structural steelwork, A. Handyside & Co.; terracotta, Burmantofts; modelling, Gilbert Seale; Farmer & Brindley; faience, Burmantofts; mosaic flooring, J. F. Ebner; iron work, Hart Son Peard & Co.; clocks, J. Moore & Sons; wallpaper, Singleton, Benda & Co.; 'Prudentia' model, F. W. Pomeroy
Cost: £28,730. Alterations £90

[585] 1894
BUSHEY, Hertfordshire: St Margaret's Clergy Orphan School [O]
School building, chapel, and lodge of brick with terracotta dressings and tiled roof
Client: Clergy Orphan Corporation
Contractors: general, J. Shillitoe & Son; terracotta, Burmantofts; modelling, Farmer & Brindley; tiles, Craven Dunnill & Co.; mosaic flooring, J. F. Ebner; iron work, Hart Son Peard & Co.
Cost: £34,235 (The list of certificates is incomplete.)
Note: The Alfred the Great millenniary memorial window was the gift of AW

[586] 1894–9
HAWKHURST, Sussex: Tongswood [O]
Work to conservatory and library of existing house; also lodge, farmhouse, cottages, smithy, poultry house, and George and Dragon Inn
Client: Misses Goldsmid
Contractors: general, F. Marshall & Son; tiles, Craven Dunnill & Co.; iron work, Hart Son Peard & Co.; plaster ornaments, G. Jackson
Cost: £12,190

[587] 1894–6
LONDON, Chelsea Bridge Road: British Institute of Preventive Medicine [O Dem]
Laboratories and offices on four floors and attics; ground floor stone, upper floors brick with stone dressings. Matching extension by Paul Waterhouse 1908
Client: British Institute of Preventive Medicine
Contractors: foundation, George Munday; superstructure, D. Charteris; structural steelwork, A. Handyside & Co.; heating and ventilating, Langfield & Co.; G. N. Haden; stone carving, Farmer & Brindley; faience, Burmantofts; mosaic flooring, J. F. Ebner; iron work, Hart Son Peard & Co.
Cost: £30,650

[588] 1894–1903
LONDON, Gower Street: University College Hospital [d D O]
300-bed hospital and nurses' home of four storeys and attics, built of red brick and terracotta with a

pavilioned slate roof. Unusual X-plan with wards of shallow cross pattern developing the 'Nightingale' plan (Pls. 154, 163). See also 1877 **[447]**
Client: Sir John Blundell Maple, as a gift to the hospital
Contractors: general, Thomas Holloway; Maple & Co.; structural steelwork, A. Handyside & Co.; heating and ventilating, G. N. Haden; Ashwell & Nesbitt; furniture/fittings, Maple & Co.
Cost: £200,000

[589] 1894–6
LONDON: 20 (25) Portman Square [O Dem]
Repairs and alterations to house. Alterations 1898–9
Client: Misses Goldsmid
Contractors: general, D. Charteris; iron work, Hart Son Peard & Co.
Cost: £355. Alterations £120

[590] 1894–6
MANCHESTER, Oxford Road: Owens College [O]
Schorlemmer Laboratory. See 1869 **[321]**, 1882 **[499]**, 1884, **[515]** 1886 **[536]**, 1891 **[569]**, 1895 **[605]**, 1898 **[627]**
Client: Building Committee of Owens College
Contractors: general, R. Neill; heating and ventilating, G. N. Haden; iron work, Hart Son Peard & Co.
Cost: £4,370

[591] 1894
NEWCASTLE, Northumberland: Royal Victoria Asylum for the Blind [O]
Conversion of Benwelldene House. See 1864 **[184]**
Contractors: general, Walter Scott; iron work, Hart Son Peard & Co.
Cost: £3,060

[592] 1894–5
STANMORE, Middlesex: Lynbrook [O]
House of red brick
Client: Frederick Gordon
Contractors: general, W. Bailey; tiles, Craven Dunnill & Co.; iron work, Hart Son Peard & Co.
Cost: £2,250

[593] 1895
ASHAMPSTEAD, Berkshire: St Clement's Church [O]
Restoration of church roof
Client: Revd V. M. Moyle (of Pembroke College). AW was among the leading subscribers
Contractors: general, J. Bottrill; wrought-iron vane, Hart Son Peard & Co.
Cost: £280

[594] 1895
CARLISLE, Cumbria: Cathedral [O]
Canopy for tomb of Bishop Harvey Goodwin. Sculptured figure by Hamo Thorneycroft
Contractors: woodwork, Wm. Southern
Cost: £75

[595] 1895–8
DUNDEE, Tayside: Prudential Assurance Company Buildings [D O]
Insurance offices and chambers of stone with a slate roof. Alterations 1899
Client: Prudential Assurance Company
Contractors: structural steelwork, Arrols Bridge & Roof Co.; modelling, Farmer & Brindley; tiles, J. &. H. Potts; faience, Burmantofts; mosaic flooring, J. F. Ebner; iron work, Hart Son Peard & Co.; clock, J. Moore & Sons
Cost: £15,475

[596] 1895–9
EDINBURGH: Prudential Assurance Company Buildings [D O]
Insurance office with chambers and shops attached of pink sandstone with a slate roof
Client: Prudential Assurance Company
Contractors: general, J. Watherston & Sons; structural steelwork, Arrol's Bridge & Roof Co.; heating and ventilating, Mackenzie & Moncur; stone carving, Farmer & Brindley; tiles, Field & Allen; faience, Burmantofts; mosaic flooring, L. Oppenheimer; iron work, Hart Son Peard & Co.
Cost: £31,155

[597] 1895–8
LEEDS, W. Yorkshire: Williams Brown & Co. Bank [O]
Bank and offices of red brick and buff terracotta with plinth of polished grey granite and slate roof
Client: Williams, Williams, Brown & Co.
Contractors: foundation, S. M. McFarlane; superstructure and finishing, Armitage & Hodgson; structural steelwork, J. Butler & Co.; carving, W. Beveridge; modelling, Farmer & Brindley; terracotta, Burmantofts; tiles, A. Whitehead; faience, Burmantofts; mosaic flooring, A. Whitehead; furniture/fittings, Marsh Jones & Cribb
Cost: £45,110

[598] 1895–7
LIVERPOOL: Liverpool University, Gossage Chemistry Laboratories [D O]
Laboratories and teaching rooms of brick with terracotta dressings and slate roof. See also 1881 **[488]**, 1884 **[512]**, 1887 **[542]**, 1888 **[549]**, 1899 **[633]**
Client: Committee of Liverpool University College
Contractors: general, Jones & Sons; structural steelwork, A. Handyside & Co.; heating and ventilating, G. N. Haden; modelling, Farmer & Brindley; furniture/fittings, Doulton; Jones & Sons; iron work, Hart Son Peard & Co.
Cost: £8,210

[599] 1895–7
LIVERPOOL: Liverpool University, Medical School [D O]
Extension of Medical School of brick and red terracotta. See Liverpool Royal Infirmary 1886 **[532]**
Client: Committee of Liverpool University College
Contractors: Jones & Sons
Cost: £1,795

[600] 1895–8
LIVERPOOL: Liverpool University, Thompson Yates Laboratory [D O]
Teaching rooms and laboratories of brick with red terracotta dressings and slate roof. See also 1881 **[488]**, 1884 **[512]**, 1887 **[542]**, 1888 **[549]**, 1899 **[633]**
Client: Committee of Liverpool University College
Contractors: general, Wm. Tomkinson & Son; structural steelwork, A. Handyside & Co.; terracotta panel, J. C. Edwards; faience, Burmantofts; mosaic flooring, J. F. Ebner; iron work, Hart Son Peard & Co.
Cost: £26,050

[601] 1895–6
LONDON: 7 Devonshire Square [O]
Alterations, decorations, and new drainage
Client: M. D. Malleson
Contractors: George House
Cost: £285

[602] 1895–1905
LONDON, Furnivals Inn: Prudential Assurance Company Offices [D O]
Extension of offices of red brick and terracotta of three and four storeys and attics with slate roofs, set around three courtyards and other light wells with public offices to Holborn, interior faience decoration. Furniture designed by Paul Waterhouse (Col. Pl. 5; Pls. 124–32, 169, 183, 190)
Client: Prudential Assurance Company
Contractors: general, Holland & Hannen; structural steelwork, A. Handyside & Co.; heating and ventilating, J. Jeffreys; terracotta, Gibbs & Canning; modelling, Farmer & Brindley; marble columns, caps, and bases, W. H. Burke & Co.; tiles, W. de Morgan; faience, Burmantofts; mosaic flooring, J. F. Ebner; chimney-pieces, Burmantofts; iron work, Hart Son Peard & Co.; model, C. N. Thwaite
Cost: to 1899 £150,155

[603] 1895 and 1901
LONDON: 44 Grosvenor Gardens [O]
Minor alterations and repairs to house including kitchen tiling
Client: Sir J. W. Pease, MP
Contractors: general, G. Harris & Son; tiles, Craven Dunnill & Co.
Cost: £90

[604] 1895–6
LONDON: 20a St James's Place [D O]
Alterations to kitchen. Alterations 1902
Client: W. Christie-Miller

Contractors: E. Harris & Son
Cost: £1,200

[605] 1895
MANCHESTER, Oxford Road: Owens College, Christie Library [D O]
Library building, stone faced with a tiled roof. See 1869 **[321]**, 1882 **[499]**, 1884 **[515]**, 1886 **[536]**, 1891 **[569]**, 1894 **[590]**, 1898 **[627]**
Client: Building Committee of Owens College
Contractors: general, Henry Vickers; structural steelwork, A. Handyside & Co.; heating and ventilating, J. Grundy; stone carving, Earp & Hobbs; tiles, D. Conway; mosaic flooring, J. F. Ebner; decoration, Heighway and Sons; iron work, Hart Son Peard & Co.
Cost: £17,565

[606] 1895–9
MANCHESTER, King Street/Tib Lane: Prudential Assurance Company Building [D O]
Extension of building. See 1886 **[535]**
Client: Prudential Assurance Company
Contractors: foundation, P. Hodkinson; superstructure and finishing, R. Neill; structural steelwork, A. Handyside & Co.; heating and ventilating, G. N. Haden; terracotta, J. C. Edwards; modelling, Farmer & Brindley; tiles, D. Conway; faience, Burmantofts; mosaic flooring, L. Oppenheimer; stained glass, R. B. Edmundson; furniture/fittings, F. Rawcliffe; G. Goodall; G. Wragge; W. Wilson; iron work, Hart Son Peard & Co.; J. Jones; clock, J. Moore & Son
Cost: to 1898 £21,950

[607] 1895–7
ST LEONARDS-ON-SEA, Sussex: Westerleigh [O]
Additions to house. Alterations 1899–1900
Client: Miss Mary Waterhouse
Contractors: general, Padgham & Hutchinson; iron work, Hart Son Peard & Co.
Cost: £3,000. Alterations £640

[608] 1895–8
SHEFFIELD, S. Yorkshire: Prudential Assurance Company Building [D O]
Offices of red brick and terracotta. Alterations 1899
Client: Prudential Assurance Company
Contractors: general, Geo. Longden & Son; structural steelwork, A. Handyside & Co.; terracotta, J. C. Edwards; modelling, Farmer & Brindley; tiles, Craven Dunnill & Co.; faience, Burmantofts; mosaic flooring, Craven Dunnill & Co.; stained glass, J. V. Rowlands; chimney-pieces, Hopton Wood Stone Co.; iron work, Hart Son Peard & Co.; clock, J. Moore & Sons
Cost: £20,820. Alterations £70

[609] 1895–1905
LOCATION UNKNOWN: Portobello House [O]
Alterations to house, including drive and laundry
Client: J. S. Harmood-Banner (Committee of the estate of E. A. R. Morris)
Contractors: E. Willcox
Cost: £2,195

[610] 1896
CHELMSFORD, Essex: Brooklands [O]
Alterations to drainage
Client: W. Christie-Miller
Contractor: George House
Cost: £415

[611] 1896
DARLINGTON, Co. Durham: Mary Pease Almshouses [O]
Almshouses of brick with slate roofs. Design work by Paul Waterhouse
Client: J. W. Pease, MP
Contractors: general, Robert Kitching; iron work, Wm. Rusell & Son
Clerk of works: J. P. Pritchett
Cost: £1,500

[612] 1896
HOLMBURY ST MARY, Surrey [O]
Monument to the late Mrs Edwin Waterhouse
Client: Edwin Waterhouse
Contractors: stone carving, Farmer & Brindley
Cost: £60

[613] 1896–8 *et seq.*
LIVERPOOL: Pearl Life Assurance Office [O]

Insurance office building faced in Portland stone, with corner turret and faience decoration to public office
Client: Pearl Life Assurance Co.
Contractors: general, E. Morrison & Sons; structural steelwork, A. Handyside & Co.; heating and ventilating, J. Grundy; modelling, Farmer & Brindley; faience, Burmantofts; mosaic flooring, J. Rust; J. F. Ebner; chimney-pieces, Shuffrey & Co.; iron work, light fittings, Hart Son Peard & Co.
Cost: estimate £23,818; cost to 1898 £20,076

[614] 1896
LONDON: Hotel Metropole [O]
Alterations to chimneys etc.
Client: ?Frederick Gordon & Co.
Contractors: general, J. Styles; iron work, Hart Son Peard & Co.
Cost: £1,195

[615] 1896
LONDON, 12 Great George Street: Royal Institution of Chartered Surveyors [D d O]
Offices, hall, library, and museum for professional institution, of red brick and Darley Dale stone
Client: Royal Institution of Chartered Surveyors
Contractors: general, Foster & Dicksee; structural steelwork, A. Handyside & Co.; heating and ventilating, J. Jeffreys; stone carving, Farmer & Brindley; tiles, Craven Dunnill & Co.; mosaic flooring, L. Oppemheimer; iron work, Hart Son Peard & Co.
Cost: £27,770

[616] 1896–7
STOUGHTON, Sussex: Parish Church [O]
Repairs to porch, new seats, and iron grilles
Contractors: stone carving, Farmer & Brindley; furniture/fittings, W. Poate & Son; iron work, Hart Son Peard & Co.
Cost: £225

[617] 1896
YATTENDON, Berkshire [O]
Alterations to cottages
Client: A. Waterhouse

[618] 1896
LOCATION UNKNOWN [O]
Granite cross
Client: Mrs Pearson Lee
Contractors: Sanders & Co.
Cost: estimate £25

[619] 1897
LONDON, Grays Inn Road: Hospital Saturday Fund [O]
Unspecified work
Client: Hospital Saturday Fund
Contractors: general, R. E. Worsley & Co.; iron work, Hart Son Peard & Co.
Cost: £1,035

[620] 1897
ROWLANDS CASTLE, Hampshire: Stansted College [O]
Alterations to drainage
Contractors: J. Edney
Cost: £80

[621] 1898–1902
CAMBRIDGE, Cambridgeshire: Girton College [D]
Chapel, two new ranges of rooms, new Hall and kitchens and swimming pool. See also 1871 **[346]**, 1875 **[414]**, 1878 **[459]** 1880 **[475]**, 1883 **[501]**, 1886 **[531]**
Client: Mistress and Fellows of Girton College
Contractors: general, ?Rattee & Kett; chimney-pieces, Shuffrey & Co.; metal window shutters, Borthwick Gate Co.

[622] 1898
LEEDS, W. Yorkshire: Yorkshire College (University) [D O]
Extensions to Dyehouse in red brick with stone dressings and a slate roof. Alterations see 1878 **[460]**, 1883 **[504]**, 1884 **[510]**, 1885 **[520]**, 1892 **[574]**, 1898 **[623, 624]**
Client: Committee of Yorkshire College

Contractors: general, J. Wood & Son; modelling, Farmer & Brindley; brass inscription, Hart Son Peard & Co.

[623] 1898
LEEDS, W. Yorkshire: Yorkshire College (University) [D O]
New Woollen Department of red brick with stone dressings. Alterations see 1878 **[460]**, 1883 **[504]**, 1884 **[510]**, 1885 **[520]**, 1892 **[574]**, 1898 **[622, 624]**
Client: Committee of Yorkshire College
Contractors: Charles Myers
Cost: tender £6,425

[624] 1898
LEEDS, W. Yorkshire: Yorkshire College (University) [D O]
New engine house, cloakroom, and cycle shed, Leather Industries Department, all of red brick with stone dressings and slate roofs. See 1878 **[460]**, 1883 **[504]**, 1884 **[510]**, 1885 **[520]**, 1892 **[574]**, 1898 **[622, 623]**
Client: Committee of Yorkshire College
Contractors: J. Wood & Son
Cost: tender £3,215

[625] 1898
LONDON: Berkeley Chapel [O]
Two desks and seats
Client: ?Albert Brassey
Contractors: furniture/fittings, E. J. Wright
Cost: £15

[626] 1898–9
LONDON, Charing Cross: Grand Hotel [D Dem]
Additional wing and reconstruction of entrance hall, Grill Room, Walnut Room, and Reading Room. Marble and alabaster facing, faience and walnut panelling
Client: not recorded
Contractors: furniture/fittings, Maple & Co.

[627] 1898–1902
MANCHESTER, Oxford Road: Owens College, Whitworth Hall [D]
Grand hall, stone faced with tiled roof, hammer beam interior, faience faced passages. Commission completed by Paul Waterhouse. See 1869 **[321]**, 1882 **[499]**, 1884, **[515]** 1886 **[536]**, 1891 **[569]**, 1894 **[590]**, 1895 **[605]**
Client: Building Committee of Owens College
Cost: £36,350

[628] 1898
NOTTINGHAM, Nottinghamshire: General Hospital, Jubilee Wing [S]
Addition of a block of three circular wards and sanitary tower in brick with Ancaster stone dressings, laundry, and out-patients' department; new staircases and lift
Contractors: W. Woodhead
Clerk of works: G. Reed

[629] 1898
OLDHAM, Lancashire: Prudential Assurance Offices [D O]
Insurance offices and chambers of red brick and pink terracotta with asymmetrically placed turret under a slate roof
Client: Prudential Assurance Company
Contractors: Whitworth & Whittaker
Cost: £20,225

[630] 1898
RHYL, Flintshire: Royal Alexandra Hospital [d O]
Hospital of red brick and terracotta
Client: Committee of the Royal Alexandra Hospital
Contractors: heating and ventilating, Ashwell & Nesbitt; modelling, Farmer & Brindley; mosaic flooring, De Grelle Hondret & Co.; iron work, Hart Son Peard & Co.; stone fender, Hopton Wood Stone Co.
Cost: £30,430

[631] 1899–1901
BRISTOL, Avon: Prudential Assurance Buildings [D O]
Insurance offices and commercial chambers of red terracotta with a slate roof
Client: Prudential Assurance Company
Cost: £15,000

[632] 1899–1901
HUDDERSFIELD, W. Yorkshire: Prudential Assurance Buildings [D O]
Offices and chambers of red brick and terracotta with a Westmorland slate roof.
Client: Prudential Assurance Company
Contractors: B. Graham & Sons
Cost: £7,995

[633] 1899–1902
LIVERPOOL, Liverpool University: New Medical School, Anatomy Building [O]
Lecture theatre and teaching rooms of brick with red terracotta dressings and slate roof. See also 1881 **[488]**, 1884 **[512]**, 1887 **[542]**, 1888 **[549]**, 1895 **[598]**
Contractors: Wm. Tomkinson
Cost: £12,740

[634] 1899
LONDON: Grosvenor Hotel [D]
Internal rebuilding of first and second floors to provide grand hall, reading room, marble and alabaster lined reception hall, billiard room, and glass covered lounge over entry

[635] 1899?
LONDON: Vauxhall Bridge [D]
Design for new Vauxhall Bridge, in association with J. M. Brydon, E. W. Mountford, W. D. Caröe, and Ernest George

[636] 1899–1903
PLYMOUTH, Devon: Prudential Assurance Buildings [D O Dem]
Office building of red brick and terracotta
Client: Prudential Assurance Company
Contractors: A. R. Lethbridge
Cost: £75,000

[637] 1899
THEYDON BOIS, Essex: Birch Hall [D]
New wing to house and farm buildings designed by E. Egan 1891
Client: Gerald Buxton

[638] 1899
LOCATION UNKNOWN [O]
Cross
Client: T. E. Lister
Contractors: stone carving, Hopton Wood Stone Co.
Cost: £7. 10*s*.

[639] 1899
LOCATION UNKNOWN [O]
Headstone
Client: M. D. Malleson
Contractors: stone carving, Hopton Wood Stone Co.
Cost: £5

[640] 1900
?HOLMBURY ST MARY, Surrey [O]
Monument to William Waterhouse
Client: Edwin Waterhouse
Contractors: stone carving, Farmer & Brindley
Cost: £30

[641] 1900
LONDON: Collingwood Street Board School [O]
Swimming bath

[642] 1900
LONDON, Islington: 34 Upper Street [O]
Alterations to shop
Client: Frank Harrison Baker

[643] 1900
NEWBURY, Berkshire [S]
Children's Hospital

[644] 1901–3
HULL, Humberside: Prudential Assurance Buildings [D O]
Insurance office and chambers, largely the work of Paul Waterhouse
Client: Prudential Assurance Company
Cost: £24,525

[645] 1901
LONDON, High Holborn: Staple Inn Buildings [O]
Sets of office chambers in pink brick and terracotta with a slate roof
Client: Prudential Assurance Company
Contractors: Holland & Hannen
Cost: £29,305

[646] 1901–4
SOUTHAMPTON, Hampshire: Prudential Assurance Company Building [D O]
Office building of red brick and terracotta with gabled façade to street
Client: Prudential Assurance Company
Contractors: Jenkins & Co.
Cost: £21,815

[647] ND
LONDON: 29 Dartmouth Road [O]
Unspecified work
Client: J. W. Pease

APPENDIX 2

GEOGRAPHICAL DISTRIBUTION OF WATERHOUSE'S WORKS

ALL Waterhouse's commissions are shown in this list, listed by number and grouped by county (or by larger areas) or by named towns where he had a number of commissions in one place.

Bedfordshire	178, 204, 273, 312, 410, 457
Berkshire	226, 378, 416, 425, 439, 452, 453, 466, 467, 474, 494, 517, 529, 562, 593, 617, 643
Bolton	11, 73, 74, 75, 555
Bristol	3, 5, 196, 330, 631
Buckinghamshire	259, 395, 564, 582
Cambridge	197, 268, 269, 289, 290, 291, 292, 332, 333, 345, 346, 347, 364, 399, 414, 415, 427, 442, 459, 475, 501, 502, 531, 540, 547, 565, 621
Cheshire	6, 12, 13, 17, 31, 48, 61, 81, 95, 96, 122, 126, 127, 137, 146, 166, 168, 174, 182, 183, 198, 199, 203, 210, 223, 231, 233, 288, 293, 314, 315, 327, 344, 352
Cleveland	78, 117, 176, 188, 272, 367, 368, 408, 409, 418, 423, 428, 429, 443, 489, 518, 541
Cornwall	271, 365, 417
Co. Durham	80, 349
Darlington	43, 44, 77, 98, 99, 100, 101, 128, 147, 175, 200, 201, 380, 381, 382, 487, 611
Derbyshire	313
Design Work	189, 230, 287, 325, 413, 455, 456, 486
Devonshire	371, 636
Dorset	361, 445
Essex	286, 331, 610, 637
Gloucestershire (including parts of Avon)	235, 396, 436
Hampshire	192, 193, 232, 266, 267, 323, 328, 329, 362, 440, 465, 483, 484, 496, 538, 620, 646
Hertfordshire	258, 309, 316, 335, 585

Ireland	55, 83, 93, 94, 236, 402
Kent	54, 76, 173, 194, 195, 234,308, 341, 557, 561
Lakeland	2, 4, 15, 16, 22, 26, 27, 28, 35, 39, 42, 45, 56, 58, 59, 92, 115, 123, 124, 130, 139, 149, 169, 257, 261, 294, 295, 351, 594
Lancashire	14, 21, 37, 72, 145, 148, 167, 191, 334, 397, 398, 410, 422, 444, 527, 629
Leeds	46, 60, 177, 336, 460, 504, 510, 520, 559, 574, 597, 622, 623, 624
Leicestershire	421, 575
Lincolnshire	342
Liverpool	29, 30, 47, 82, 103, 104, 131, 132, 133, 150, 206, 237, 274, 296, 317, 337, 353, 405, 406, 430, 446, 488, 497, 512, 521, 532, 542, 549, 598, 599, 600, 613, 633
London	179, 180, 181, 207, 208, 209, 238, 239, 240, 241, 242, 243, 244, 245, 275, 276, 277, 278, 297, 298, 299, 300, 301, 302, 318, 319, 338, 354, 355, 356, 357, 370, 384, 385, 386, 419, 420, 431, 432, 433, 447, 448, 461, 470, 471, 479, 480, 481, 490, 491, 492, 498, 505, 506, 513, 514, 522, 523, 533, 534, 543, 544, 550, 556, 560, 567, 568, 576, 577, 578, 583, 587, 588, 589, 601, 602, 603, 604, 614, 615, 619, 625, 626, 634, 635, 641, 642, 645, 647
Manchester	1, 7, 18, 19, 20, 32, 33, 34, 49, 50, 51, 52, 62, 63, 64, 65, 66, 67, 68, 69, 70, 83, 84, 85, 86, 87, 88, 89, 90, 105, 106, 107, 108, 109, 110, 111, 112, 113, 114, 134, 135, 136, 151, 152, 153, 154, 155, 156, 157, 158, 159, 160, 161, 162, 163, 164, 165, 211, 212, 213, 214, 215, 216, 217, 218, 219, 220, 221, 222, 246, 247, 248, 249, 250, 251, 252, 253, 279, 280, 281, 282, 283, 284, 303, 304, 305, 306, 320, 321, 339, 358, 359, 372, 373, 387, 388, 407, 422, 434, 435, 493, 499, 507, 515, 535, 536, 545, 551, 569, 570, 571, 590, 605, 606, 627
Middlesbrough	389, 390, 516, 524, 553
Middlesex	412, 438, 573, 592
Newcastle	184, 572, 591
Norfolk	476
Northamptonshire	363
Nottingham	584, 628
Overseas	129, 311, 375, 391, 400, 495, 552
Oxford (including Oxfordshire)	172, 254, 350, 392, 462, 463, 464, 477, 525, 537
Reading	71, 116, 225, 255, 285, 307, 340, 360, 376, 377, 393, 394, 424, 437, 449, 450, 472, 482, 526, 563, 579, 580
Scotland	366, 369, 383, 473, 503, 508, 530, 548, 595, 596
Shropshire	57, 411, 500

Somerset	8, 36, 186, 227, 458
Staffordshire	118, 511
Surrey	185, 202, 256, 326, 451, 528, 612, 640
Sussex	102, 205, 270, 322, 404, 478, 509, 546, 586, 607, 616
Unidentified	9, 10, 23, 38, 40, 41, 53, 119, 120, 121, 141, 142, 143, 170, 171, 187, 228, 229, 262, 263, 264, 265, 324, 343, 379, 454, 468, 485, 519, 539, 609, 618, 638, 639
Wales	138, 140, 144, 224, 310, 566, 630
Warwickshire	25, 260, 348, 403, 554, 558
Wiltshire	374
Worcestershire	441
Yorkshire (including Humberside)	24, 79, 91, 97, 125, 190, 426, 469, 581, 608, 632, 644

Appendix 3

PAINTINGS BY WATERHOUSE EXHIBITED AT THE ROYAL ACADEMY

THE information in this list is taken from A. Graves, *The Royal Academy of Arts: A Complete Dictionary of Contributors and their Work from its Foundation in 1796 to 1904* (London, 1906).

Date	*Cat. no.*	Title
1857	1094	*Cottage erecting in Cumberland*
1868	865	*Manchester New Town Hall*
1869	975	*Gonville and Caius College, Cambridge*
	986	*Manchester Town Hall*
	999	*Manchester Town Hall*
1870	792	*St. Matthew's, Blackmoor*
1871	943	*Eaton Hall, Cheshire*
	948	*Clock Tower, Manchester Town Hall*
1872	1173	*Liverpool Seamen's Orphan Institution*
	1213	*Owens College, Manchester*
1874	1095	*Court Chambers, Lincoln's Inn*
	1172	*Eaton Hall, Cheshire*
1875	945	*Eaton Hall, Cheshire*
1876	1004	*Natural History Museum, South Kensington*
	1068	*Eaton Hall, Cheshire*
1877	1093	*Clock Tower, Manchester Town Hall*
1878	Elected ARA	
	1054	*Iwerne Minister, Dorset*
	1094	*Yattendon, near Newbury*
	1109	*Central Hall, Natural History Museum*
1879	1094	*Interior, New Court, Carey Street*
	1101	*Prudential Insurance Company, Holborn*
	1102	*Pembroke College, Cambridge*
1880	1119	*Corner of Old Bond Street and Piccadilly*
	1175	*Interior of Heythrop Hall, Oxon*
	1177	*St Paul's School, Kensington*
1881	1010	*Clock Tower in memory of the late C. Turner, Esq., M.P., on the Liverpool pierhead*

Date	*Cat. no.*	Title
	1043	*Library at Eaton Hall*
	1084	*Central Institution of the City and Guilds of London Institute, Exhibition Road, South Kensington*
1882	1130	*Interior of St. Elisabeth's, Reddish, near Manchester*
	1139	*Interior of St. Elisabeth's, Reddish, near Manchester*
	1198	*St. Mary's, Twyford, Hants*
	1204	*St. Mary's, Twyford, Hants*
1883	1156	*Owens College, Manchester*
	1191	*Turner Memorial, Dingle Head, Liverpool, Home of Rest for Chronic Sufferers*
1884	1048	*Warkworth*
	1258	*Owens College, Manchester*
	1283	*Cambridge University Union Society*
	1296	*Turner Memorial Home*
	1314	*Allerton, Cannes*
1885	Elected RA	
	1348	*The Theatre, Taormina, Sicily*
	1803	*National Liberal Club, Whitehall Place*
1886	1168	*Chartres Cathedral porch, north Transept*
	1588	*Prudential Insurance Offices, Dale Street, Liverpool*
	1639	*National Liberal Club*
1887	1133	*The Lahn, from Kloster Arnstein*
	1596	*Girton College, Cambridge*
	1656	*Royal Infirmary, Liverpool*
	1680	*Manchester Town Hall* (Diploma Work)
1888	1217	*Girgenti*
	1230	*The Aventine from the Ponte Rotto*
1889	1492	*A Berkshire Lane, Easter*
	1902	*Liverpool University College*
	1917	*Hotel Metropole, Brighton*
	1934	*The Weigh-house Chapel, Mayfair*
1890	1204	*Evening near Zermatt*
	1828	*Prudential Assurance Company, Glasgow*
1891	1256	*Granada from the Gardens of the Generalife*
	1832	*National Provincial Bank of England, Manchester*
	1849	*In the Garden of the Hotel Metropole, Brighton*
1892	1191	*In Salvator Rosa's Country*
	1240	*The Tiber in 1873*
	1751	*Examination Hall, Yorkshire College, Leeds*
1893	1498	*National Provincial Bank of England, Piccadilly*
	1499	*Refuge Assurance Co., Manchester*
1894	1714	*Prudential Assurance Buildings, Nottingham*
1895	688	*On the Erme*

Date	*Cat. no.*	Title
	1463	*Prudential Assurance Company, Edinburgh*
1896	1677	*Christie Library, Owens College, Manchester*
	1687	*University College Hospital, Gower Street*
	1695	*Bank, Park Row, Leeds*
1897	715	*Bamborough Castle*
	1793	*The Surveyor's Institution*
1898	77	*Arisaig, N.B.*
	866	*Ashampstead Common*
1899	52	*On the Clunie, Braemar*
	766	*March in Granada*
	1853	*Prudential Assurance Company, Holborn Bars*
1900	599	*A Cotswold manor house*
1901	462	*At my gate*

Appendix 4

COMPETITIONS ASSESSED BY WATERHOUSE

Date	*Competition*	*Winner*
1864	Congleton Town Hall	E. W. Godwin
1865	Middlesbrough Exchange	C. J. Adams
1866	Bristol Assize Courts (2nd Competition)	E. W. Godwin
1866	Wolverhampton Town Hall	E. Bates
1867	London, Watford Orphan Asylum	—
1869	Cambridge Corn Exchange	R. R. Rowe
1869	Dumfries Infirmary	J. Starforth
1869	Plymouth Guildhall	Norman & Hine (with E. W. Godwin as consultant)
1869	Sheffield Baptist Chapel	—
1870	Birmingham, Midland Counties Idiot Asylum	—
1870	Saltburn Convalescent Home	—
1871	Birmingham Council House	R. Yeoville Thomason
1871	Blackburn Free Library and Museum	Woodzell & Collcutt
1872	Cambridge Congregational Chapel	Fuller & Cubitt
1872	Leeds Exchange and Newsroom	—
1872	St. Helen's Town Hall	H. Summers
1873	Cambridge, The Leys School Hall and Chapel	—
1873	Nottingham Temperance Hall	—
1873	Woodford Congregational Church	Rowland Plumbe
1874	London, Islington Union Chapel	J. Cubitt
1876	Nottingham University	Lockwood & Mawson
1877	Barrow-in-Furness Town Hall	W. H. Lynn
1878	Dublin City Markets	Lockwood & Mawson
1878	Southport Covered Market	—
1879	Birmingham Wesleyan College	Ball & Goddard
1879	Over Darwen Town Hall	Charles Bell
1879	Southport Convalescent Home	H. J. Paull & A. A. Bonella
1881	Newbury Grammar School	—
1882	Middlesbrough Town Hall	G. G. Hoskins
1883	Nottingham Municipal Buildings	F. Holden
1884	Burnley Municipal Buildings	Henry Holtom
1885	Birmingham Assize Courts	Aston Webb & Ingress Bell
1886	Harrogate Hospital	Thomas Worthington & J. Elgood

1886	Sunderland Municipal Buildings	Brightwen Binyon
1887	Belfast, MacArthur Hall	T. N. Deane
1887	Edinburgh Municipal Buildings	Leeming & Leeming
1887	Milan Cathedral, completion of façade	Brentano (died shortly after the competition)
1888	London, Institute of Chartered Accountants	J. Belcher
1889	Gloucester Municipal Offices	G. H. Hunt
1889	Sheffield Municipal Buildings	E. W. Mountford
1890	Halifax Infirmary	Thomas Worthington & J. Elgood
1890	London, Whitfirle Tabernacle	—
1890	Worcester, Victoria Institute	J. W. Simpson & E. J. Milner Allen
1891	Glasgow Art Gallery	J. W. Simpson & E. J. Milner Allen
1891	London, LCC Lodging Houses	Gibson and Russell
1891	London, South Kensington Museum	Aston Webb & Ingress Bell
1892	Birmingham General Hospital	W. Henman
1892	Manchester Technical School	Spalding & Cross
1892	Manchester Whitworth Institute	T. W. & R. F. Beaumont
1893	Bath Pumproom extension	J. M. Brydon
1894	London, Metropolitan Asylums Board	—
1894	North Riding of Yorkshire, County Council Offices	—
1895	Newcastle (N. Ireland) Hotel	J. T. Farrall
1896	Belfast City Hall	A. Brumwell Thomas
1896	Edinburgh, North Bridge Street reconstruction	J. H. Williams
1896	Manchester Royal Infirmary	J. W. Simpson & E. J. Milner Allen
1897	Cardiff City Hall	Lanchester, Stewart & Rickards
1897	Newcastle upon Tyne Infirmary	W. L. Newcombe & H. Percy Adams
1897	Wolverhampton Library	H. T. Hare
1899	Bradford Cartwright Hall	J. W. Simpson & E. J. Milner Allen

Appendix 5

DRAUGHTSMEN AND ASSISTANTS IN THE WATERHOUSE OFFICE

It is not possible to recover a complete list of those who worked for Waterhouse. Many were in the office for only a short period; and the office ledgers do not record names of draughtsmen over the whole period of his practice. However, this list is a useful indication of the size of the office in its early years and in the busy decade from 1865 to 1875. Several of the names are also of interest, because they relate to other well-known practitioners. The list is compiled from all the surviving ledgers, and includes clerks and office boys, who were drawn in occasionally to help with tracings. Dates show the period for which each draughtsman worked for Waterhouse. Those in brackets indicate that the draughtsman worked for a period of less than six months. Often the period was only a matter of weeks. Dates followed by ff. indicate that the draughtsman is known, from other sources, to have worked for Waterhouse after the period covered by the office ledgers. Several of these worked almost their whole career in the office (see Pl.159). Draughtsmen who set up in practice on their own as architects or surveyors are indicated by *.

Name	*Date(s)*	*Further information*
Aitken	1859–60	
Ardley	1874–8	
Atkins	1868	
Baker	(1862), 1863	
Battye, J.	1859–63	later a clerk of works for AW
Bellringer	1863–4 ff.	
Belshaw	(1871)	Belshaw worked for AW for only 19 days, but T. O. Belshaw was later a frequent subcontractor, and is recorded as attending AW's funeral
Binyon, Brightwen*	1866–71	
Bond	(1869)	
Brace	(1873)	
Brydon*	1869	
Burgess	1867–72	
Byrne	(1867)	
Clarke	(1863)	
Cooke	1875–6	
Cooper, T.*	1865–76 ff.	
Cotton	1861	
Deane	1869–76 ff.	
Dorman	(1869)	

Name	*Date(s)*	*Further information*
Entwistle	1860–6 ff.	
Evans, C.	1861–76 ff.	
Evans, F. W.	1865–6	
Farmer	(1861)	
Foster	1871–2	
Froude	1867–70	
Fry	1865–7	
Fuller	1859–60	
Geldart, Ernest*	1865–72	later clerk in holy orders
Gibbs	1870–1	
Gordon	1874–5, 1876, 1878–79	
Grosse	(1886)	
Haigh	1874–6, 1879–80	
Hansom	1866–7	
Hardcastle	1871–5	
Harrison	1871–5, 1877	
Harvey, W.*	(1863)	
Hendry	1868	
Hodgson	1865–7	
I'Anson	1867–8, 1869	
Jones	(1871)	
King	1866–9	
Krolow, Thysson	1863–4	
Kurtson	(1877)	
Lamb	1874–5	
Lee	1872–3	
Lester	(1862–3)	
Littler, Henry	1860	1860 and later a clerk of works for AW
Liversage	(1862)	
Maitland	1871–2	
Marnock	(1871–2)	
Masey	(1878) (1879)	
Mee	1862–4	
Moore	(1878)	
Murray	(1867)	
Nicolay, G. W.	1870–6 ff.	also perspectivist
Peabody	(1869)	
Pennington	(1876–7)	
Philips	(1877)	
Popplewell	1863–8	
Redmayne, Giles T.*	1859–64	
Reynolds	1867–9	
Rhodes	1863–4	
Robinson, Michael	1860–75 ff.	quantity surveyor
Scott, C. H.	1859–75	

Name	*Date(s)*	*Further information*
Sherwin	(1868)	
Showell	(1861) (1862)	
Sly	(1863)	
Speechly	(1869)	
Spice	1870–5 ff.	
Spurrell	1872–4	
Steane, George	1861–72	
Steane, James	1863–4	
Sutton	(1878) (1879)	
Tate	1865–7	
Tatham	1871–5, 1877–8	
Thomas	1870–1	
Thornicraft	1864–75 ff.	
Turner	1866–7	
Wainwright	1863–4	
Watson	1871–4	
Wheeler	1866–7	
Willey, John	1859–65 ff.	chief clerk
Willinck, Charles	(1873), 1874–5	
Wilson	(1864), 1867–75	
Wilson, J. K.	(1879)	
Windsor	1867–75	

BIBLIOGRAPHY

A. MANUSCRIPT SOURCES AND DRAWINGS

'Scraps from France, Switzerland and Italy', album with mounted sketchbook leaves, 1853–4. Private collection.

Sketchbooks

1851–1853. Private collection.

10 September 1855–19 February 1858. Private collection.

13 August 1857–6 October 1859. Private collection.

1860–1864. Private collection.

1861–1865. Private collection.

1866. Private collection.

1868–1873. Private collection.

1873–1880. Private collection.

1879–1885. Private collection.

1884–1887. Private collection.

1887–1892. Private collection.

1890–1891. Private collection.

1891–1895. Private collection.

1896–1899. Private collection.

Office Letter book.

Volume I, 18 September 1856–14 October 1862. Private collection.

Volume II, 16 November 1864–9 May 1865. RIBA Library.

Volume III, 23 November 1866–21 March 1867. RIBA Library.

The Ledger. Private collection.

Long Ledger. Private collection.

The Register No. 1. Private collection.

Certificate Register. Private collection.

'Architectural Fragments Sketches etc.', scrapbook assembled by AW, 1848 *et seq*. Private collection.

Drawings

The largest holding of drawings from the Waterhouse Office is in the RIBA Drawings Collection. This covers the majority of his schemes from the early 1850s to 1901 and later; it has recently been enlarged by the drawings archive from the Prudential Assurance Corporation. There are a number of other important holdings, in particular: Public Record Office (Drawings for the Natural History Museum), Victoria & Albert Museum (selected perspectives), Manchester University School of Architecture (perspectives and some elevations for Manchester Town Hall and one or two other buildings), Balliol College, Oxford (a number of drawings for works at Balliol College). Many of AW's drawings for his own house, and a number of his perspectives are held in the collection of the Waterhouse family, as are a number of his sketches and watercolours. There are also a large number of smaller archives in the hands of colleges, institutions, and owners of Waterhouse buildings.

B. PUBLISHED ADDRESSES OF ALFRED WATERHOUSE

1865. *Building News* 23 June, 441, 'A Short Description of the Manchester Assize Courts'. Address to an ordinary meeting of the RIBA (also published in *Papers Read at the RIBA 1864/5* (1865), 165–78).

1877. *RIBA Sessional Papers*, 12 (1876–7), 117 ff., 'The New Town Hall in Manchester'. Address to an ordinary meeting of the RIBA.

1878. *Builder* 31 Aug., 905. Address to the Manchester Society of Architects.

1879. *Berkshire Chronicle* 15 Nov., 6. Address at the distribution of prizes at Reading School of Science and Art

1883. *Building News* 2 Mar., 245 Address to the Annual Meeting of the Birmingham School of Art (read in AW's absence).

1885. *Builder* 7 Mar., 331 and 14 Mar., 373, 'Westminster Abbey'. Address to students of the Royal Academy.

1888. *Builder* 10 Nov., 336. Inaugural Address as President of the RIBA (also published in *RIBA Journal of Proceedings* 5. 17–29, and *Transactions of the RIBA*, NS 5 (1889), 1.

1889. *Builder* 2 Feb., 90. Presidential address to students, RIBA (also published in *RIBA Journal of Proceedings*, 5. 125–9, and *Transactions of the RIBA*, NS 5 (1889), 6.

1889. *Builder* 9 Nov., 328. Presidential address, RIBA (also published in the *RIBA Journal of Proceedings*, 6. 21–34, and *Transactions of the RIBA*, NS 6 (1890), 1.

1890. *Builder* 25 Jan., 57. Presidential address to students, RIBA (also published in the *RIBA Journal of Proceedings*, 6. 121–6, and *Transactions of the RIBA*, NS 6 (1890), 8.

1890. *Builder* 8 Nov., 359. Presidential address, RIBA (also published in *RIBA Journal of Proceedings*, 7. 21–35, and *Transactions of the RIBA*, NS 7 (1891), 1.

1891. *Builder* 24 Jan. 64. Presidential address to students, RIBA (also published in *RIBA Journal of Proceedings*, 7. 121–6, and *Transactions of the RIBA*, NS 7 (1891), 28.

Note: References for AW's Presidential addresses are given to *Builder*, although they were also reported in *Building News*. The *Builder* carried fuller reports.

C. SELECTED NINETEENTH-CENTURY PUBLICATIONS

AXON, W. E. A., *An Architectural and General Description of the Town Hall, Manchester, &c.* (Manchester and London, 1878).

BLAND, W., *Experimental Essays of the Principles of Construction in Arches, Piers, Buttresses, Etc. Useful to the Practical Builder* (London, 1862).

BOWMAN, H., and CROWTHER, J. S., *Churches of the Middle Ages* (London, 1848).

BROTHERS, ALFRED, FRAS, with an introduction by James Crostow, *Manchester As It Was: A Series of Views* (Manchester, 1878).

CATTOIS, F., and VERDIER, A., *Architecture civile et domestique au Moyen-age à la Renaissance* (Paris, 1855–7).

CHAMBERS, Sir WILLIAM, *A Treatise on the Decorative Part of Civil Architecture*, ed. J. Gwilt (London, 1825).

CRANE, WALTER, *The Bases of Design* (London, 1898).

DARBYSHIRE, A., *An Architect's Experiences: Professional, Artistic, Theatrical* (Manchester, 1897).

DONALDSON, J., *Farm Buildings* (London, 1851).

DONALDSON, T. L., *Handbook of Specifications* (London, 1860).

EASTLAKE, C. L., *The History of the Gothic Revival* (London, 1872). Facsimile ed. with an introduction by J. Mordaunt Crook (Leicester, 1970).

FERGUSSON, J. J., *The History of the Modern Styles of Architecture* (2nd edn., London, 1862).

FERREY, B., *Recollections of A. W. N. Pugin and his father Augustus Pugin* (London, 1861; repr. 1978).

HARDING, J. D., *Principles and Practice of Art* (London, 1845).

HOSKINS, G. G., *The Clerk of Works. A Vade Mecum for all Engaged in the Superintendence of Building Operations* (London, 1876).

HOWARD, ELIOT (ed.), *The Eliot Papers* (privately published, 1895).

INWOOD, H. W., *Of the Resources of Design in the Architecture of Greece etc., Egypt, and other Countries obtained by the Studies of the Architects of those Countries from Nature* (London, 1834).

JONES, OWEN, *The Grammar of Ornament* (London, 1856).

KERR, R., *The Gentleman's House or how to Plan English Residences from the Parsonage to the Palace* (London, 1864; repr. 1972).

—— *Seven Other Lamps of Architecture: Letters from Architecturus to his Son, Builder*, 9 (1851) and 10 (1852).

LAXTON, F. W., *The Builder's Price Book for 1855* (London, 1855).

LOUDON, J. C., *An Encyclopaedia of Cottage Farm and Villa Architecture* (London, 1833; 2nd edn., 1846).

MOLESWORTH, Sir G. L., *Pocket Book of Useful Formulae and Memoranda for Civil and Mechanical Engineers* (London, 1863).

NIGHTINGALE, B., *Lancashire Non-conformity: Or Sketches, Historical and Descriptive of the Congregational and Old Presbyterian Churches in the County*, 6 vols. (Manchester, 1890–3).

PALEY, F. A., *Gothic Mouldings* (London, 1845).

PROUT, SAMUEL, FSA, *Hints on Light, Shadow and Composition etc. as applicable to Landscape for Juvenile Inquirers* (London, 1838).

PUGIN, A. W. N., *Contrasts; or, A Parallel between the Noble Edifices of the Fourteenth and Fifteenth Centuries, and Similar Buildings of the Present Day; Shewing the Present Decay of Taste* (London, 1836).

—— *The True Principles of Pointed or Christian Architecture* (London, 1841).

RICKMAN, T. M., *An Attempt to Discriminate the Styles of Architecture in England* (London, 1817).

RUSKIN, J., *Lectures on Architecture and Painting* (New York, 1854; repr., 1978).

—— *The Seven Lamps of Architecture* (London, 1849).

—— *The Stones of Venice* (London, 1851–3).

St Anne's Church, Manchester, 'Description of the Restoration by A. Waterhouse', May Day Bazaar Catalogue (1889).

SCOTT, G. G., *Personal and Professional Recollections* (London, 1879).

—— *Remarks on Secular and Domestic Architecture* (London, 1858).

SHARPE, E. B., *Seven Periods of English Architecture Defined and Illustrated* (London, 1851).

SHAW, R. N., *Architectural Sketches from the Continent* (London, 1858).

—— and JACKSON, T. G. (eds.), *Architecture: A Profession or an Art?* (London, 1892).

STREET, A. E., *A Memorial of G. E. Street, RA* (London, 1888).

STREET, G. E., *An Urgent Plea for the Revival of the True Principles of Architecture in the Public Buildings of Oxford* (Oxford and London, 1853).

—— *Brick and Marble Architecture in North Italy* (London, 1855).

VIOLLET-LE-DUC, E. E., *Dictionnaire raisonné de l'architecture française du* XIe au XVIe siècle (Paris, 1854–68).

—— *Dictionnaire raisonné du mobilier français de l'époque carlovingienne à la Renaissance*, 6 vols. (Paris, 1858–75).

WATERHOUSE, A., *Courts of Justice Competition: General Description of Design Submitted by Alfred Waterhouse, Architect* (London, 1867).

—— *Description of the Hundred of Salford Assize Courts of the County of Lancs* (Manchester, 1859).

WATERHOUSE, EDWIN (ed.), *Extracts from the Journal of Mary Waterhouse 1825–1880* (Privately published, 1907).

Theodore Waterhouse 1838–1891. Notes on his Life and Extracts from his Letters and Papers (Privately published, London, 1894).

D. SELECT BIBLIOGRAPHY OF TWENTIETH-CENTURY WORKS

ALLAN, A. R., 'The Victoria Building: Its Furnishing and Use' in *The University of Liverpool Recorder*, 83 (Apr. 1980).

AMES, WINSLOW, *Prince Albert and Victorian Taste* (London, 1967).

ANDERS, M., 'Burmantofts Faience', B.Arch. diss., Newcastle University, 1983.

—— and GARLICK, ALAN, 'Burmantofts Pottery', Bradford Art Galleries and Museums and Leeds City Museums exhibition catalogue (1983).

ANSON, P. F., *Fashions in Church Furnishings 1840–1940* (London, 1960).

ARCHER, J. H. G. (ed.), *Art and Architecture in Victorian Manchester* (Manchester, 1985).

—— 'A Civic Achievement: The Building of Manchester Town Hall, Part One: The Commissioning' in *Transactions of the Lancashire and Cheshire Antiquarian Society*, 81 (1982).

ARNOLD, H. G., *Victorian Architecture in Reading* (Reading, 1976).

ASLIN, E., *The Aesthetic Movement* (London, 1969; 2nd ed., 1981).

—— *E. W. Godwin: Furniture and Interior Decoration* (London, 1986).

—— *Nineteenth Century English Furniture* (London, 1962).

BEATTIE, M., 'Alfred Waterhouse and Allerton Priory', B.Arch. diss., Newcastle University, 1989.

BEDARIDA, F., *A Social History of England 1851–1975* (London, 1979).

BETJEMAN, J., *First and Last Loves* (London, 1952).

BINFIELD, CLYDE, *So Down to Prayers: Studies in English Nonconformity 1780–1920* (London, 1977).

BLAU, E., *Ruskinian Gothic: The Architecture of Deane and Woodward 1845–61* (Princeton, NJ, 1982).

BRIGGS, A., *Victorian Cities* (London, 1963).

BROOKS, M. W., *John Ruskin and Victorian Architecture* (London, 1987).

BROWNLEE, D. B., *The Law Courts: The Architecture of George Edmund Street* (Cambridge, Mass., 1984).

CHAPMAN, V., 'George Gordon Hoskins, J.P., F.R.I.B.A.: A Darlington Architect and his Work 1864–1907', *Durham Archaeological Journal* 4 (1988), 61–8.

—— *Rural Darlington: Farm, Mansion and Suburb*, Durham County Library, Local History Publication, 8 (1975).

CHAPPLE, J. E. V., and POLLARD, F. S. (eds.), *Letters of E. Gaskell* (Manchester, 1966).

CLARK, KENNETH, *The Gothic Revival* (London, 1928; new edn., 1950).

CLARKE, B. F. L., *Church Builders of the Nineteenth Century: A Study of the Gothic Revival in England* (London, 1938; repr. 1969).

COLE. D., *The Work of Sir Gilbert Scott* (London, 1980).

COLLINS, PETER, *Changing Ideals in Modern Architecture* (London, 1965).

COLVIN, H. M., *A Biographical Dictionary of British Architects 1600–1840* (London, 1978).

—— *Unbuilt Oxford* (New Haven, Conn., 1983).

Country Life, 2 (1897), 182 ff.; 9 (1901), 496 ff.; 47 (1920), 724 ff.; and 156 (1974) 614 ff.

CREIGHTON, L., *Life and Letters of Thomas Hodgkin Longmans* (London, 1918).

CROOK, J., MOURDANT, *William Burges and the High Victorian Dream* (London, 1981).

—— *The Dilemma of Style: Architectural Ideas from the Picturesque to the Post Modern* (London, 1987).

CUNNINGHAM, COLIN, *Victorian and Edwardian Town Halls* (London, 1981).

—— MALTBY, SALLY, and MACDONALD, SALLY, *Alfred Waterhouse 1830–1905* (London, 1983).

—— *Prudential Buildings, Holborn Bars* (London, 1987).

DELLHEIM, C., *The Changing Face of the Past* (Cambridge 1982).

DIXON, R., and MUTHESIUS, S., *Victorian Architecture* (London, 1978).

DOIG, A., *The Architectural Drawings Collection of King's College, Cambridge: A Catalogue and Historical Synopsis of the Major Project Drawings of the Eighteenth and Nineteenth Centuries* (Amersham, 1979).

DUTTON, RALPH, *The Victorian Home: Some Aspects of Nineteenth Century Taste and Manners* (London, 1954).
ENSOR, Sir ROBERT, *England 1870–1914* (Oxford, 1936).
FAWCETT, J. (ed.), *Seven Victorian Architects* (London, 1976).
FRANKLIN, JILL, *The Gentleman's Country House and its Plan, 1834–1914* (London, 1981).
FURNEAUX JORDAN, R., *Victorian Architecture* (Harmonsdworth, 1966).
GIROUARD, M., *Alfred Waterhouse and the Natural History Museum* (New Haven, Conn., 1981).
—— *Life in the English Country House* (New Haven, Conn., 1978).
—— *Sweetness and Light: The 'Queen Anne' Movement 1860–1900* (Oxford, 1977).
—— *The Victorian Country House* (New Haven, Conn., 1979).
GLOAG, J., *Victorian Taste: Some Aspects of Architecture and Industrial Design 1820–1900* (London, 1962; repr., 1972).
GOODHART-RENDEL, H. S., *English Architecture Since the Regency: An Interpretation* (London, 1953).
GOSDEN, P. H. J. H., and TAYLOR, A. J. (eds.), *Studies in the History of a University 1874–1974* (Leeds, 1975).
GRAY, A. STUART, *Edwardian Architecture: A Biographical Dictionary* (London, 1985).
HANDLEY-REED, C., 'Sculpture in High Victorian Architecture' in *The High Victorian Cultural Achievement* (London, 1967) (2nd Conference Report of the Victorian Society).
HARDMAN, M., *Ruskin & Bradford: An Experiment in Cultural History* (Manchester, 1986).
HARPER, R., *Victorian Architectural Competitions* (London, 1983).
—— *Victorian Building Regulations* (London, 1985).
HENDERSON, PHILIP, *Letters of William Morris* (London, 1950).
HERSEY, G. L., *High Victorian Gothic: A Study in Associationism* (Baltimore, 1972).
HITCHCOCK, HENRY-RUSSELL, *Architecture: Nineteenth and Twentieth Centuries* (3rd edn., Harmondsworth, 1968).
HUXLEY, GERVAS, *Victorian Duke* (Oxford, 1967).
ISICHEI, E. A., *Victorian Quakers* (Oxford, 1970).
JACKSON, B. H. (ed.), *Recollections of T. G. Jackson* (Oxford, 1950).
JARVIS, P., 'Waterhouse in Darlington', B.Arch. diss., Newcastle University, 1983.
JENKINS, FRANK, *Architect and Patron* (Oxford, 1961).
JERVIS, SIMON, *Victorian Furniture* (London, 1968).
JONES, E. (ed.), *The Memoirs of Edwin Waterhouse* (London, 1988).
KELLY, T., *For the Advancement of Learning: The University of Liverpool 1881–1981* (Liverpool, 1981).
LETHABY, W. R., *Architecture: An Introduction to the History and Theory of the Art of Building* (London, 1911).
—— *Philip Webb and his Work* (London, 1935).
LEVER, J., *Architects' Designs for Furniture* (London, 1982).
—— 'Great George Street: Why choose Waterhouse', in *Charter Centenary*, supplement to the *RICS Journal* (Sept. 1981).
MACLEOD, ROBERT, *Style and Society: Architectural Ideology in Britain 1840–1914* (London, 1971).
MANNING, ELFRIDA, *Marble and Bronze: The Art and Life of Hamo Thorneycroft* (London, 1982).
MILLS, ERNESTINE, *Life and Letters of Frederic Shields* (London, 1912).
MUTHESIUS, HERMANN, *Das Englische Baukunst und Gegenwart* (Berlin, 1900–4).
—— *Das Englische Haus* (Berlin, 1904).
MUTHESIUS, S., *The High Victorian Movement in Architecture 1850–1870* (London, 1972).
NAPIER, M., and LAING, A. (eds.), *The London Oratory: Centenary 1884–1984* (London, 1985).
NAYLOR, G., *The Arts and Crafts Movement: A Study of its Sources, Ideals and Influences on Design Theory* (London, 1971).
OLLEY, J., and WILSON, C., 'The Natural History Museum' in D. Cruikshank (ed.), *Timeless Architecture* (London, 1985).

ORMOND, L., and ORMOND, R., *Lord Leighton* (New Haven, Conn., 1975).
PEVSNER, N., *Some Architectural Writers of the Nineteenth Century* (Oxford, 1972).
—— *et al.*, *The Buildings of England* (Harmondsworth, 1951 *et seq.*).
—— *Pioneers of Modern Design from William Morris to Walter Gropius*, first published as *Pioneers of the Modern Movement* (revised edn., Harmondsworth, 1960).
PHYSICK, J., *The Victoria and Albert Museum: The History of its Building* (London, 1982).
—— and DARBY, M., *Marble Halls: Drawings and Models for Victorian Secular Buildings* (Victoria & Albert Museum (London, 1973)).
Prudential Assurance Company, *A Century of Service: The Story of the Prudential 1848–1948* (London, 1948).
QUINEY, A., *John Loughborough Pearson* (New Haven, Conn., 1979).
REED, B., *Late Eighteenth and Nineteenth Century Sculpture in the British Isles* (London, 1976).
SAINT, A., *Richard Norman Shaw* (New Haven, Conn., 1976).
—— *The Image of the Architect* (New Haven, Conn., 1983).
—— 'Whatever Happened to Jonathan Carr?', *The London Journal*, 12 1 (Summer 1986) (Commemorative volume for Francis Sheppard).
SHEPPARD, F. H. W. (ed.), *Survey of London*, xxxviii. *The Museums Area of South Kensington and Westminster* (London, 1975).
SMITH, S. A., 'Alfred Waterhouse', Ph.D thesis, London University, Courtauld Institute, 1970.
SPIERS, MAURICE, *Victoria Park Manchester* (Manchester, 1976).
SPIERS, R. P., *Architectural Drawing* (London, 1887; 2nd edn., 1902).
STAMP, G., *The Great Perspectivists* (London, 1982)
STANTON, P., *Pugin* (London, 1971).
STEEGMAN, J., *Victorian Taste: A Study of the Arts and Architecture from 1830 to 1870* (Cambridge, Mass., 1970).
STEWART, CECIL, *The Stones of Manchester* (London, 1956).
SUMMERSON, J. N., *Heavenly Mansions and Other Essays on Architecture* (London, 1949).
—— *The London Building World of the Eighteen Sixties* (London, 1973).
—— *Victorian Architecture: Four Studies in Evaluation* (New York, 1970).
TAYLOR, J., 'Circular Hospital Wards: Professor John Marshall's Concept and its Exploration by the Medical Profession in the 1880s', *Medical History*, 32 (1989).
THOMPSON, P., *William Butterfield 1814–1900* (London, 1971).
THWAITE, ANN, *Edmund Gosse: A Literary Landscape* (Oxford, 1984).
VELUET, P., 'Alfred Waterhouse', B.Arch. diss., Newcastle University, 1970.
VENN, J., *Biographical History of Gonville and Caius College 1349–1897*, iii (Cambridge, 1901).
WAINWRIGHT, CLIVE, and GERE, CHARLOTTE, *Architect-Designers from Pugin to Mackintosh* (London, 1981).
Paul Waterhouse: A Collection of Certain of his Papers and Addresses (London, 1930).
WATERSON, M. (ed.), *The Country House Remembered* (London, 1985).

INDEX

Bold numbers indicate Colour Plates and numbers in Italic indicate Plates. Figures are referenced by page number only. Buildings are referenced by location as well as by architect.

Plates

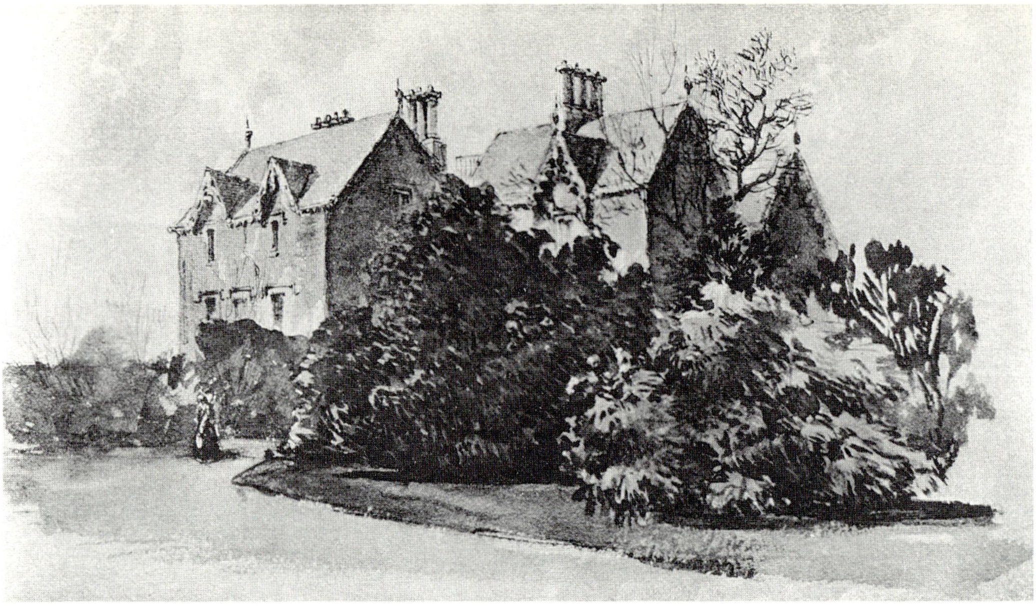

2. Oakfield, Aigburth, Liverpool, from a sketch, *c.*1848

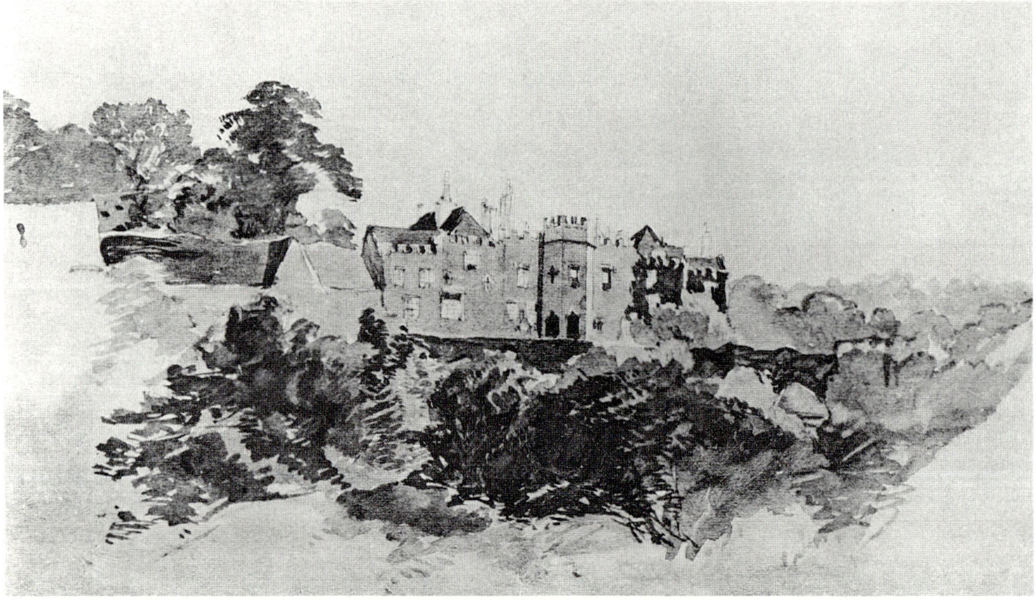

3. Sneyd Park, Bristol, from a sketch by William Waterhouse, *c.*1854

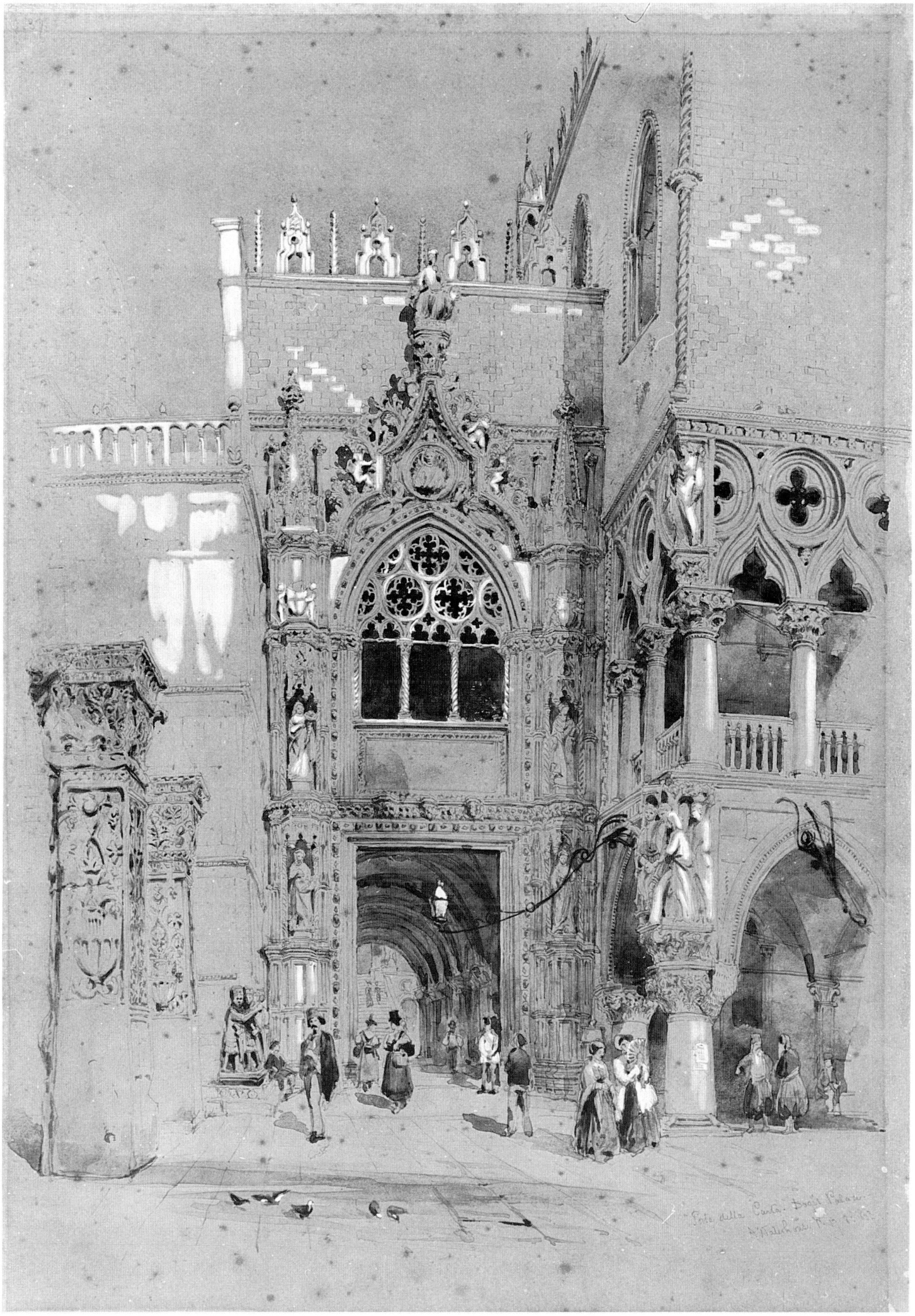

4. Venice, Porta della Carta, dated 1853

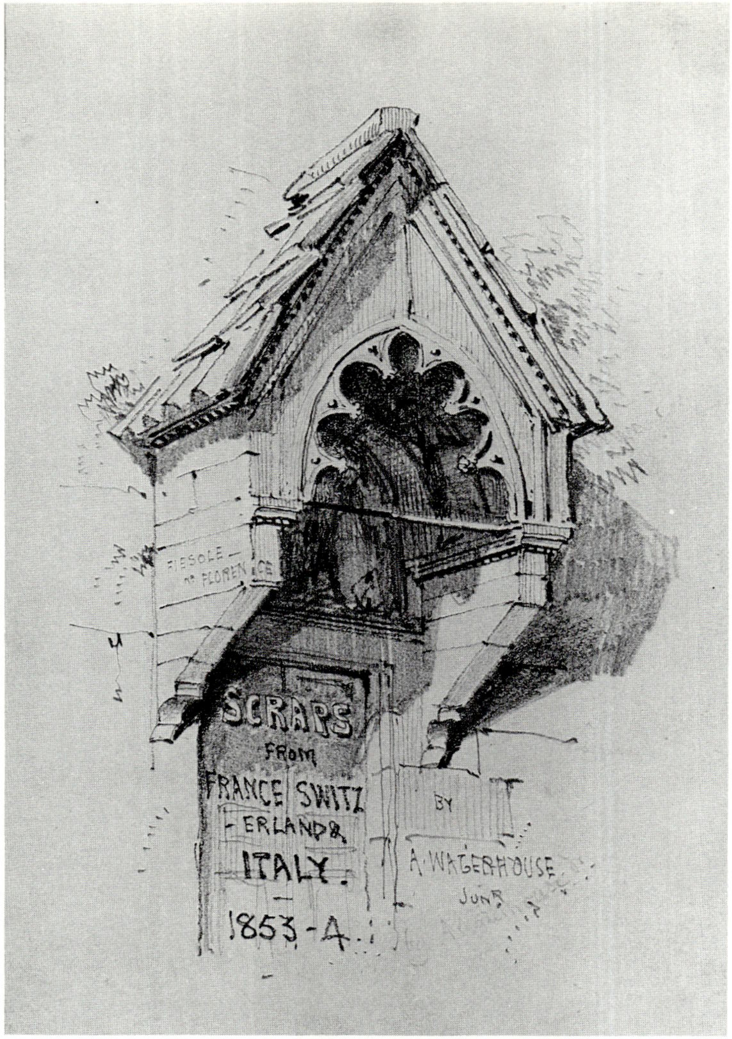

5. 'Scraps from France, Switzerland and Italy': frontispiece of Presentation Volume (1853–4)

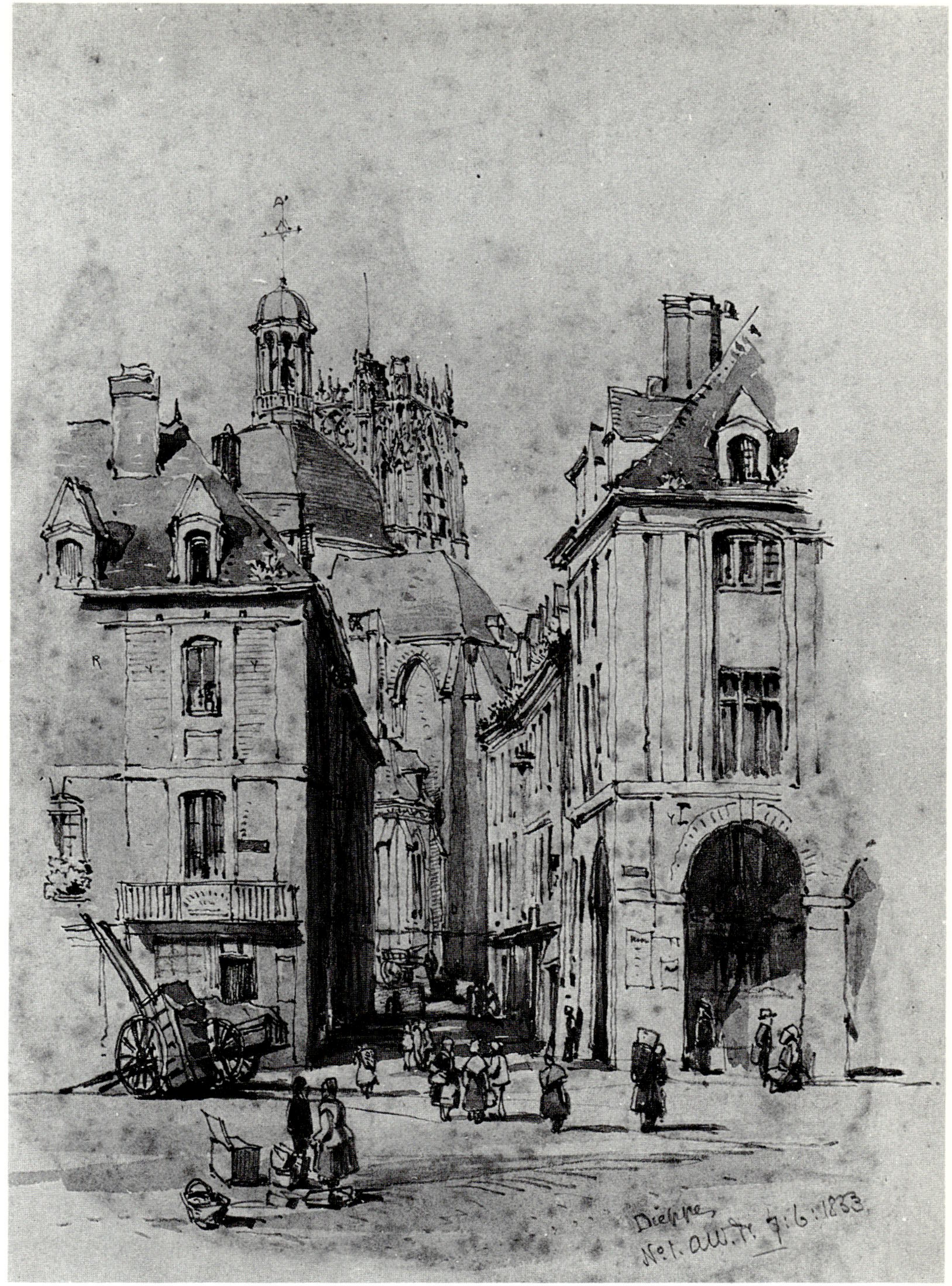

6. Right, Dieppe: watercolour dated 1853

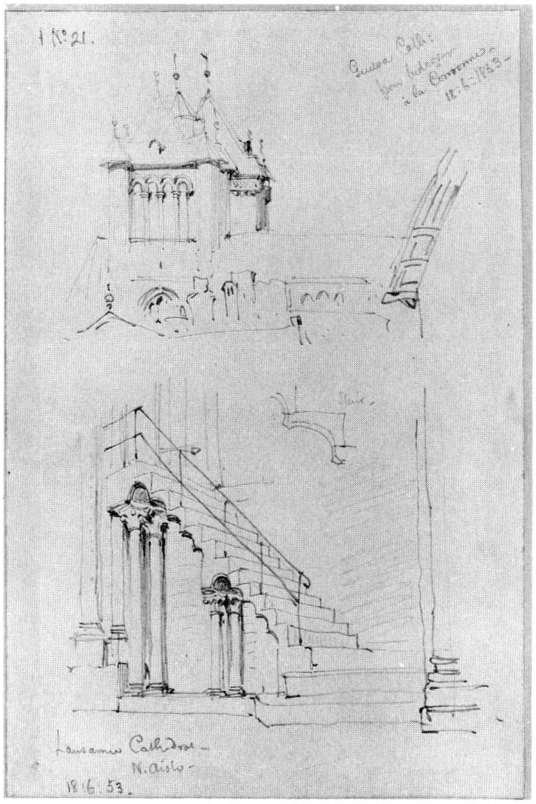

7. Sketchbook 1853: Geneva Cathedral, Lausanne Cathedral

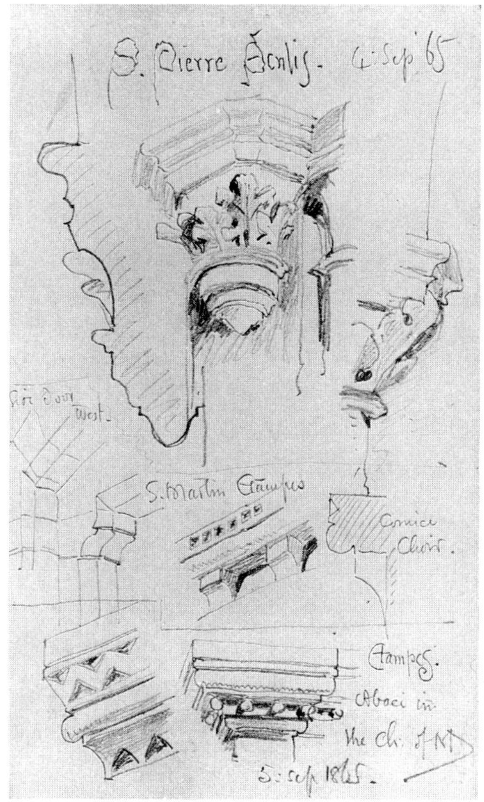

8. Sketchbook 1865: details from Senlis and Étampes

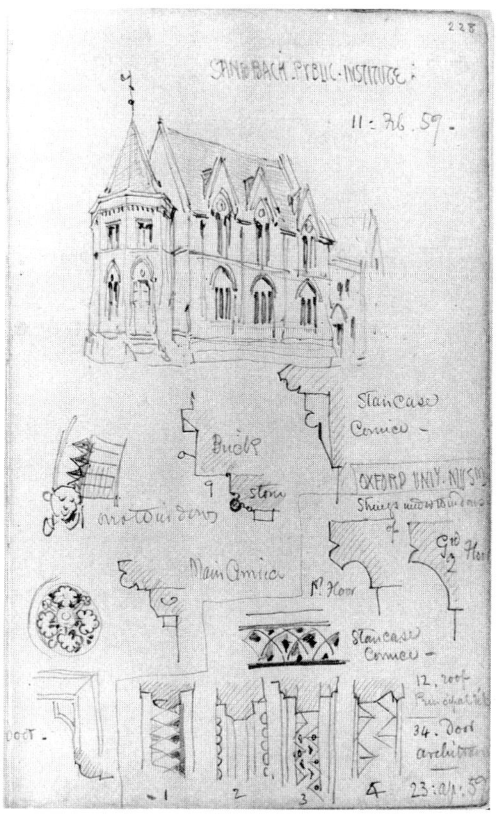

9. Sketchbook 1859: Sandbach Public Institute and details from Oxford University Museum

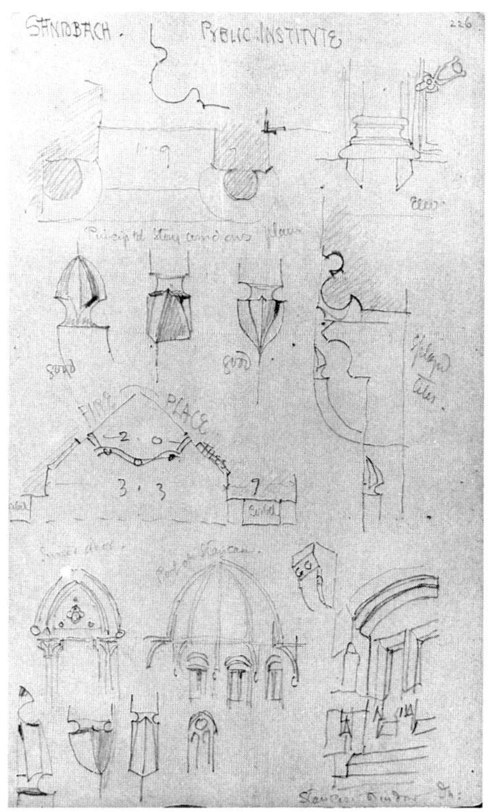

10. Sketchbook 1859: details from Sandbach Public Institute

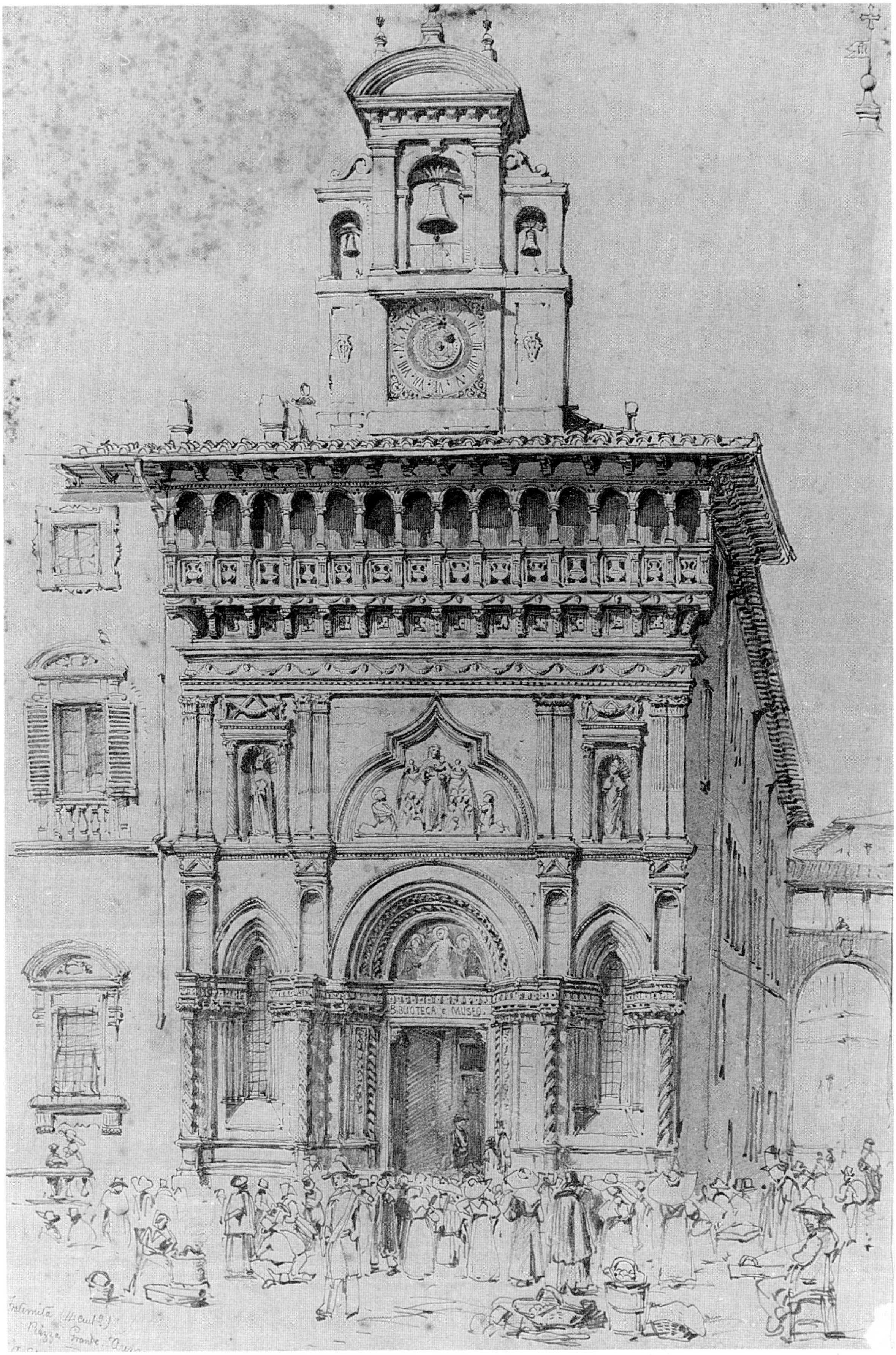

11. Piazza Grande, Arezzo, dated 1853: pencil and wash

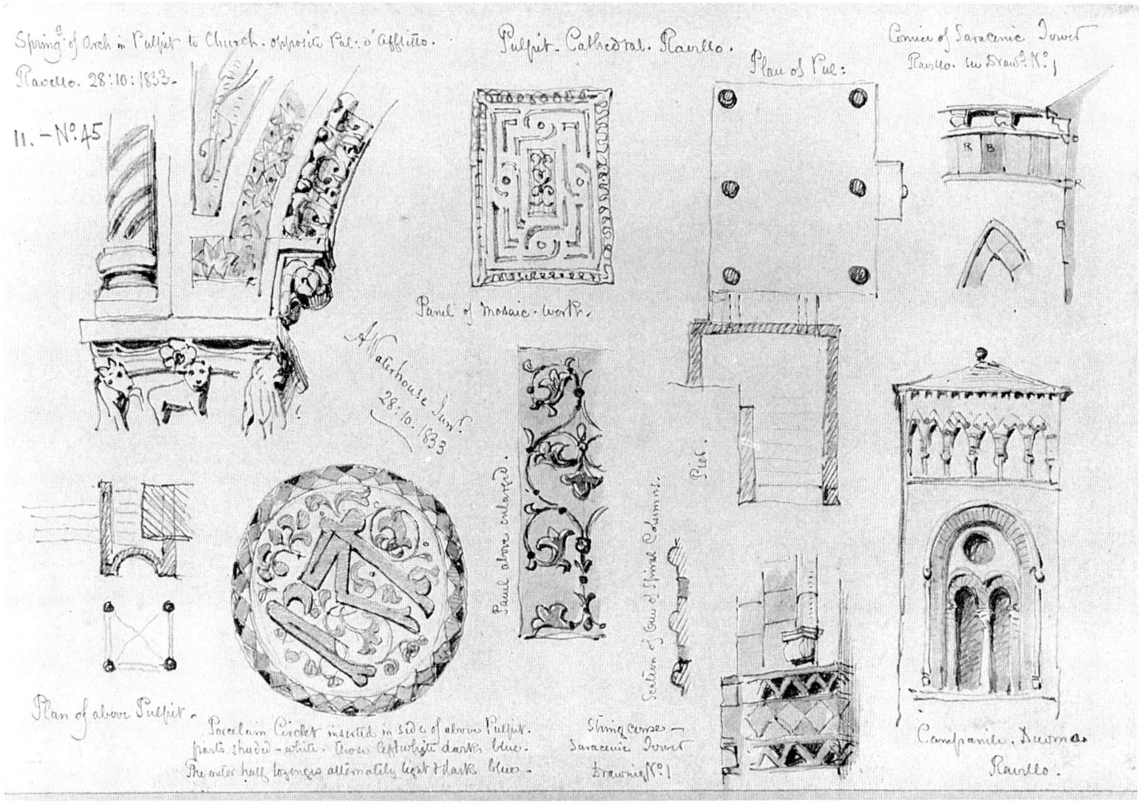

12. Sketchbook 1853: details from Ravello

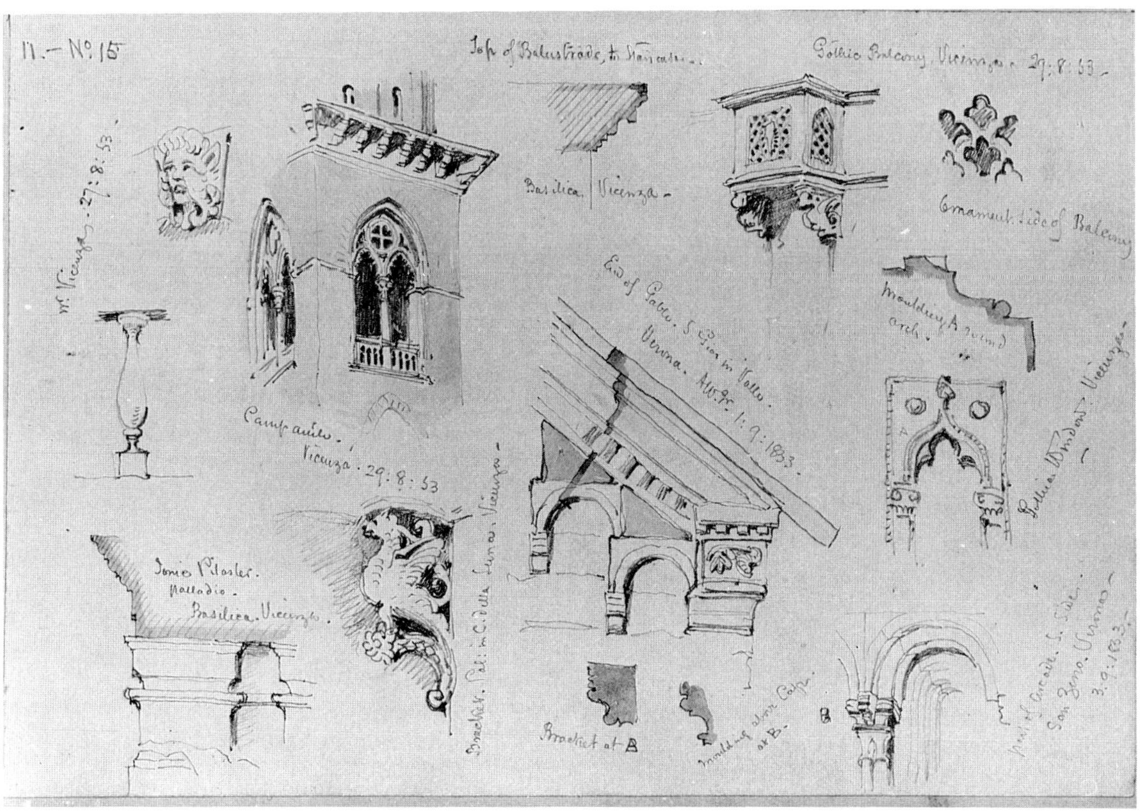

13. Sketchbook 1853: details from Vicenza and Verona

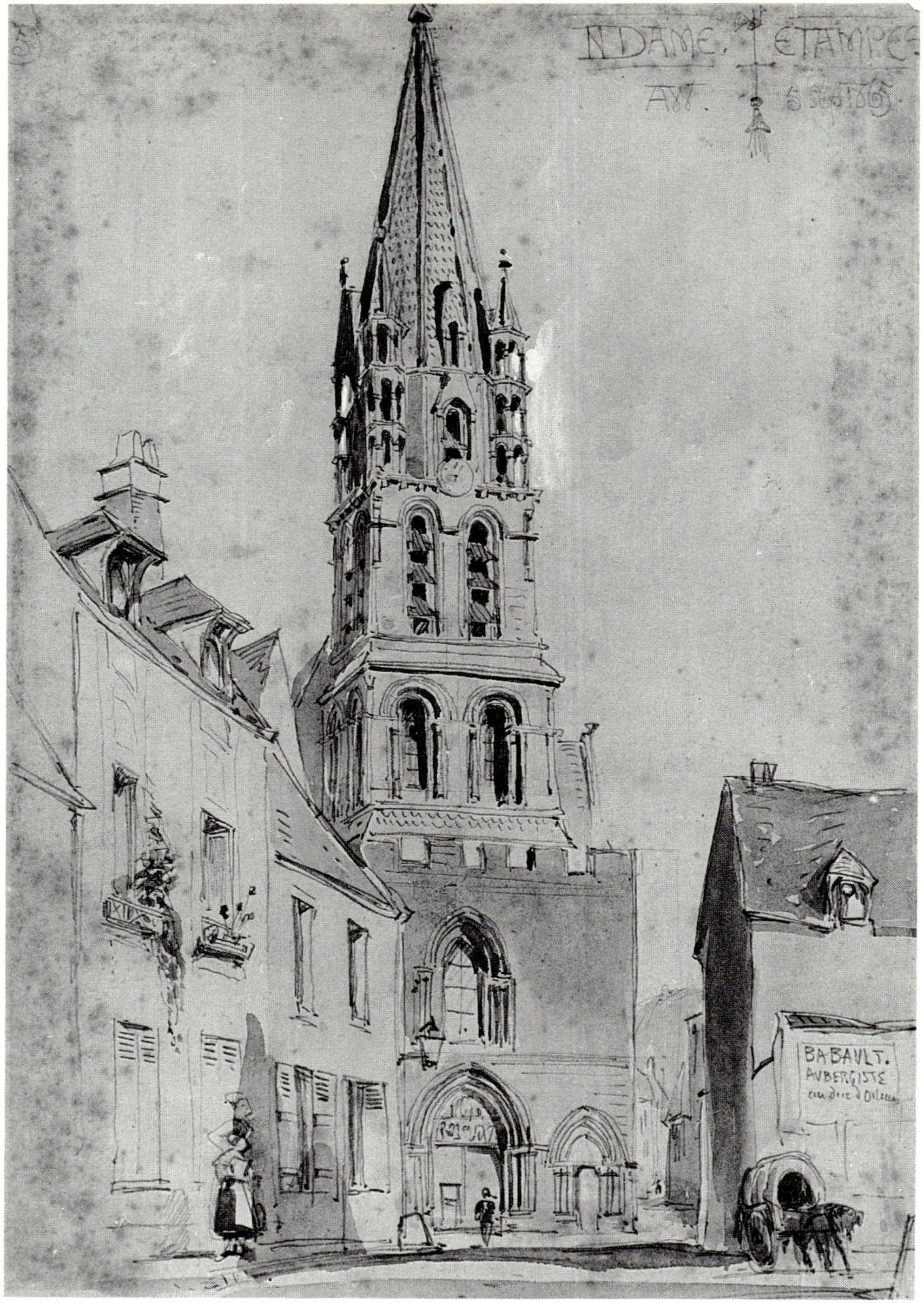

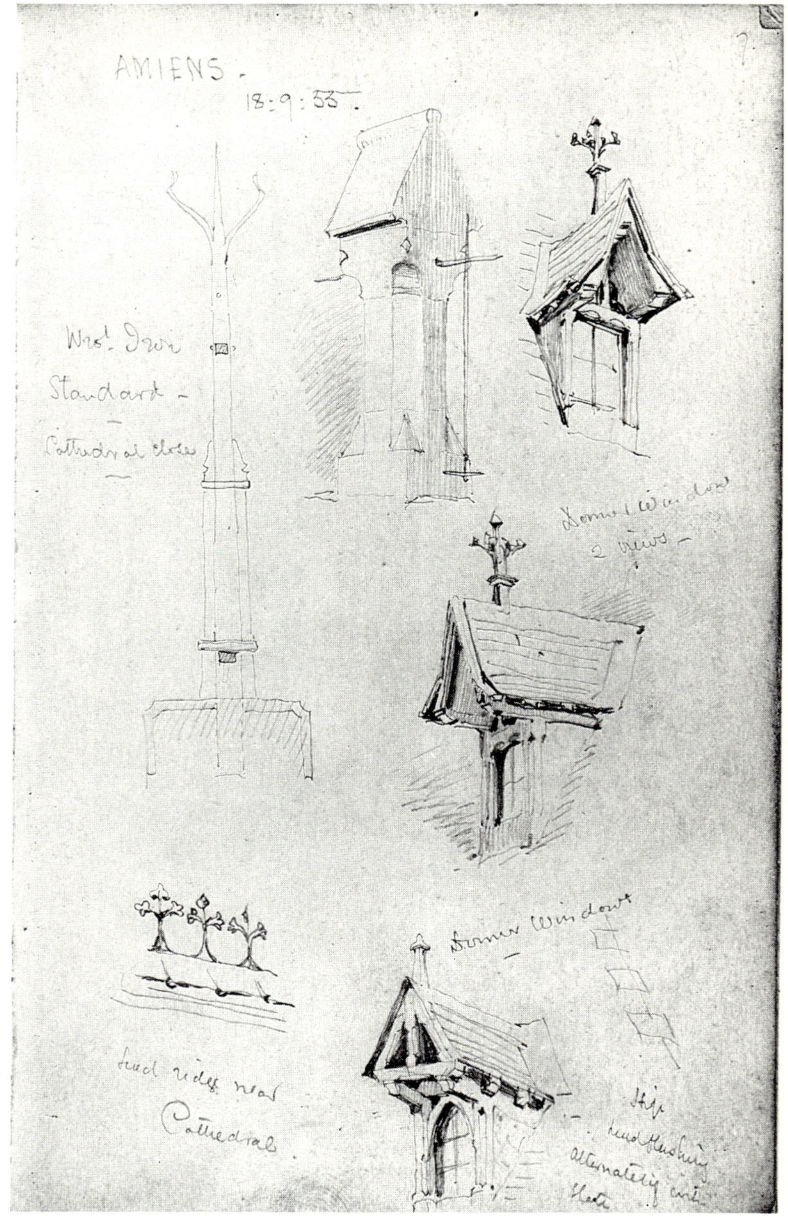

15. Sketchbook 1855: details from Amiens

14. Left, Notre Dame, Étampes: pencil and wash dated 1865

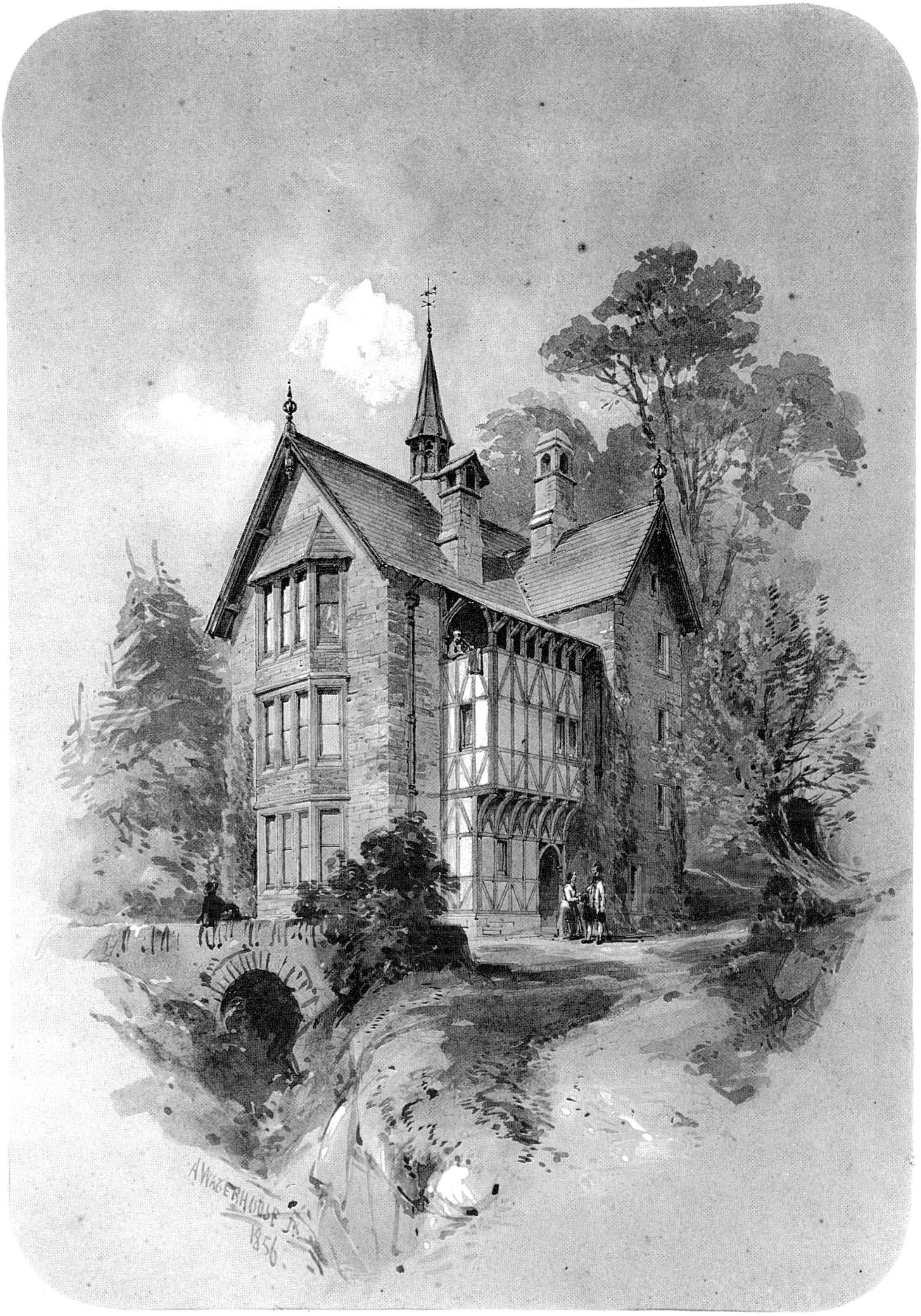

16. Fawe Park, Keswick: perspective dated 1856, exh. Royal Academy (1857) **[45]**

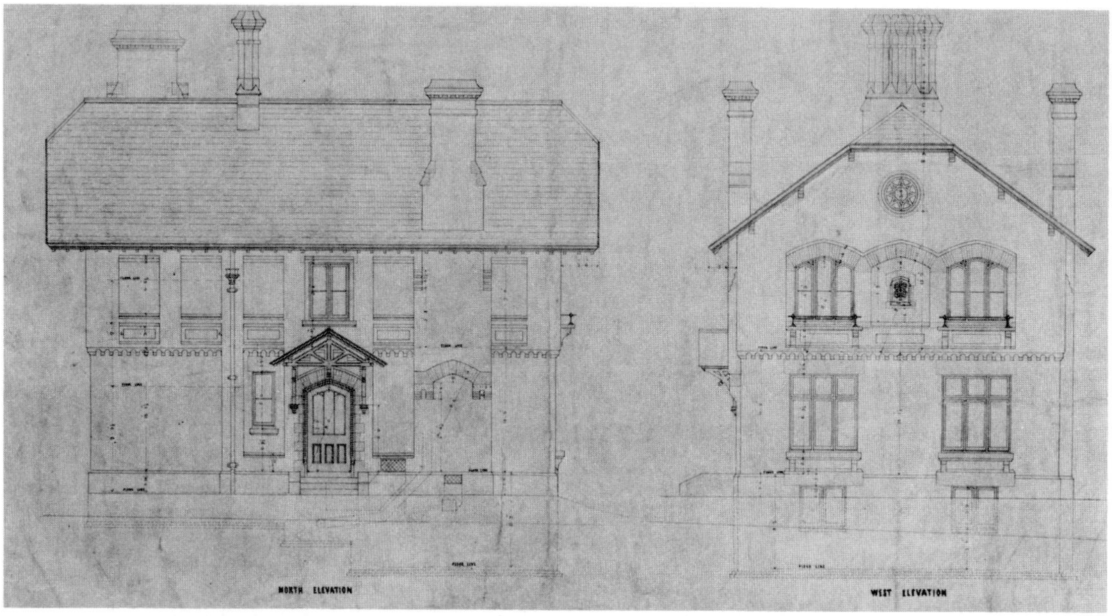

17. Manchester, Barcombe Cottage, elevations, *c.*1858 **[52]**

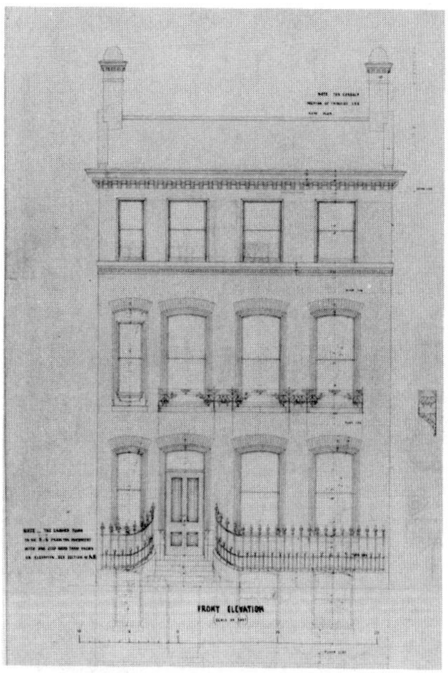

18. Liverpool, house for Sebastian Waterhouse, elevation, 1857 **[29]**

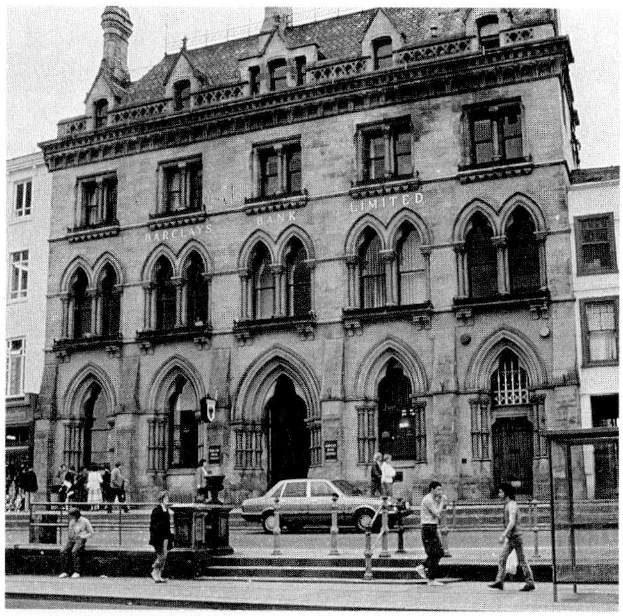

20. Darlington, Backhouse's Bank (1864) **[175]**

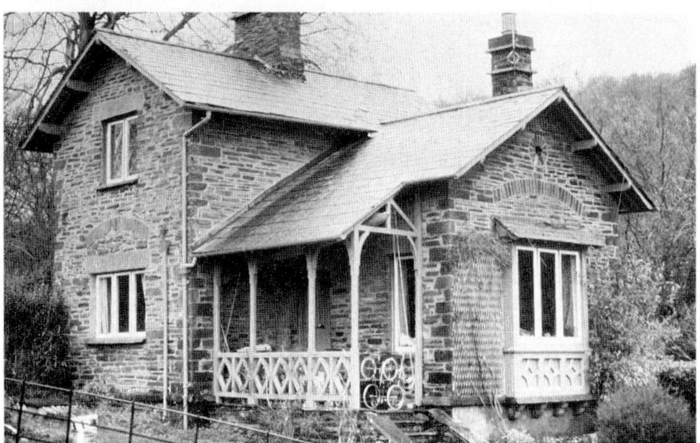

19. Brathay Hall, Lodge (1858) **[42]**

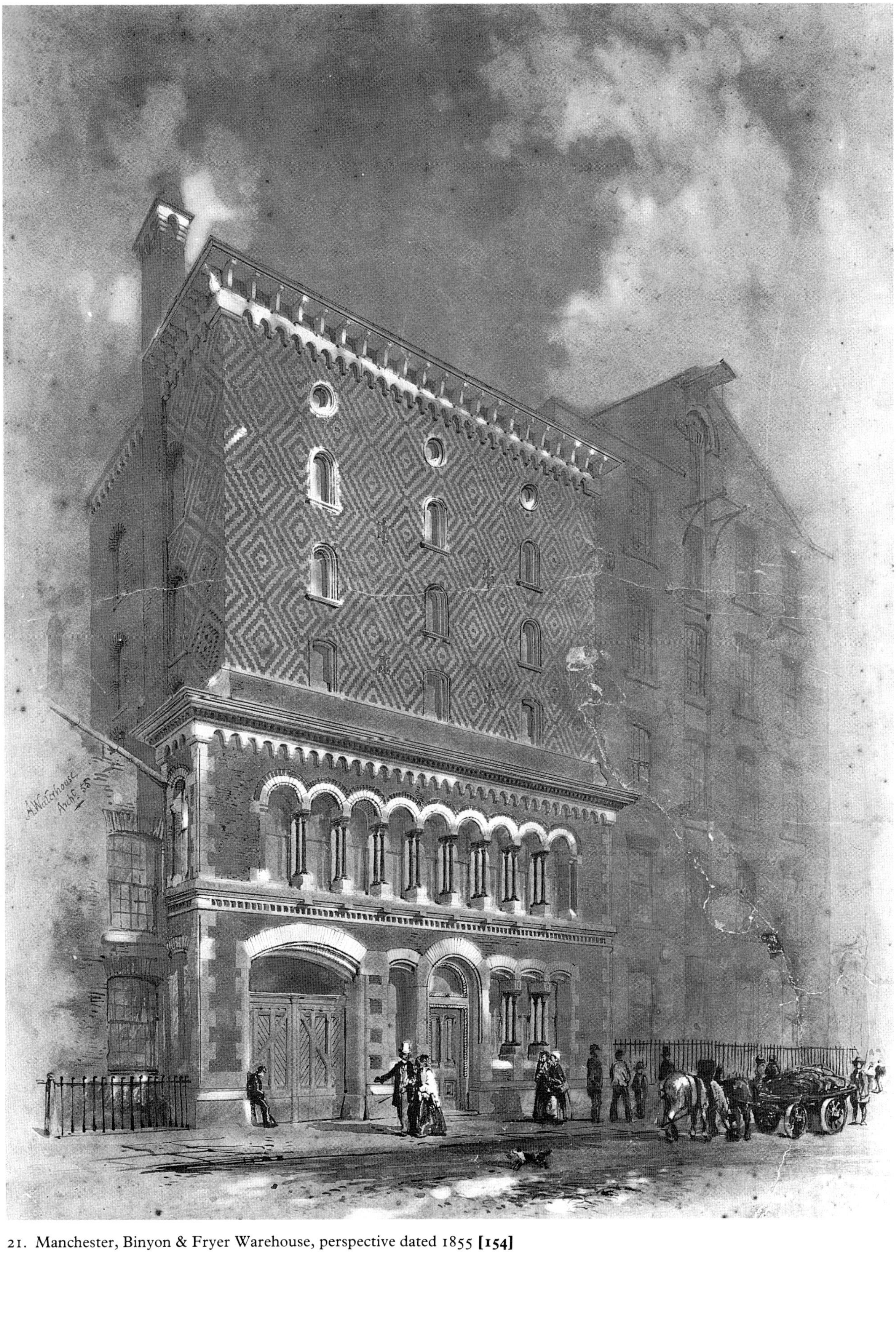

21. Manchester, Binyon & Fryer Warehouse, perspective dated 1855 **[154]**

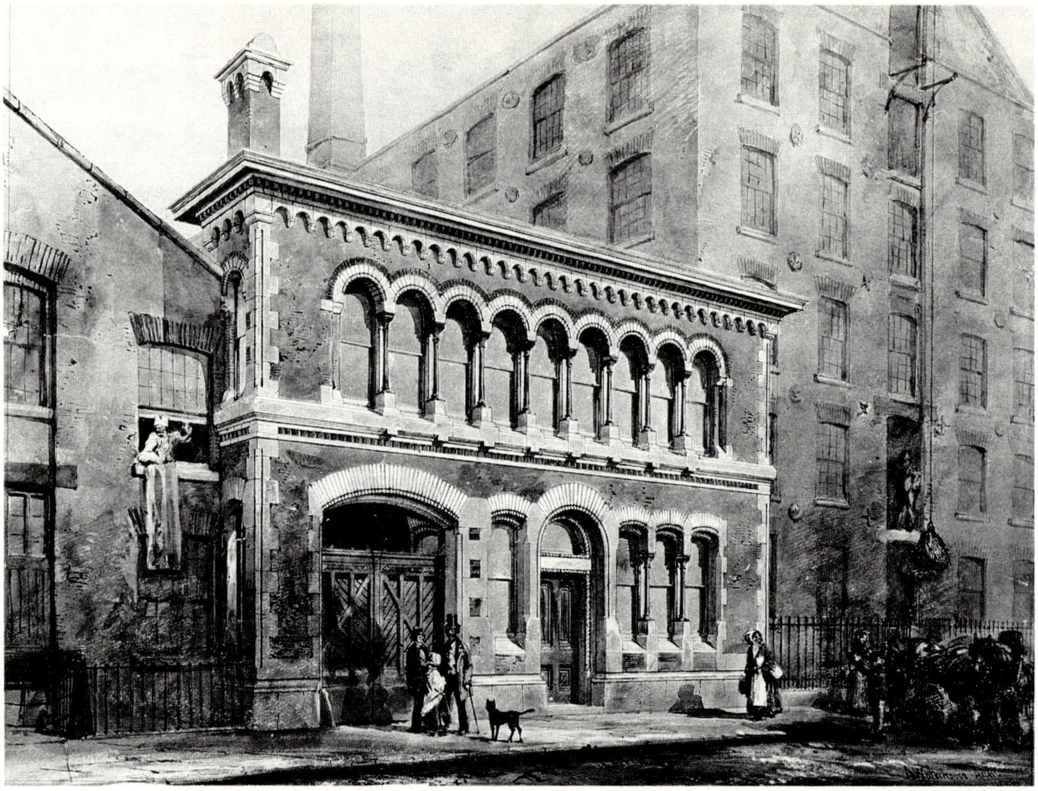

22. Manchester, Binyon & Fryer Warehouse, perspective dated 1855 **[7]**

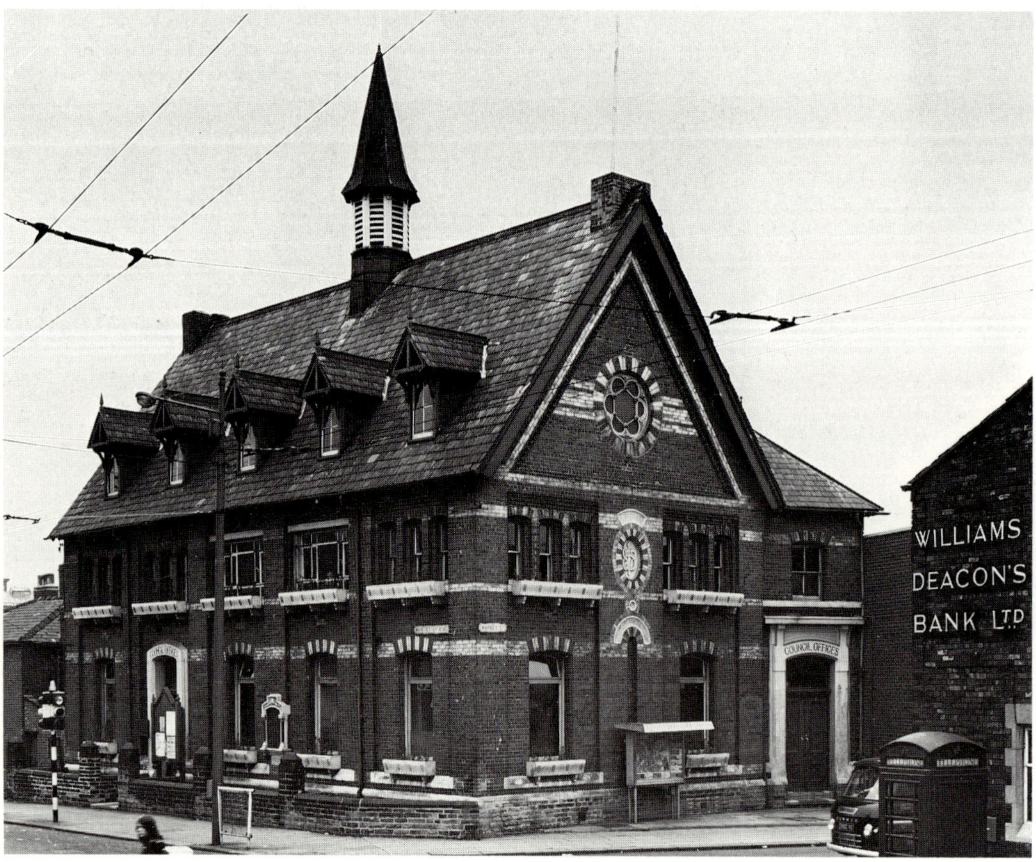

23. Manchester, Droylsden Institute (1858) **[50]**

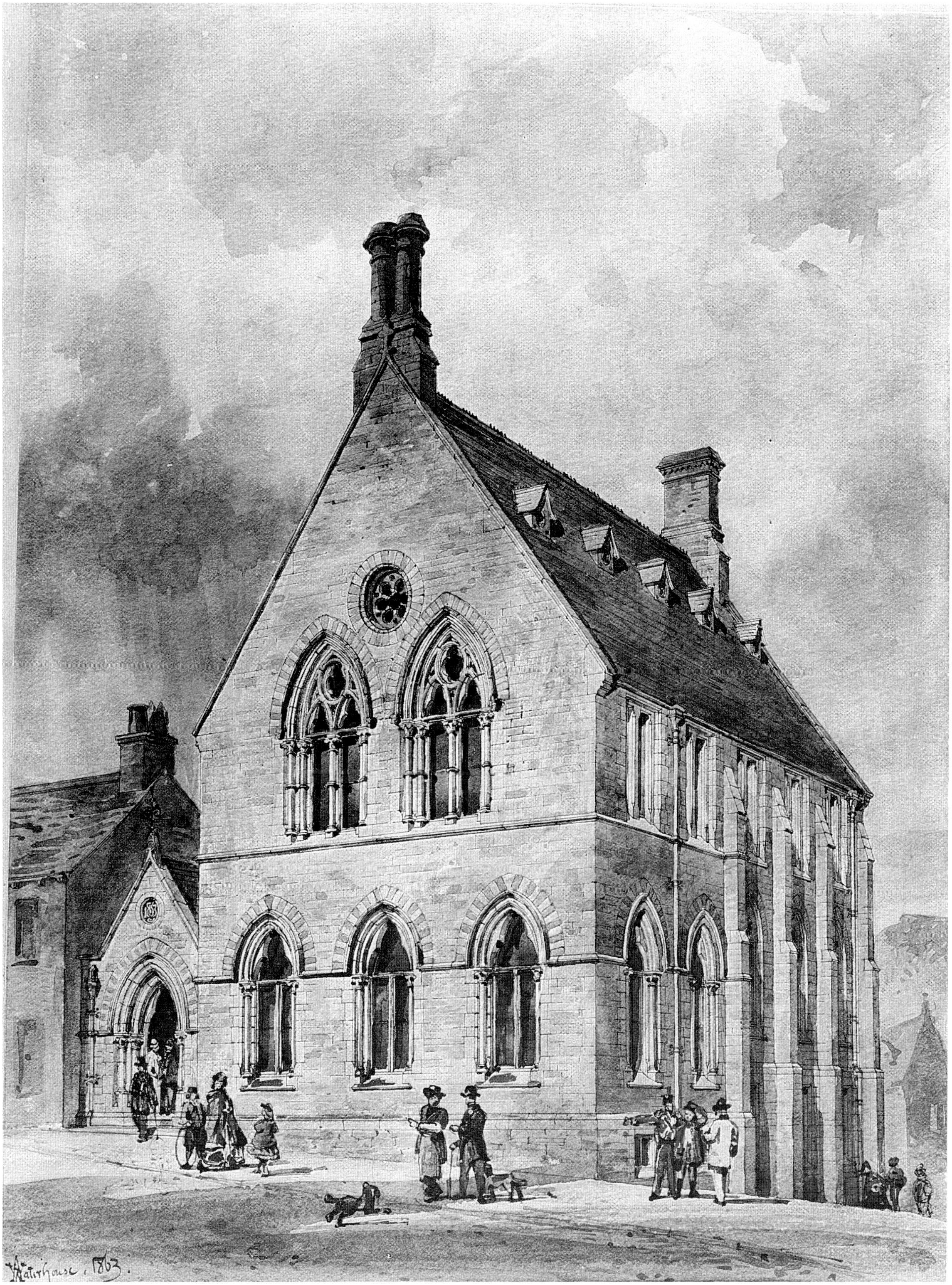

24. Bingley, Mechanics Institute, perspective dated 1863 **[125]**

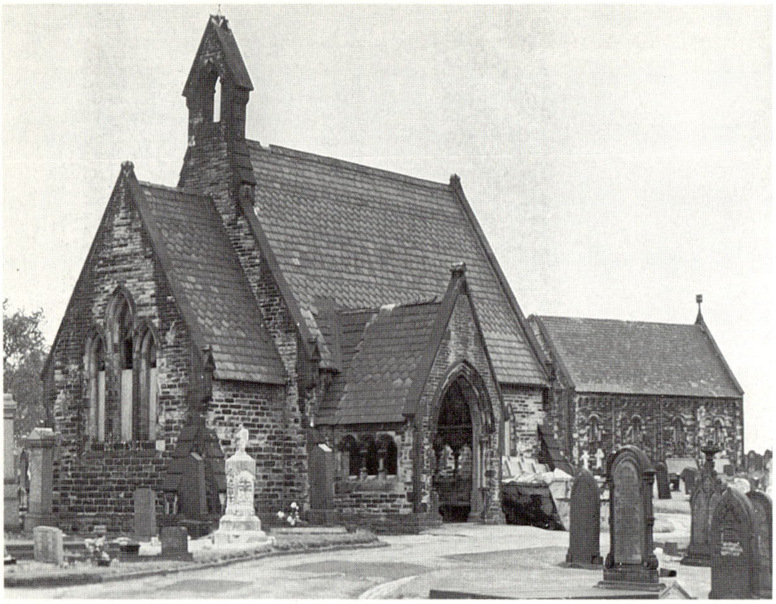

25. Ince-in-Makerfield, Cemetery Chapels (1856) **[14]**

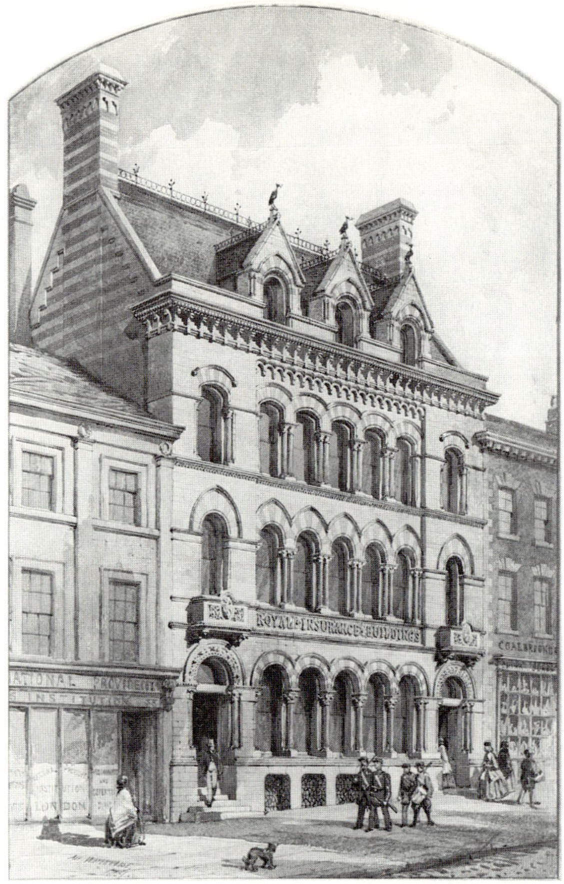

28. Manchester, Royal Insurance Building, perspective dated 1861 **[108]**

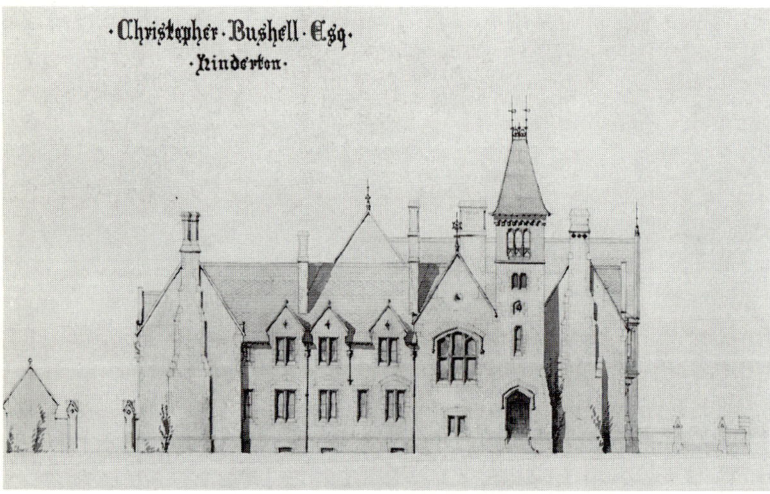

26. Cheshire, Hinderton Hall, sketch elevation, *c.*1856 **[13]**

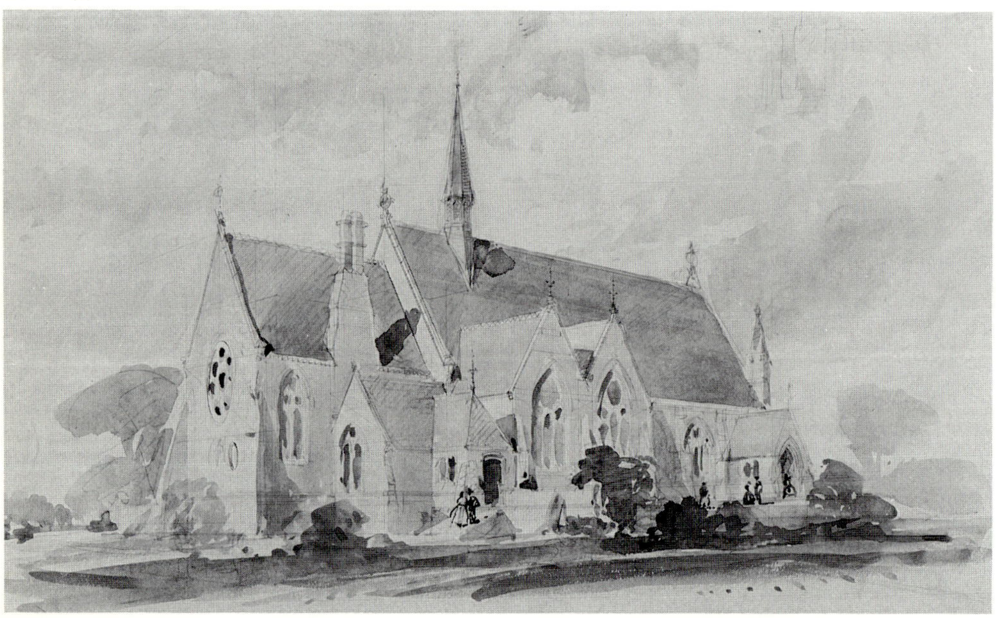

27. Manchester, St John, Brooklands, perspective sketch, *c.*1863 **[152]**

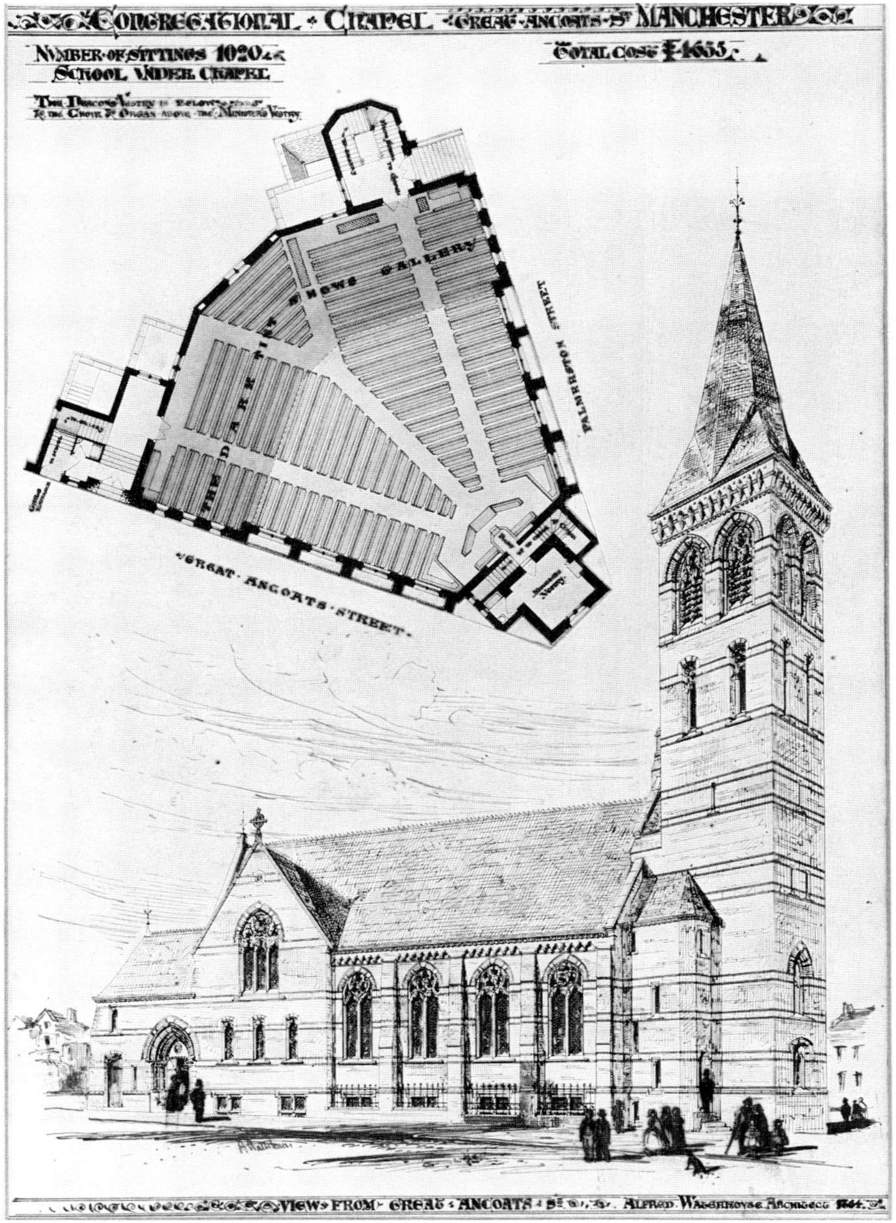

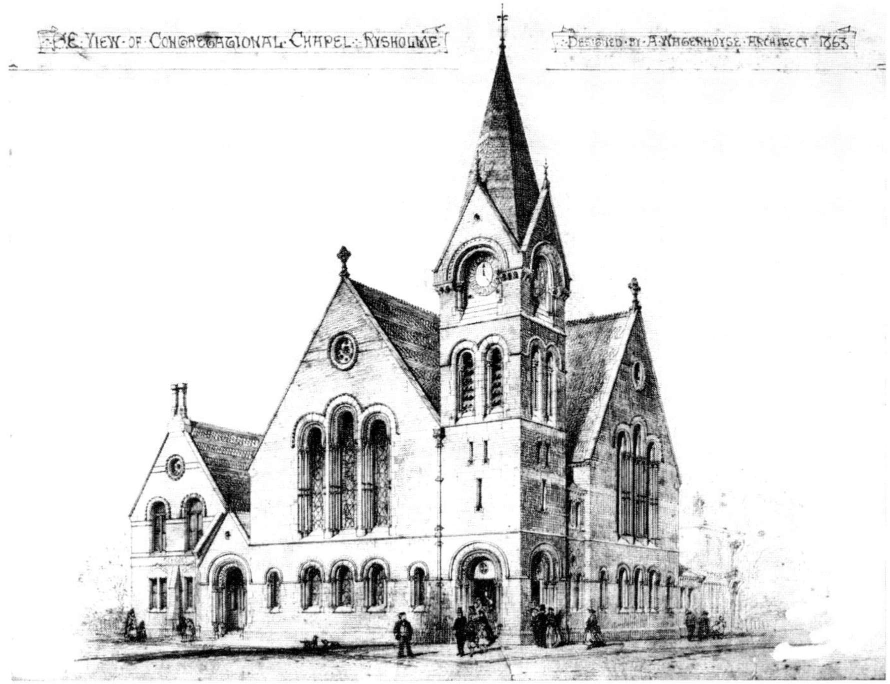

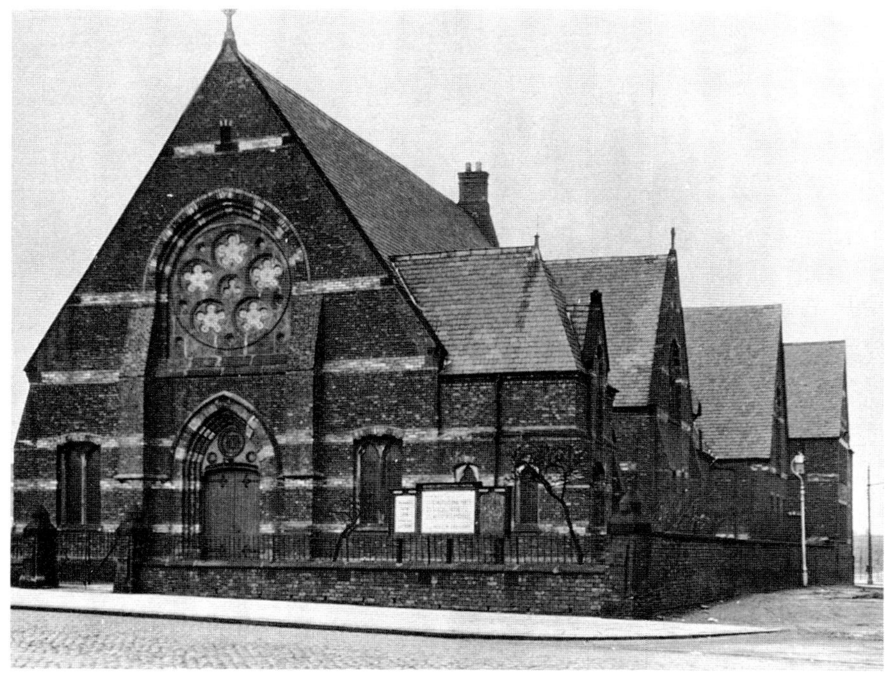

29. Above, Manchester, Ancoats Congregational Chapel, ink perspective dated 1864 **[105]**

30. Above right, Manchester, Rusholme Congregational Chapel, ink perspective dated 1863 **[164]**

31. Right, Manchester, Besses o' th' Barn Congregational Chapel (1863), *c.*1900 **[163]**

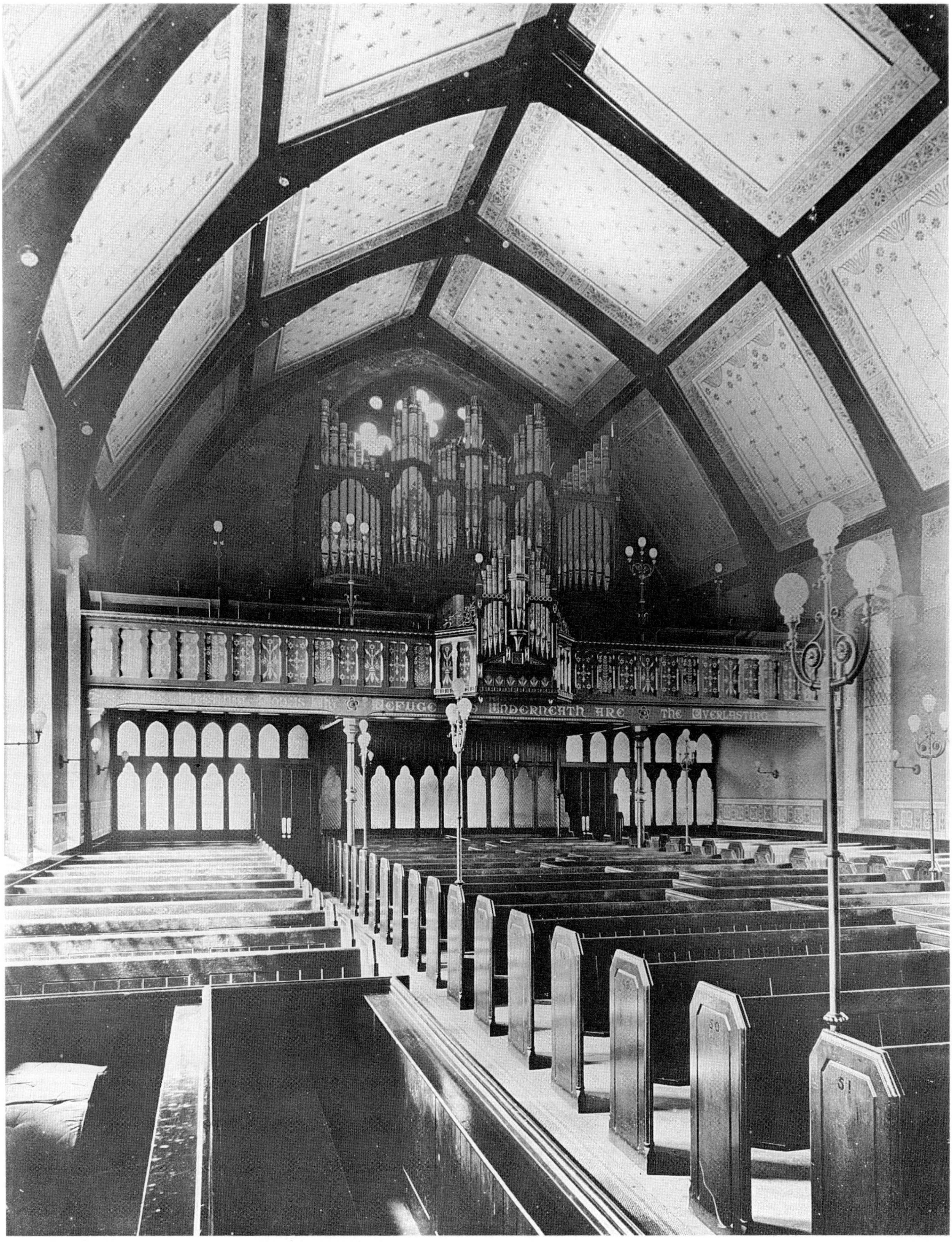

32. Manchester, Besses o' th' Barn Congregational Chapel (1863), interior, *c.*1900 **[163]**

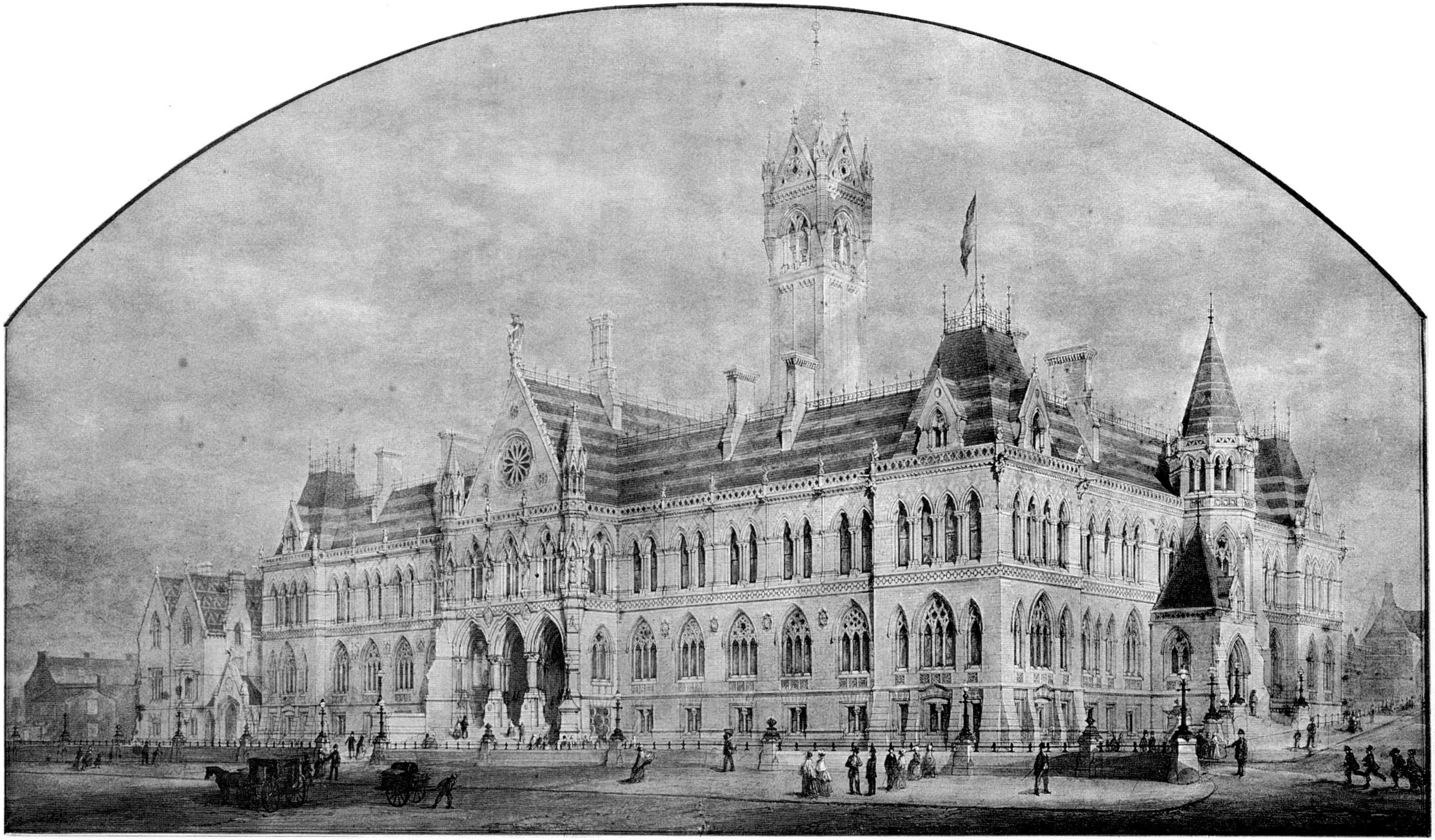

33. Manchester Assize Courts, perspective by J. Battye, *c.*1865 **[67]**

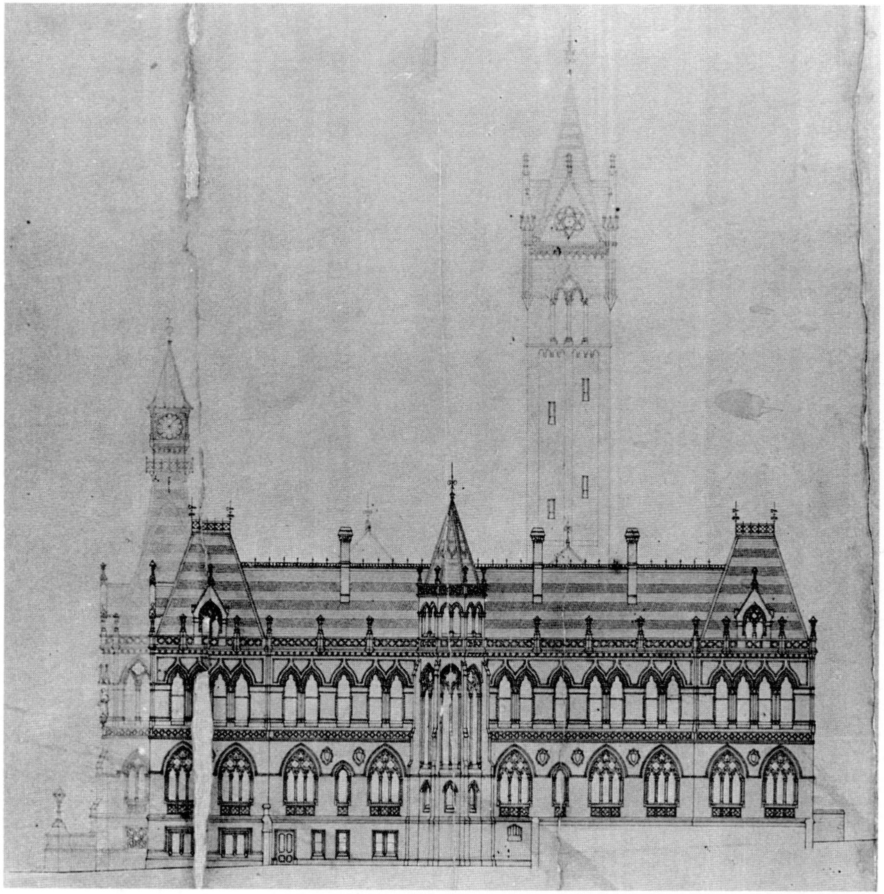

34. Manchester Assize Courts, side elevation, competition drawing, 1859 **[67]**

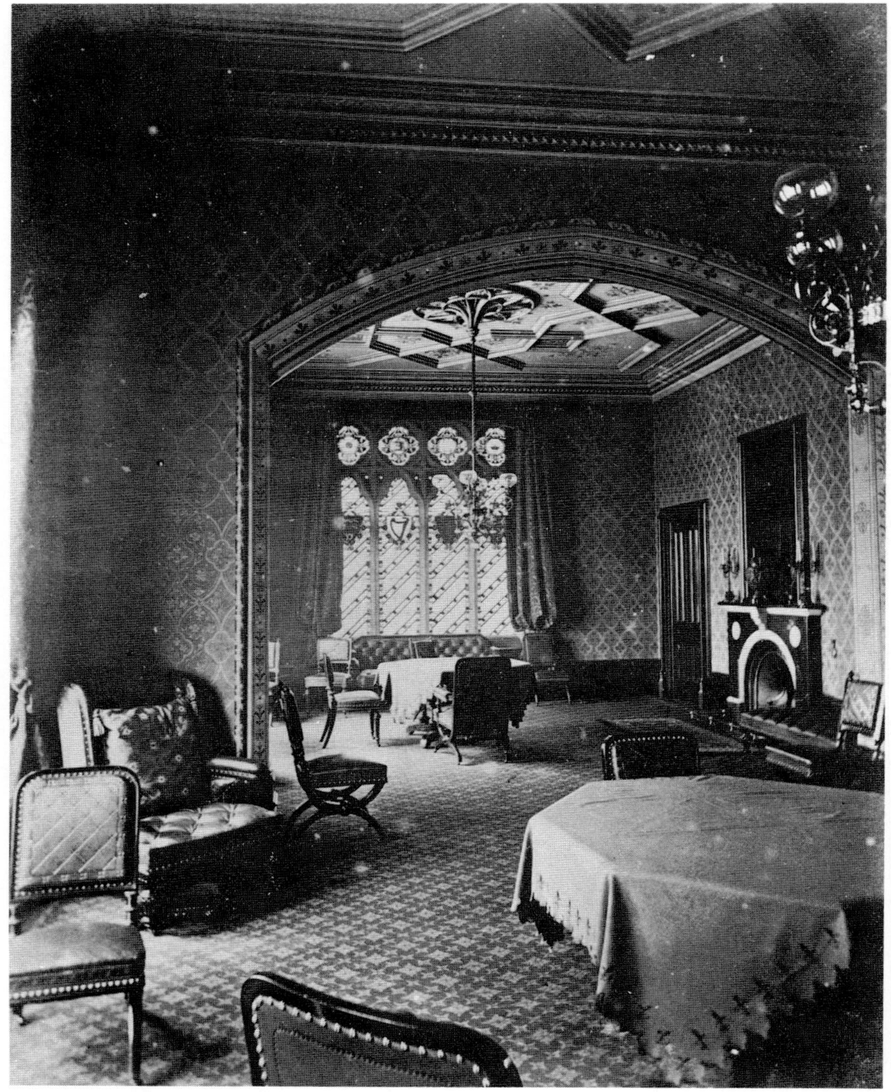

35. Manchester Assize Courts, sitting-room in Judges' Lodgings, *c.*1865 **[67]**

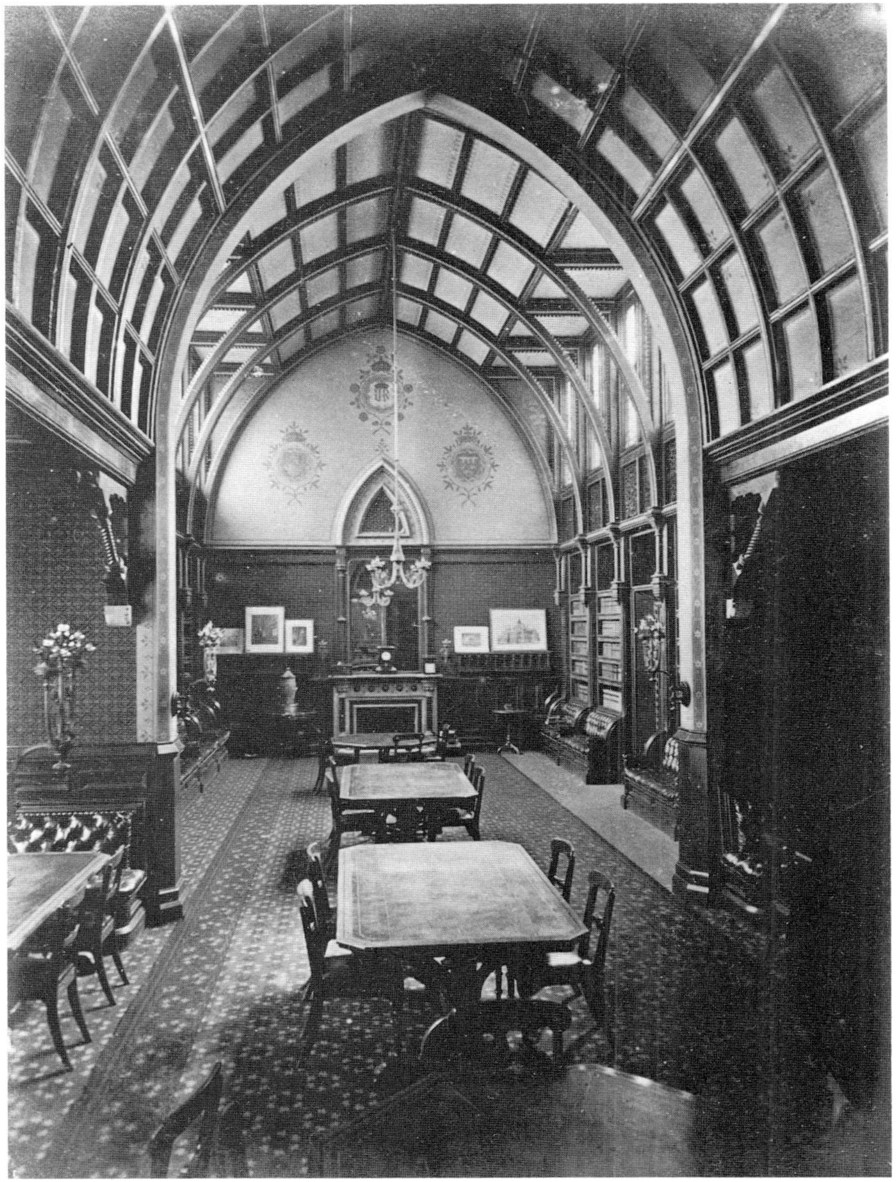

36. Manchester Assize Courts, Barristers' Library, *c.*1865 **[67]**

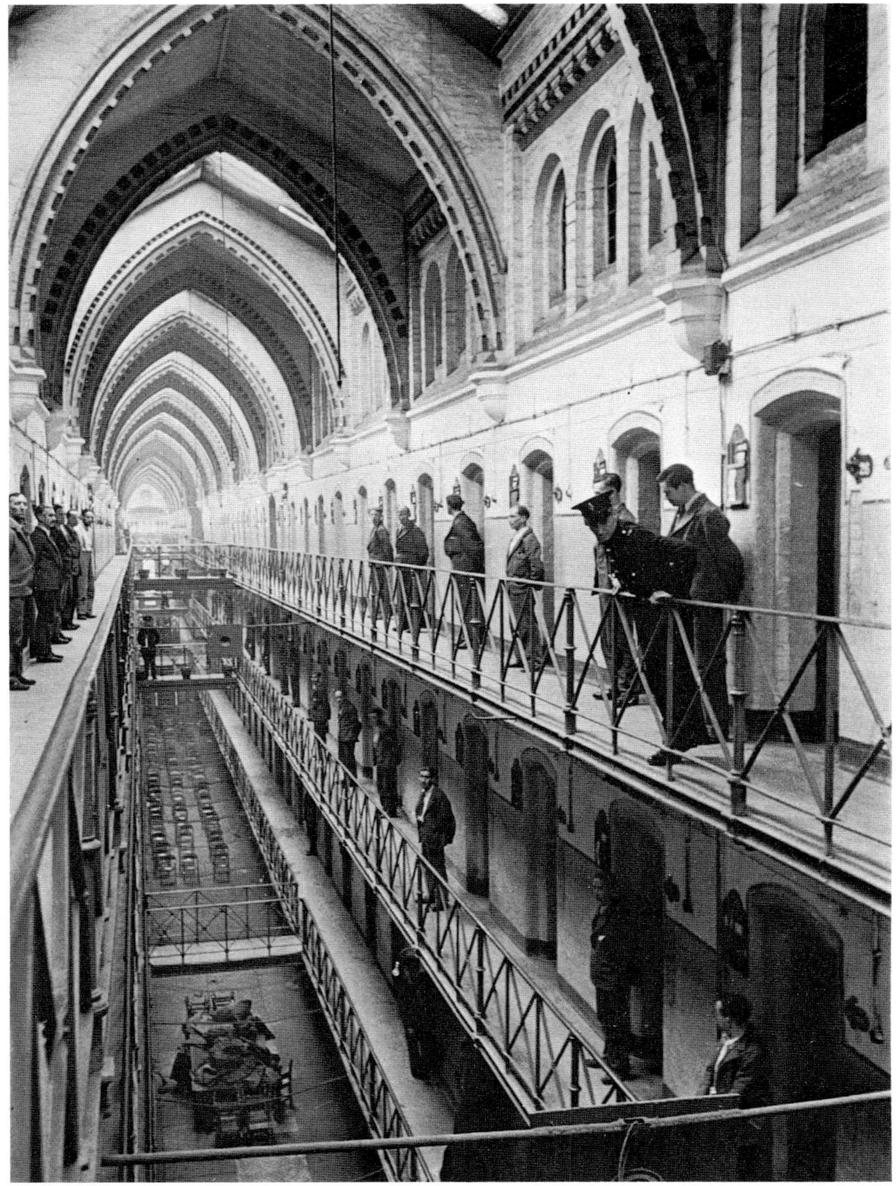

37. Manchester, Strangeways Gaol (1861), interior, *c.*1940 **[113]**

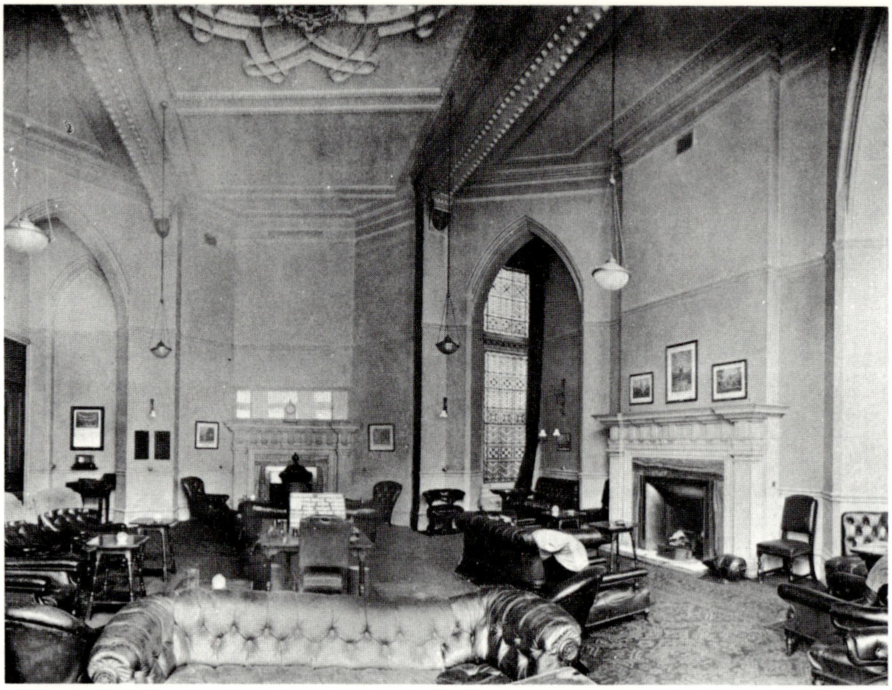

38. London, New University Club (1865), Morning Room, *c.*1920 **[208]**

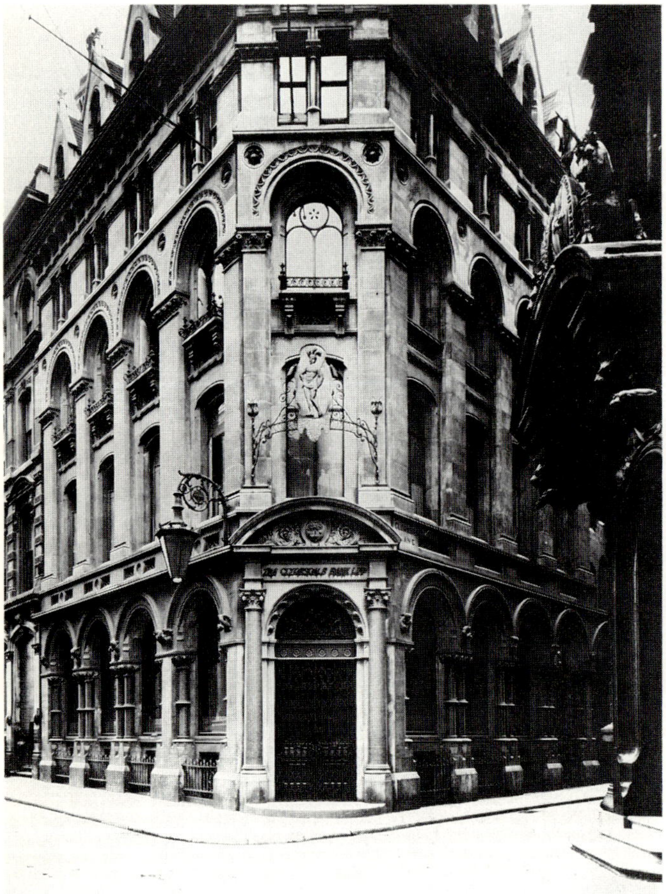

39. London, Alexander and Cunliffe's Bank (1864), *c.*1900 **[179]**

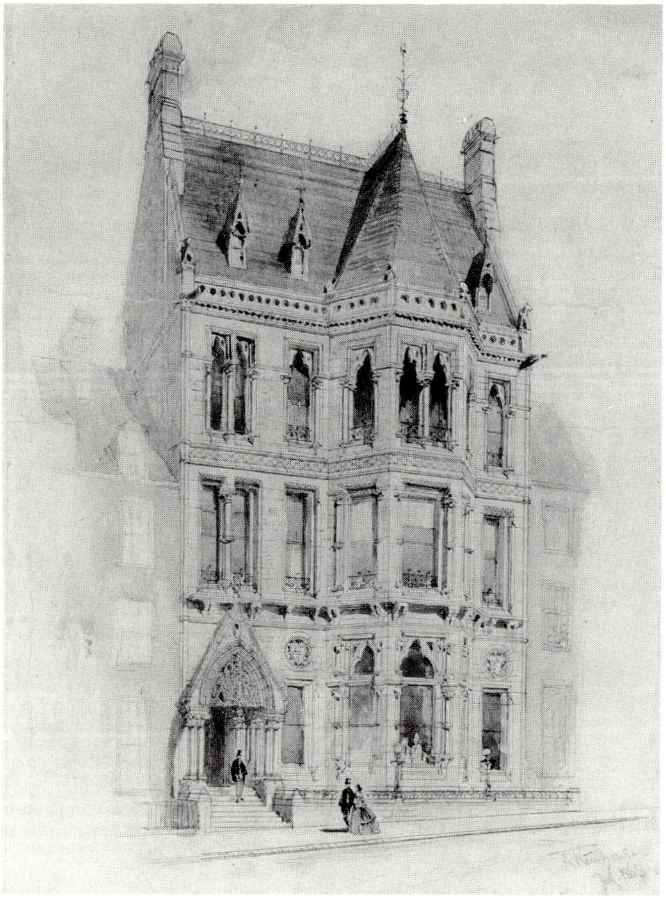

40. London, New University Club, sketch perspective dated 1865 **[208]**

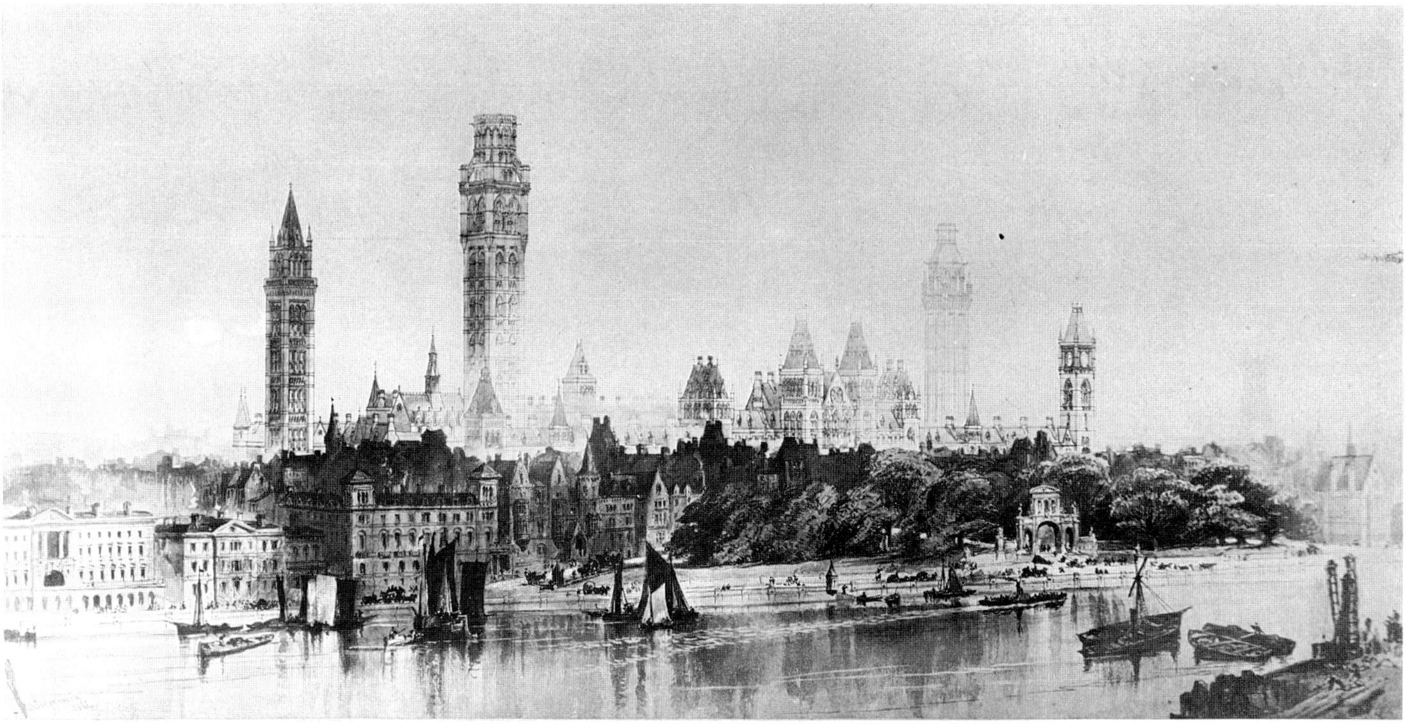

41. London, Law Courts Competition, perspective from the south bank of the River Thames, 1867 **[278]**

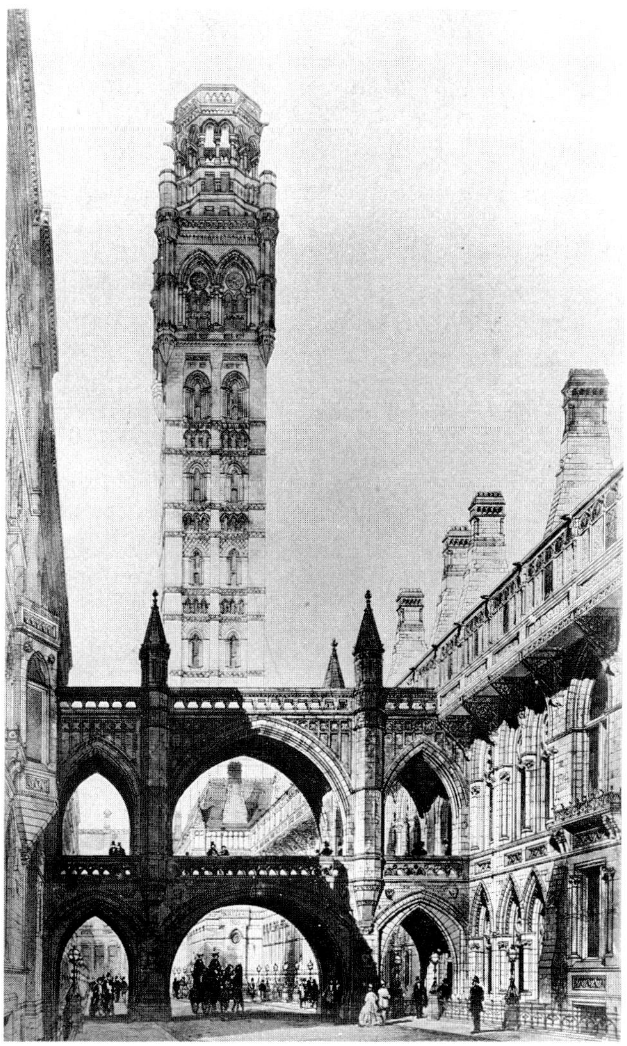

42. London, Law Courts Competition, perspective of interior street, 1867 **[278]**

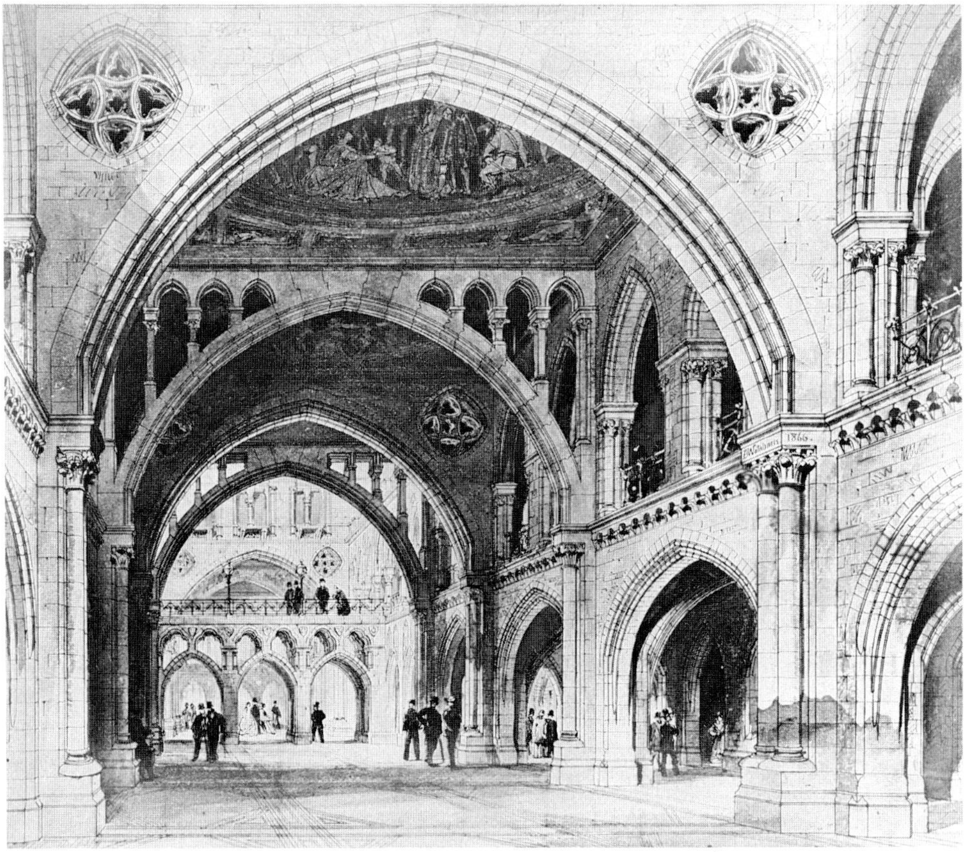

43. London, Law Courts Competition, perspective of transverse hall, 1867 **[278]**

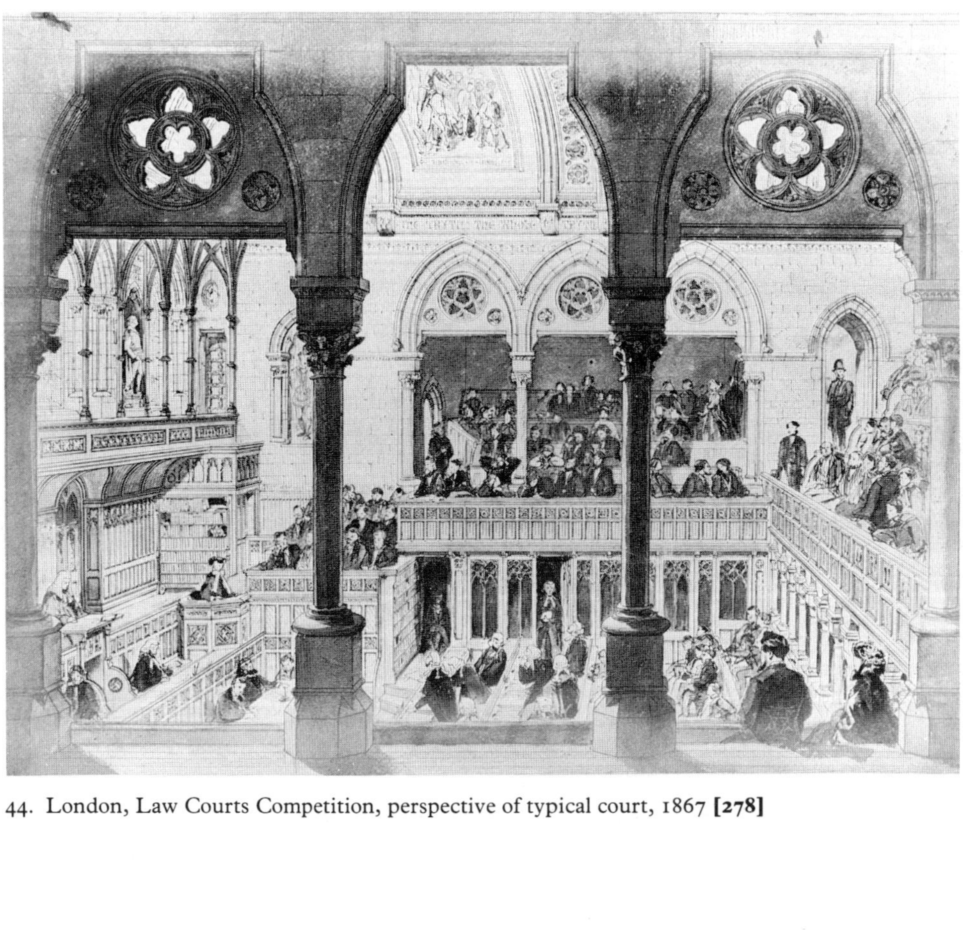

44. London, Law Courts Competition, perspective of typical court, 1867 **[278]**

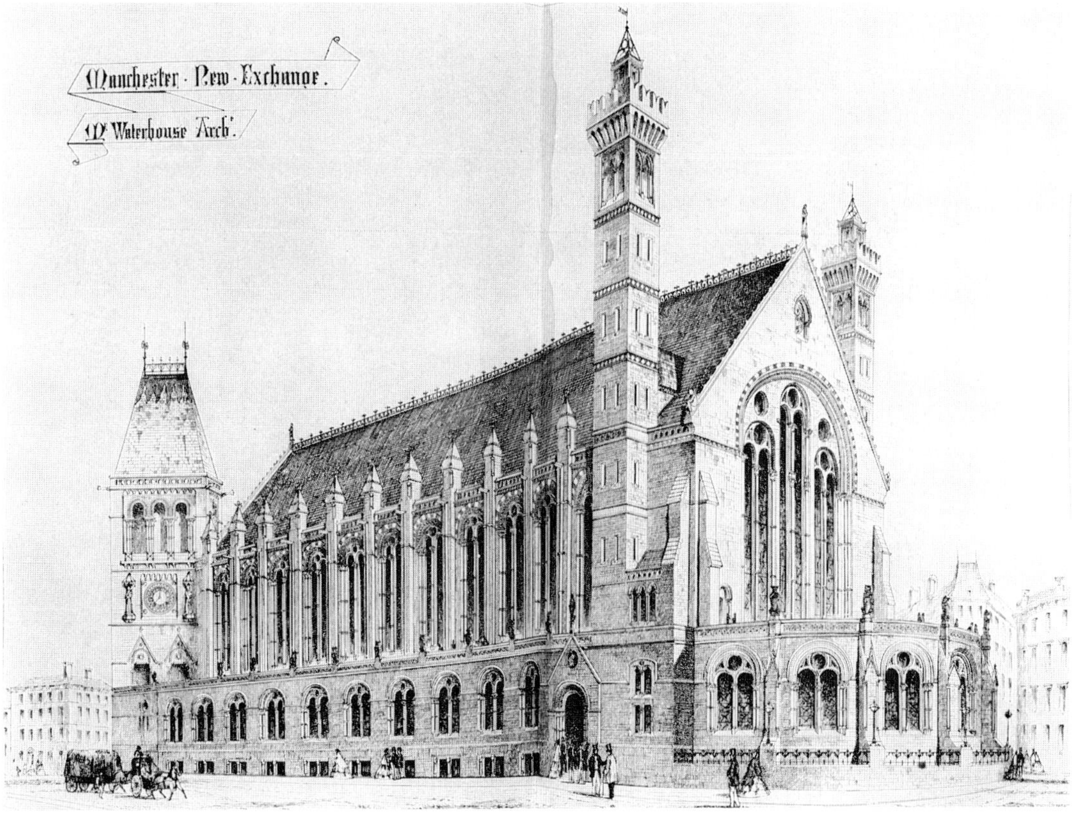

45. Manchester, design for New Exchange, from *Building News* (1865) **[215]**

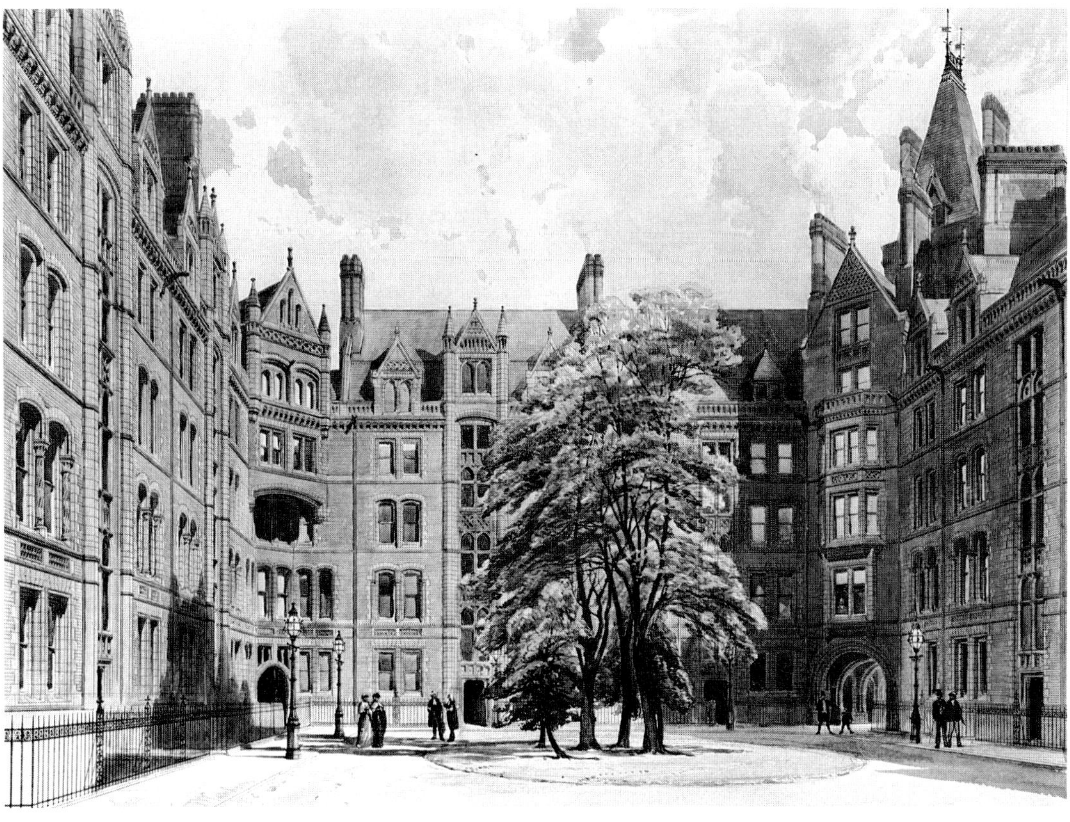

46. London, New Court Carey Street, perspective dated 1879 **[370, 470]**

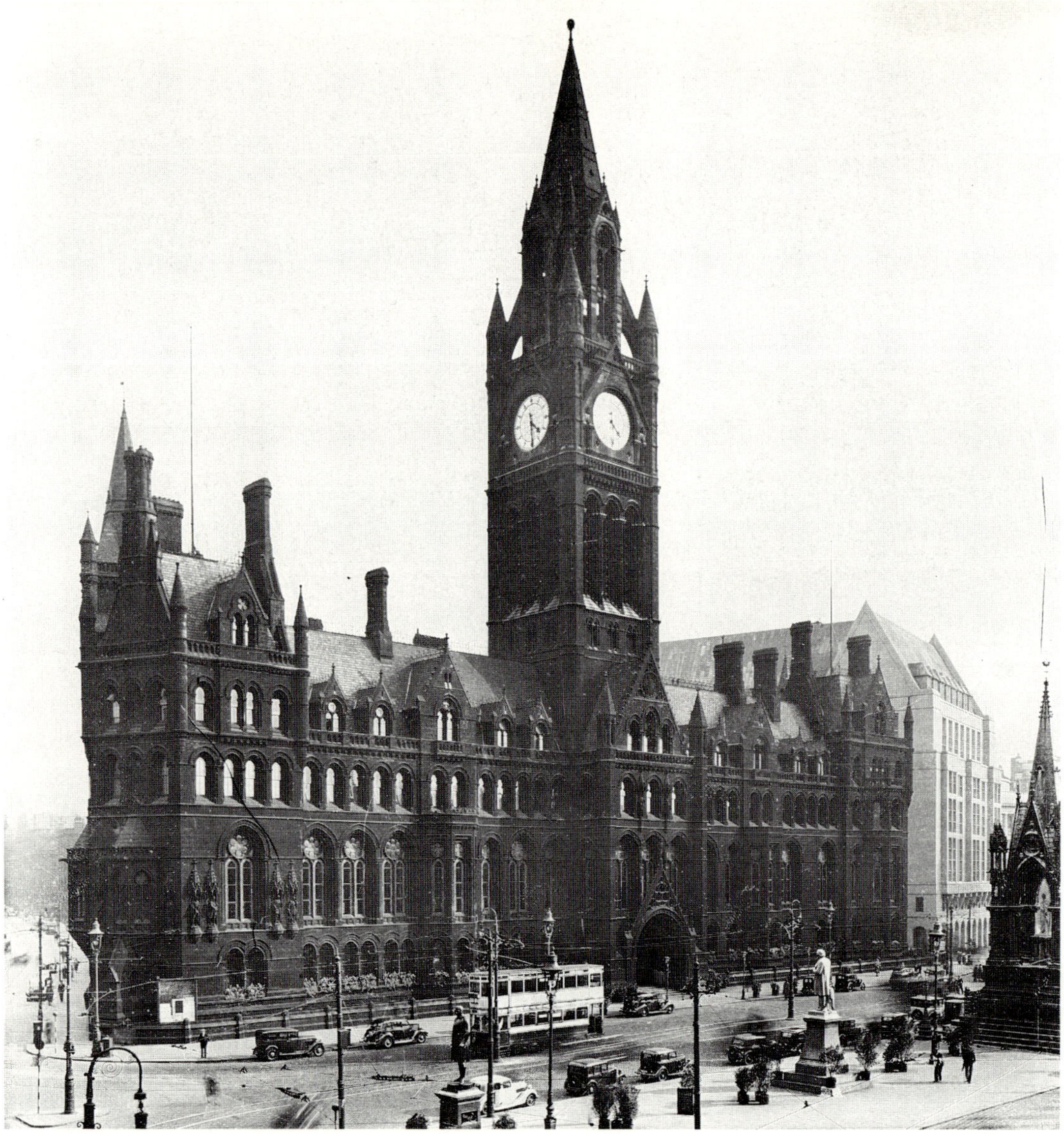

47. Manchester Town Hall (1867), from Albert Square, *c.*1945? **[284]**

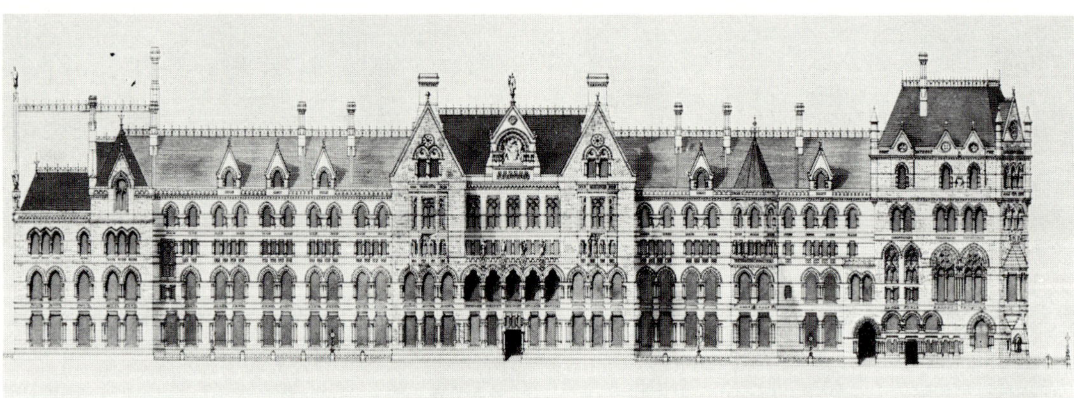

48. Manchester Town Hall, Princess Street elevation, first competition, 1867 **[284]**

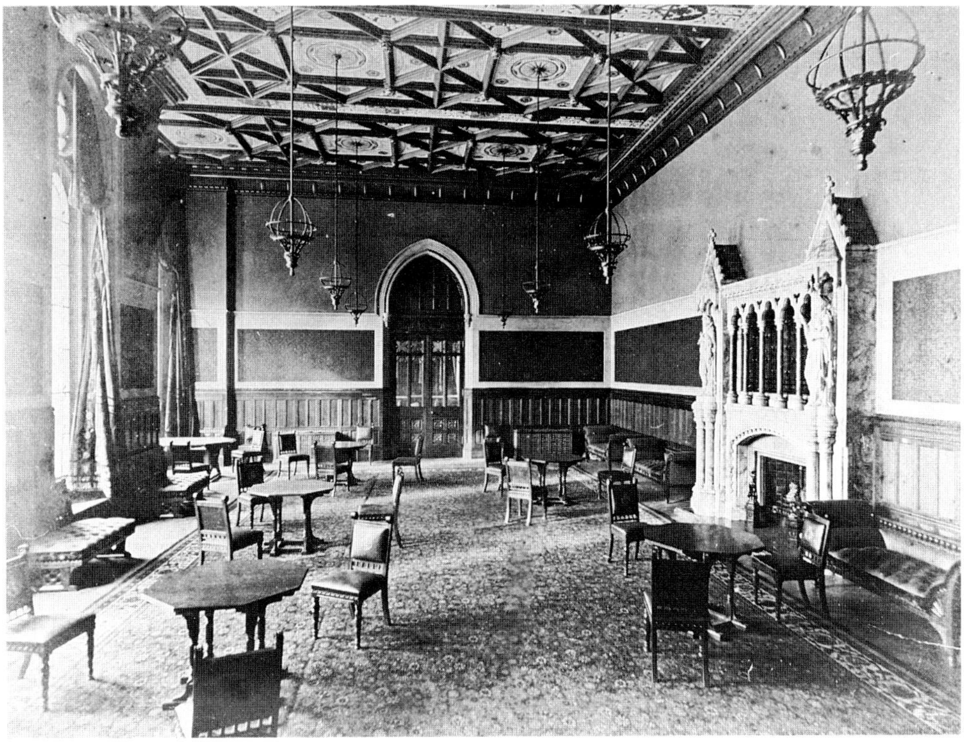

49. Manchester Town Hall (1867), reception room, 1877 **[284]**

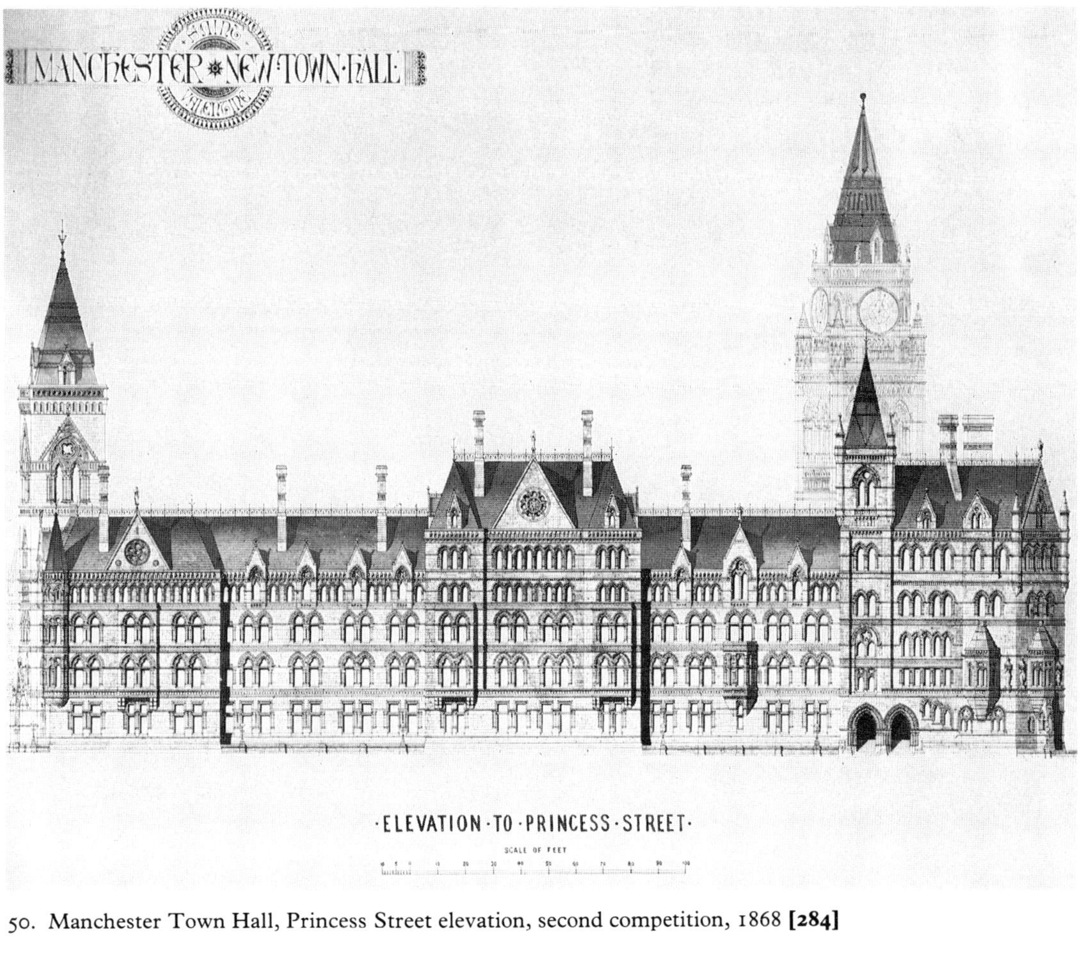

50. Manchester Town Hall, Princess Street elevation, second competition, 1868 **[284]**

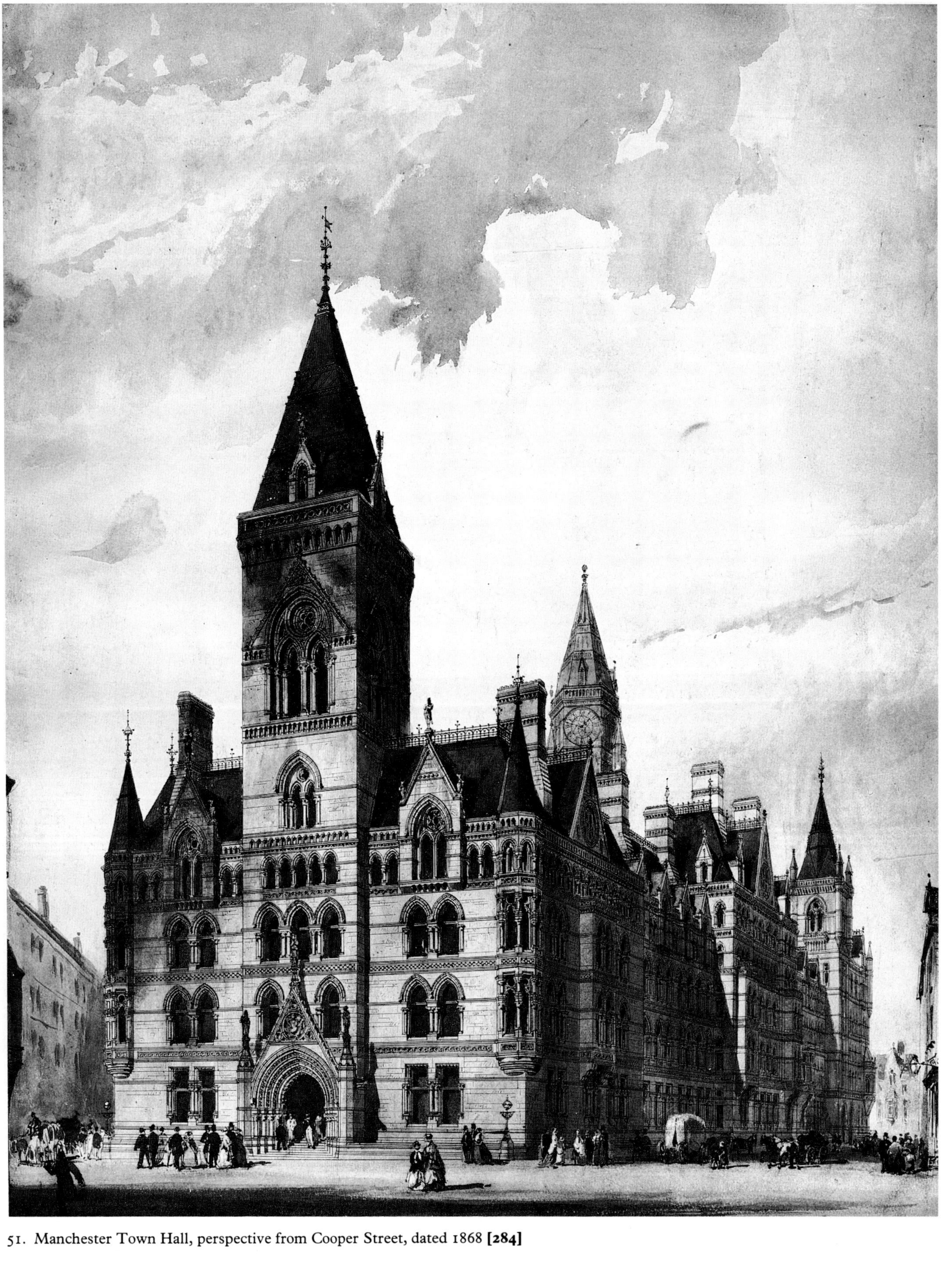

51. Manchester Town Hall, perspective from Cooper Street, dated 1868 **[284]**

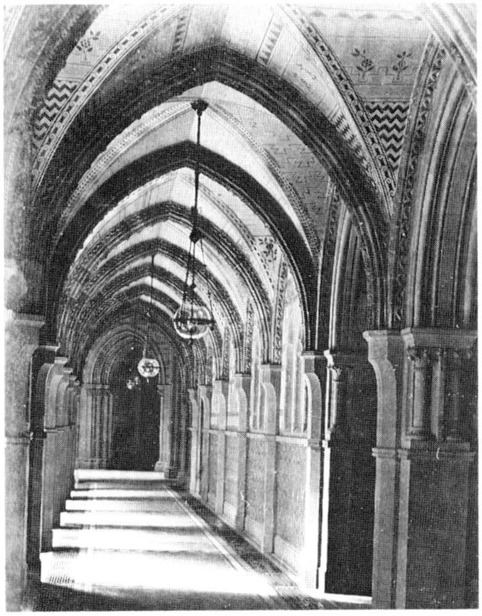

52. Manchester Town Hall (1867), ground-floor corridor, 1877 **[284]**

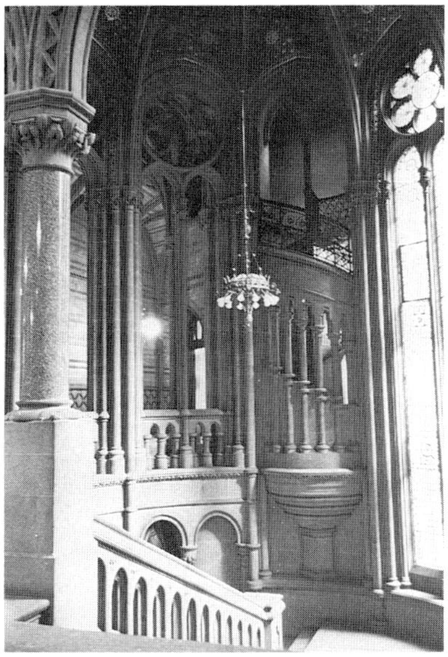

53. Manchester Town Hall (1867), grand staircase **[284]**

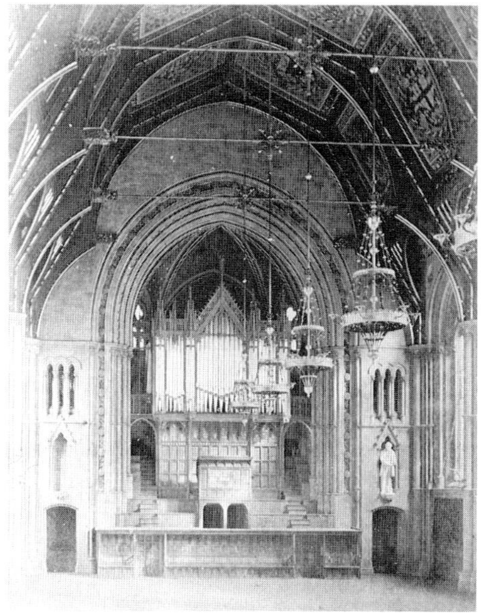

54. Manchester Town Hall (1867), Great Hall, 1877 **[284]**

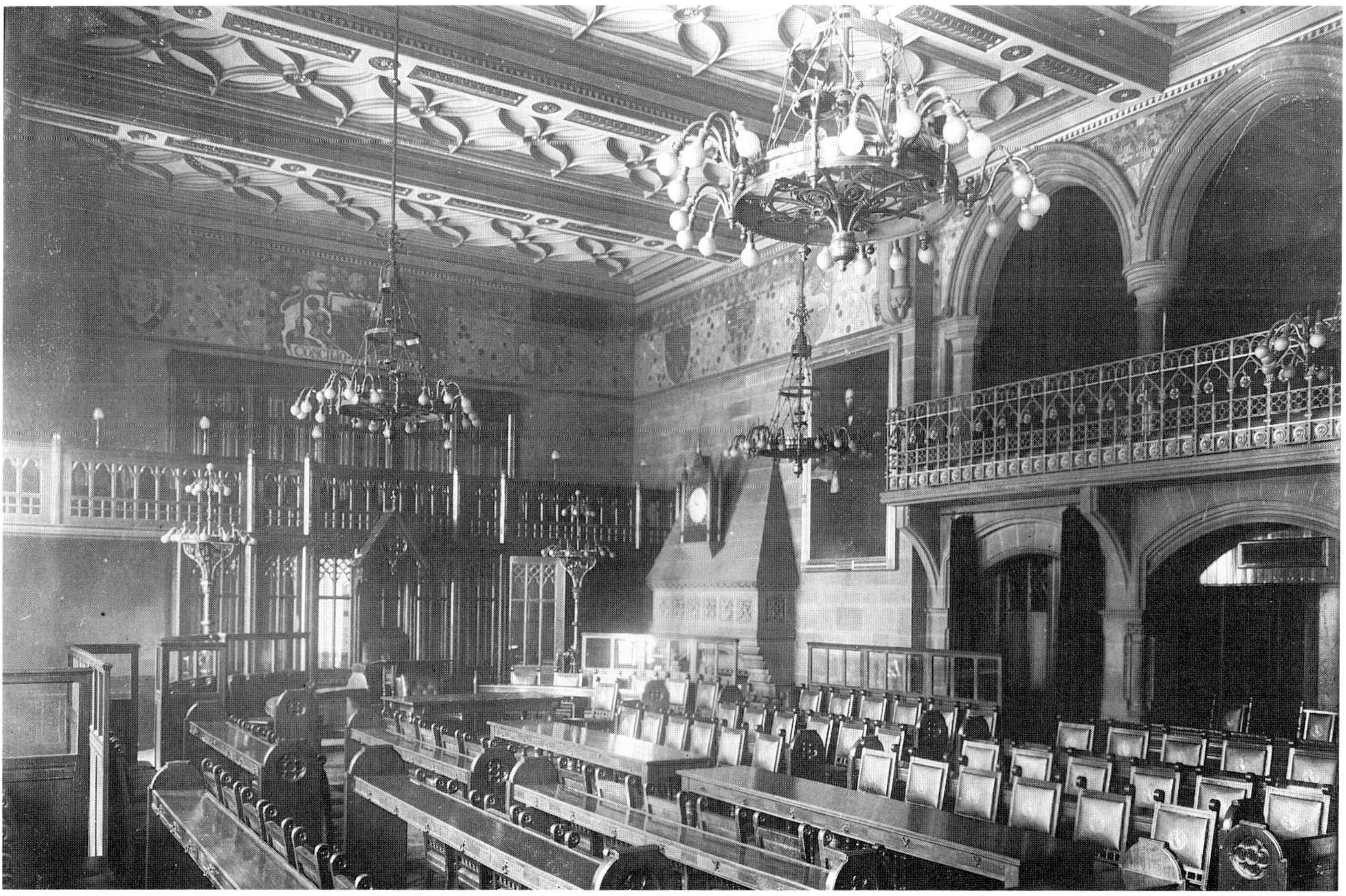

55. Manchester Town Hall (1867), Council Chamber, *c.*1900 **[284]**

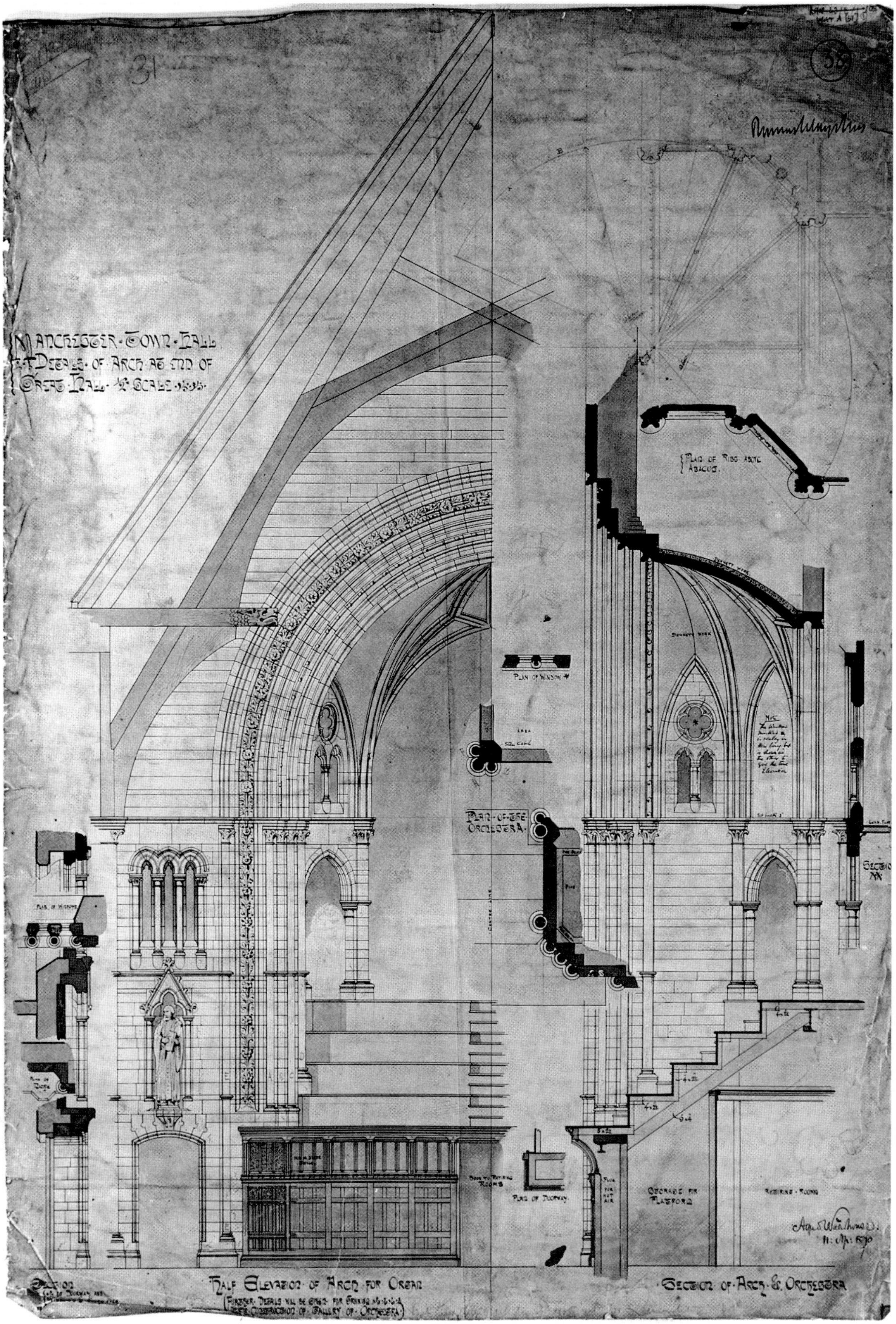

56. Manchester Town Hall, details of arch at rear of Great Hall, dated 1870 **[284]**. This recess was subsequently enlarged and filled with the town hall organ

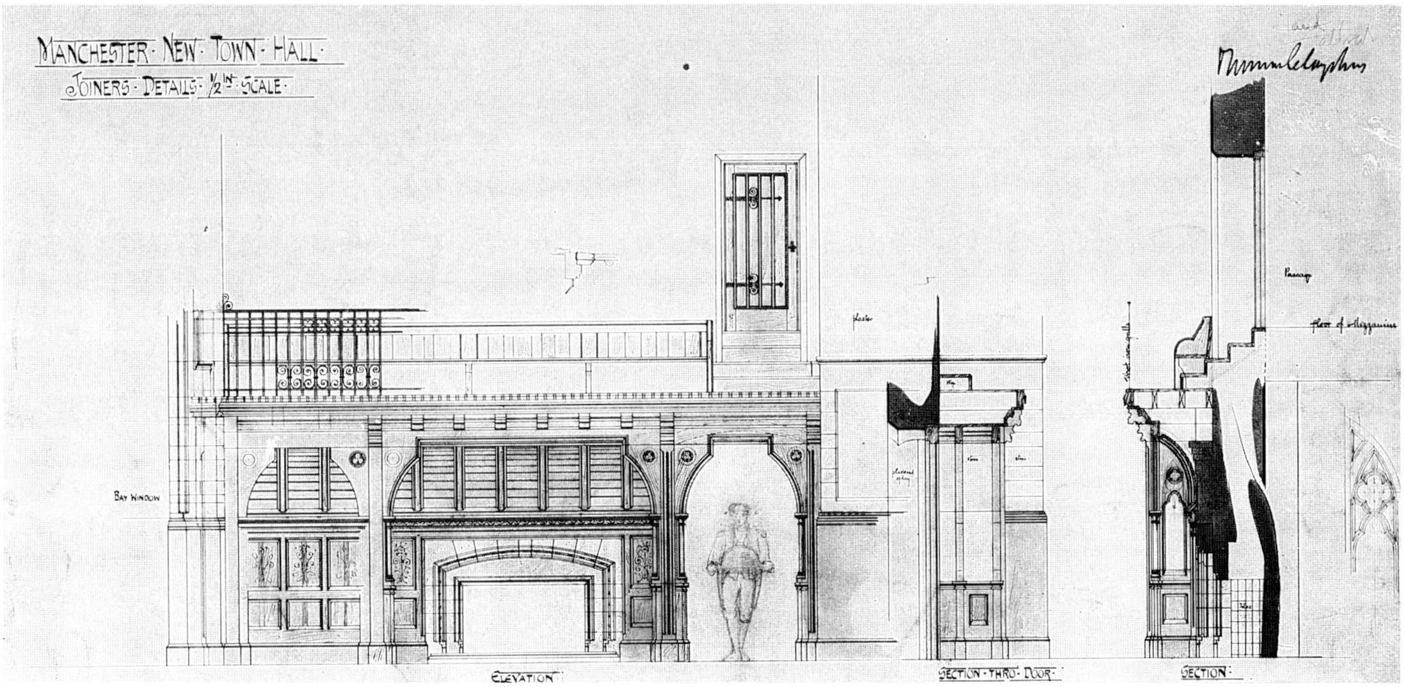

57. Manchester Town Hall, joiners' details of fireplace, minstrels' gallery, and serving entry in Banqueting Hall, *c.*1875 **[284]**

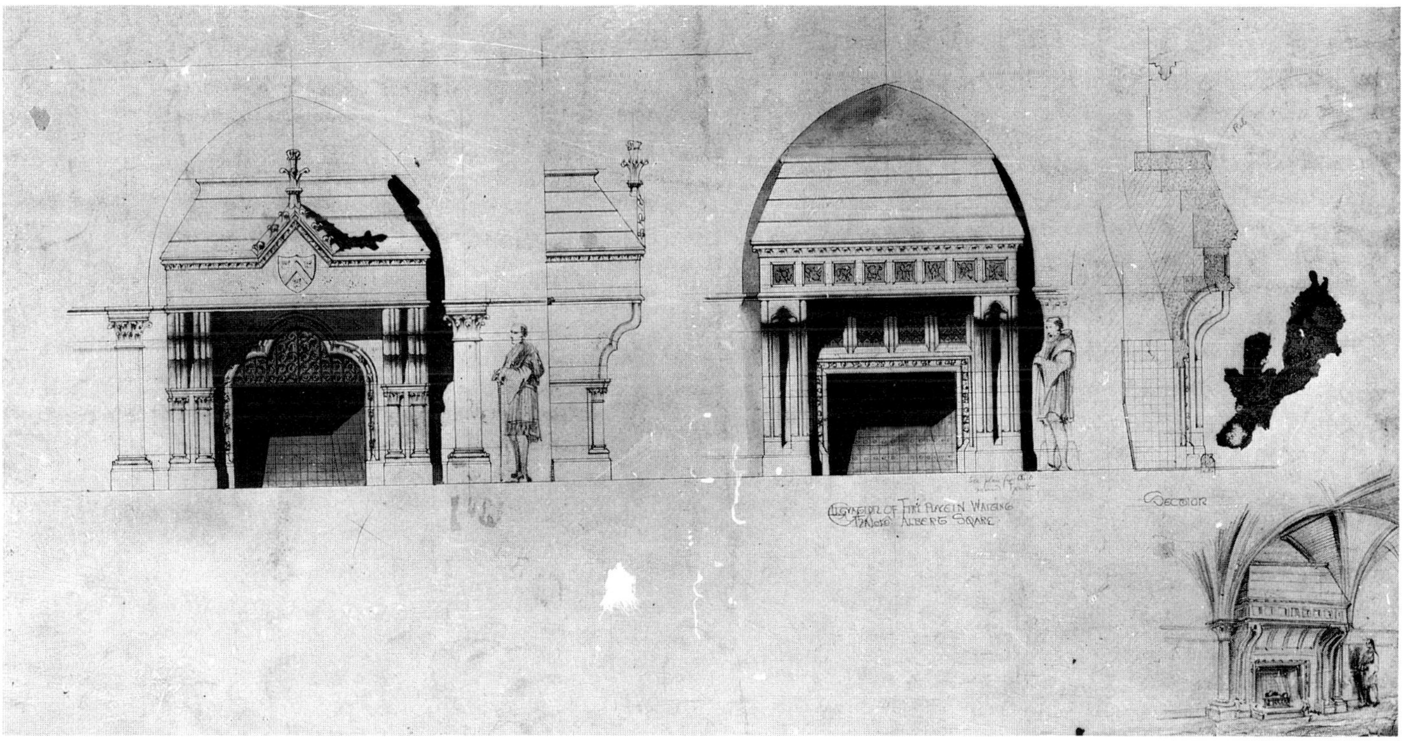

58. Manchester Town Hall, two designs for stone fireplaces, *c.*1869 **[284]**

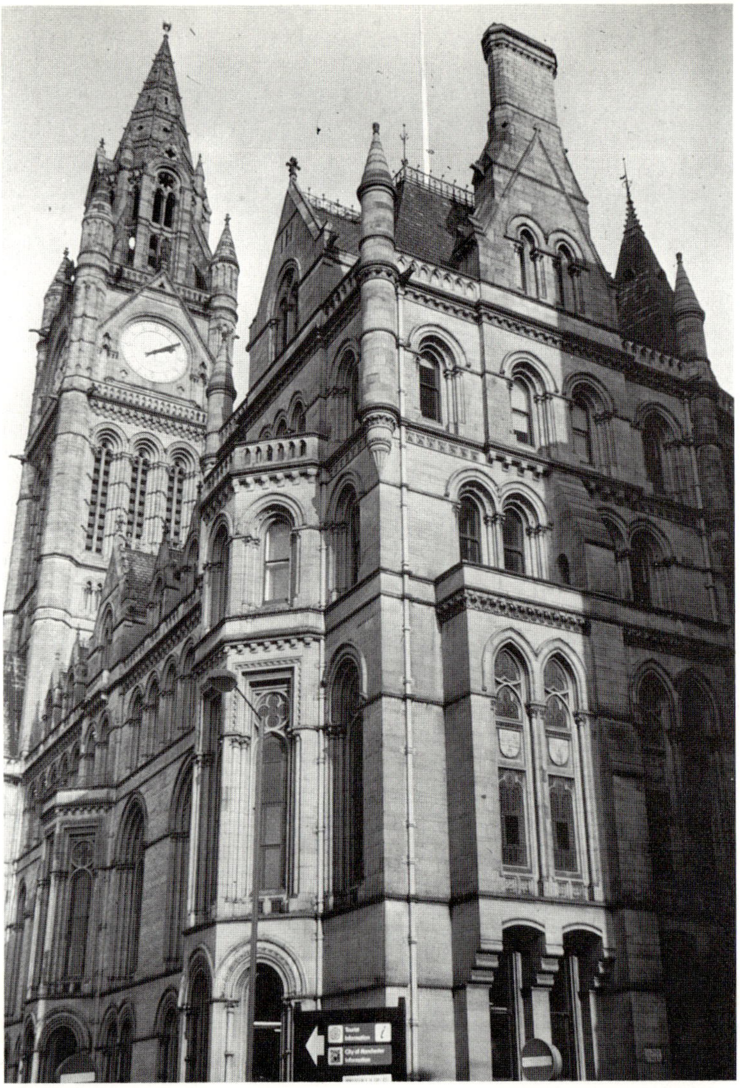

59. Manchester Town Hall (1867), corner of Albert Square and Lloyd Street **[284]**

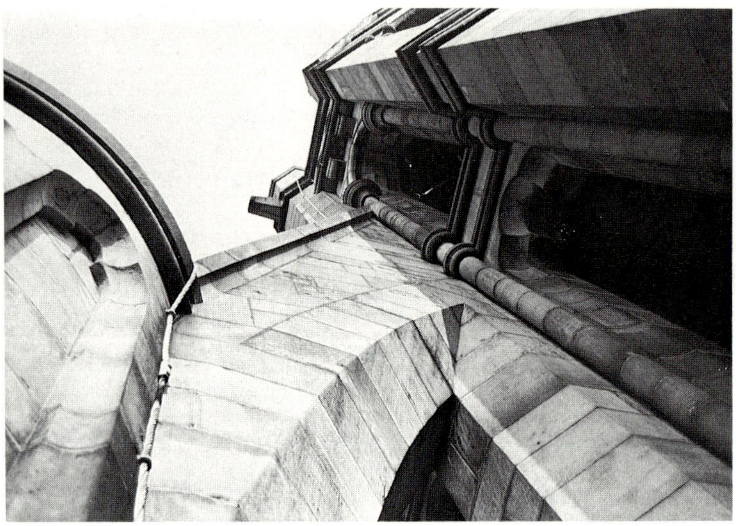

60. Manchester Town Hall (1867), detail of stonework of tower **[284]**

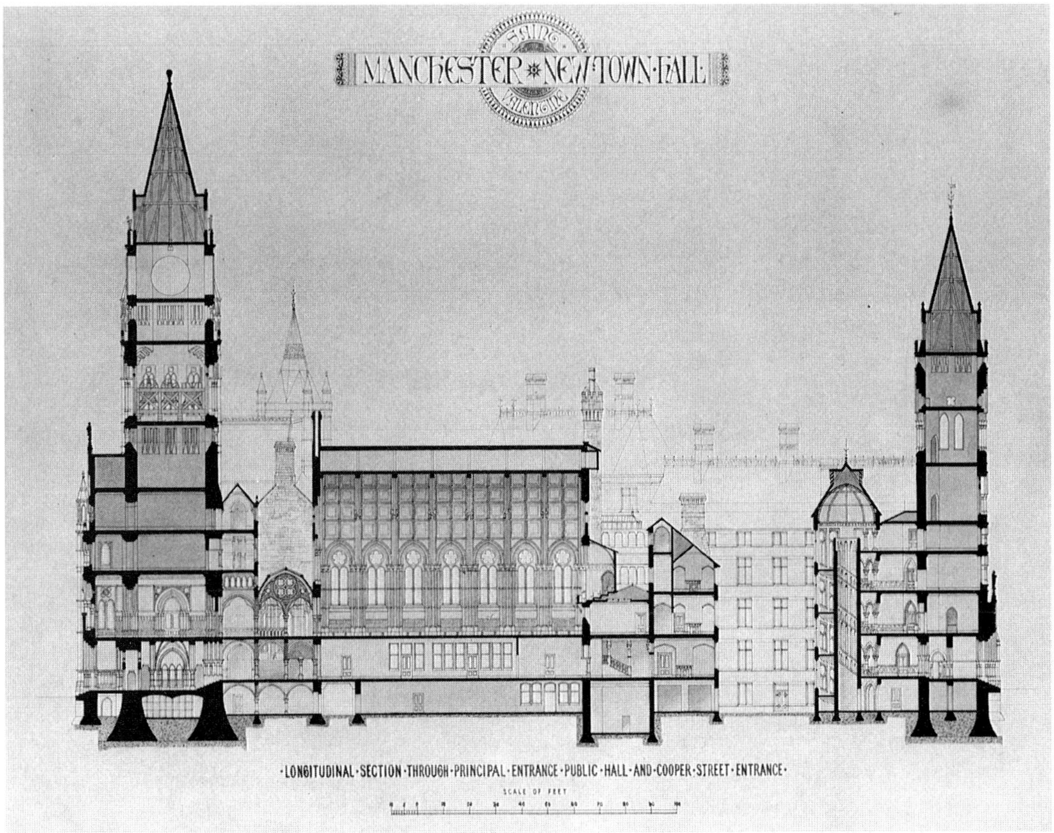

61. Manchester Town Hall, longitudinal section, competition drawing, second competition, 1868 **[284]**

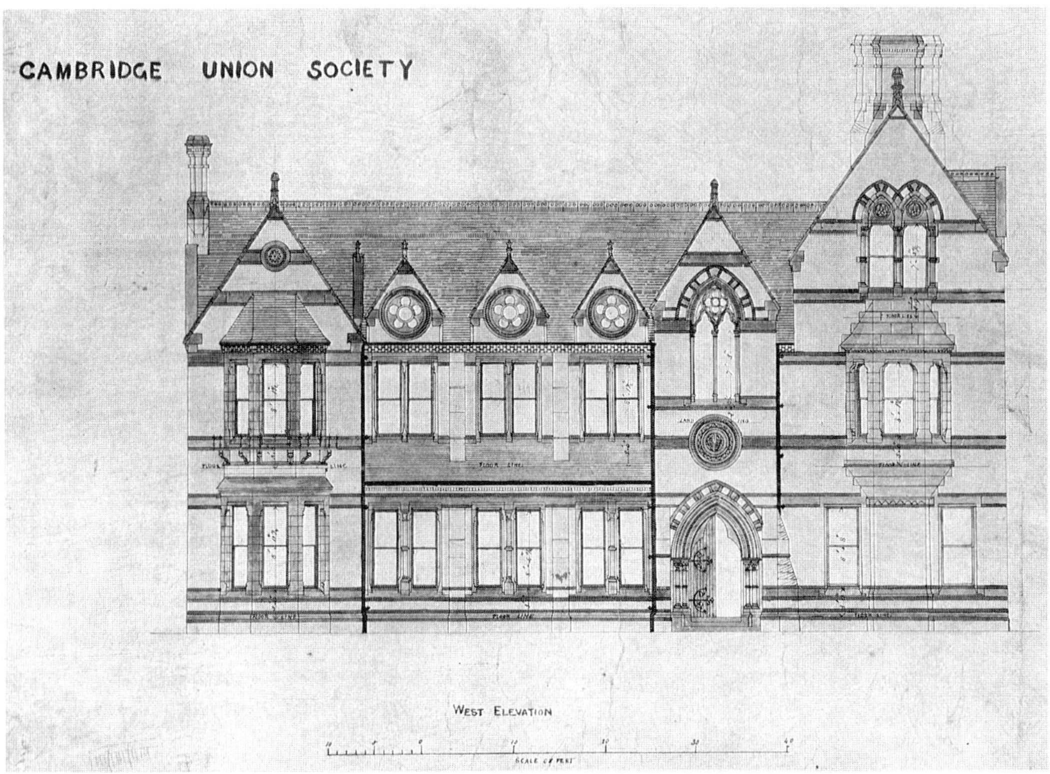

62. Cambridge Union Society, front elevation, *c.*1865 **[197]**

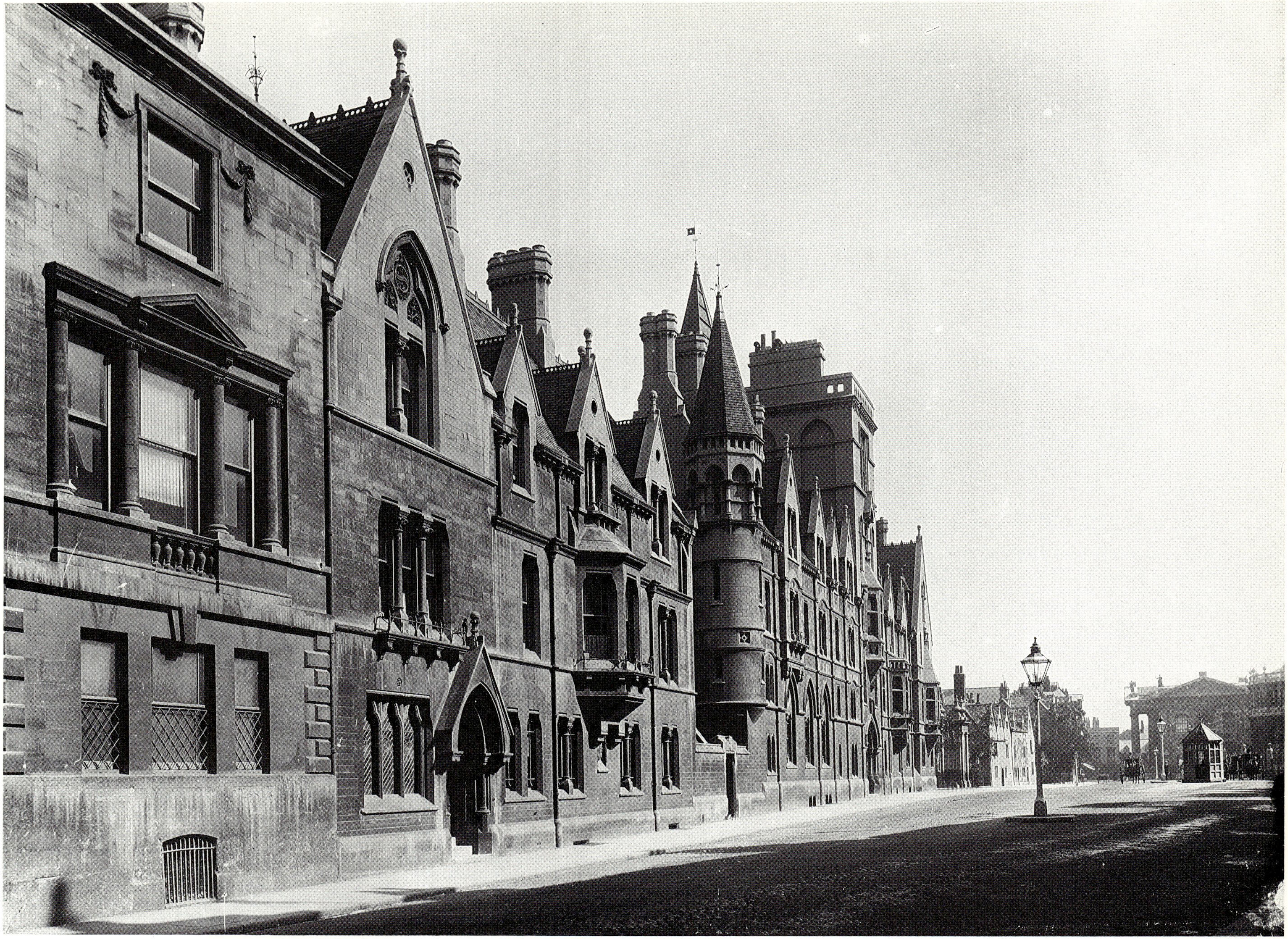

63. Oxford, Balliol College (1866), Broad Street front **[254]**

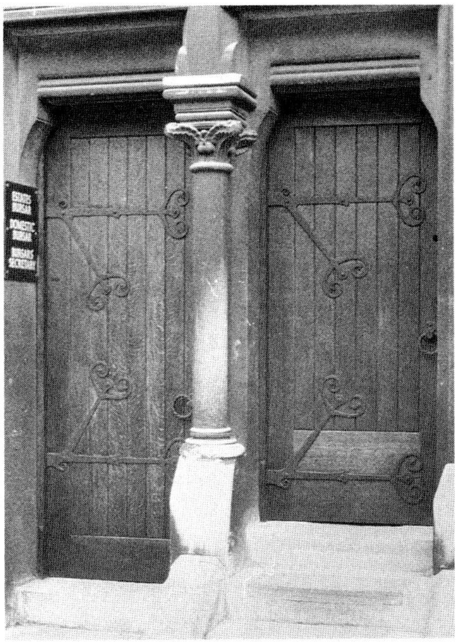

64. Oxford, Balliol College (1866), double doorway at the junction of the first two sections of the Brackenbury Buildings **[254]**

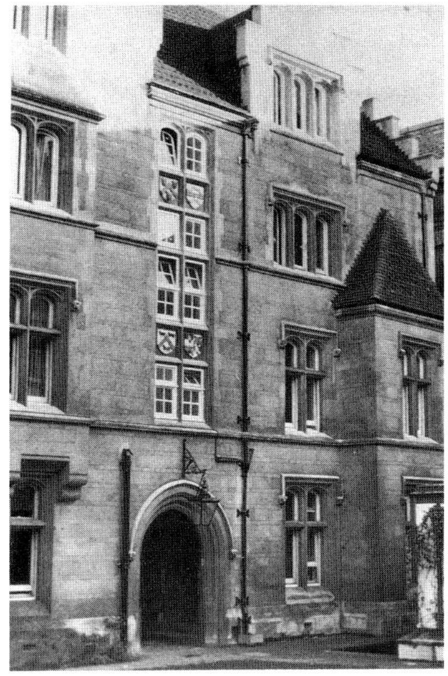

65. Cambridge, Trinity Hall, South Court (1870) **[332]**

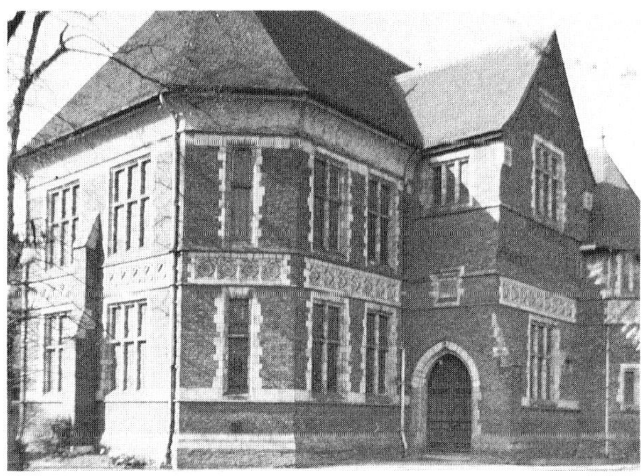

66. Oxford Union Society (1878), new debating hall **[464]**

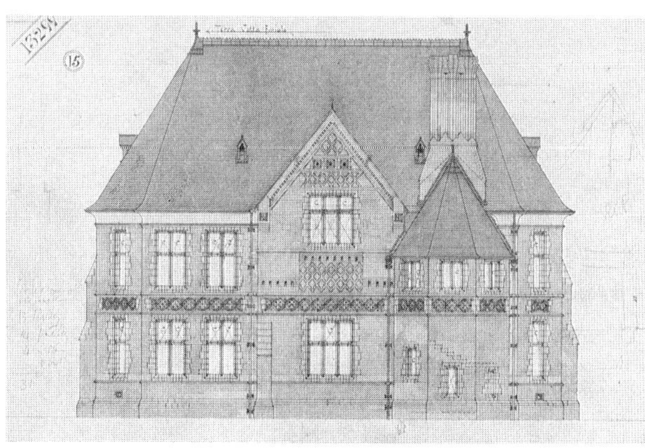

67. Oxford Union Society, contract elevation of new debating hall dated 1878 **[464]**

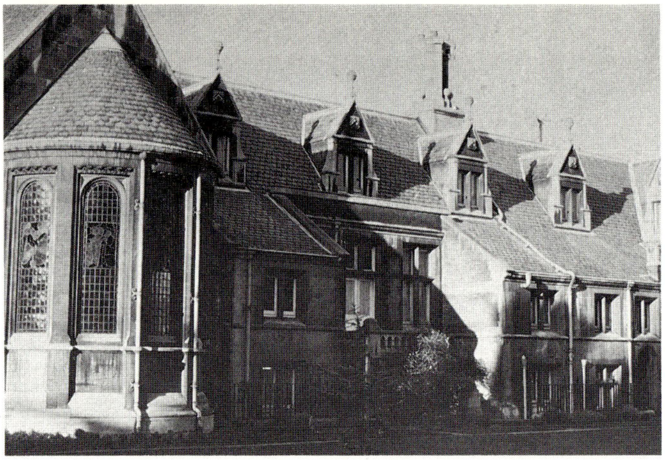

68. Cambridge, Gonville and Caius College (1867), rebuilt section of Tree Court and new apse to Chapel **[268]**

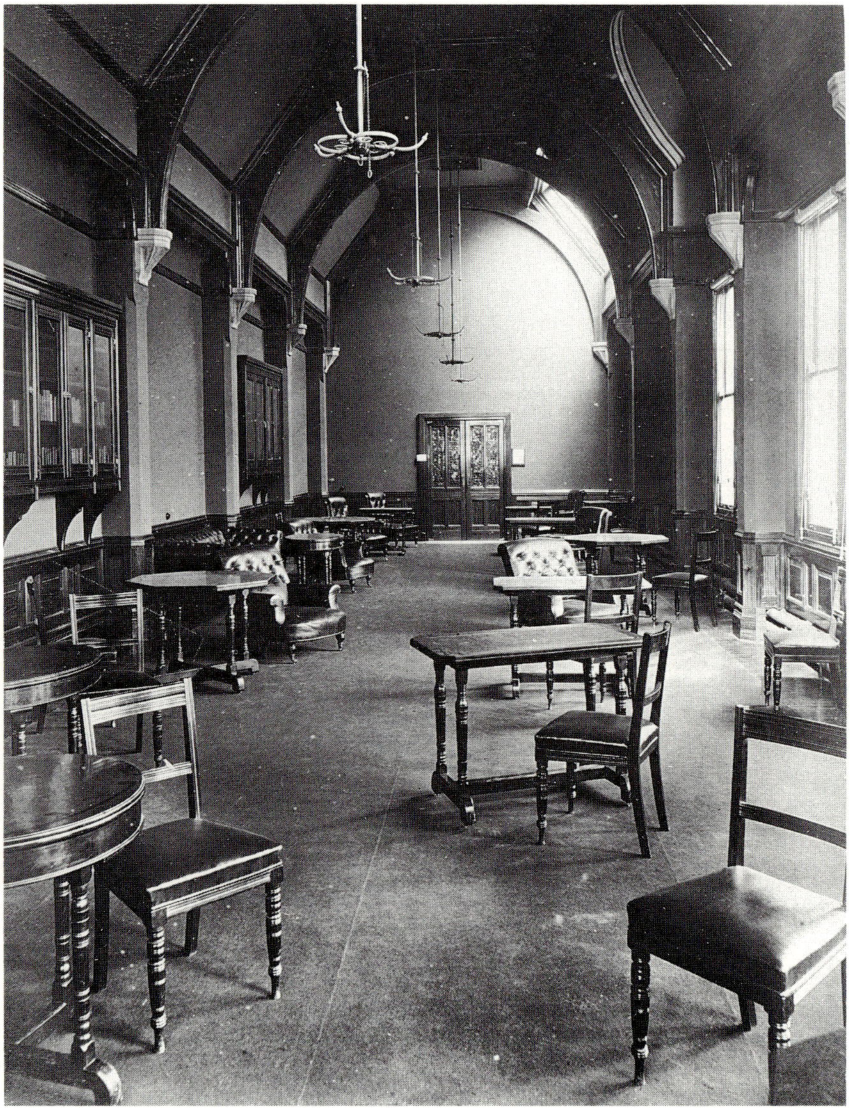

69. Cambridge Union Society, reading room in the new wing, *c.*1884 **[197]**

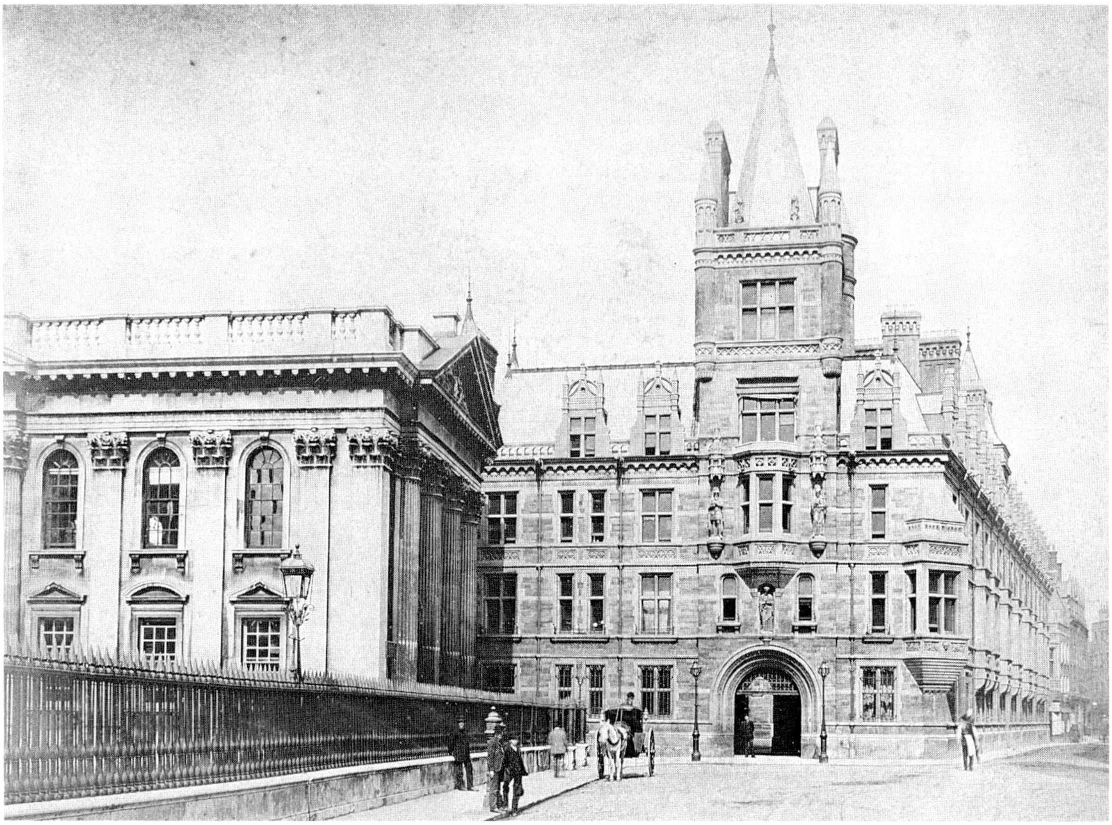

70. Cambridge, Gonville and Caius College (1867), from King's Parade, *c*.1876 **[268]**

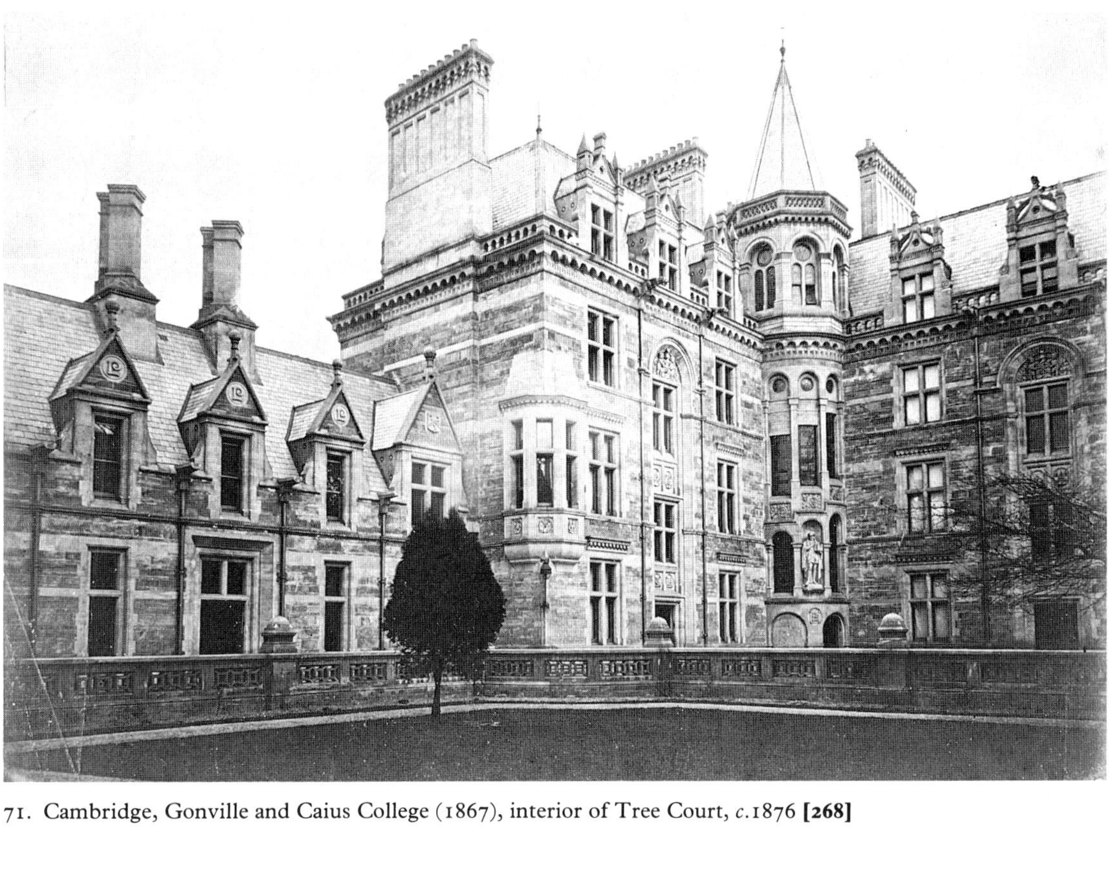

71. Cambridge, Gonville and Caius College (1867), interior of Tree Court, *c*.1876 **[268]**

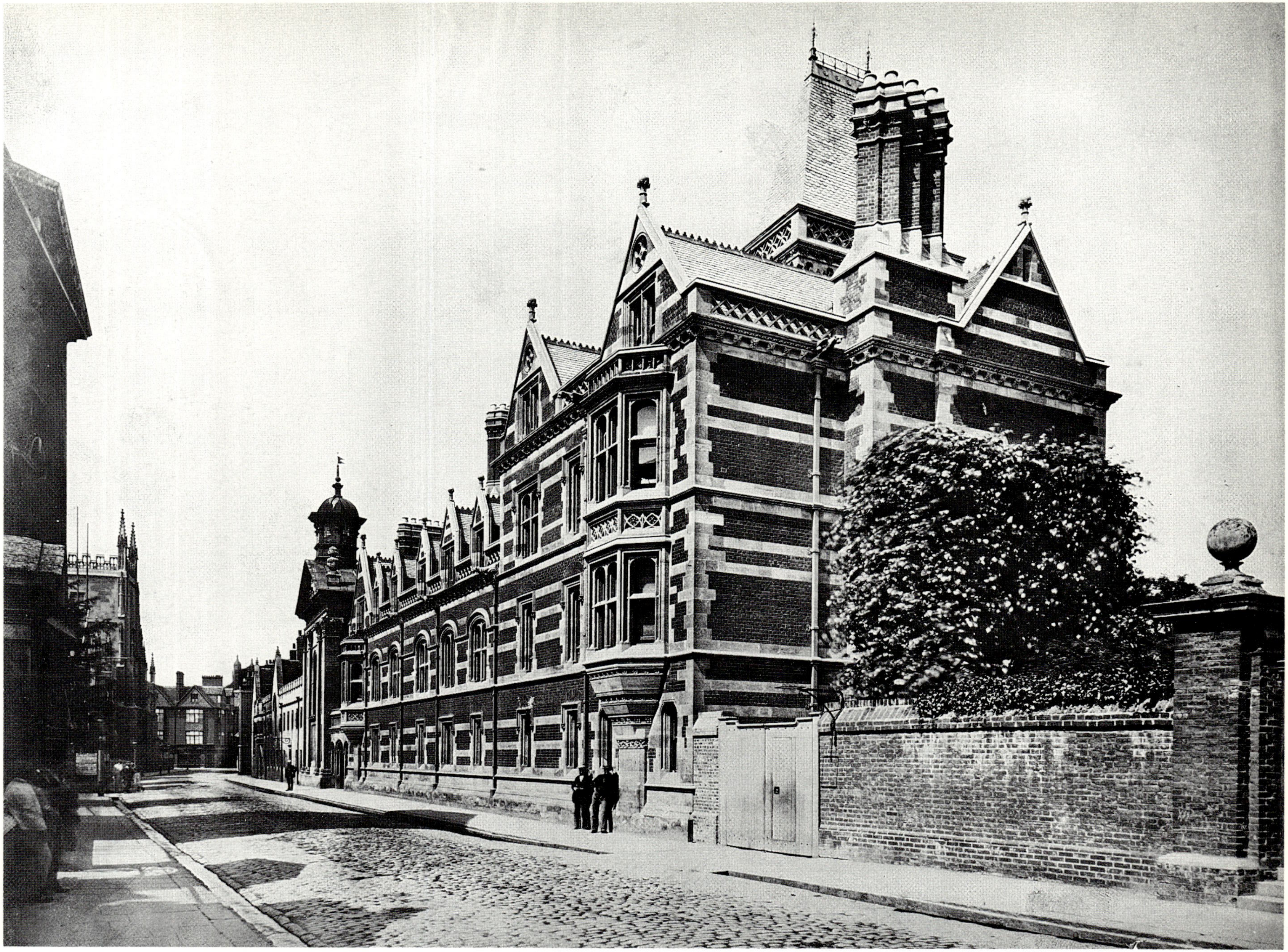

72. Cambridge, Pembroke College (1870), new sets of rooms from Trumpington Street, *c.*1900 **[333]**

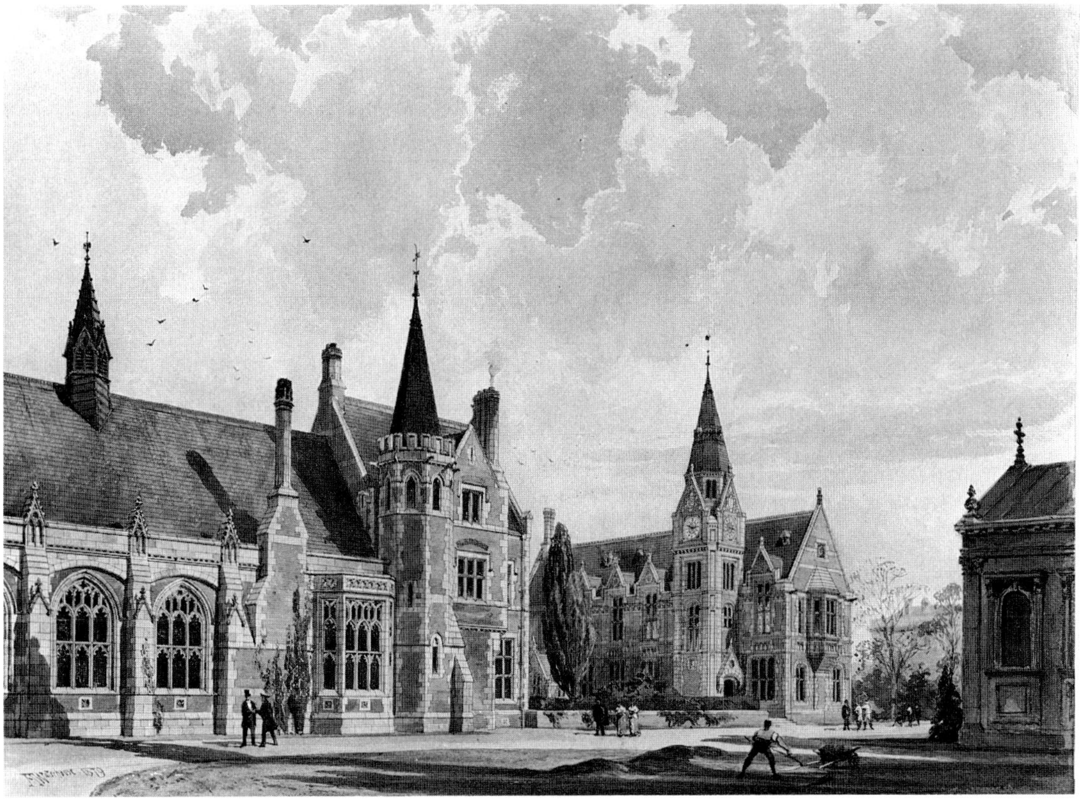

73. Cambridge, Pembroke College, new Hall and Library, perspective dated 1879 **[415]**

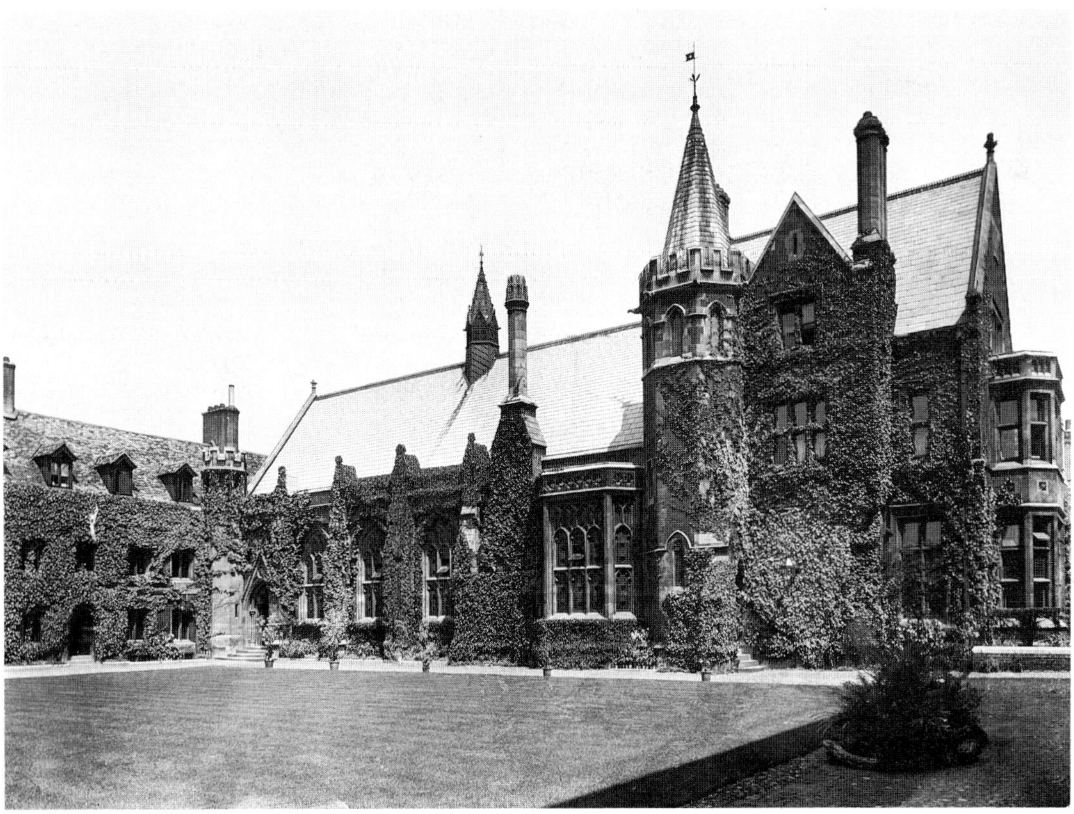

74. Cambridge, Pembroke College (1875), new Hall, *c.*1900 **[415]**

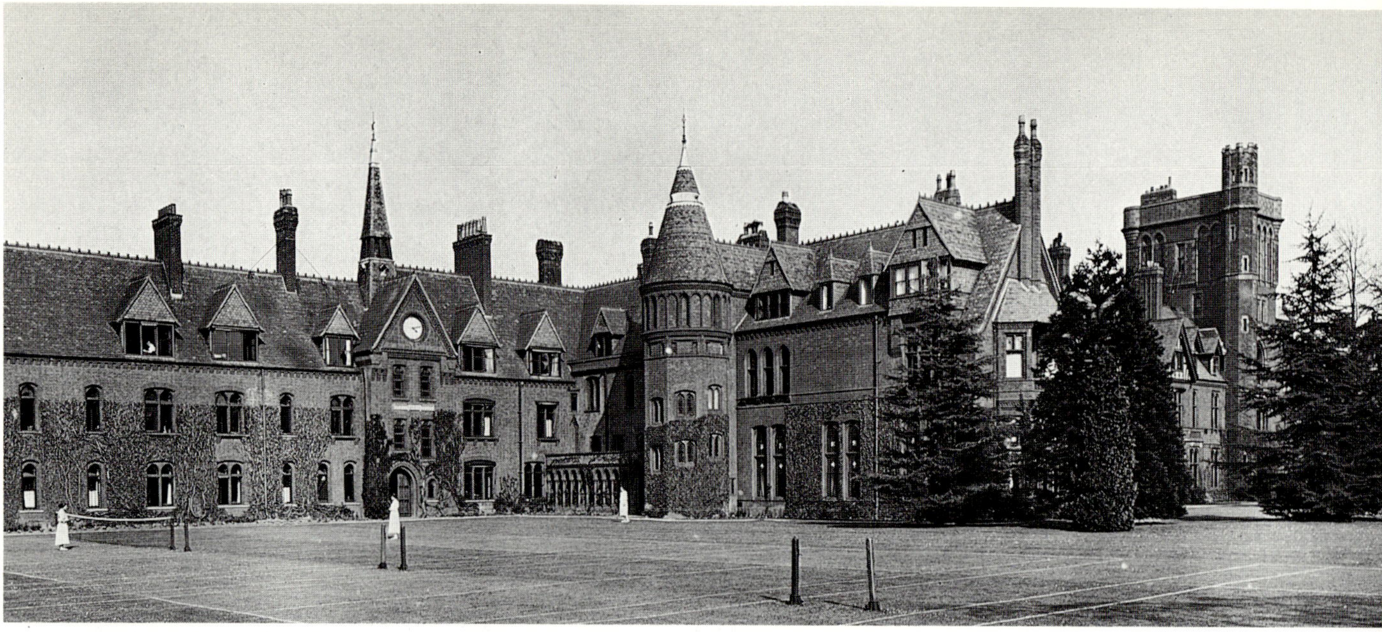

75. Cambridge, Girton College (1871 *et seq.*), *c.*1900. The original buildings are in the centre, with the stair turret and Stanley Library to the right and gate tower beyond **[346, 414, 501, 531]**

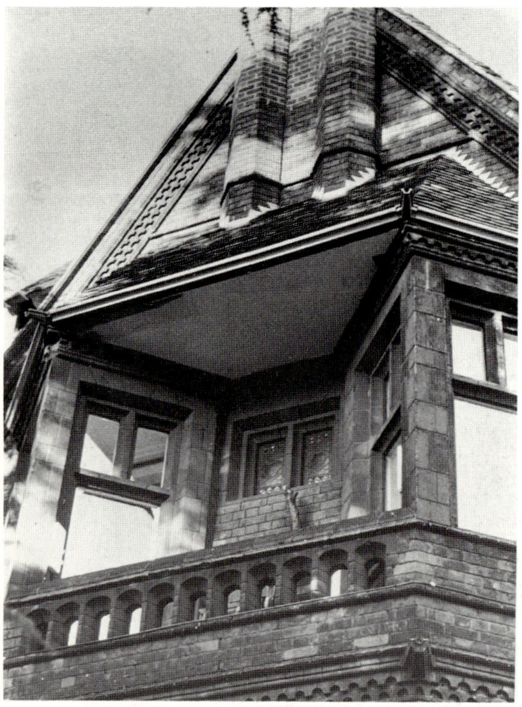

76. Cambridge, Girton College (1883), detail of loggia to the Mistress's Lodgings **[501]**

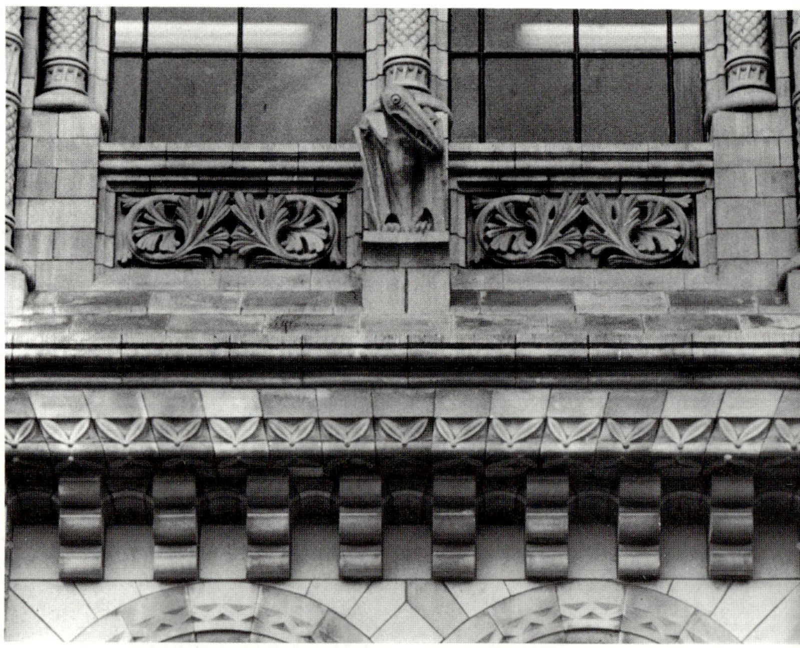

77. London, Natural History Museum (1870), detail of terracotta ornament at first-floor level **[243]**

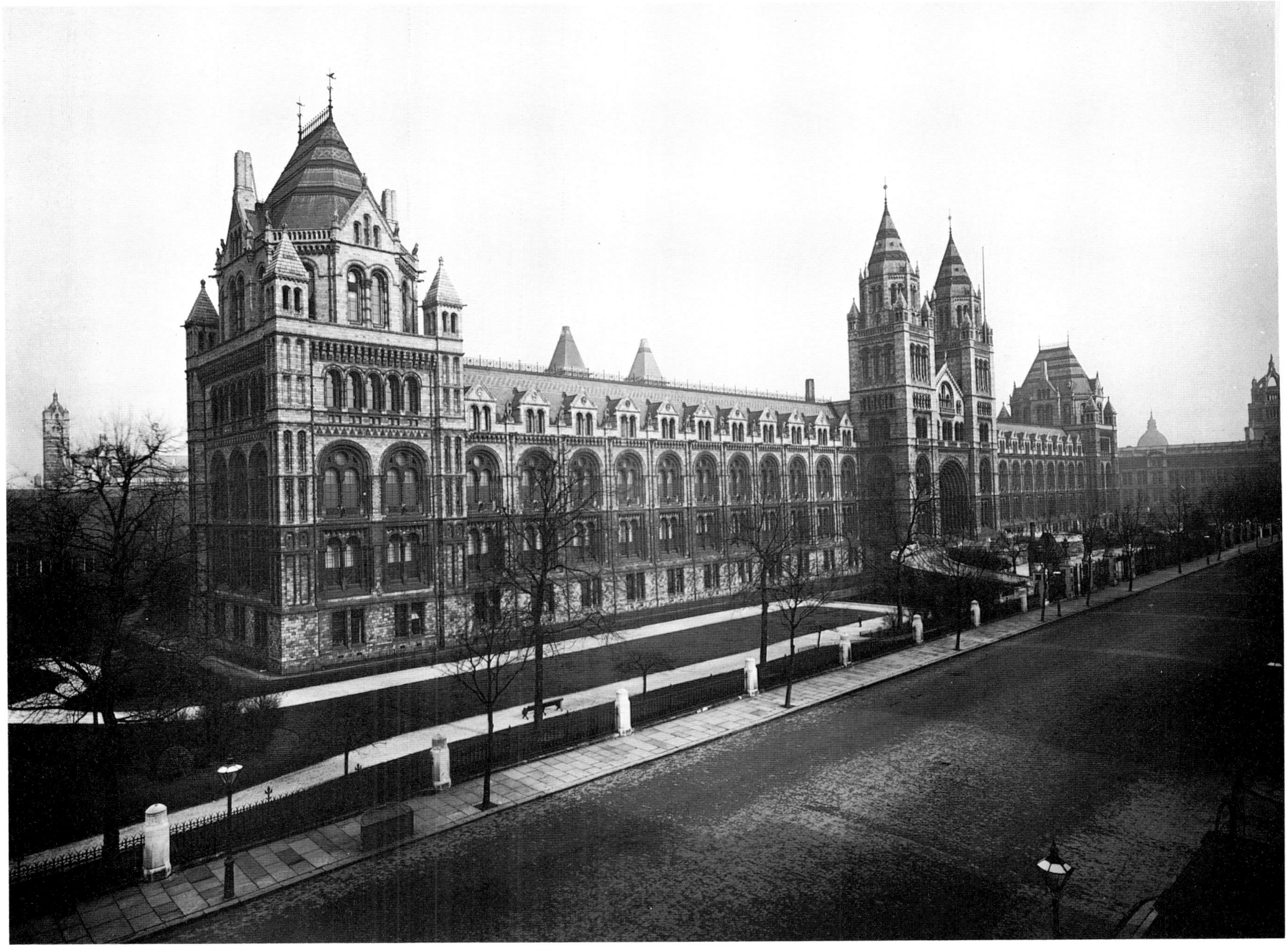

78. London, Natural History Museum (1870), *c.*1910 **[243]**

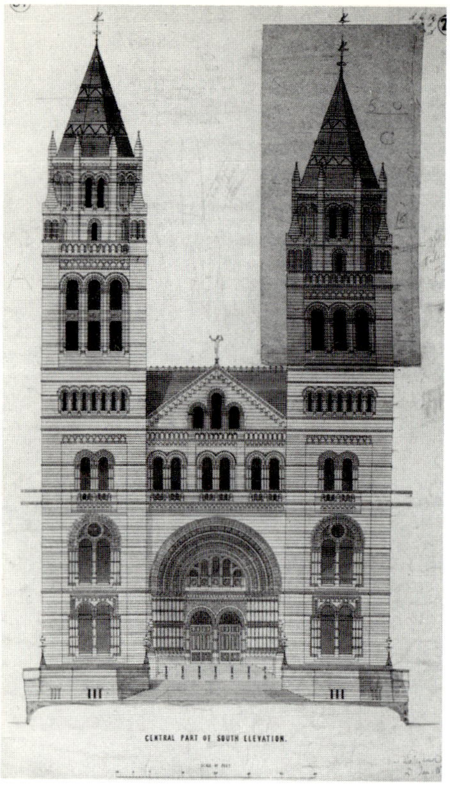

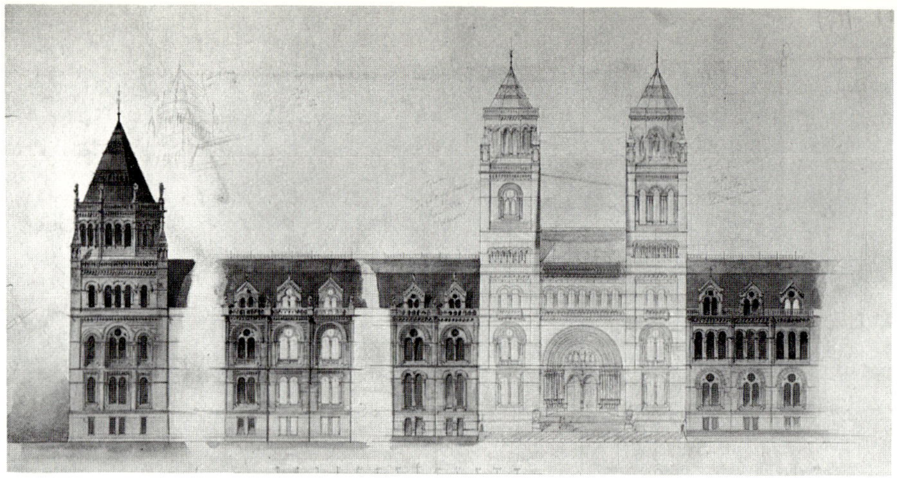

80. London, Natural History Museum, sketch design for south façade, *c.*1870 **[243]**

79. Left, London, Natural History Museum, central part of south elevation with alternative designs for towers, dated 1872 **[243]**

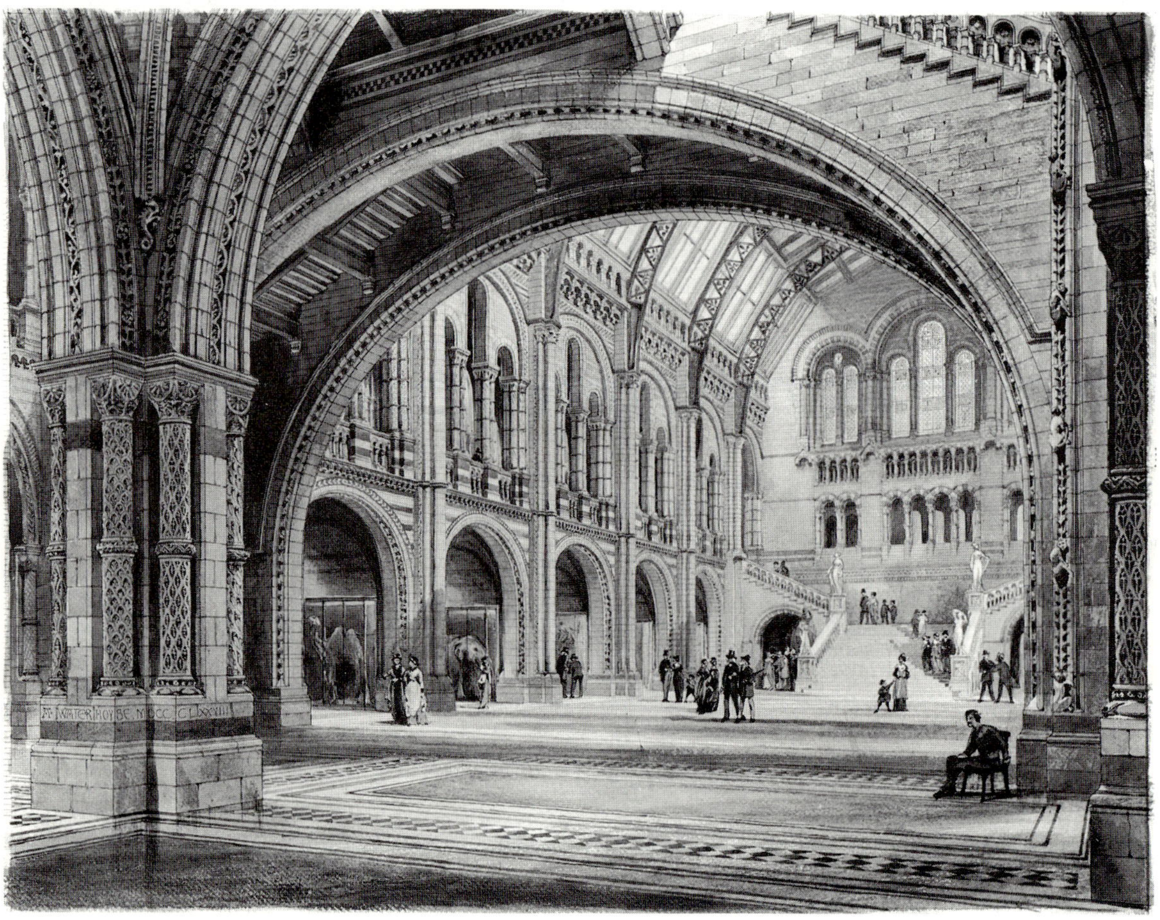

81. London, Natural History Museum, perspective of the index hall from entrance, dated 1878 **[243]**

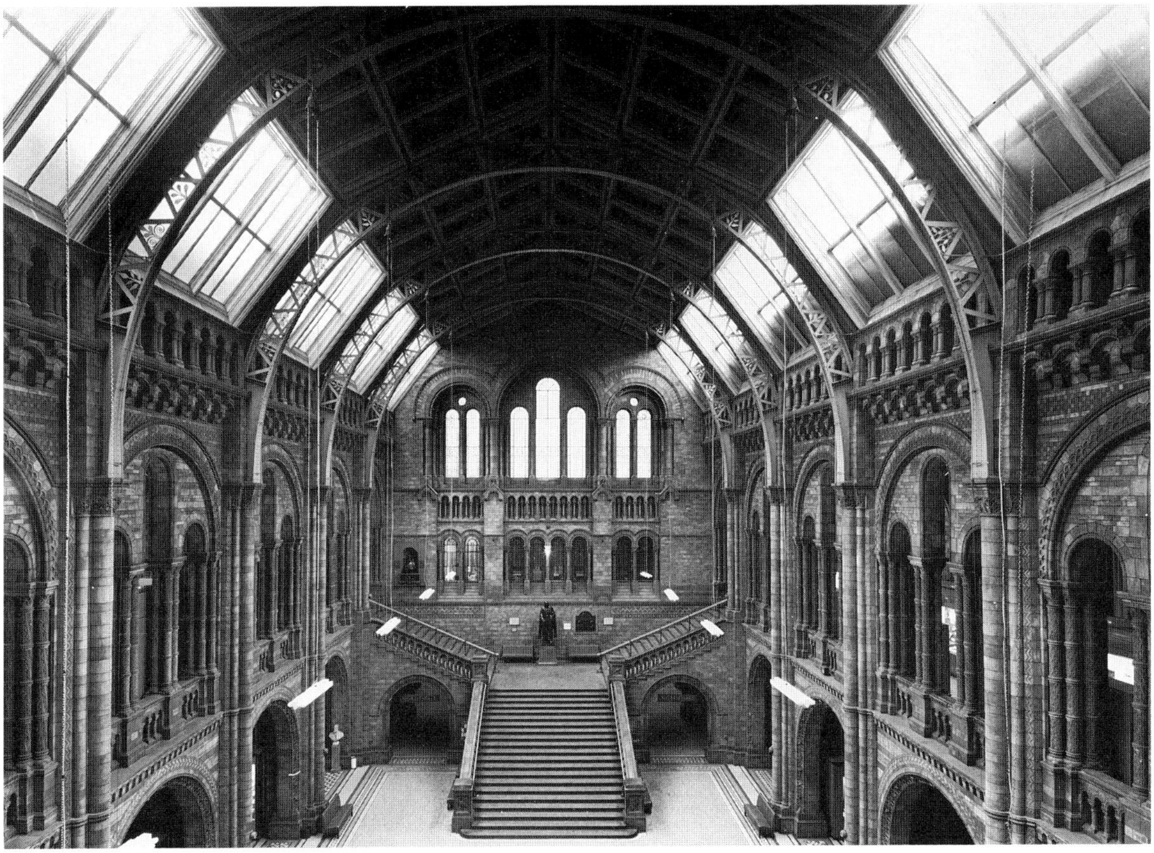

82. London, Natural History Museum (1870), the index hall from the flying stair **[243]**

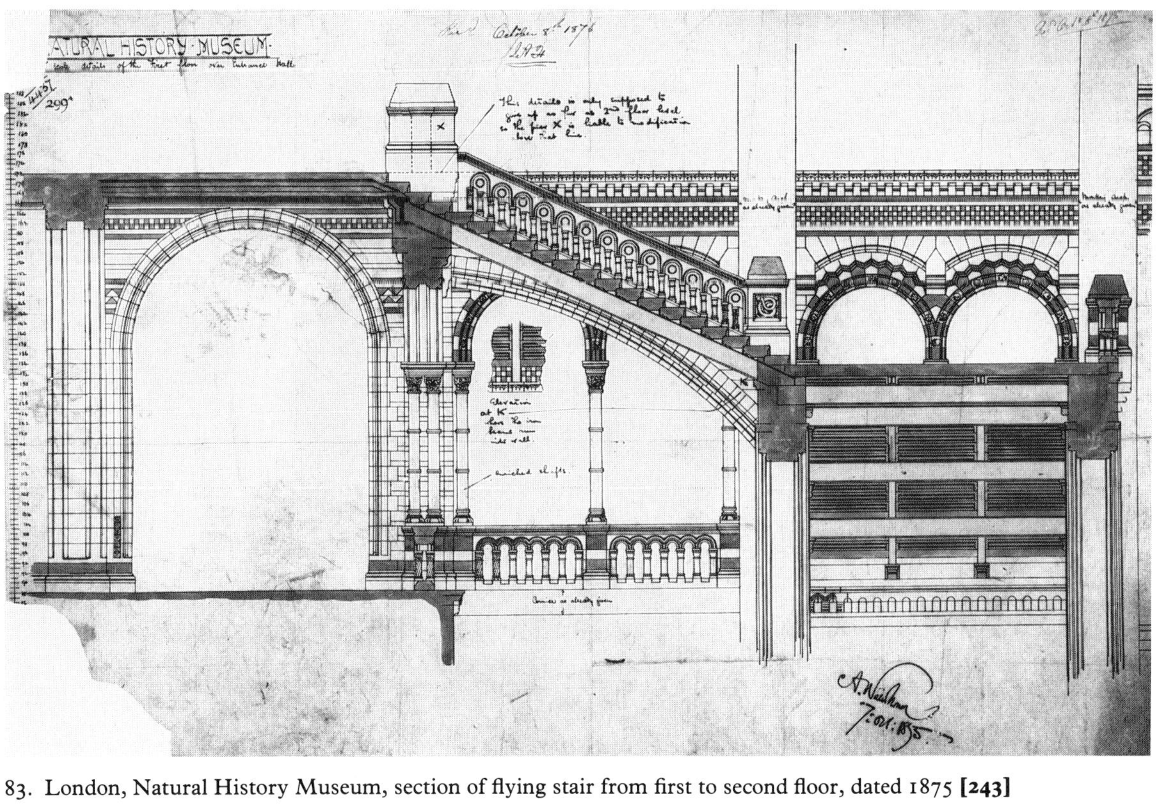

83. London, Natural History Museum, section of flying stair from first to second floor, dated 1875 **[243]**

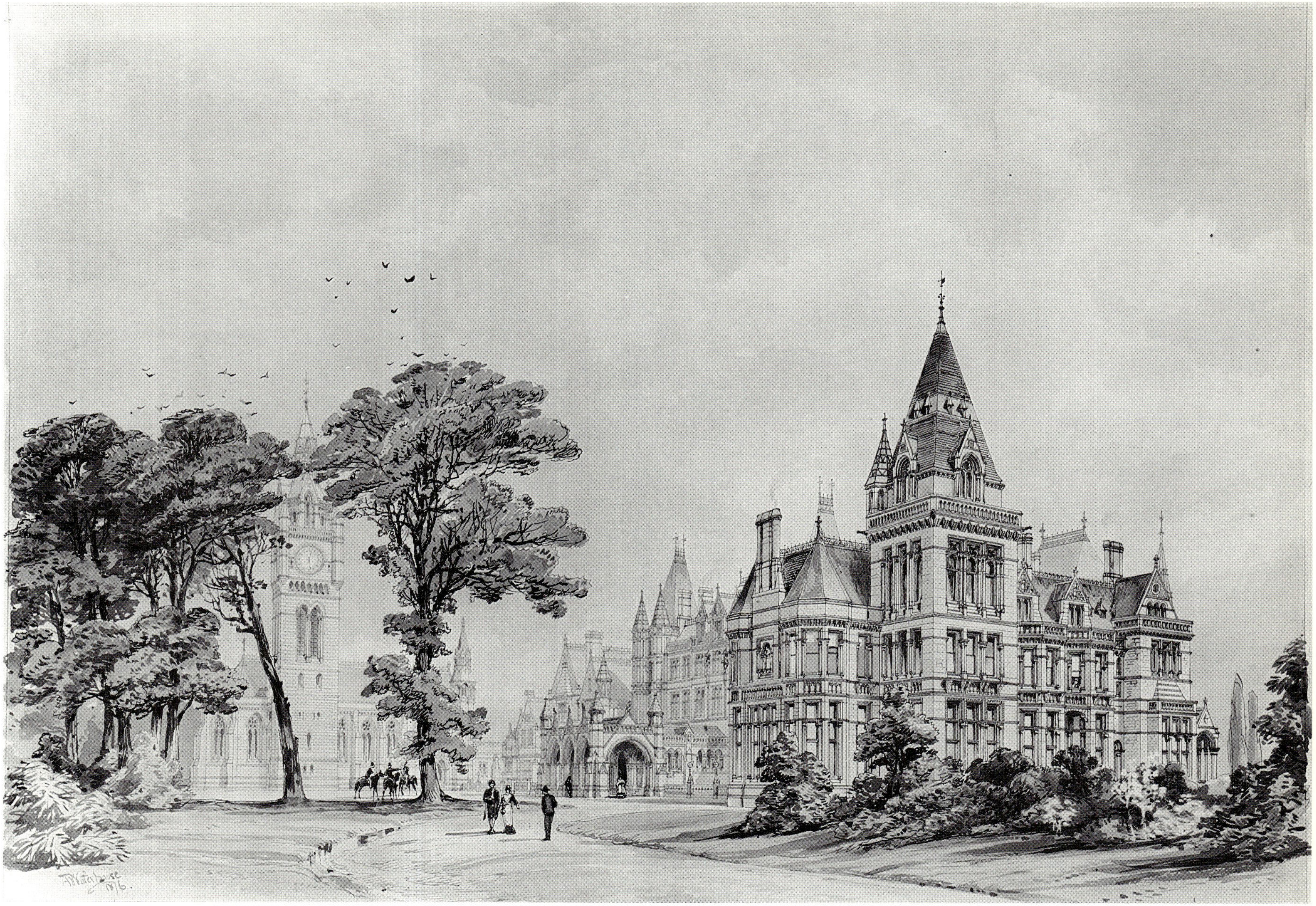

84. Eaton Hall, perspective of Library wing, Hall, and Chapel from the south-west, dated 1876 **[315]**

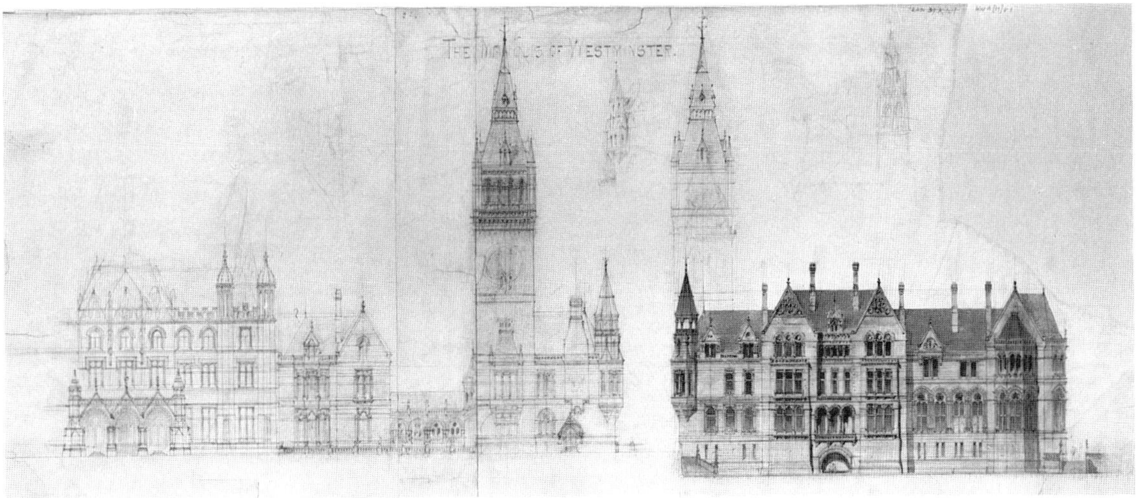

85. Eaton Hall, sketch design for Library wing, *c.*1869 **[315]**

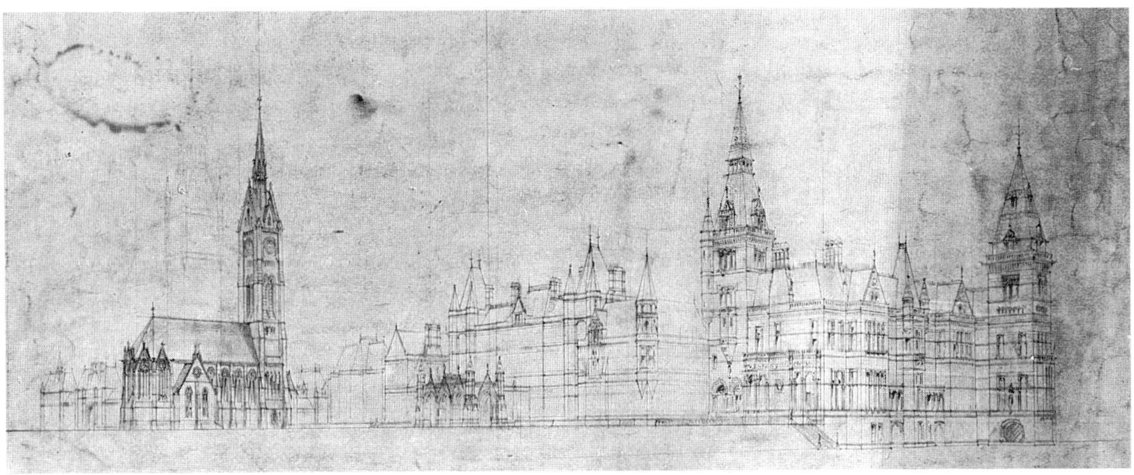

86. Eaton Hall, revised sketch design for Library wing, *c.*1870 **[315]**

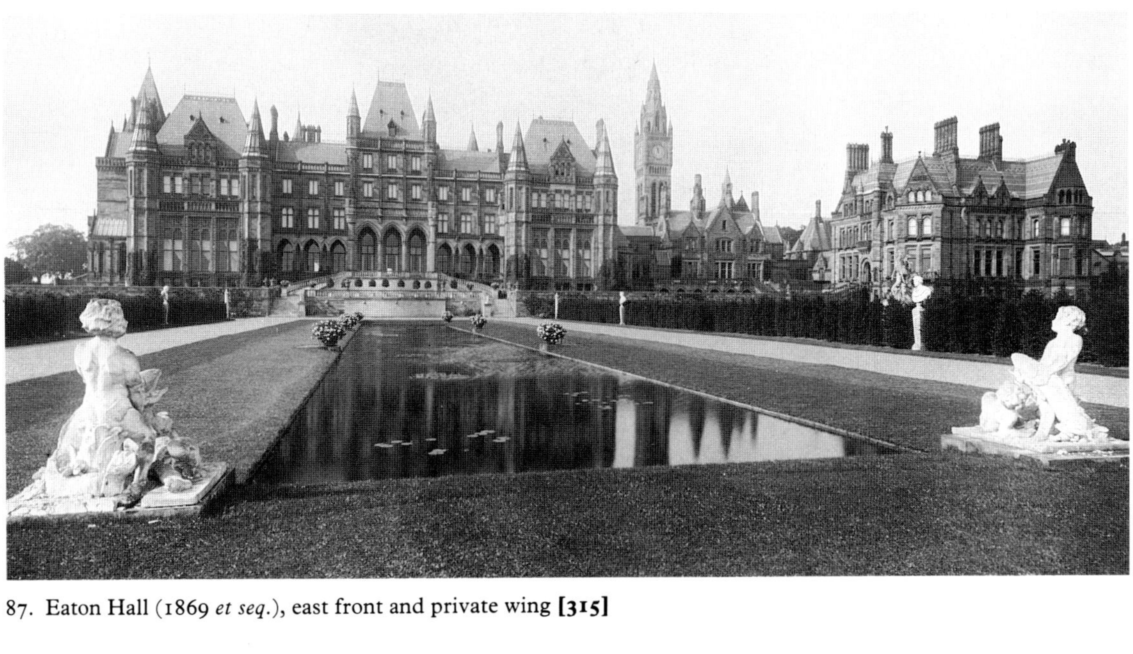

87. Eaton Hall (1869 *et seq.*), east front and private wing **[315]**

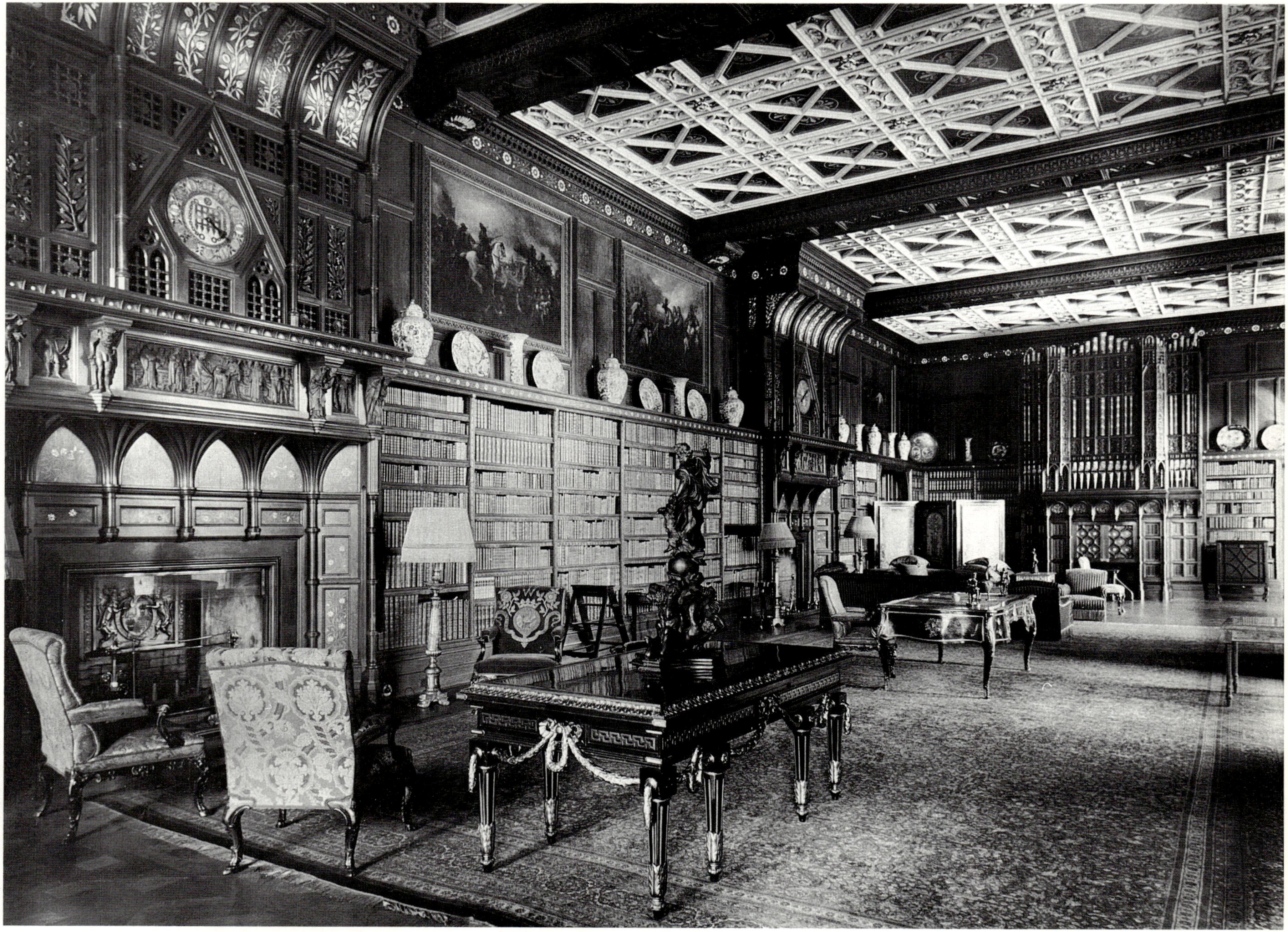

88. Eaton Hall (1869 *et seq.*), Library about 1890 **[315]**

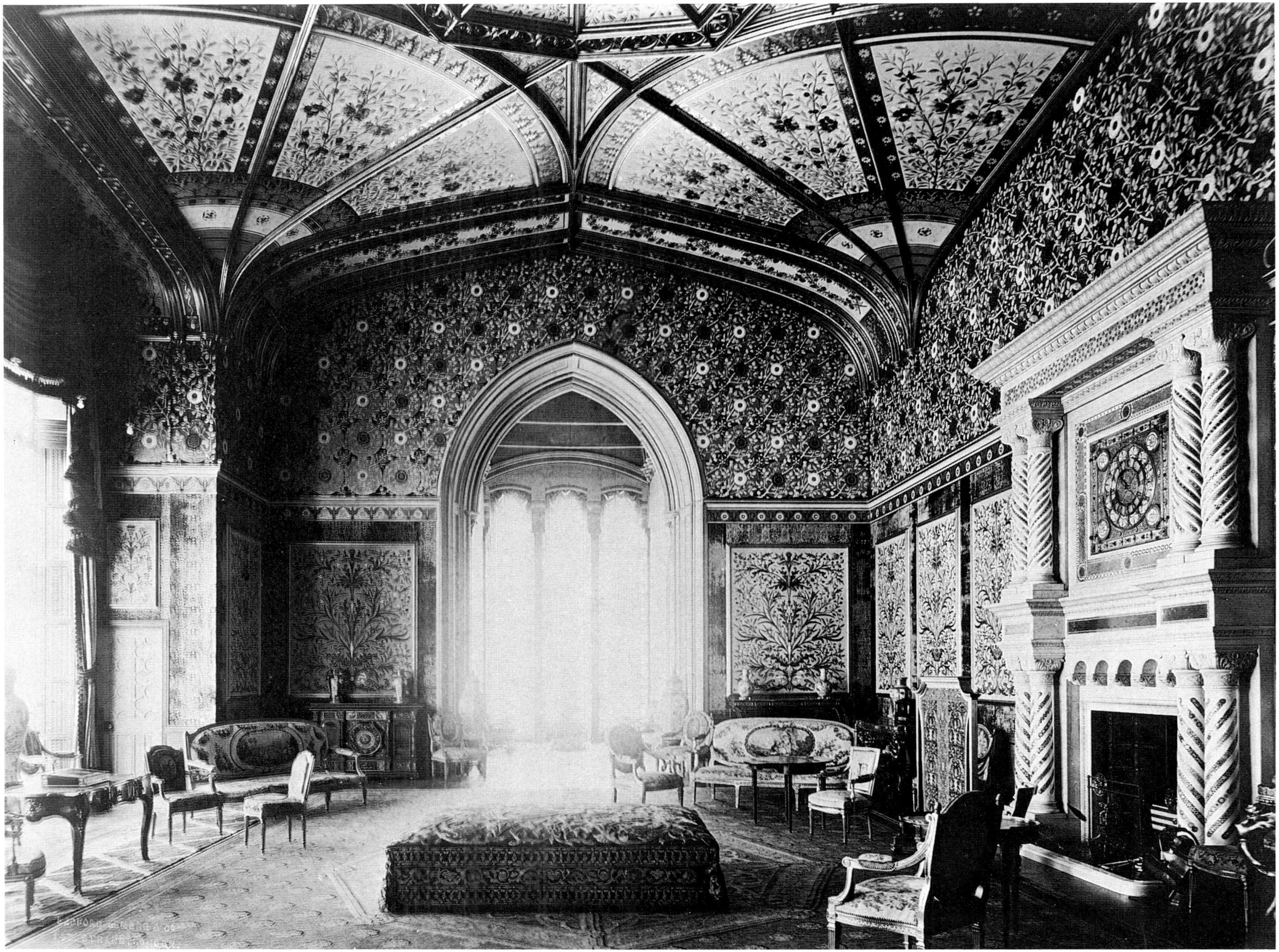

89. Eaton Hall (1869 *et seq.*), Great Drawing-Room about 1887 **[315]**

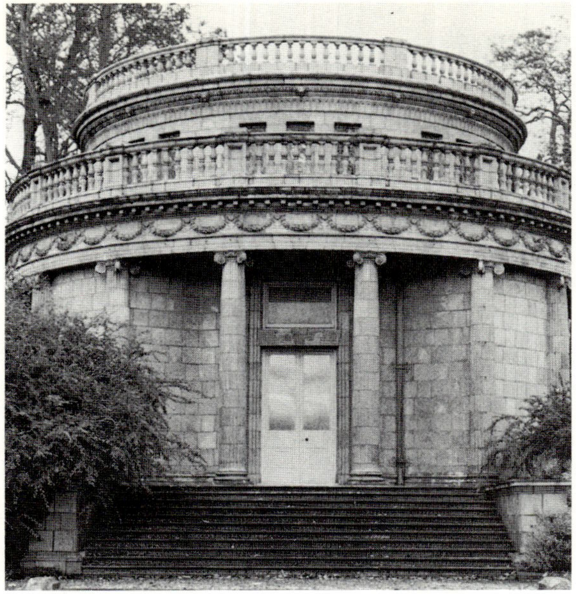

90. Eaton Hall, terracotta parrot house, *c.*1880 **[315]**

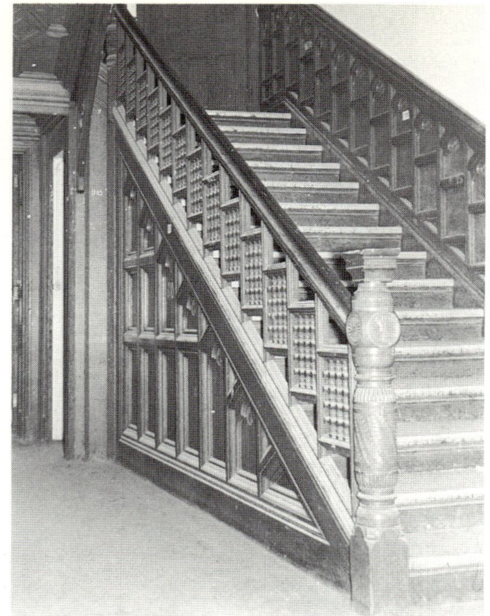

91. Eaton Hall, Library wing (1869 *et seq.*), staircase from first to second floor in 1959 shortly before demolition **[315]**

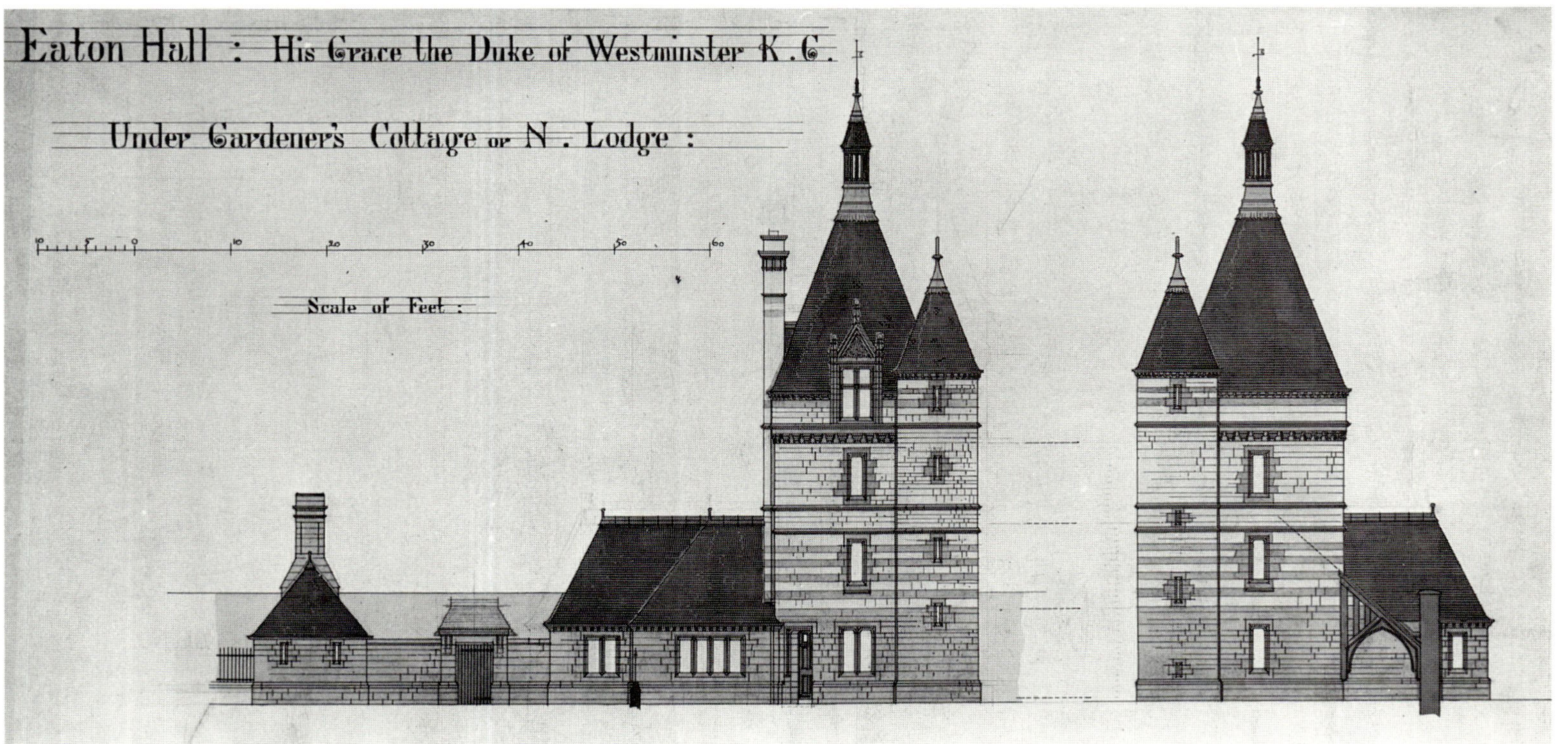

92. Eaton Hall, elevation of north lodge, dated 1881 **[315]**

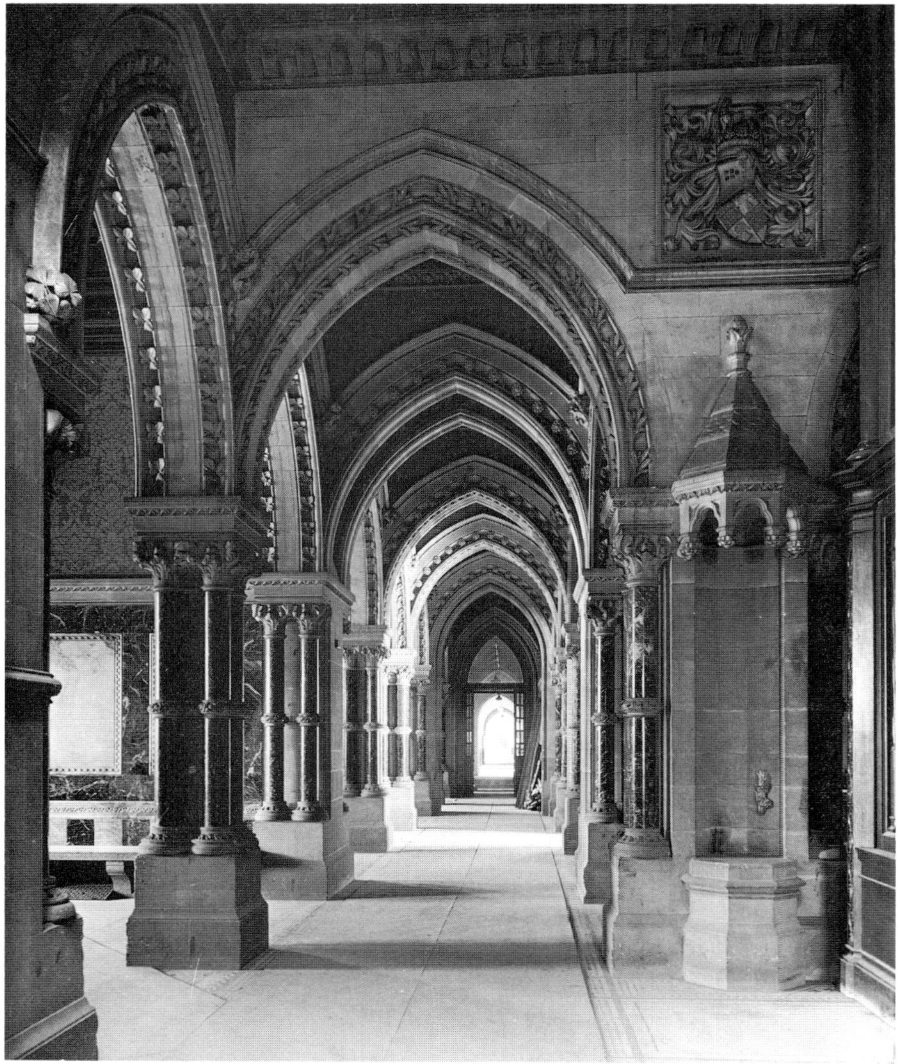

93. Eaton Hall (1869 *et seq.*), entrance hall and great corridor shortly before demolition in 1959 **[315]**

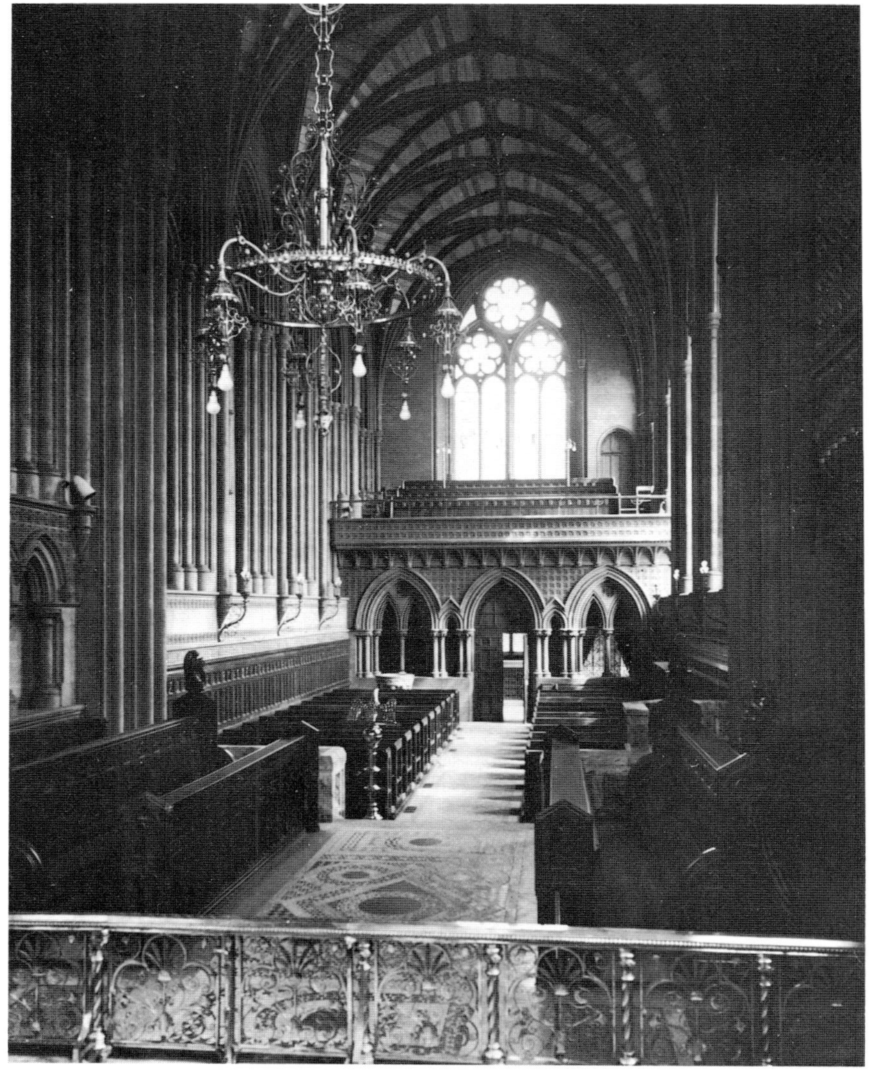

94. Eaton Hall (1869 *et seq.*), interior of Chapel in 1959 **[315]**

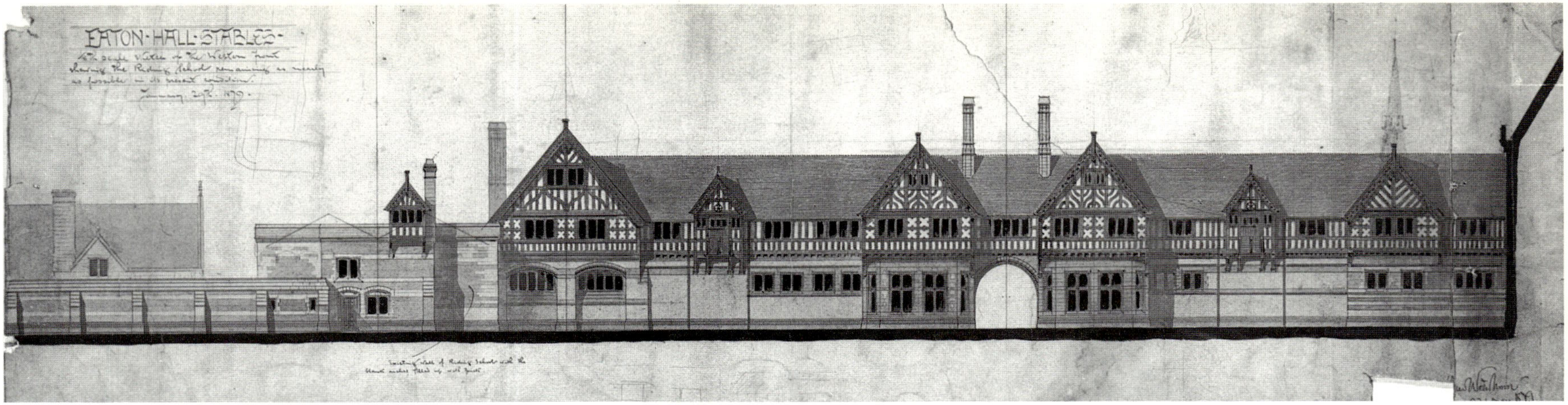

95. Eaton Hall, west elevation of covered riding school and stable court, dated 1879 **[315]**

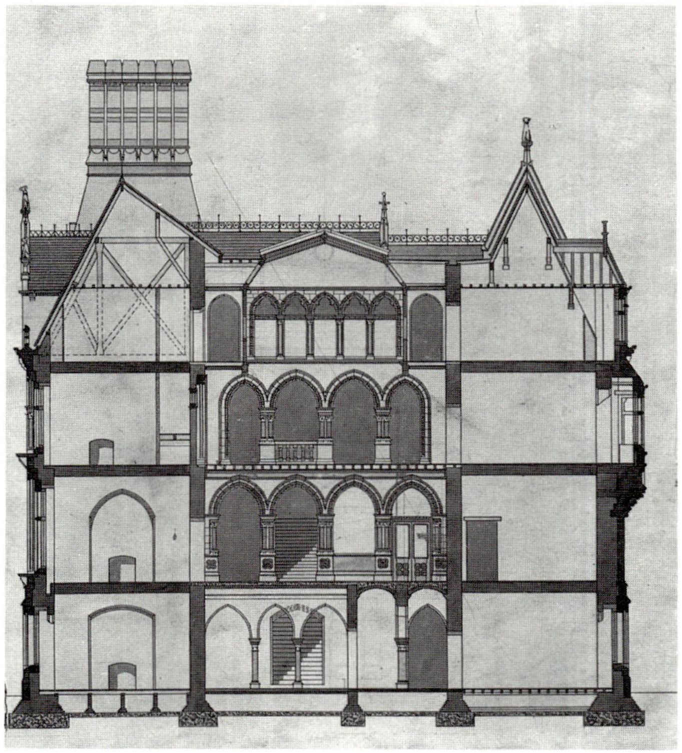

96. Eaton Hall, section of private wing facing north, *c.*1870 **[315]**

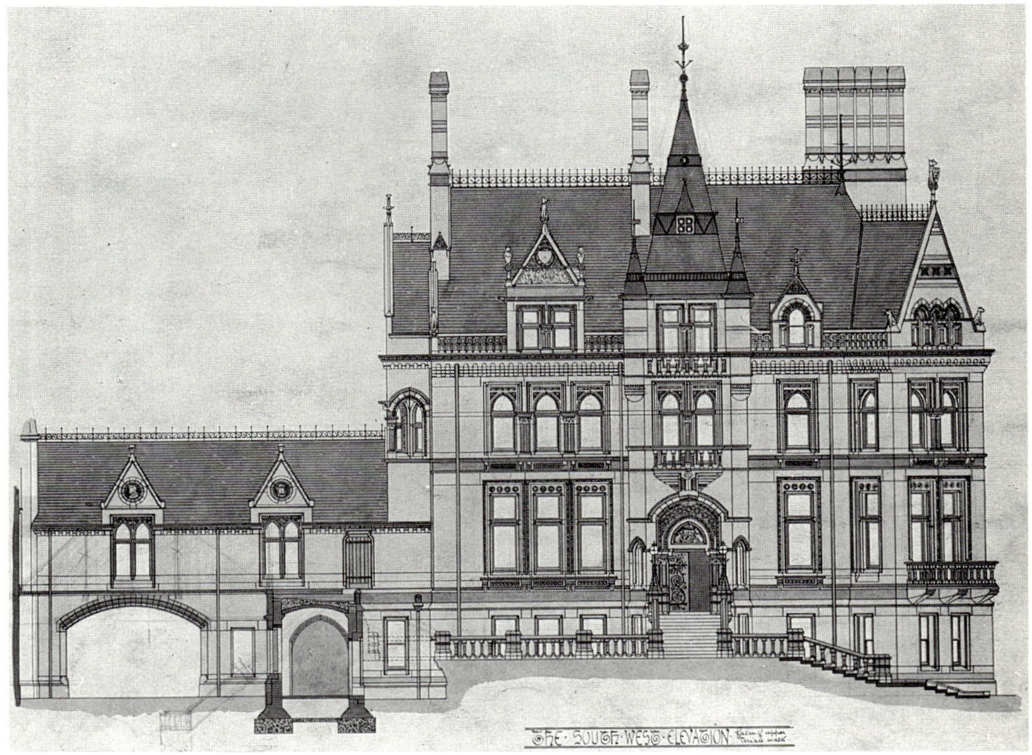

97. Eaton Hall, south elevation of private wing, *c.*1870 **[315]**

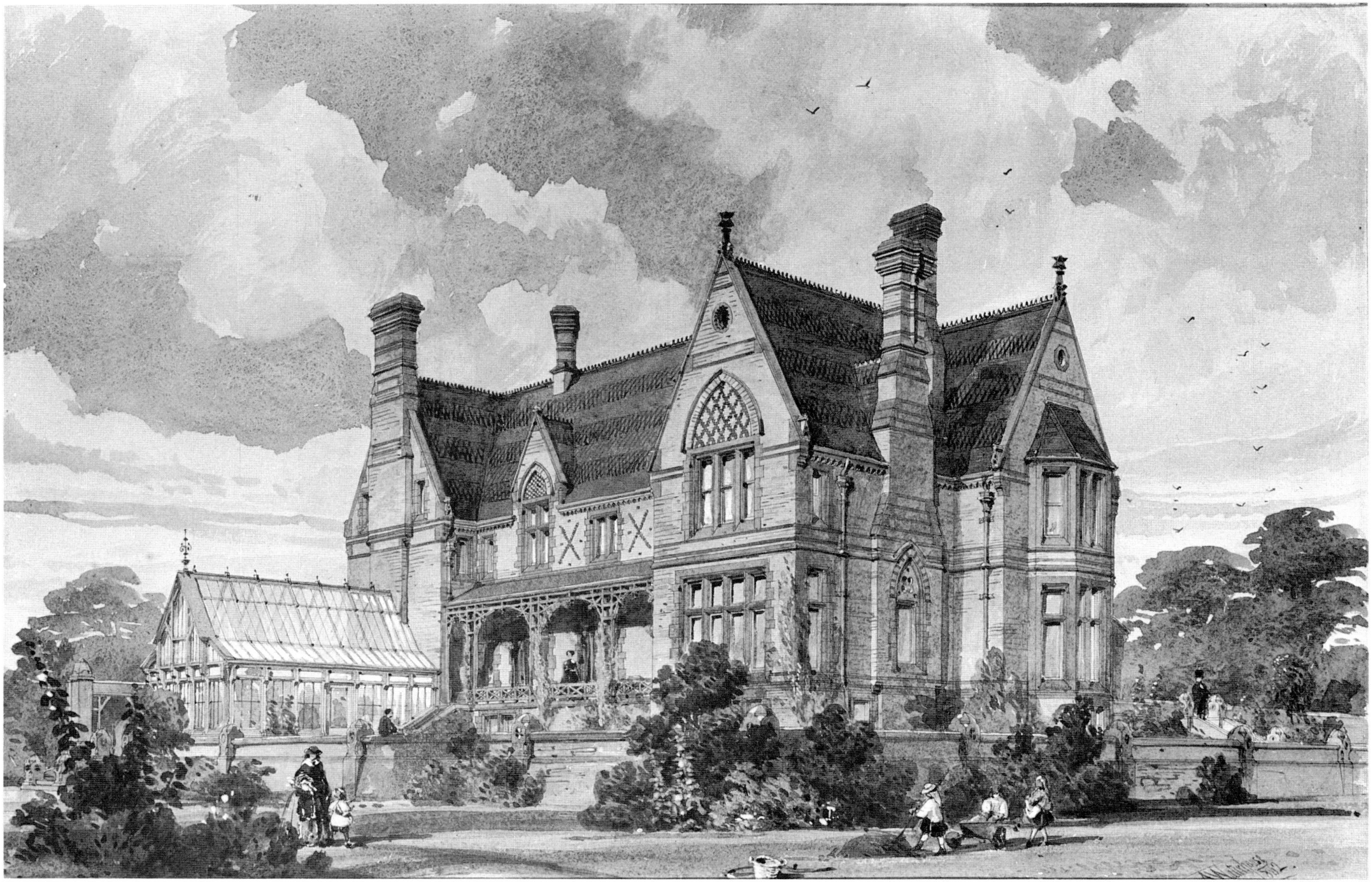

98. Liverpool, New Heys, watercolour perspective dated 1862 **[103]**

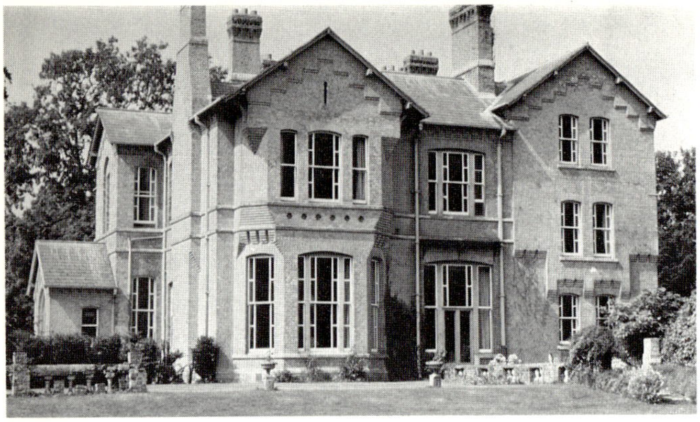

99. Crimplesham Hall (1880), garden front **[476]**

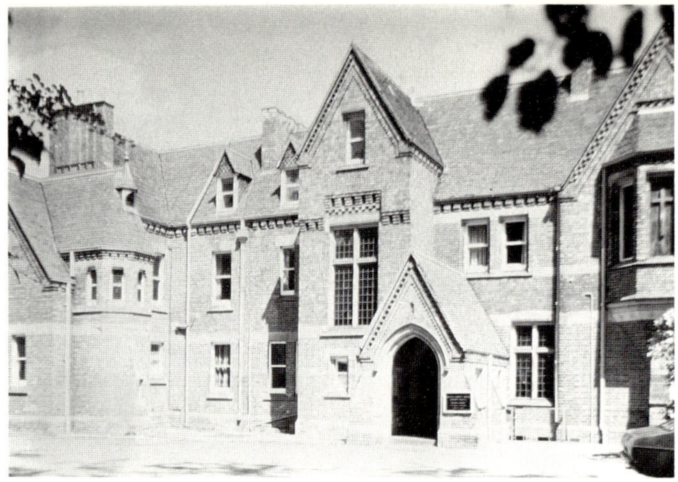

100. Darlington, Hurworth Grange (1873), entrance front **[381]**

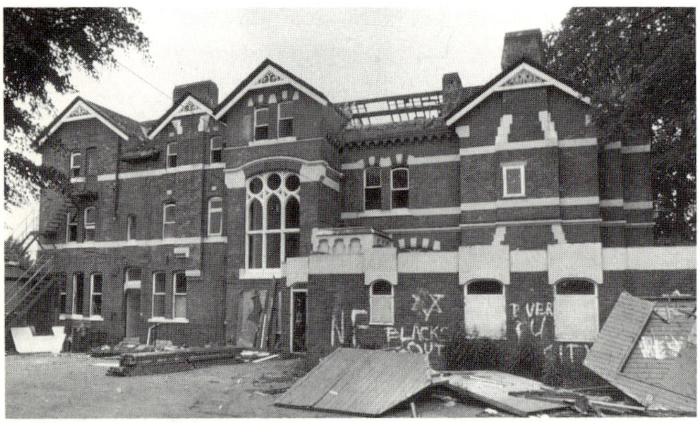

101. Manchester, West Bank House (1867), entrance front during demolition, 1984 **[281]**

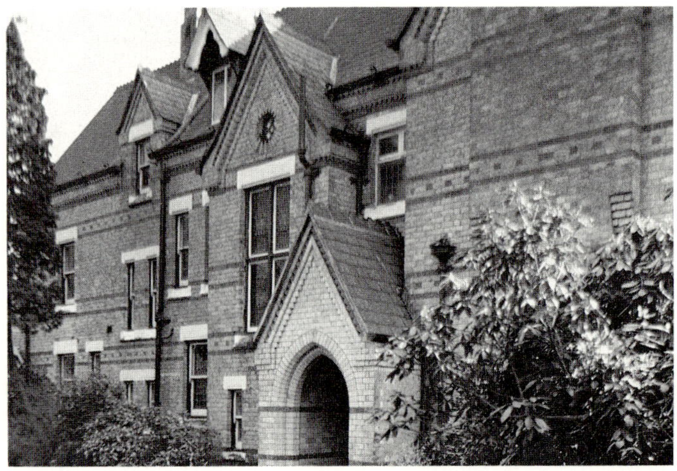

102. Manchester, Bowdon, Merle Bank (1869), entrance front **[314]**

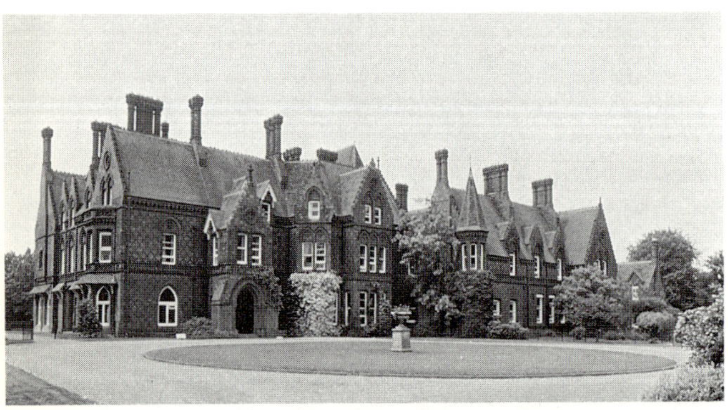

103. Stanstead Abbots, Easneye Park (1866), entrance front **[258]**

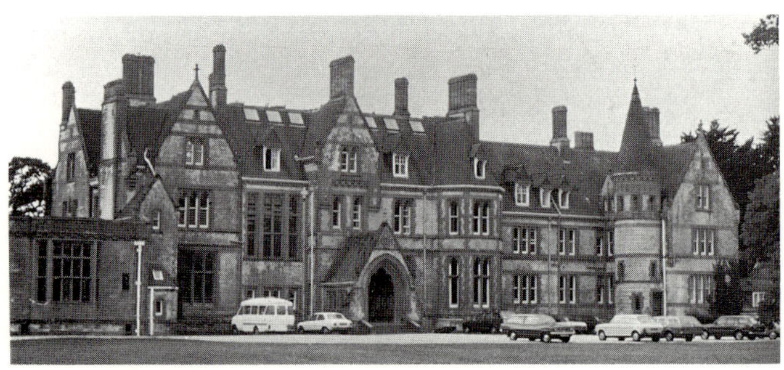

104. Iwerne Minster (1877), entrance front **[445]**

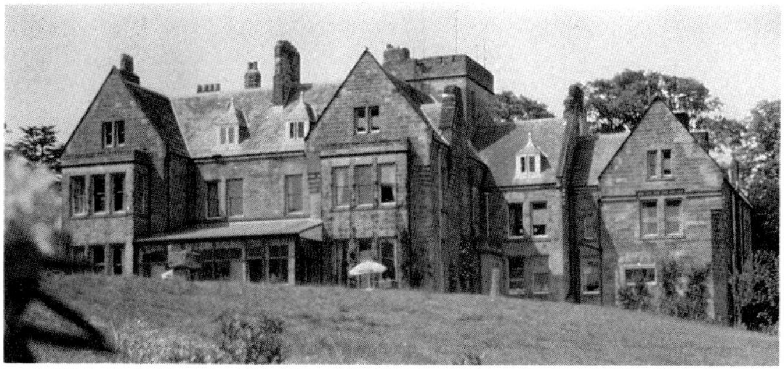

105. Grinkle Park (1881), garden front **[489]**

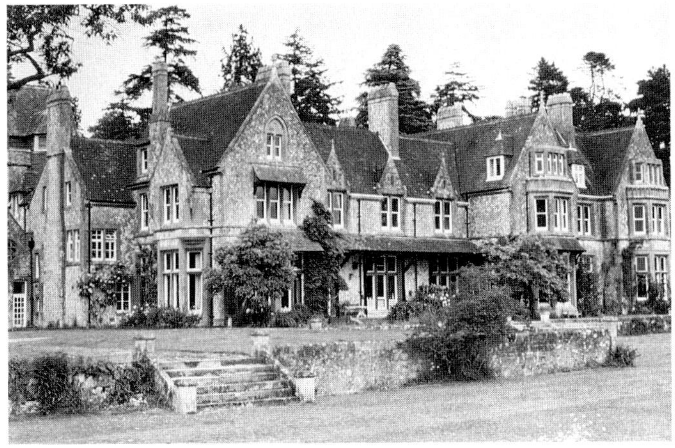

106. Blackmoor House (1868), garden front **[192]**

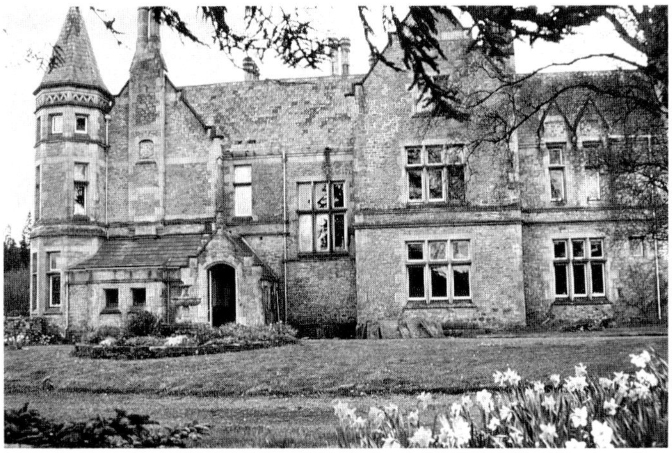

107. Hamsterley, Dryderdale (1871), entrance front **[349]**

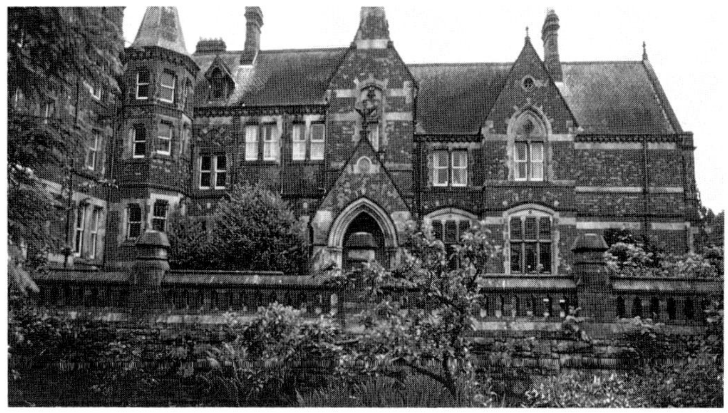

108. Guisborough, Hutton Hall (1864), entrance front **[176]**

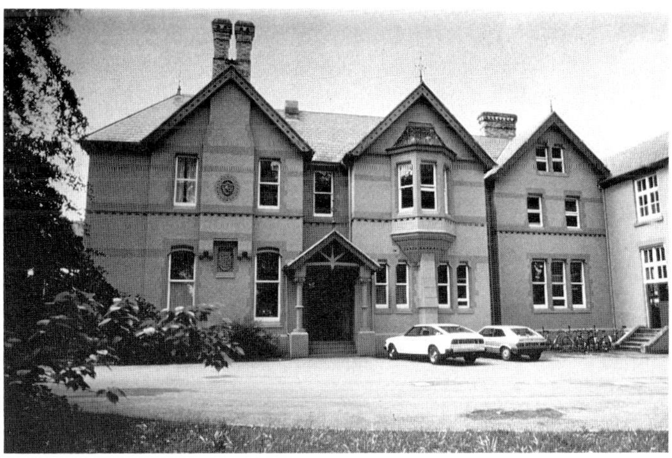

109. Darlington, Hummersknott (1863), entrance front **[147]**

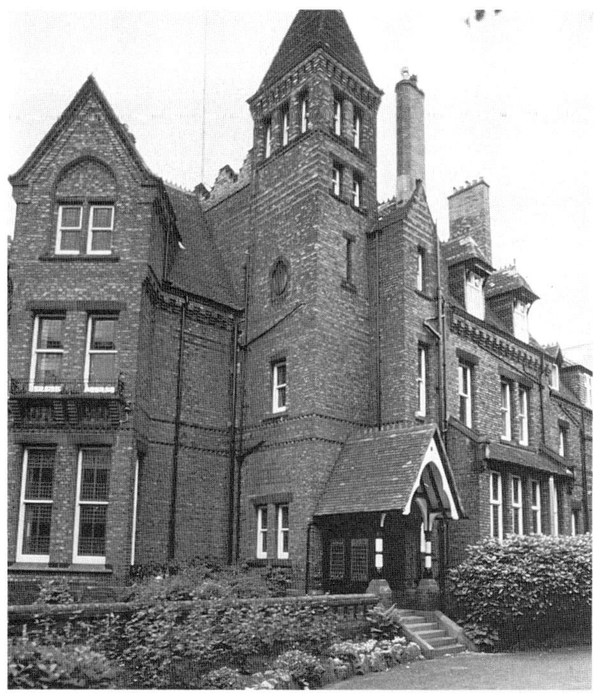

110. Manchester, Victoria Park, T. R. Hetherington house (1871) **[358]**

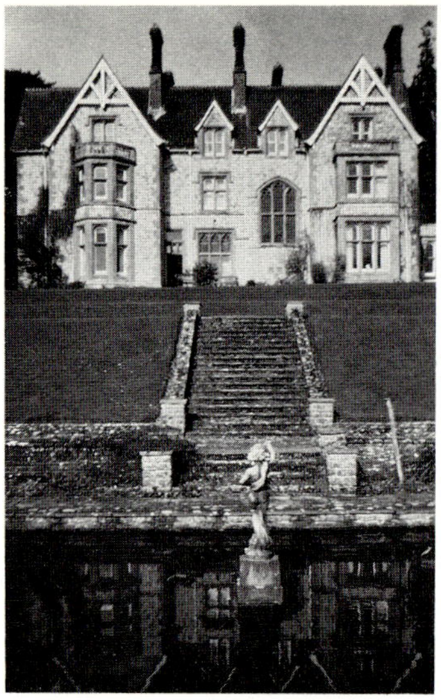

111. Coldhayes (1869), garden front **[323]**

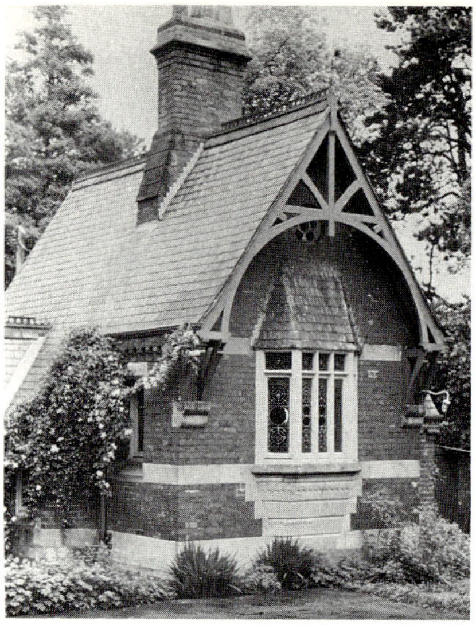

112. Guisborough, Hutton Hall (1864), lodge **[176]**

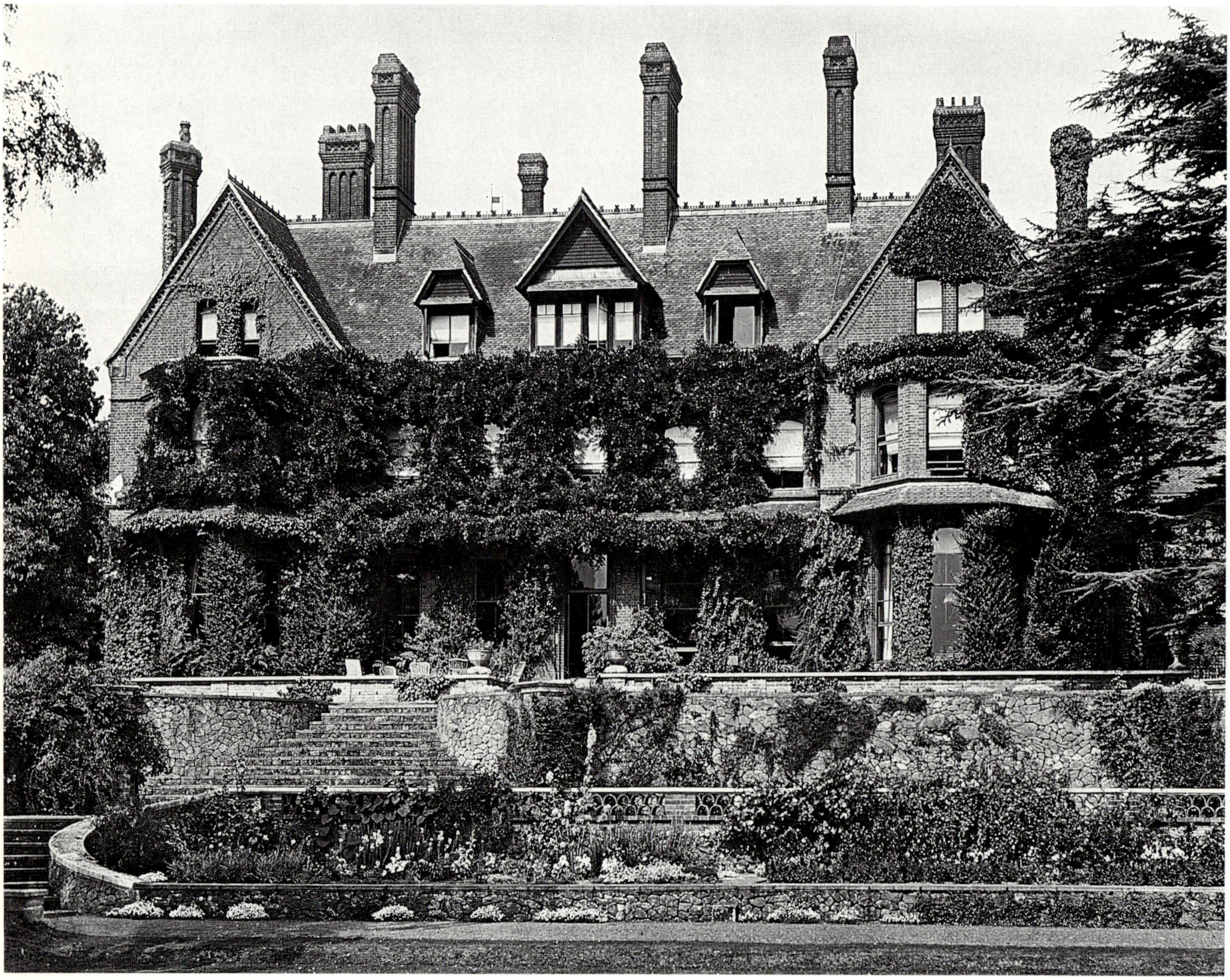

113. Abinger Hall (1870), garden front, *c.*1900 **[326]**

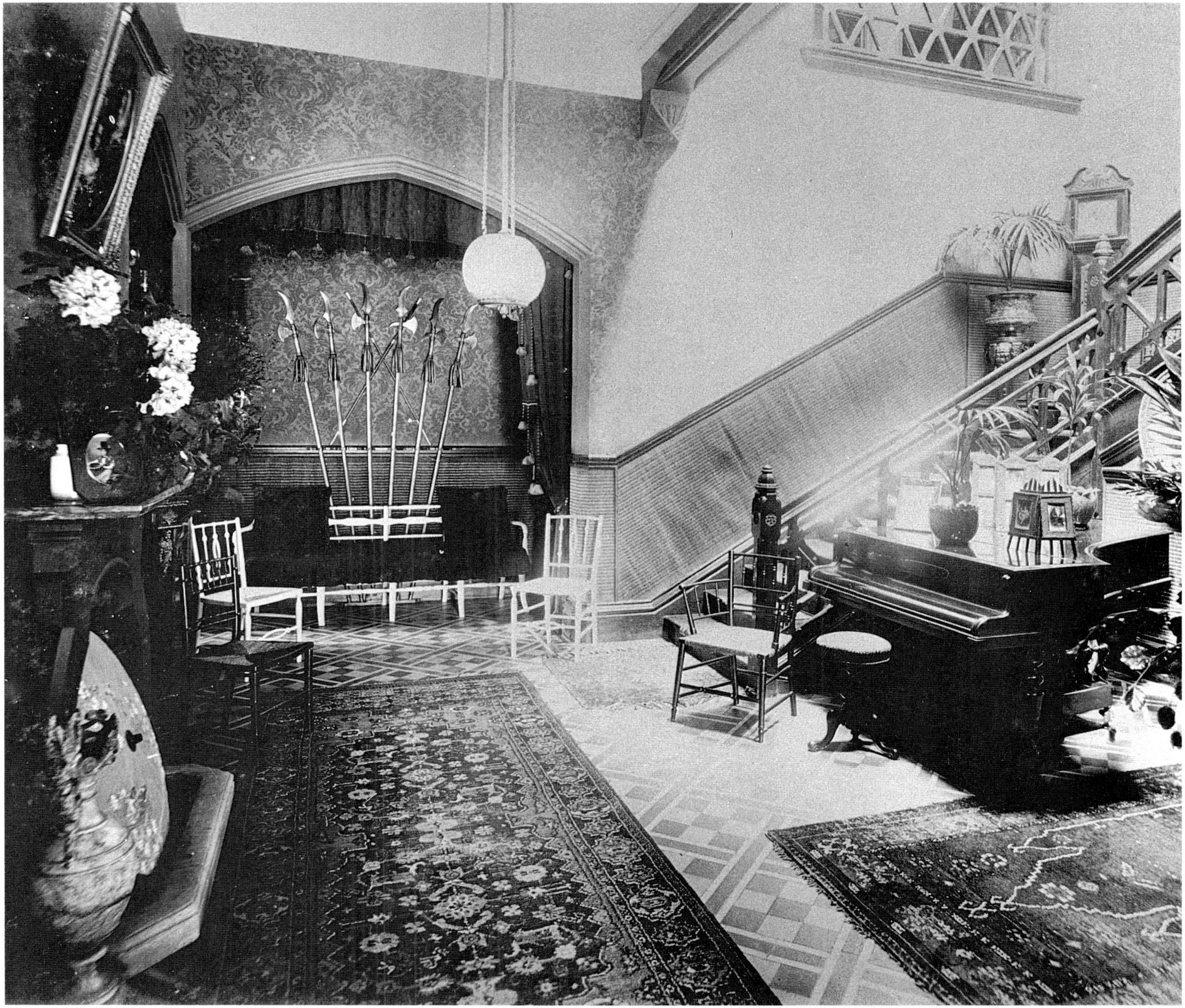

114. Leighton Buzzard, The Heath (1864), staircase hall, *c.*1890 **[178]**

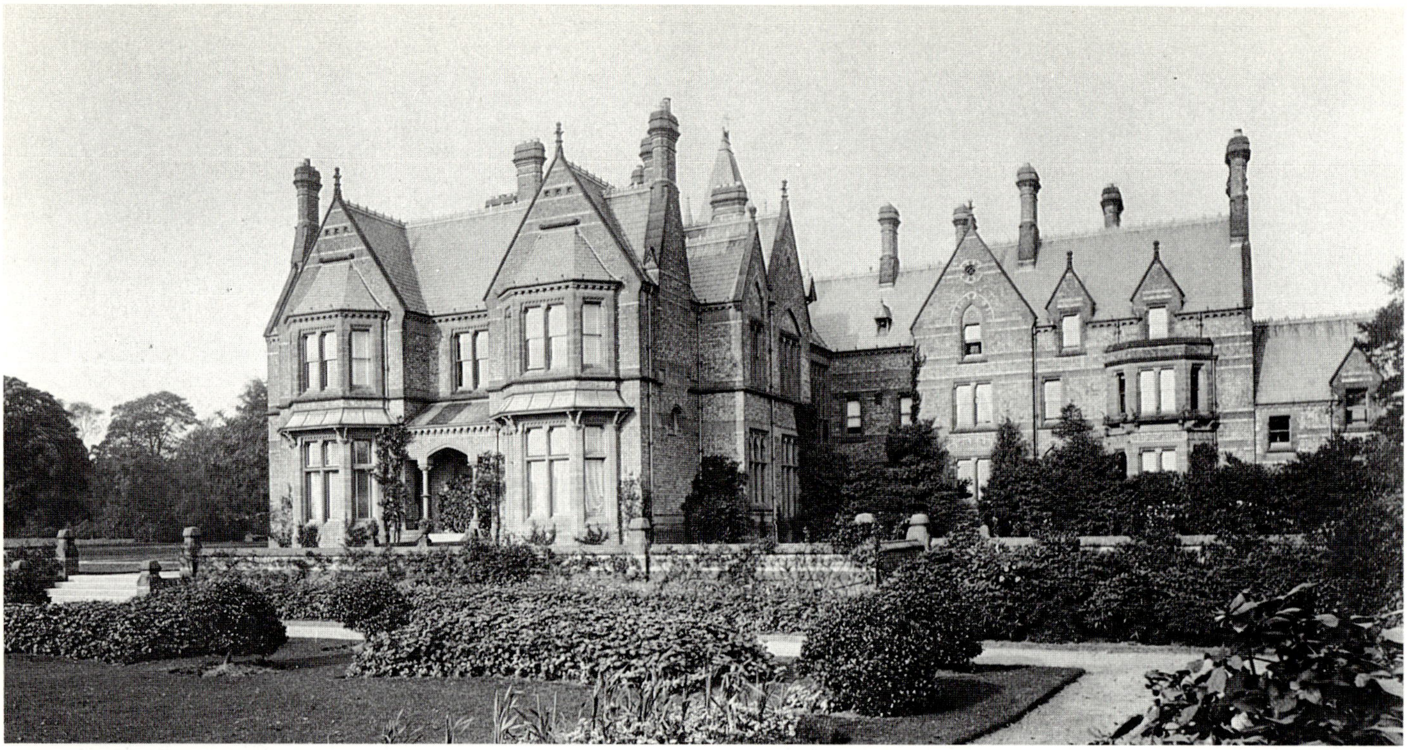

115. Liverpool, Allerton Priory (1866), garden front, *c.*1880 **[237]**

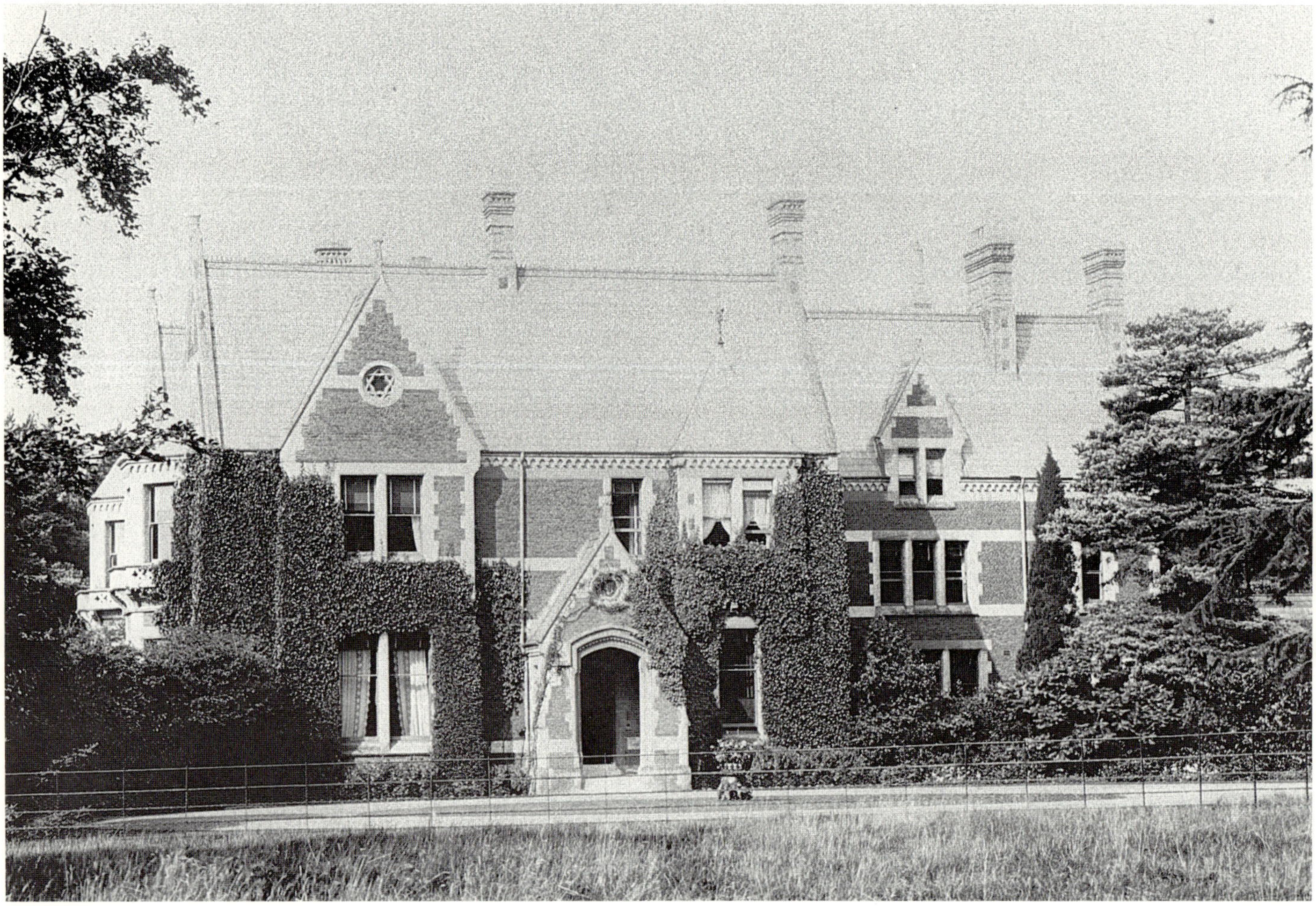

116. Leighton Buzzard, The Heath (1864), entrance front, *c.*1890 **[178]**

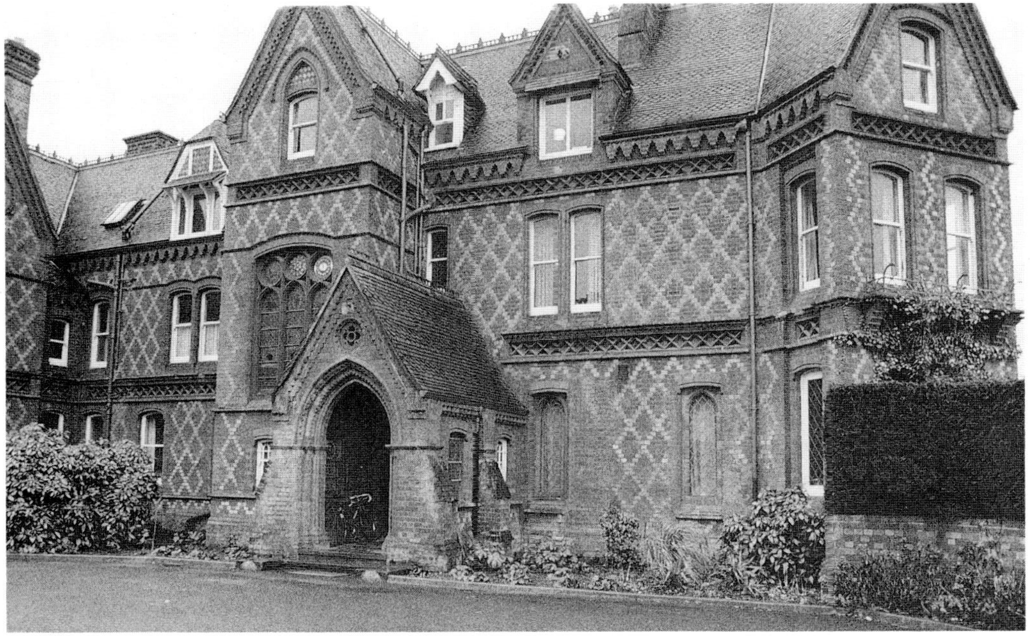

117. Reading, Foxhill (1867), entrance front **[285]**

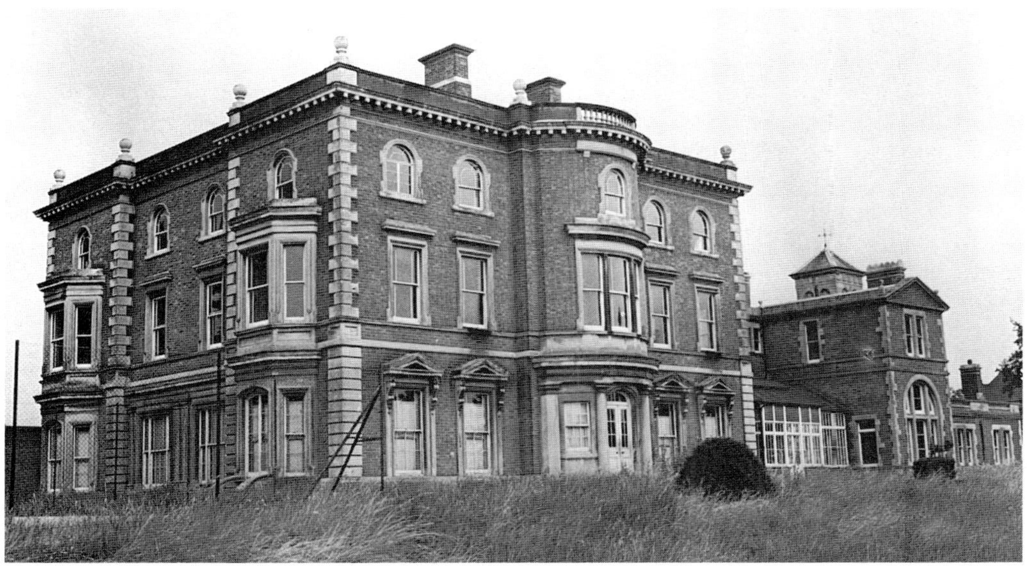

118. Easenhall, Town Thorns (1871), garden front and billiard room/service wing **[348]**

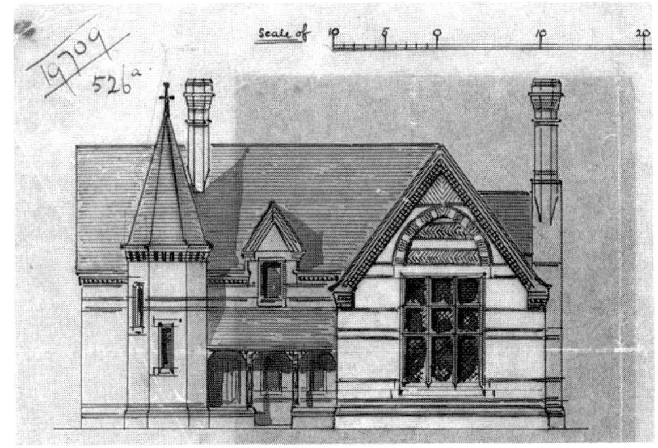

119. Guisborough, Hutton Hall, design for village reading room, dated 1872 **[368]**

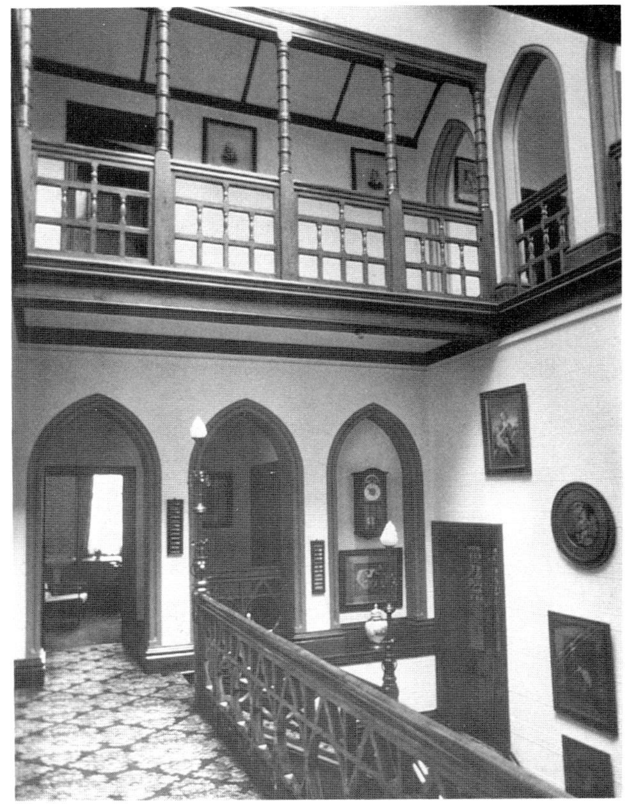

120. Manchester, Victoria Park, T. R. Hetherington house (1871), staircase hall **[358]**

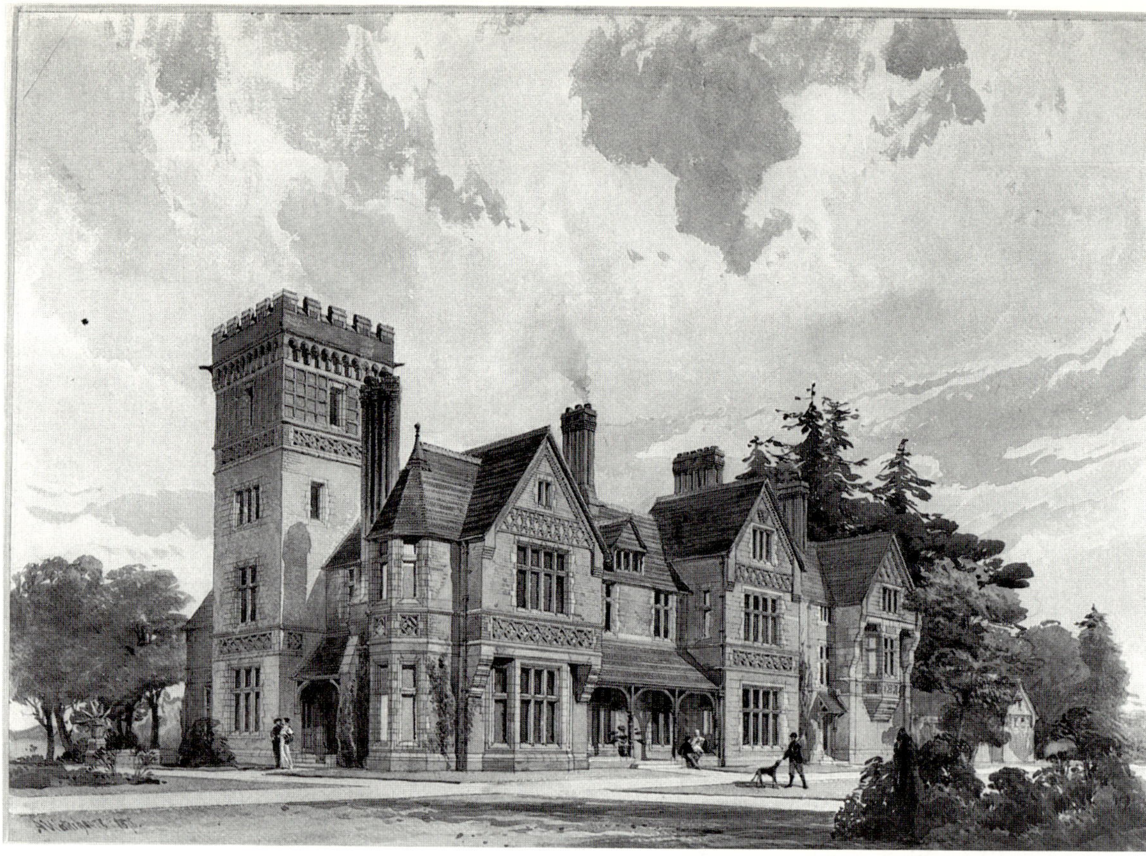

121. Yattendon Court, perspective dated 1878 **[453]**

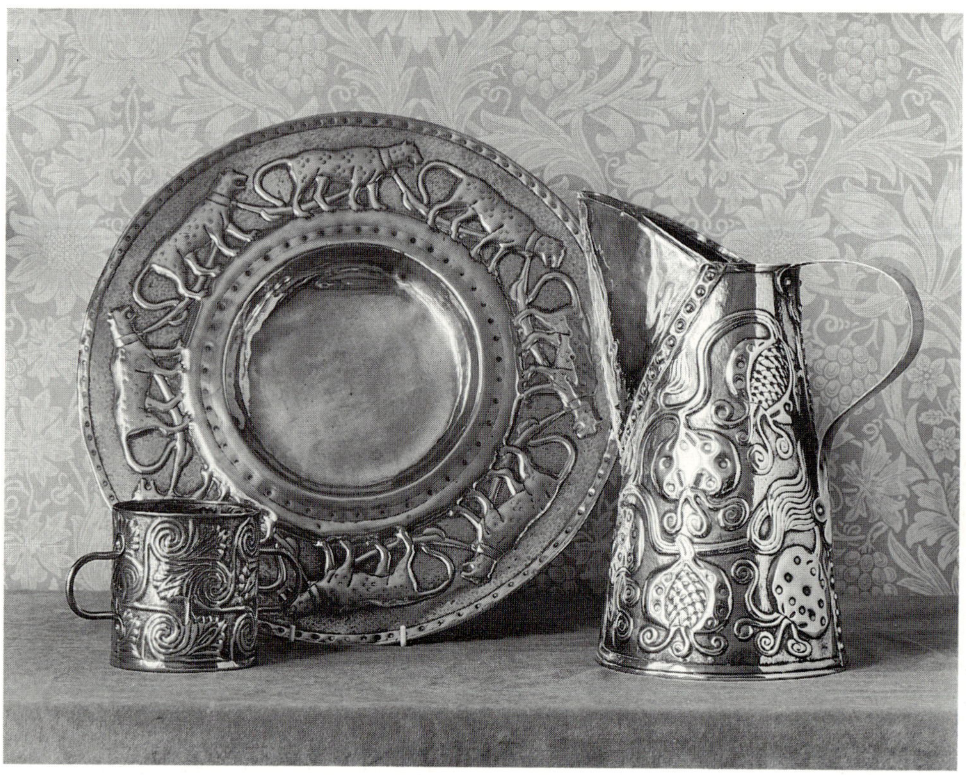

122. Examples of Yattendon Guild work in repoussé copper

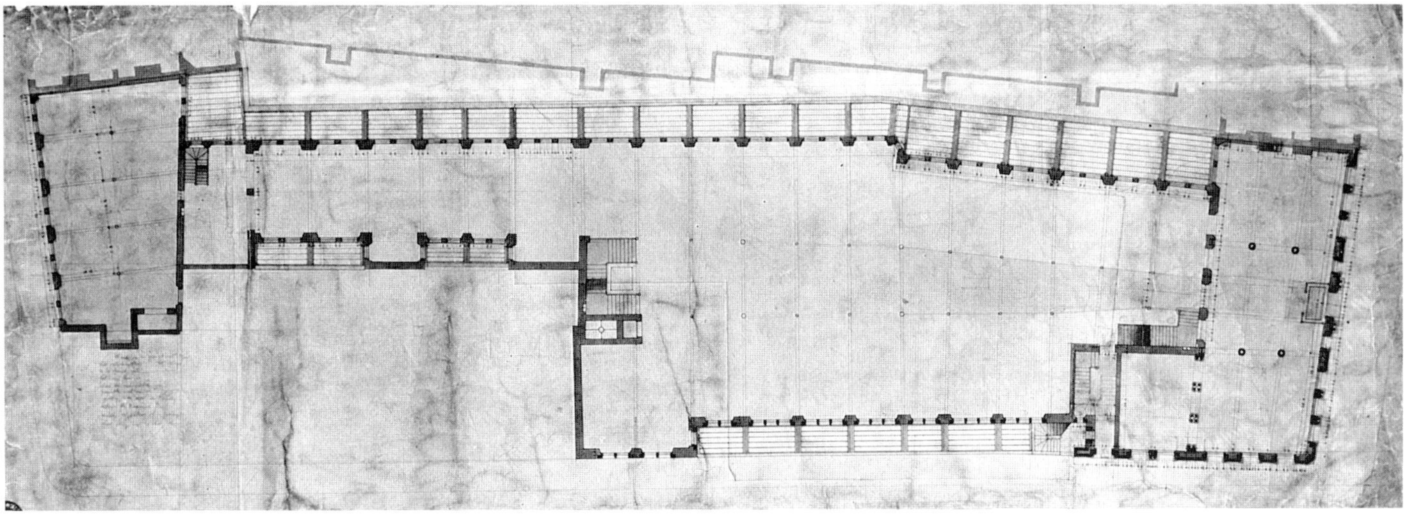

123. London, Holborn, Prudential Assurance building, ground plan, 1876 **[431]**

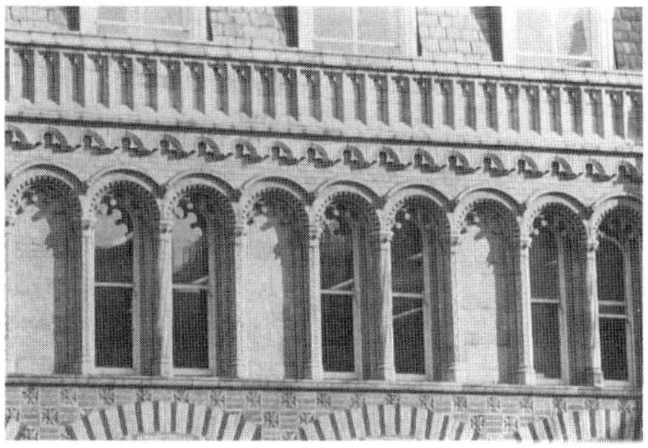

124. London, Holborn, Prudential Assurance Buildings (1895), terracotta, with brick and tile, in the north court **[602]**

125. Below, London, Holborn, Prudential Assurance Buildings (1895), principal façade, 1905 **[602]**

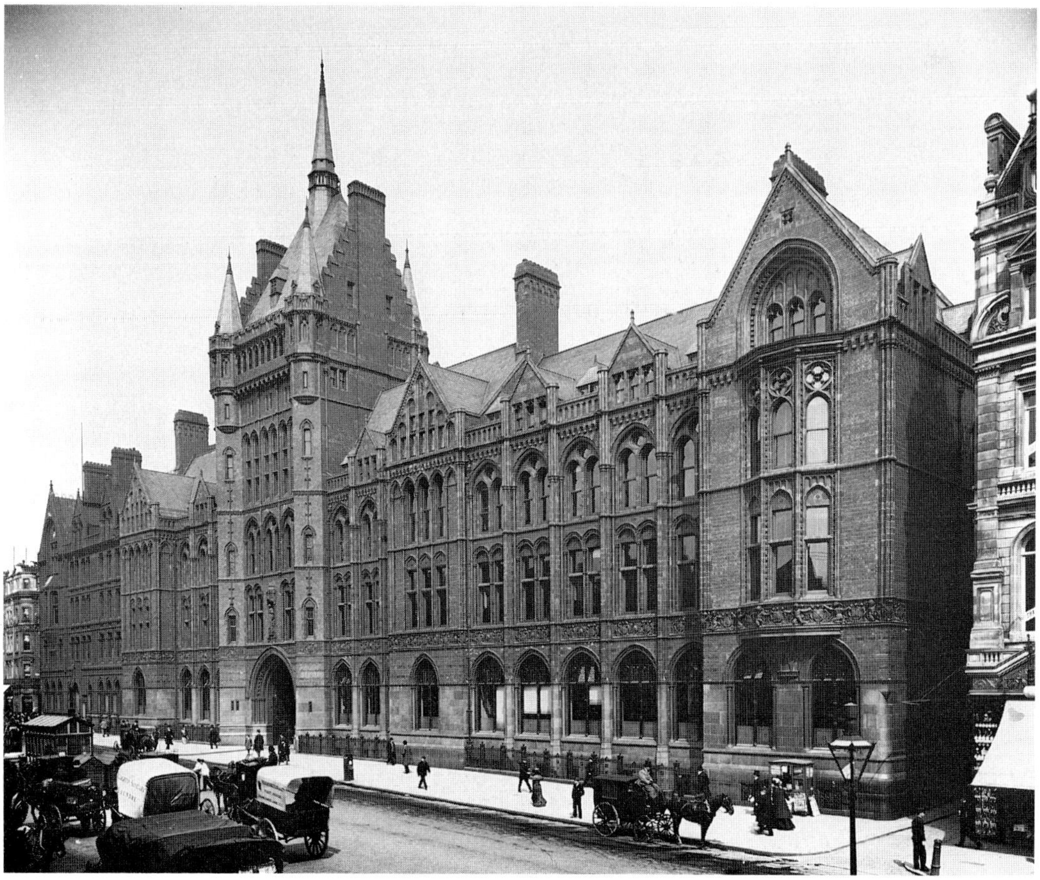

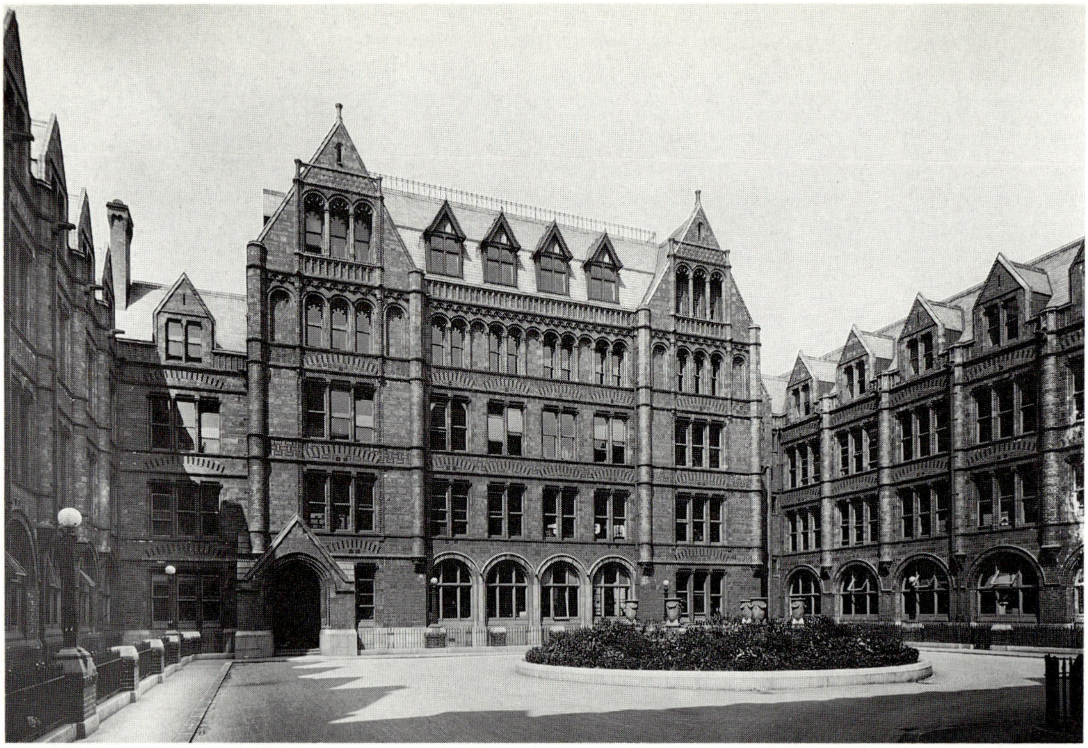

126. London, Holborn, Prudential Assurance Buildings (1895), north court, 1905 **[602]**

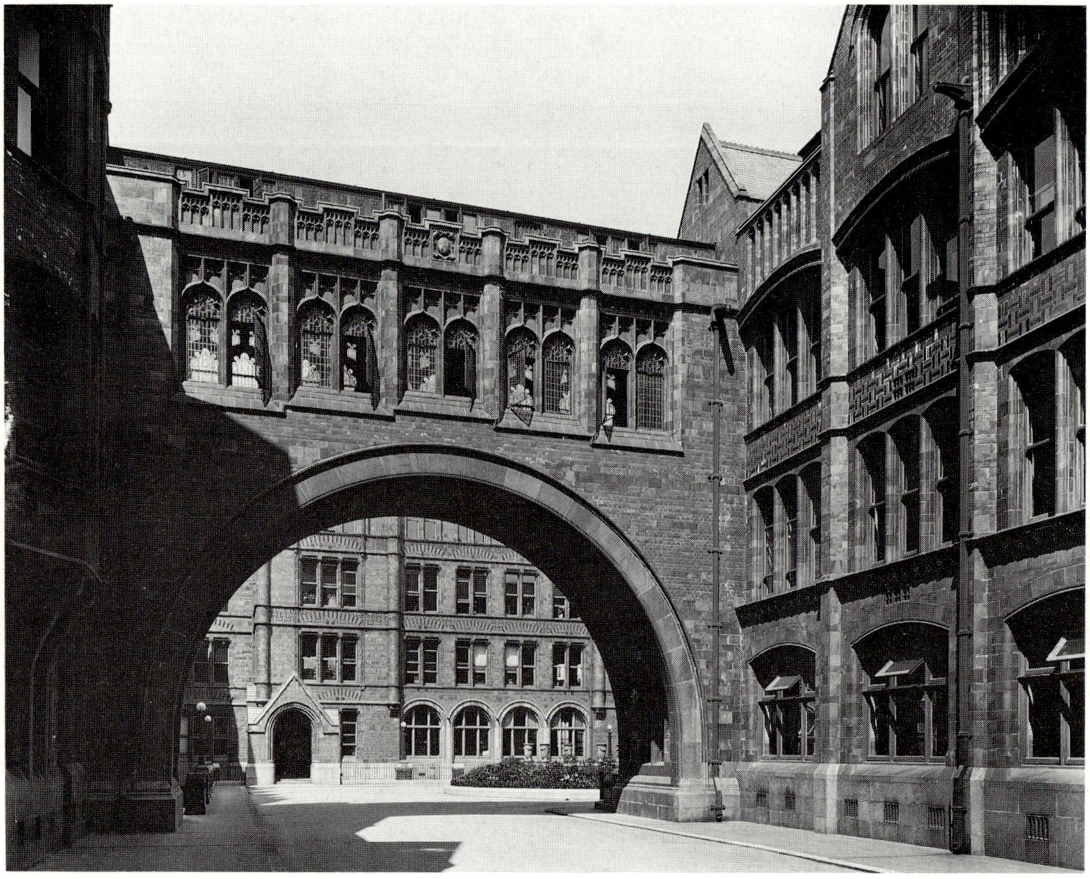

127. London, Holborn, Prudential Assurance Buildings (1895), central court and archway to north court, 1905 **[602]**

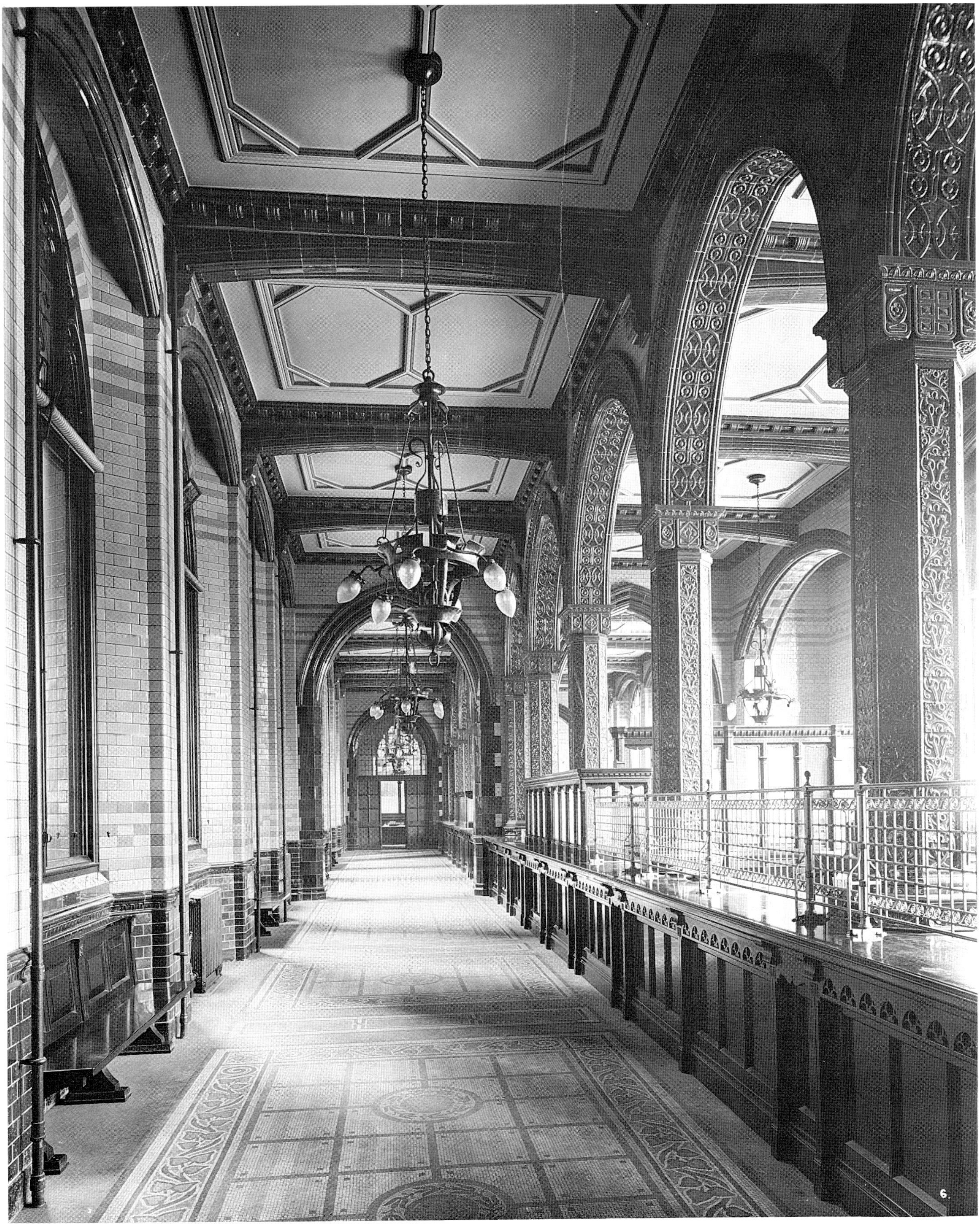

128. London, Holborn, Prudential Assurance Buildings (1895), cashiers' office, 1905 **[602]**

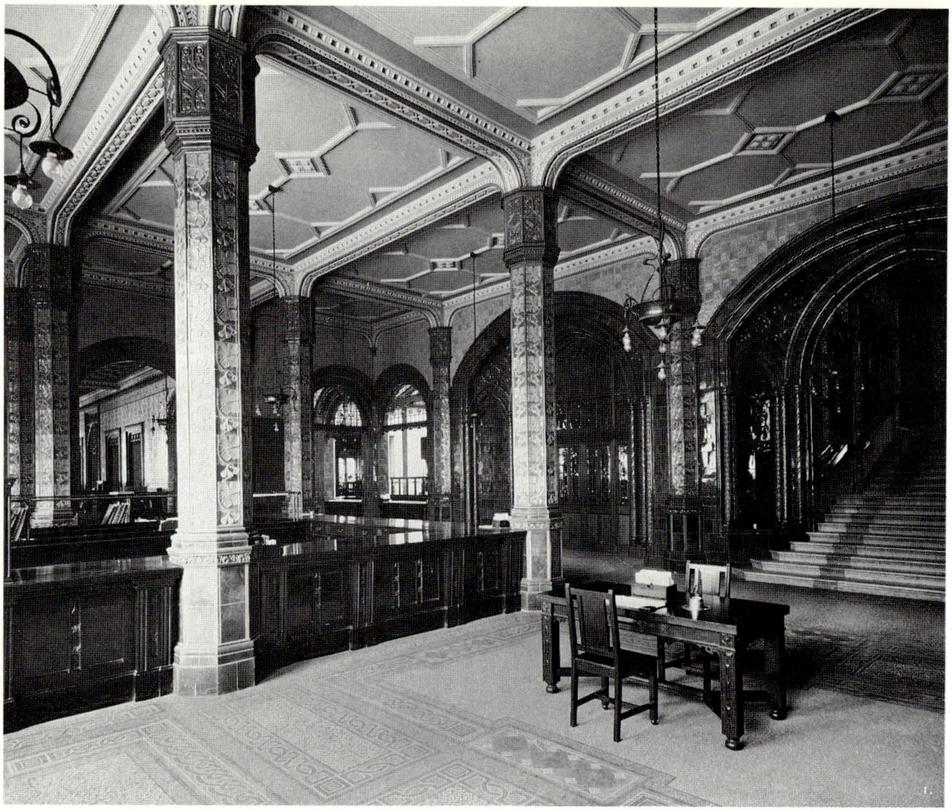

129. London, Holborn, Prudential Assurance Buildings (1895), main entrance foyer, 1905 **[602]**

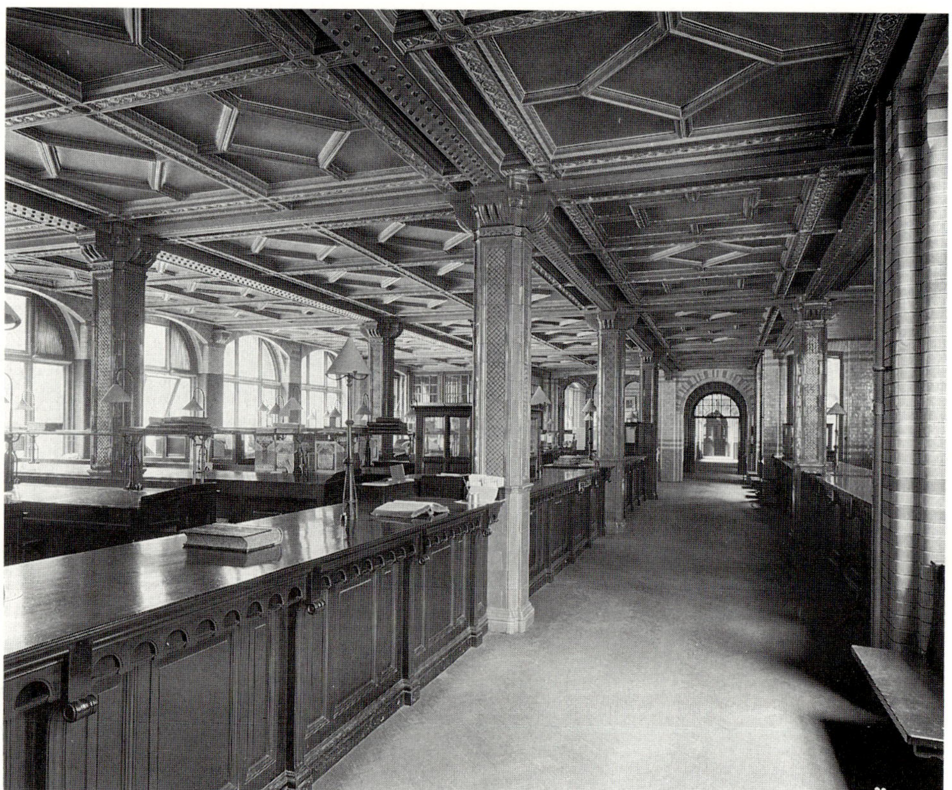

130. London, Holborn, Prudential Assurance Buildings (1895), clerks' public office, 1905 **[602]**

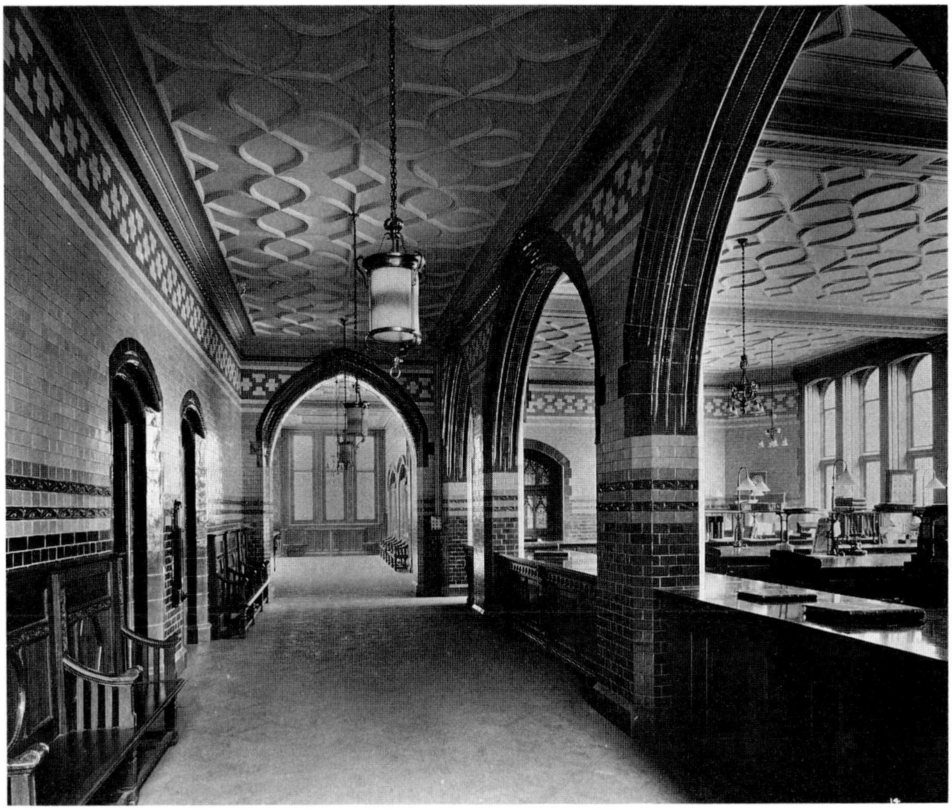

131. London, Holborn, Prudential Assurance Buildings (1895), managers' clerks' office, 1905 **[602]**

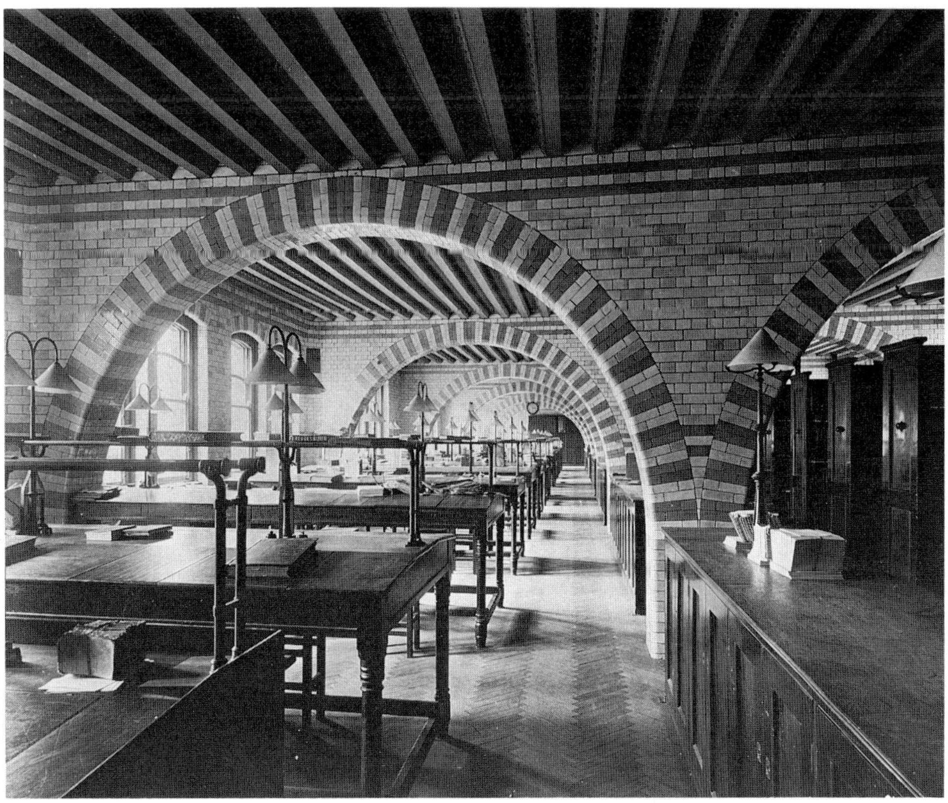

132. London, Holborn, Prudential Assurance Buildings (1895), clerks' office, 1905 **[602]**

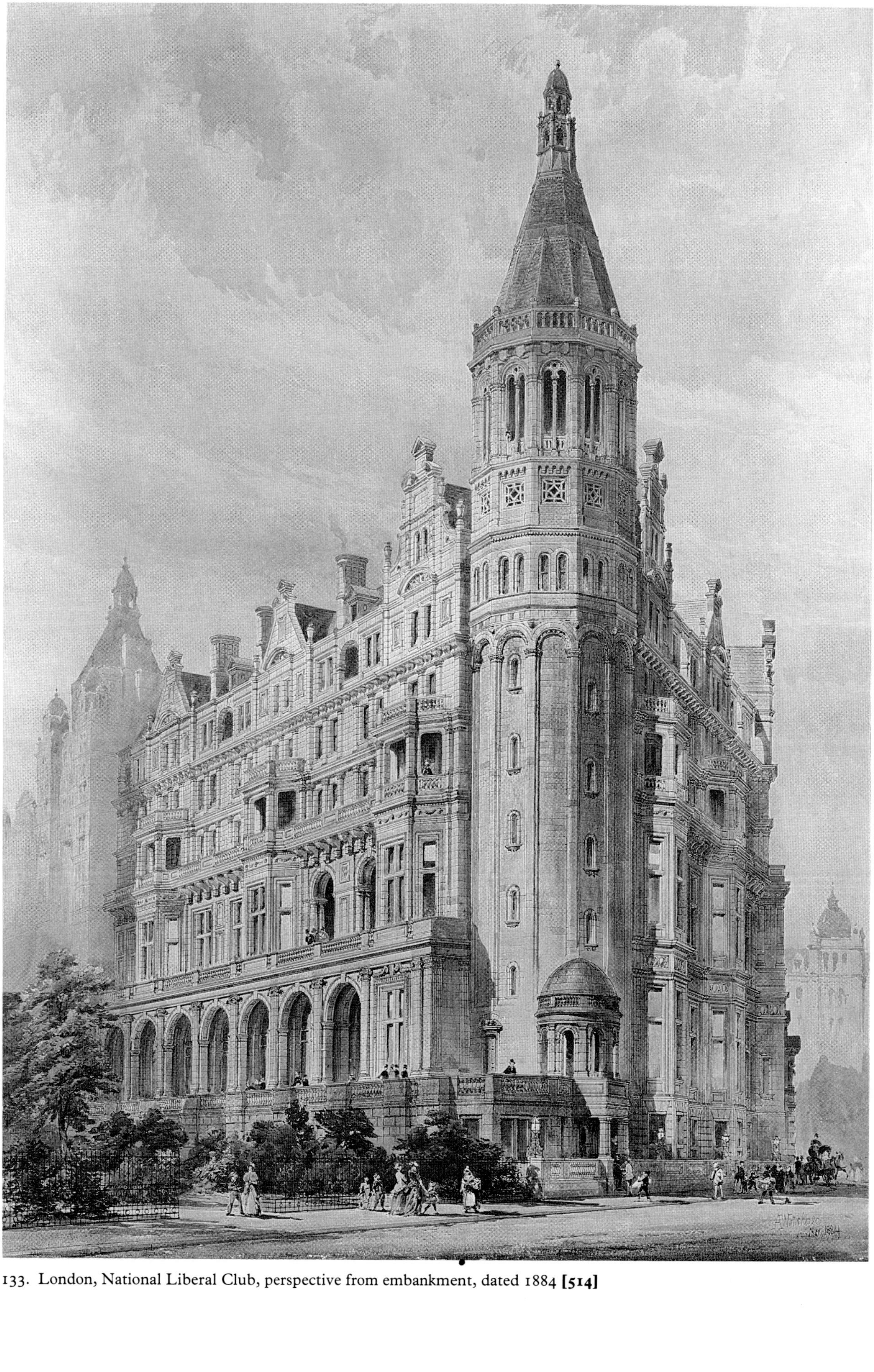

133. London, National Liberal Club, perspective from embankment, dated 1884 **[514]**

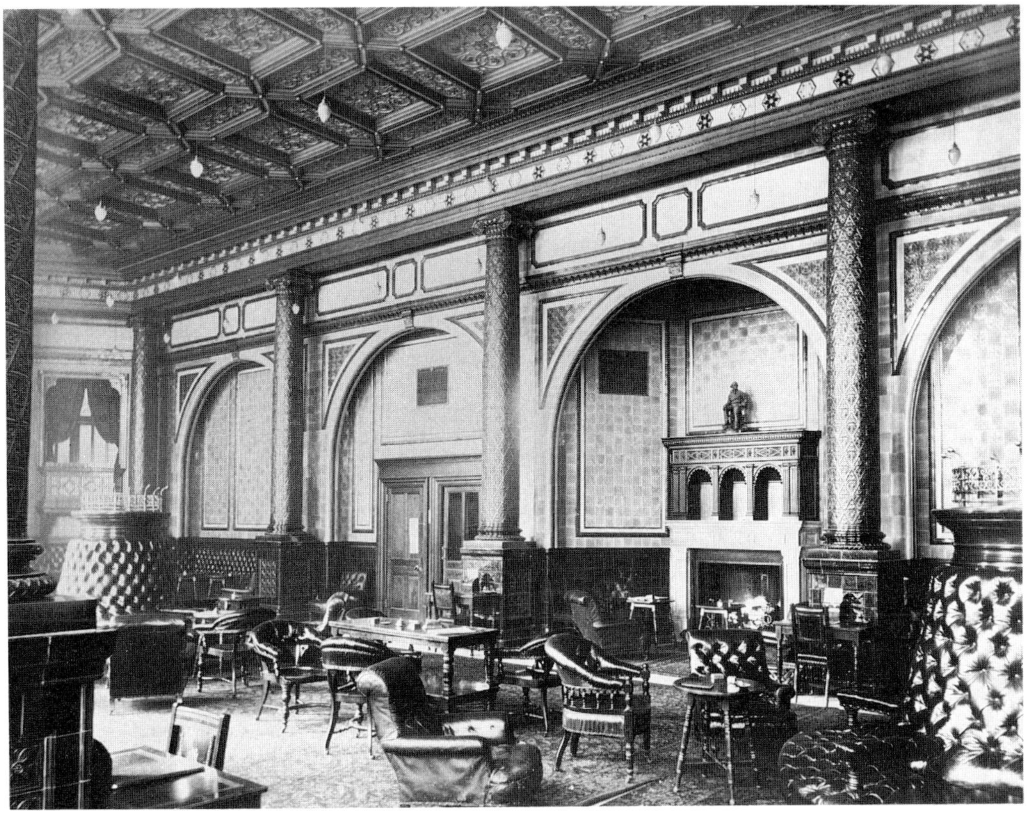

134. London, National Liberal Club (1884), basement smoking-room, 1887 **[514]**

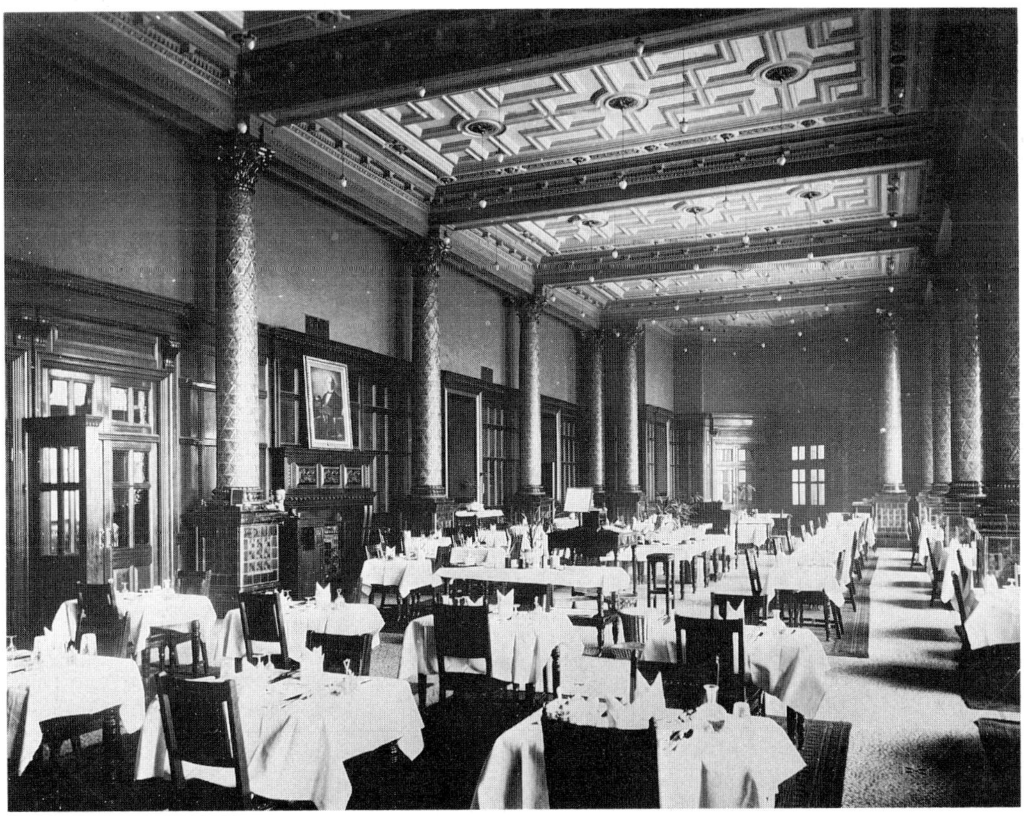

135. London, National Liberal Club (1884), dining-room, 1887 **[514]**

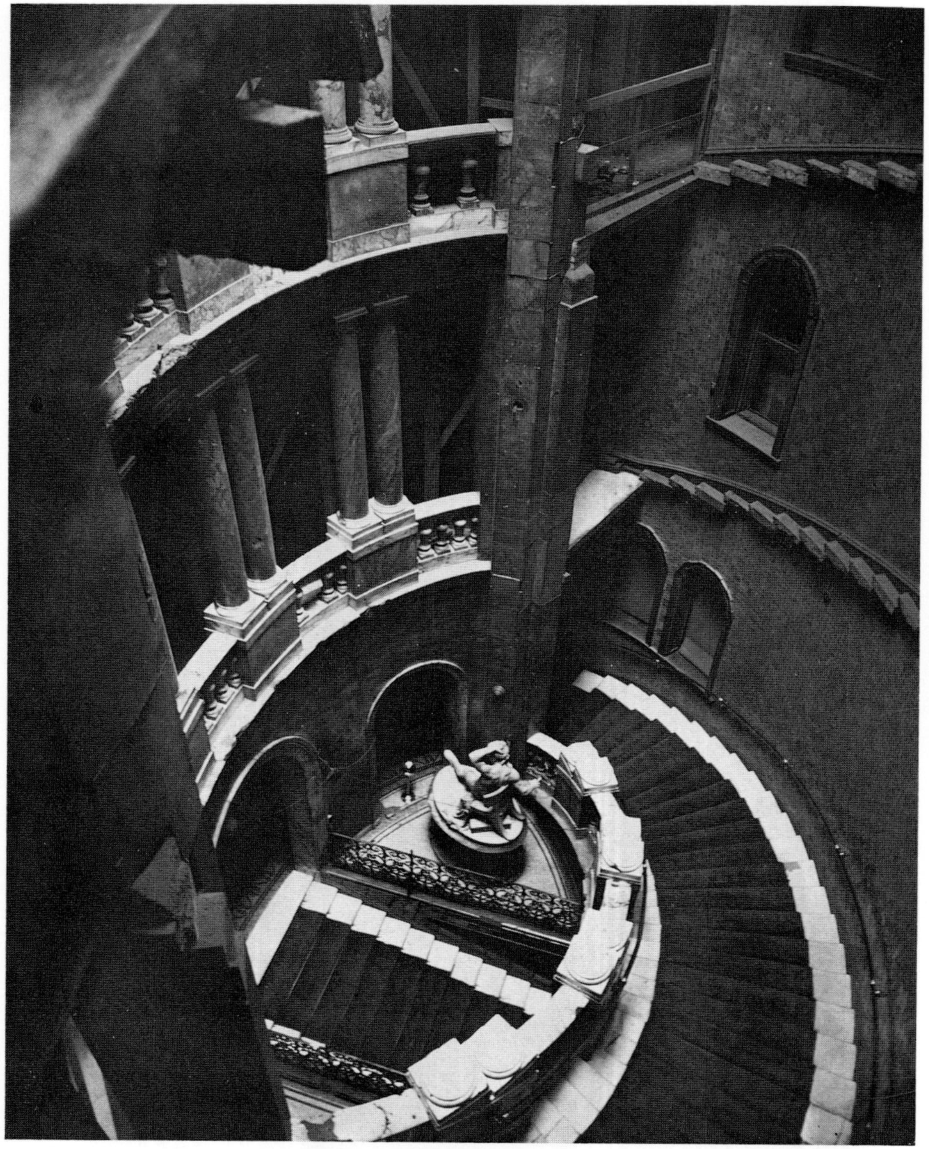

136. London, National Liberal Club (1884), principal staircase after bombing in 1941 **[514]**

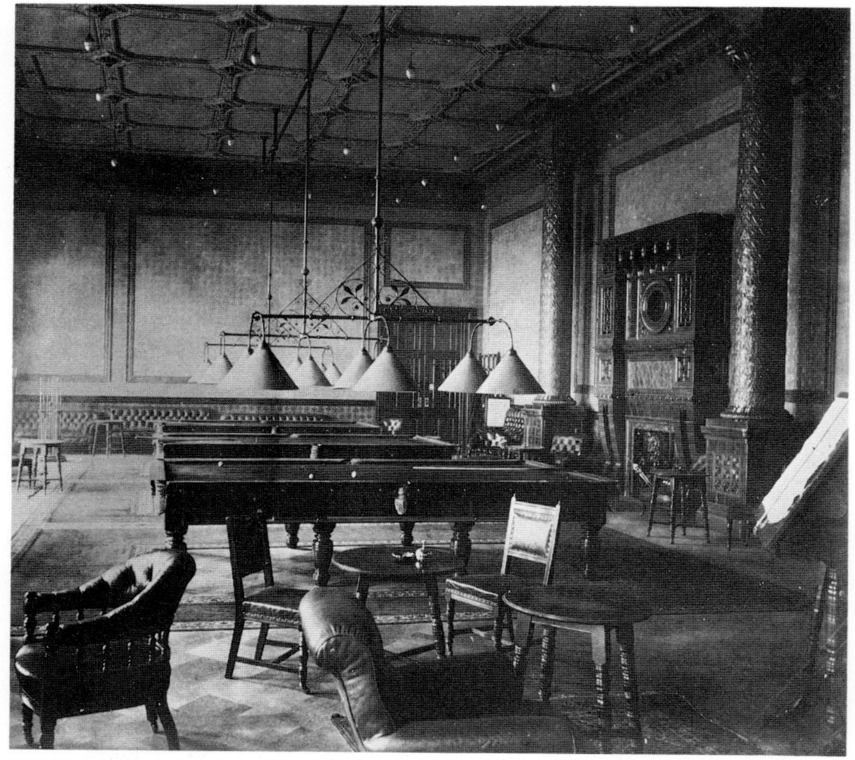

137. London, National Liberal Club (1884), billiards room on the first floor, 1887 **[514]**

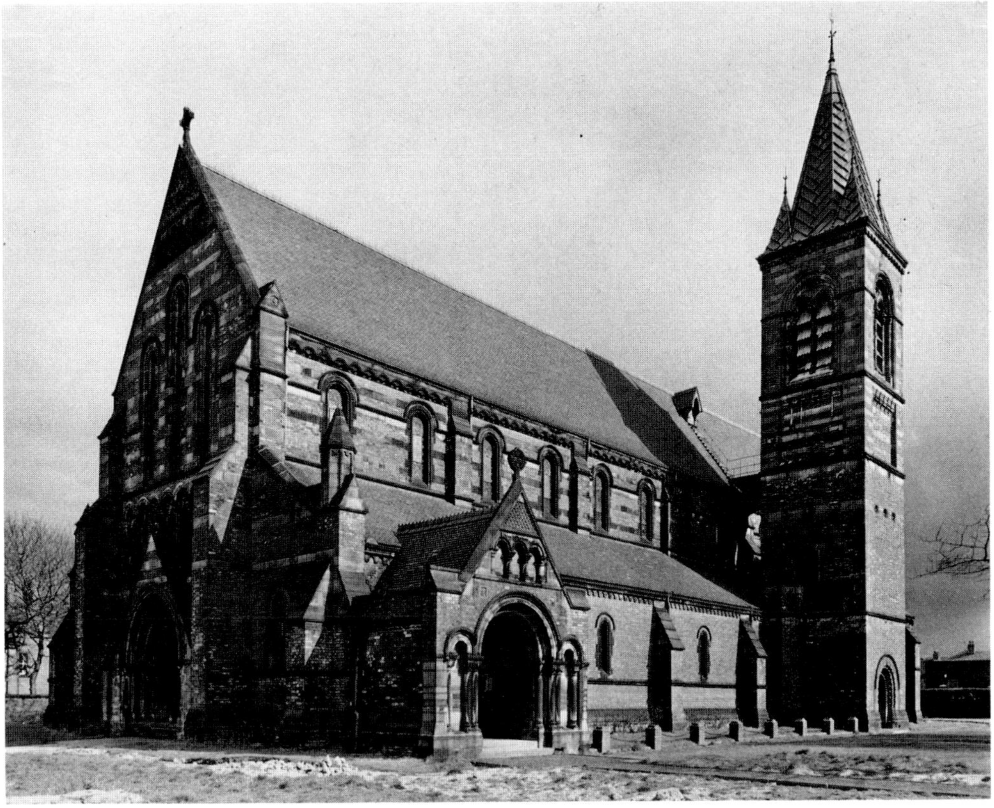

138. Manchester, Reddish, St Elisabeth's Church (1883) **[507]**

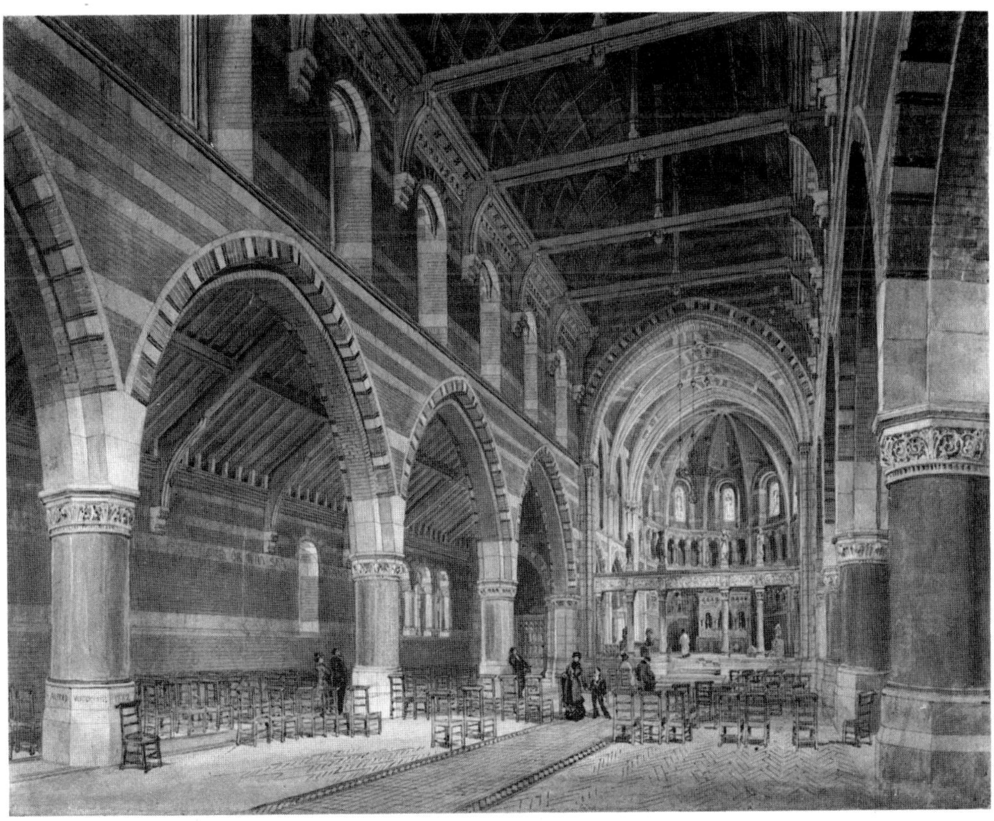

139. Manchester, Reddish, St Elisabeth's Church, interior perspective, dated 1882 **[507]**

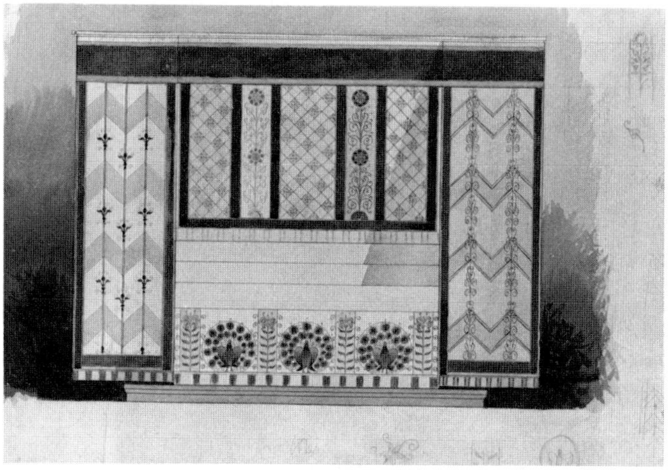

140. Manchester, Reddish, St Elisabeth's Church, sketch design for white altar frontal and triptych reredos, *c.*1885 **[507]**

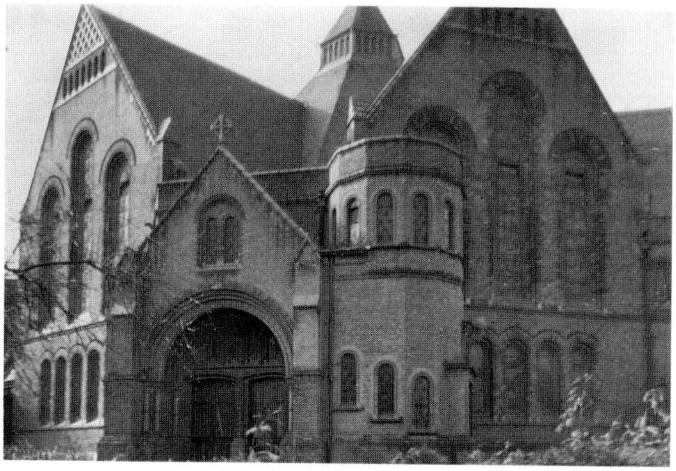

141. London, Lyndhurst Road Congregational Church (1883) **[505]**

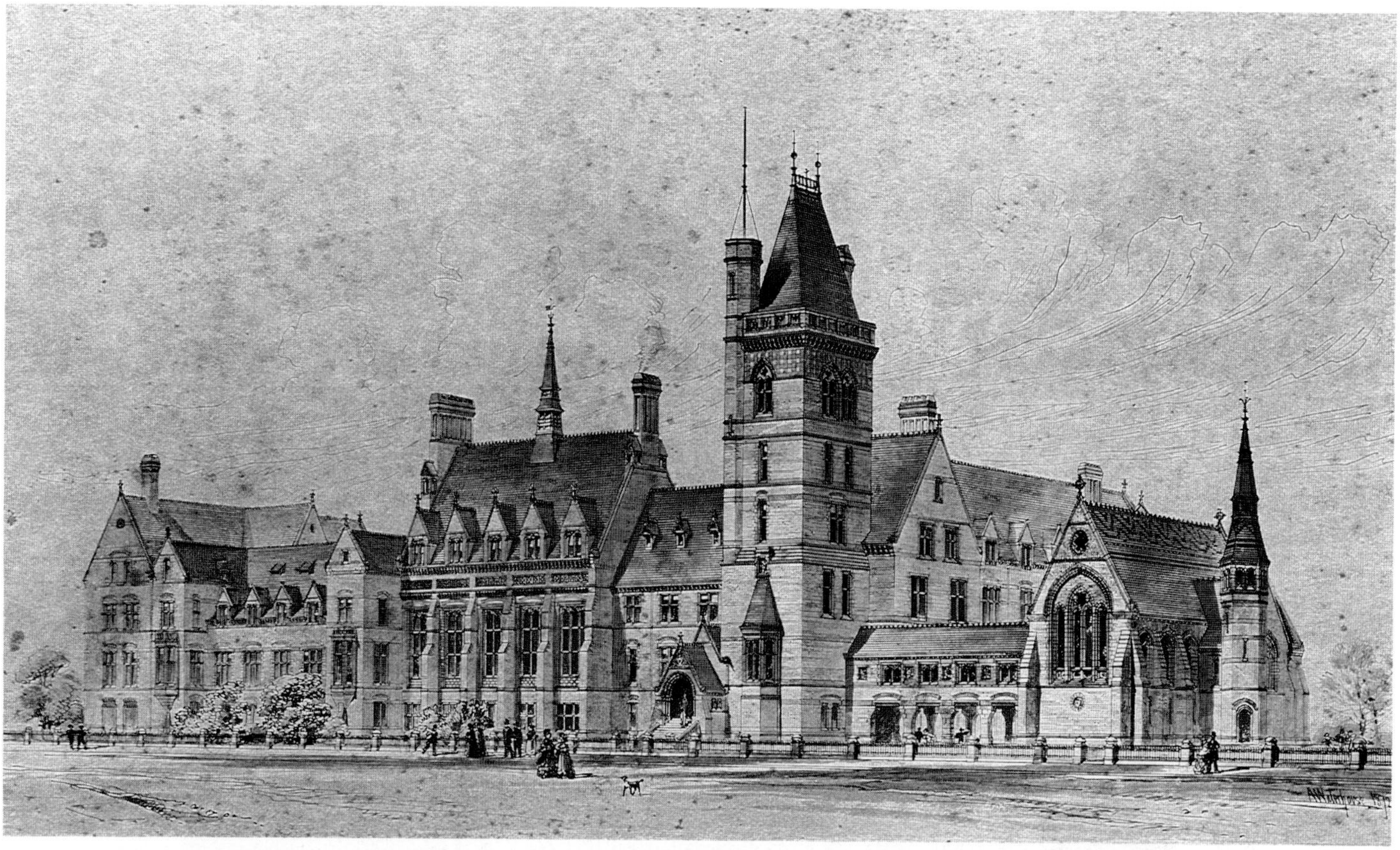

142. Liverpool Seamen's Orphans' Institution (1870), perspective, dated 1872. The chapel to the right was not built **[337]**

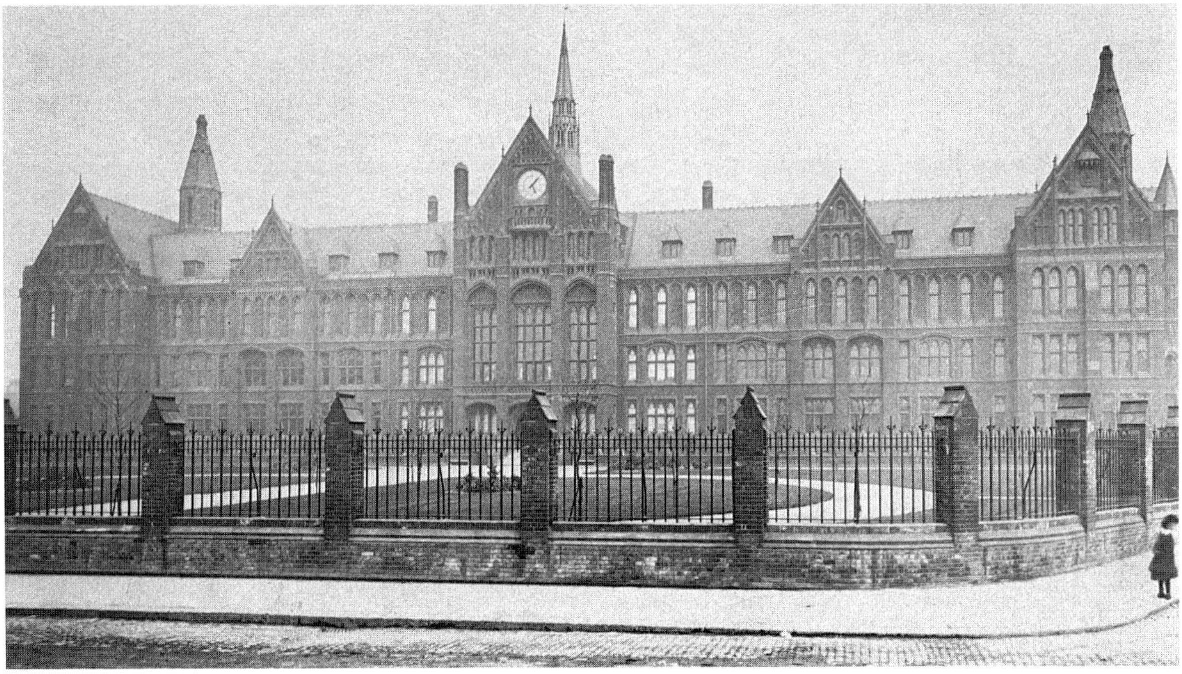

143. London, Hammersmith, St Paul's School (1881), *c.*1887 **[492]**

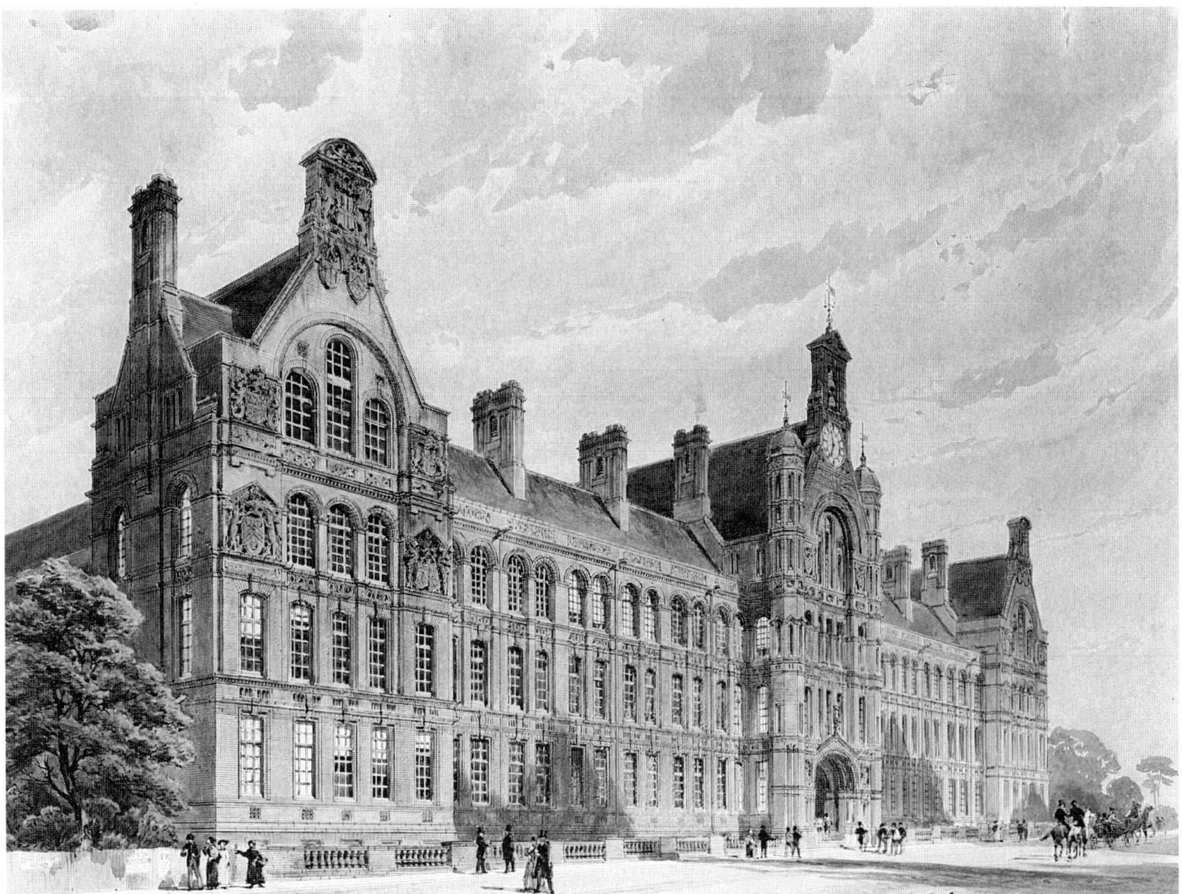

144. London, Exhibition Road, City and Guilds Technical Institute, perspective dated 1881 **[490]**

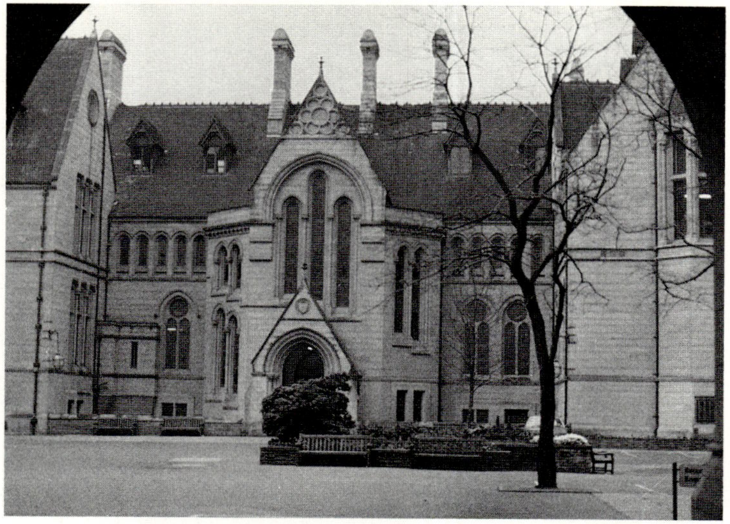

145. Manchester, Owens College (1869), entrance to the original building **[321]**

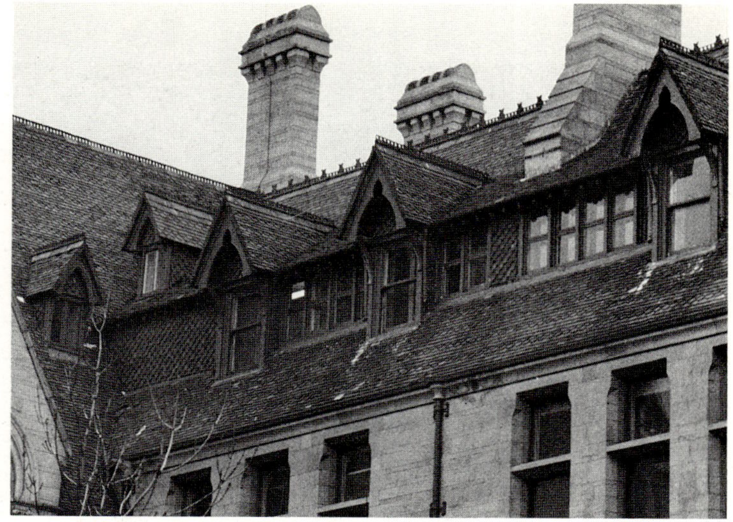

146. Manchester, Owens College (1869), detail of roof of original building **[321]**

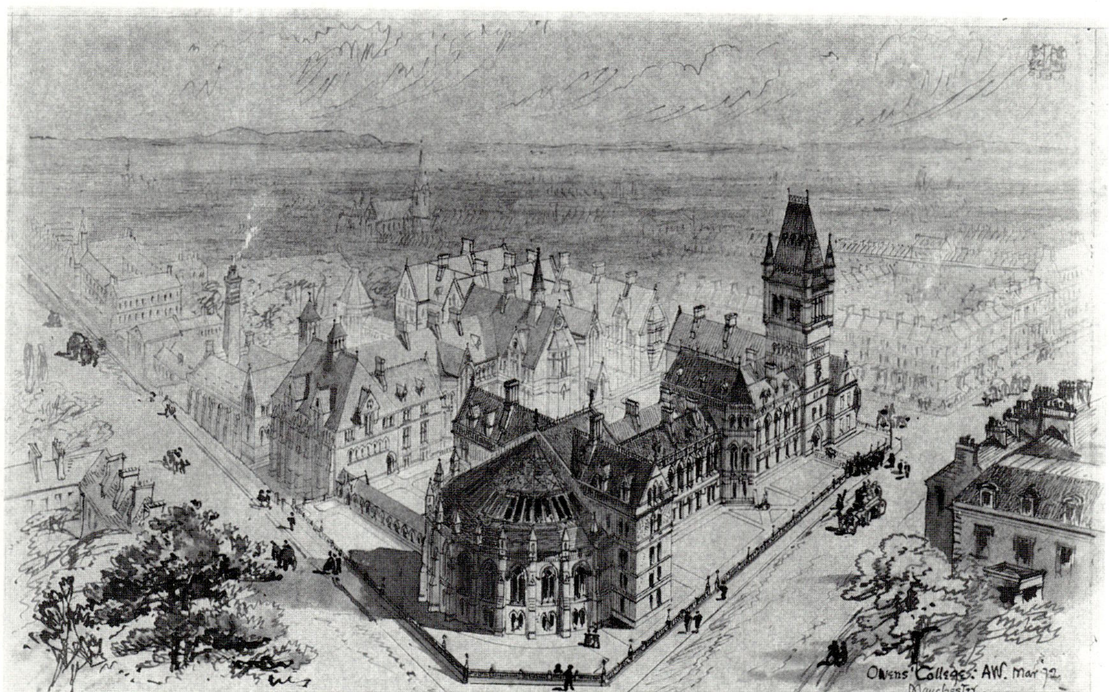

147. Manchester, Owens College (1869), bird's-eye perspective, dated 1872 **[321]**

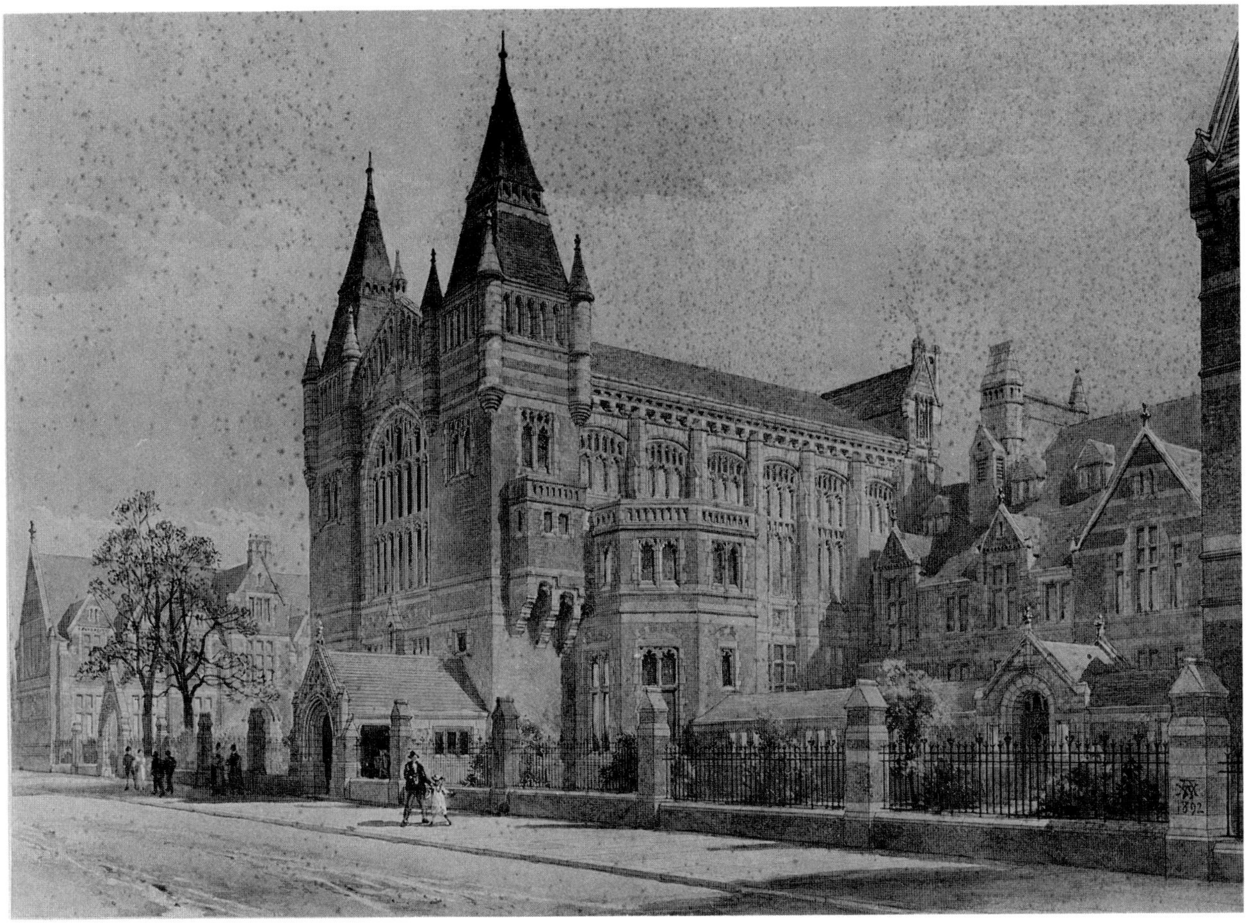

148. Leeds, Yorkshire College (University), perspective of Great Hall, dated 1892 **[574]**

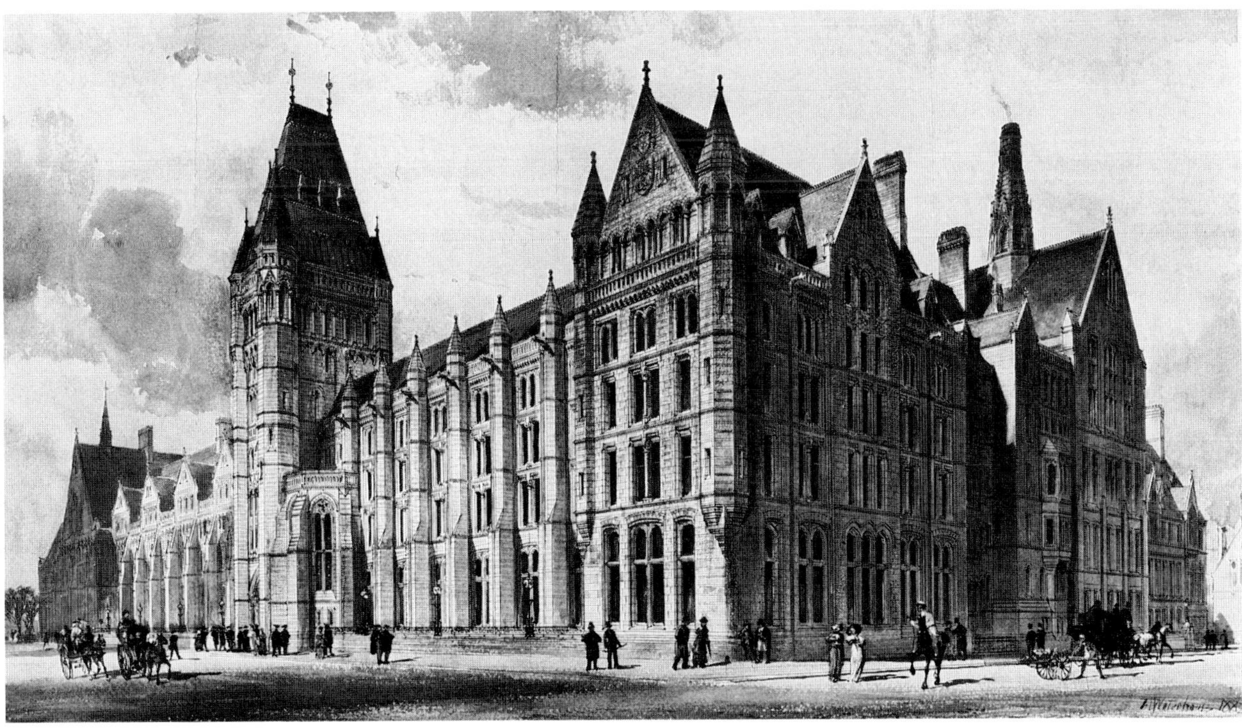

149. Manchester, Owens College, Oxford Road façade, perspective, dated 1882 **[515]**

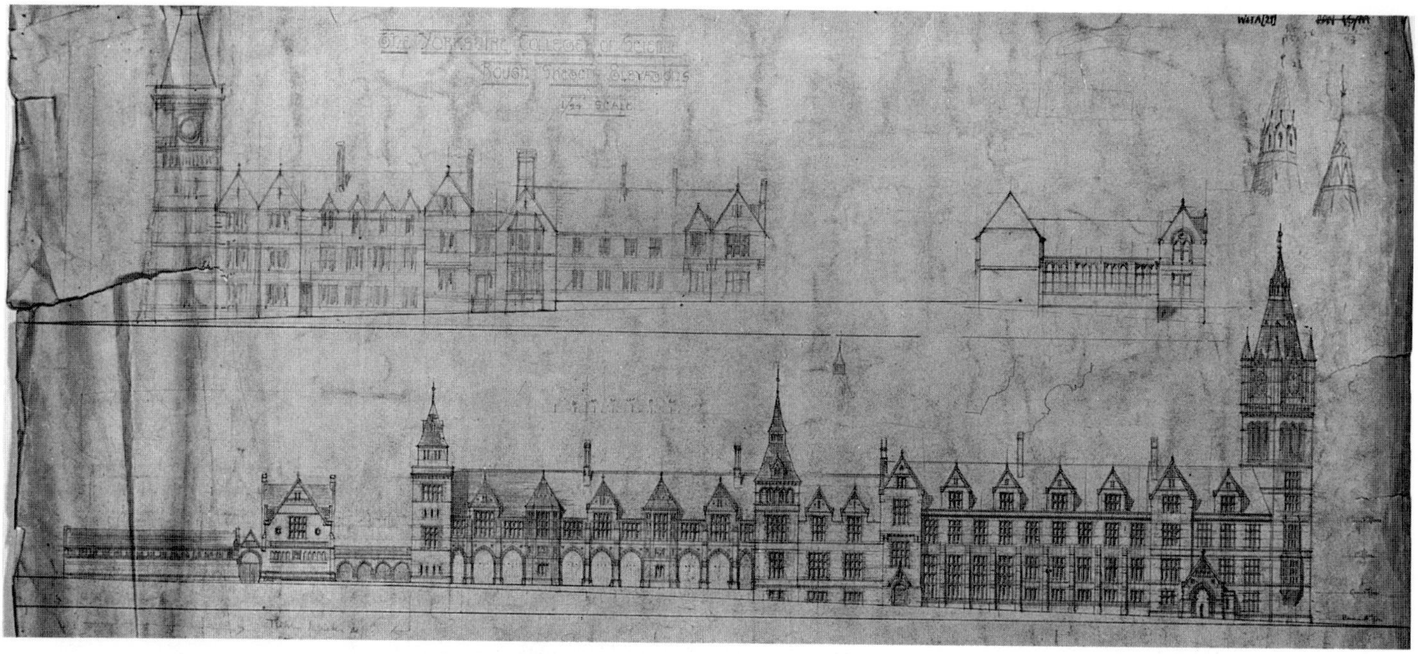

150. Leeds, Yorkshire College, proposed elevation, *c.*1878 **[460]**

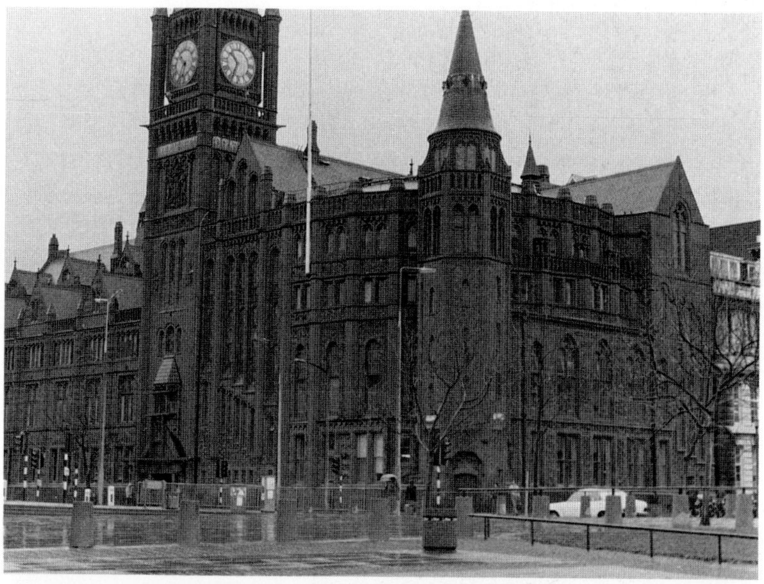

151. Liverpool, University (1888), Victoria Building and Jubilee Tower **[549]**

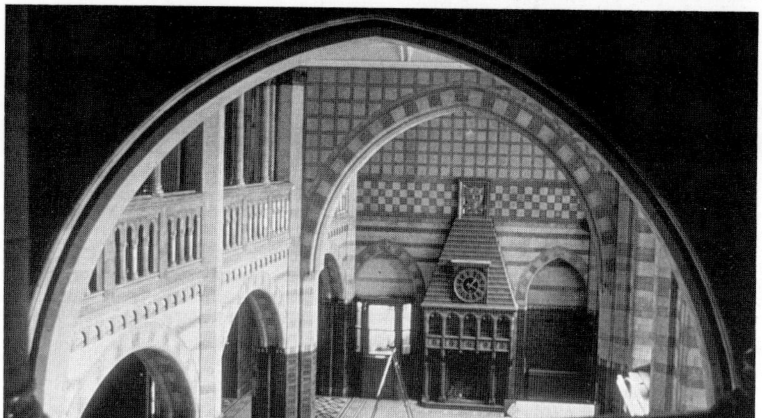

152. Liverpool, University (1888), assembly space in the Victoria Building **[549]**

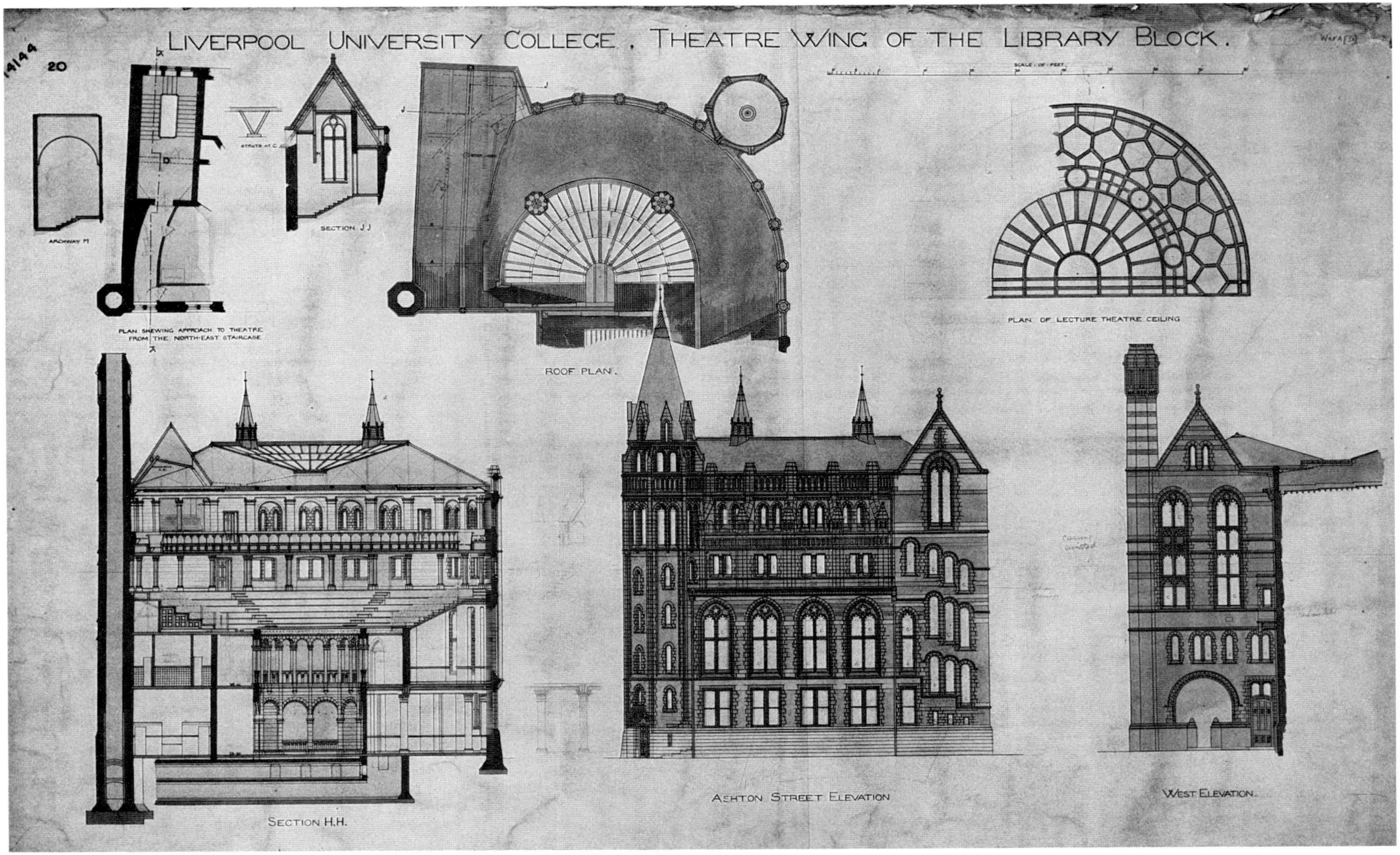

153. Liverpool, University, Victoria Building (1888), section, elevation, and roof plan, shewing the assembly space beneath the lecture theatre **[549]**

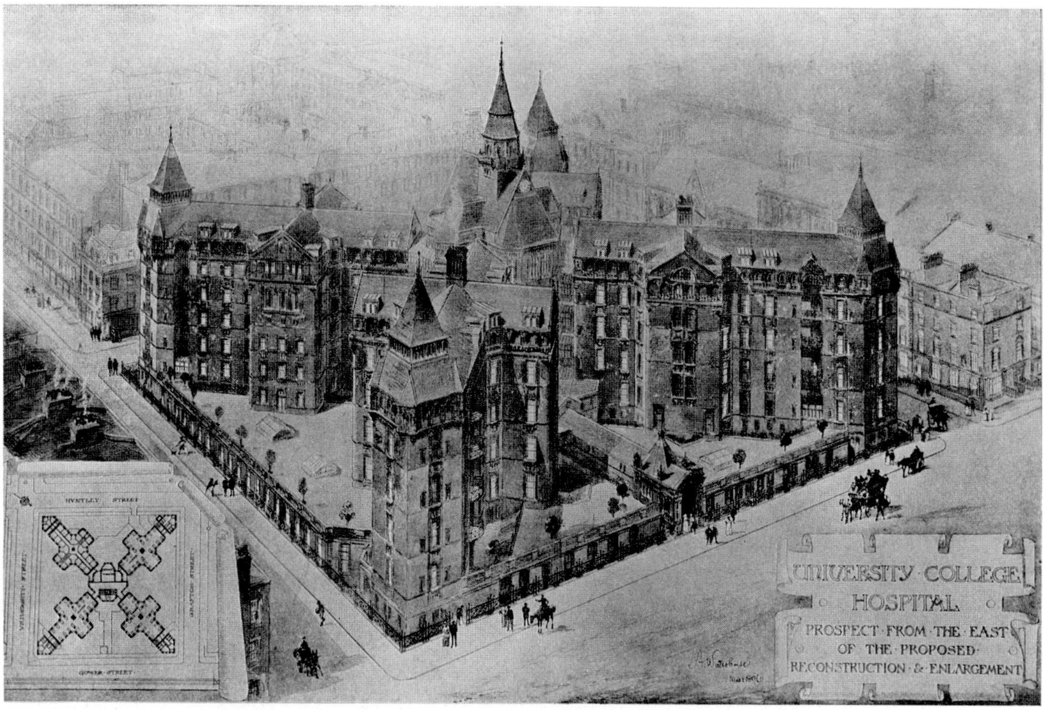

154. London, University College Hospital, bird's-eye perspective, dated 1896 **[588]**

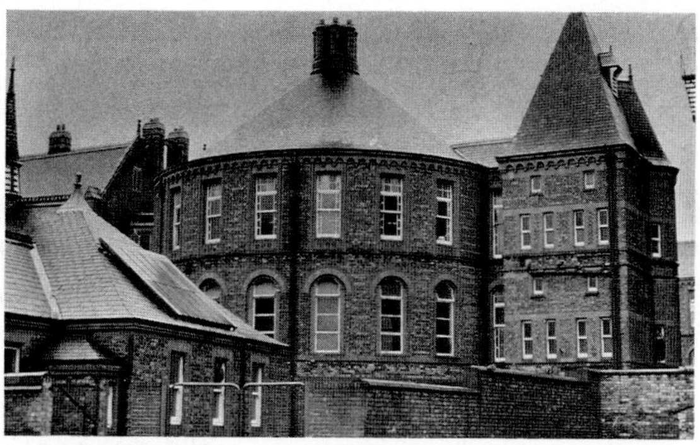

155. Liverpool, Royal Infirmary (1886), circular ward wing **[532]**

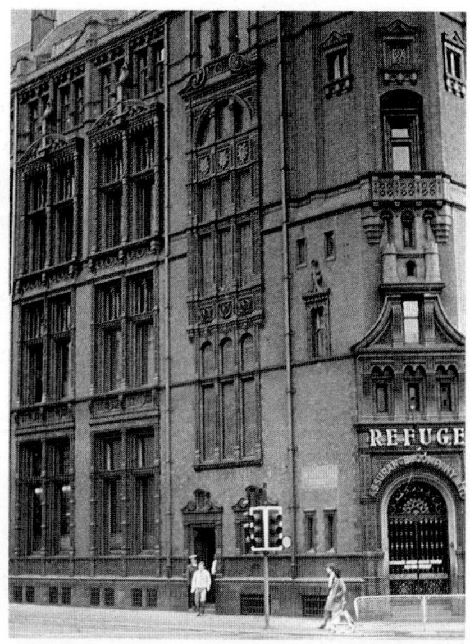

157. Manchester, Refuge Assurance Building (1891), public entrance, clerks' entrance and staircase tower, and offices **[570]**

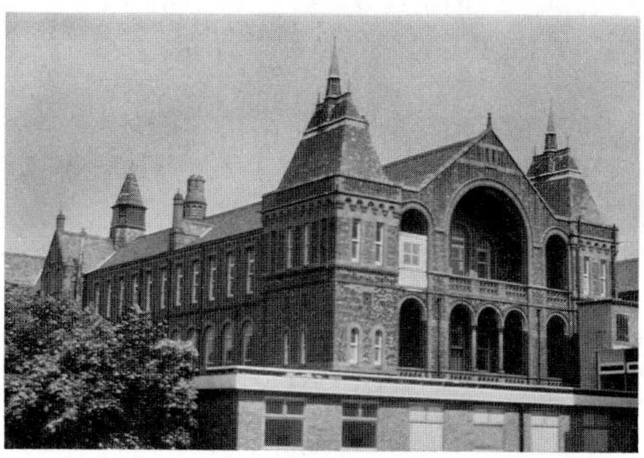

156. Liverpool, Royal Infirmary (1886), ward wing **[532]**

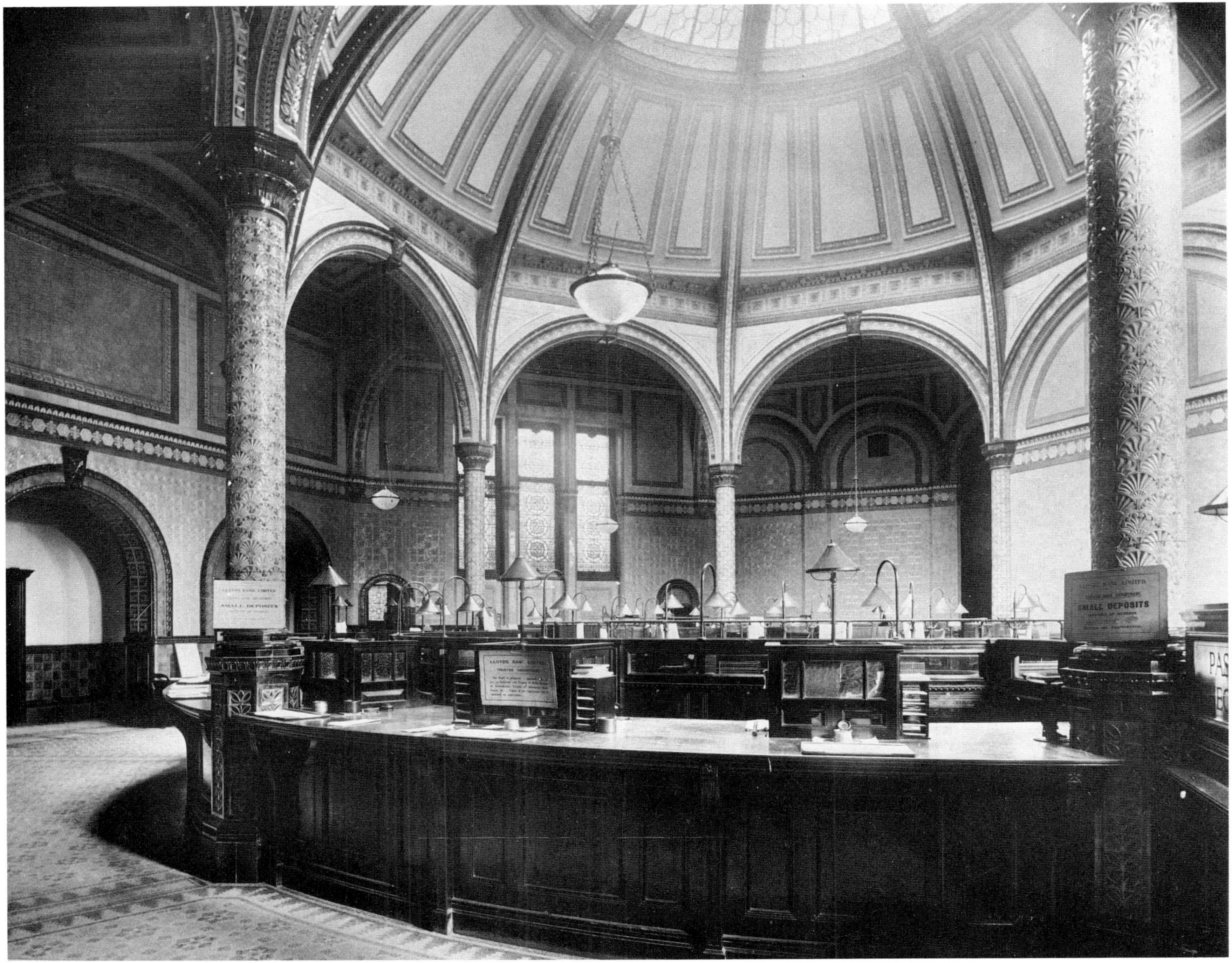

158. Cambridge, Fosters' Bank (1891), interior of banking hall, *c.*1900 **[565]**

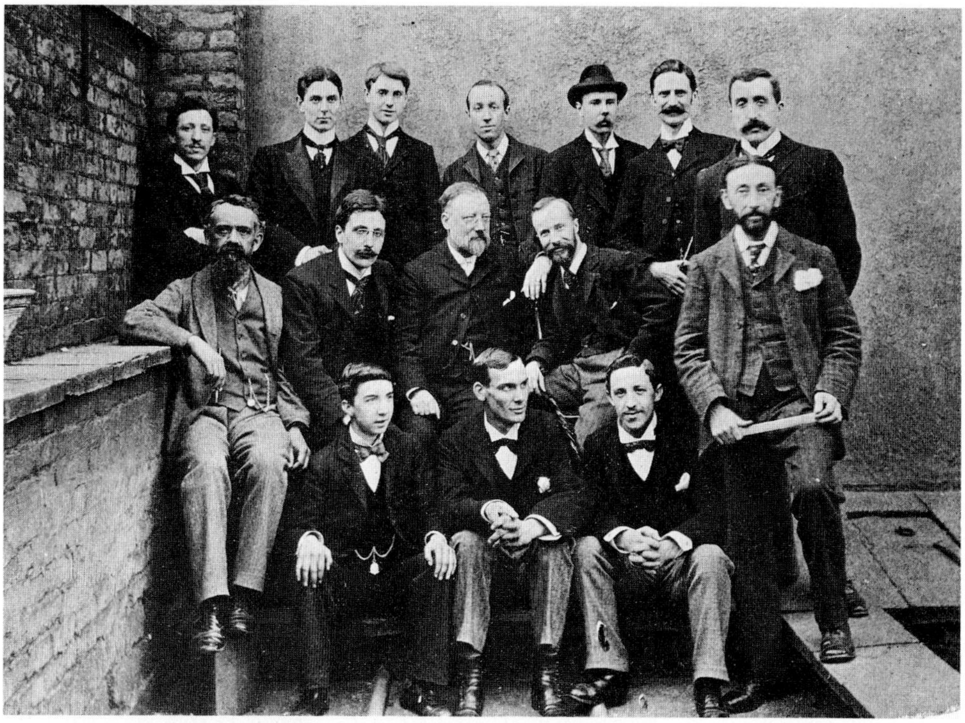

159. Staff of Waterhouse's office on the roof of 8 New Cavendish Street, *c.*1900. *Front row l. to r.*, William Minter, Alan George (son of Sir Ernest George), A. Grosse; *middle row l. to r.*, Pennington, Crum Watson, Cooper (chief draughtsman), Robert Ardley, A. Spice; *back row l. to r.*, James Grosse, Whitwell, Masey, Mansford, E. J. Clark, S. Powell, Gridley

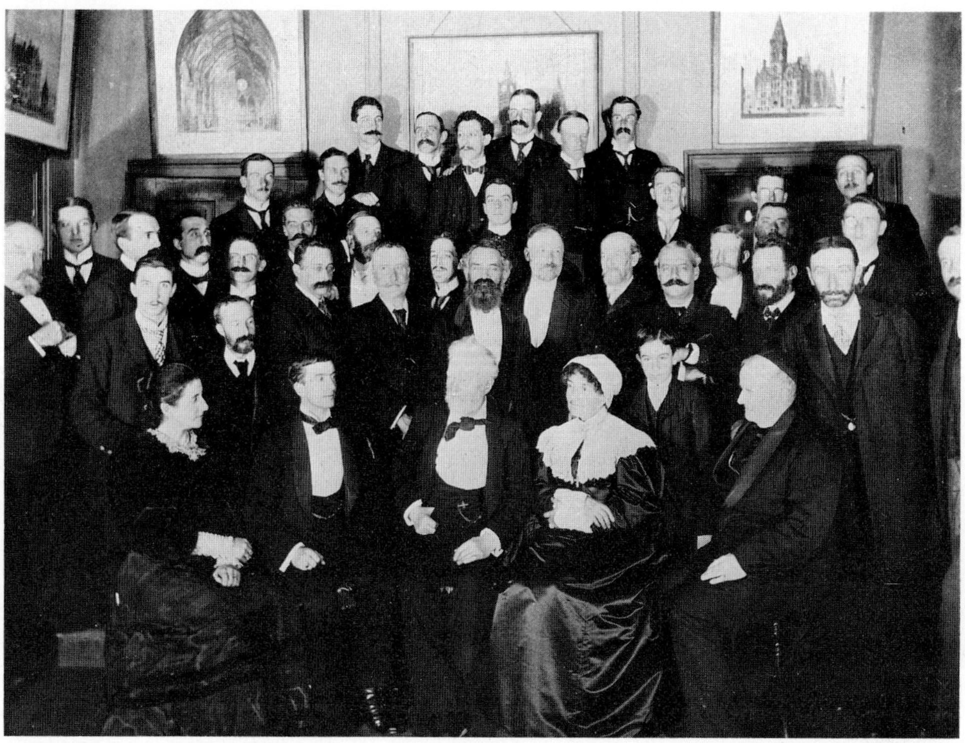

160. Past and present staff of the office assembled for Alfred Waterhouse's seventieth birthday, 1900. *Seated l. to r.*, Mrs Paul Waterhouse, Paul Waterhouse, Alfred, Mrs Waterhouse (Bessie), Revd Ernest Geldart; *standing l. to r.*, G. T. Redmayne, Whitwell, ?, office boy, Gridley, Robert Ardley, Powell, ?, ?, ?, ?, Alfred Grosse, Pennington, Cooper (chief draughtsman), J. M. Brydon, Fellowes Prynne, 'Willie' Bloomfield (*in front*), ?, ?, Spice, Shackle; *at rear l. to r.*, Hodgson, Mount, Crum Watson, Mansell, James Grosse, Ballardie, ?, McGareth Hogg, Elsley Smith, Stroud, L. S. Sullivan, ?, Mansford, Fred Macey. (On the wall perspectives of Manchester Town Hall, Manchester Assize Courts, Liverpool University, and Manchester Town Hall)

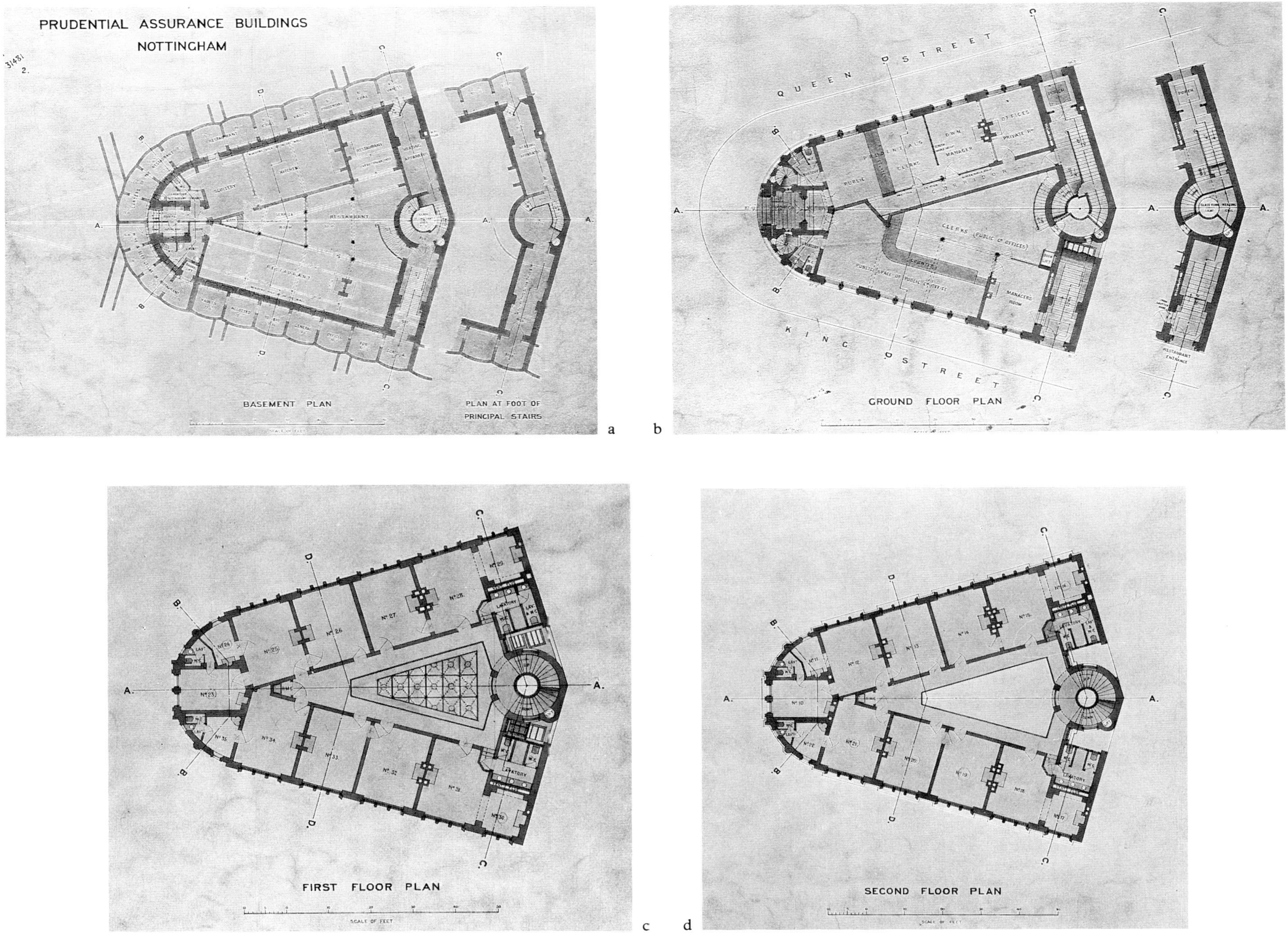

a b c d

161. Nottingham, Prudential Assurance Building (1893): plans: *a*. basement, *b*. ground floor, *c*. first floor, *d*. second floor **[584]**

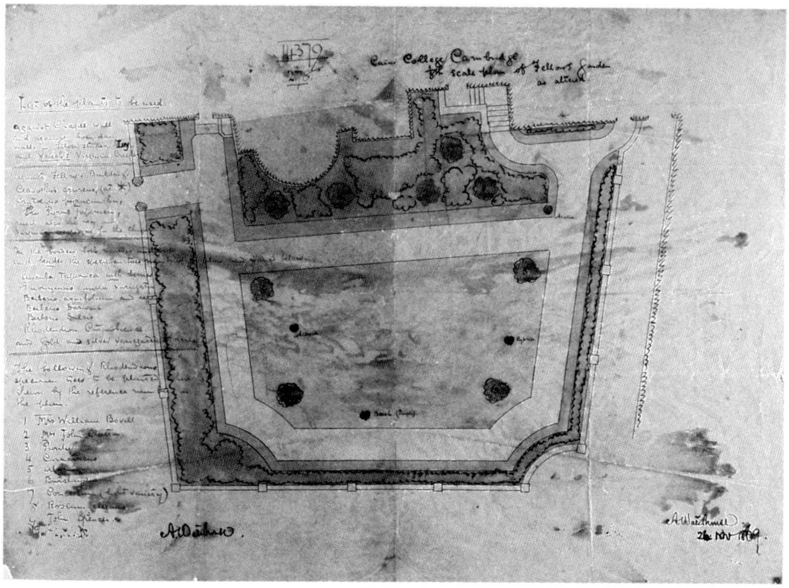

162. Cambridge, Gonville and Caius College (1867), plan for planting the Fellows' garden in Tree Court, dated 1869. The three trees named are Acacia, purple beech, and cypress; and the notes name nine specimen rhododendron bushes **[268]**

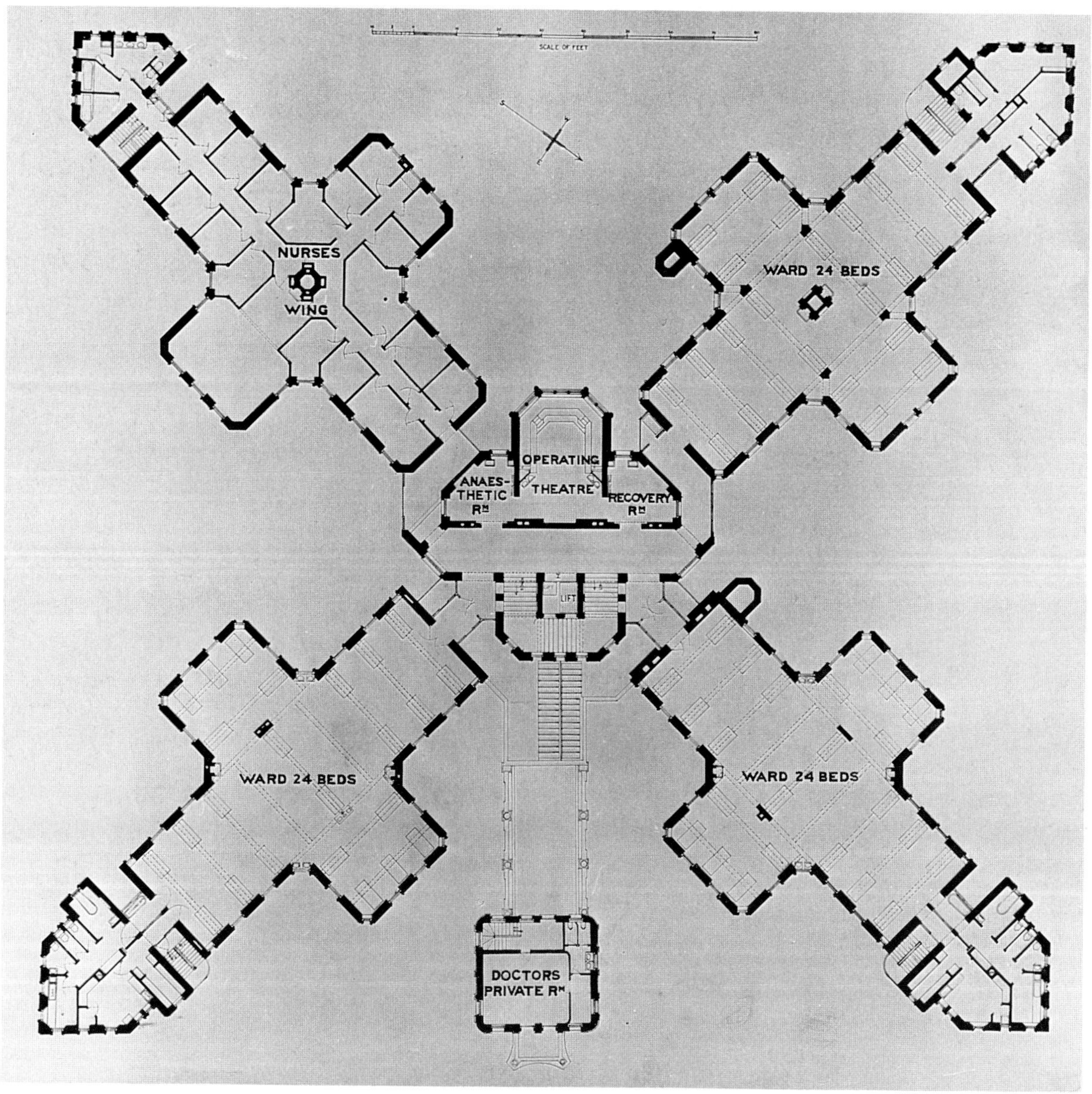

163. London, University College Hospital (1894), first-floor plan **[588]**

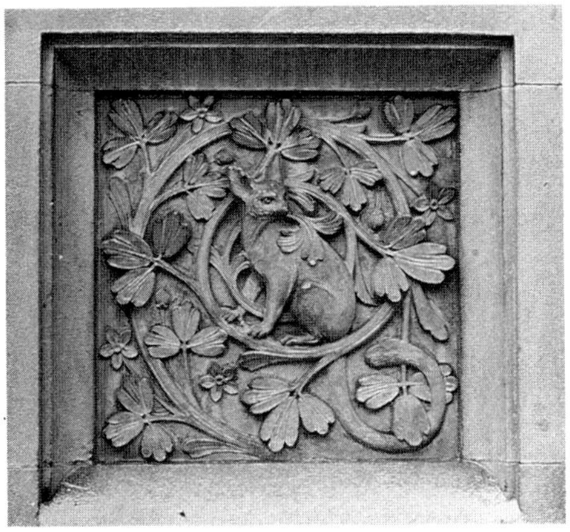

164. London, Natural History Museum (1870), terracotta panel from gate piers **[243]**

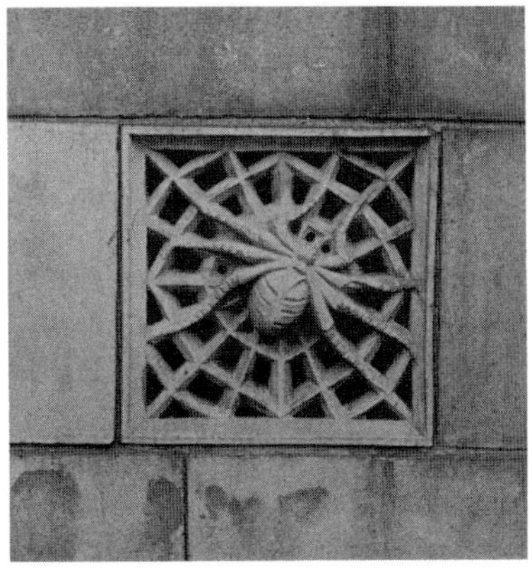

165. London, Natural History Museum (1870), terracotta air brick **[243]**

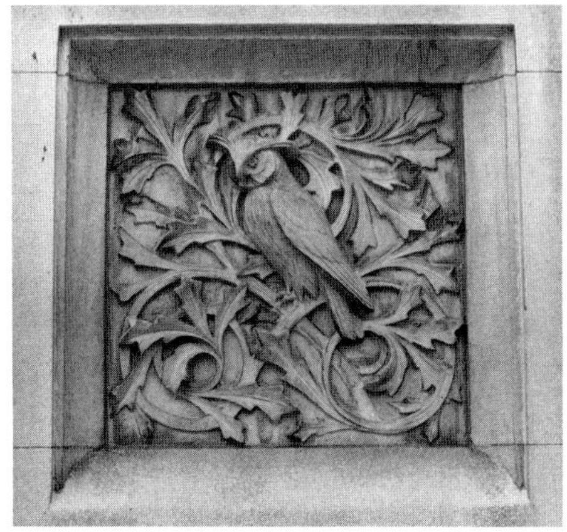

166. London, Natural History Museum (1870), terracotta panel from gate piers **[243]**

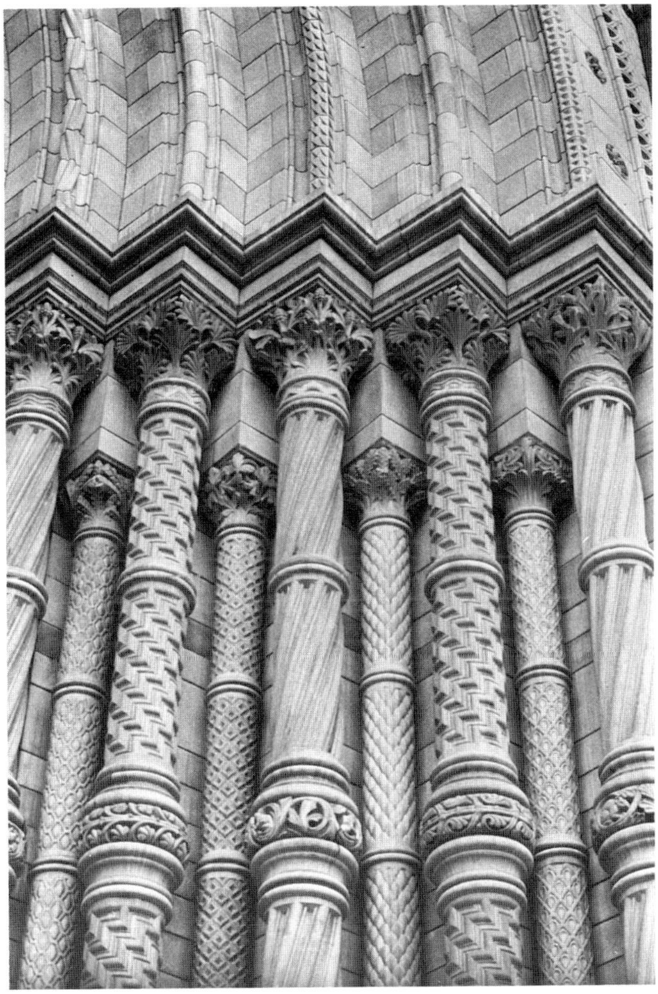

167. London, Natural History Museum (1870), terracotta columns of principal entrance **[243]**

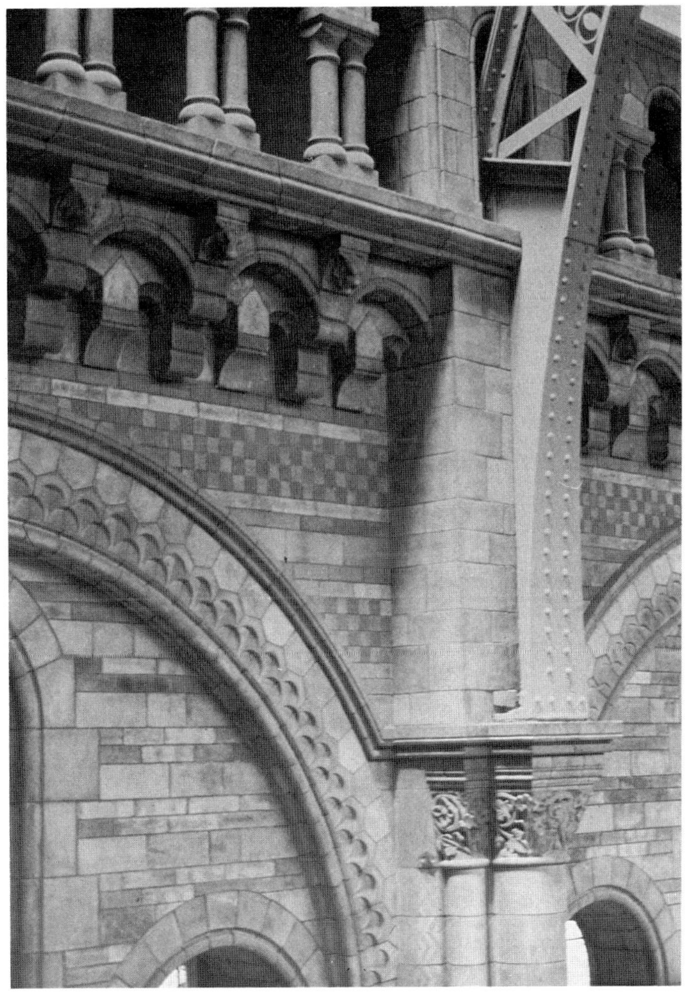

168. London, Natural History Museum (1870), springing of iron roof to index hall **[243]**

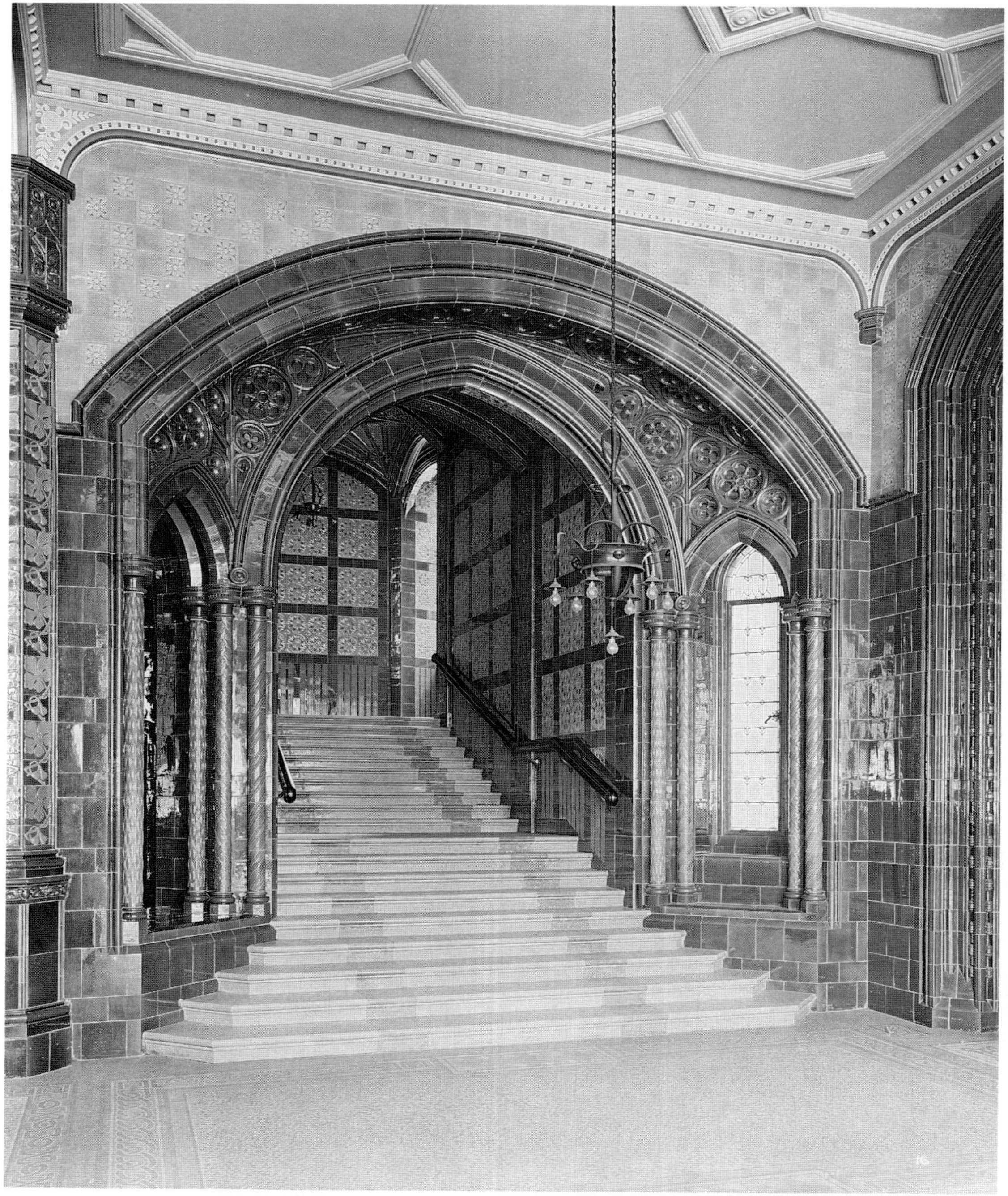

169. London, Holborn, Prudential Assurance Buildings (1895), faience decoration of the principal staircase in olive green and cream, 1905 **[602]**

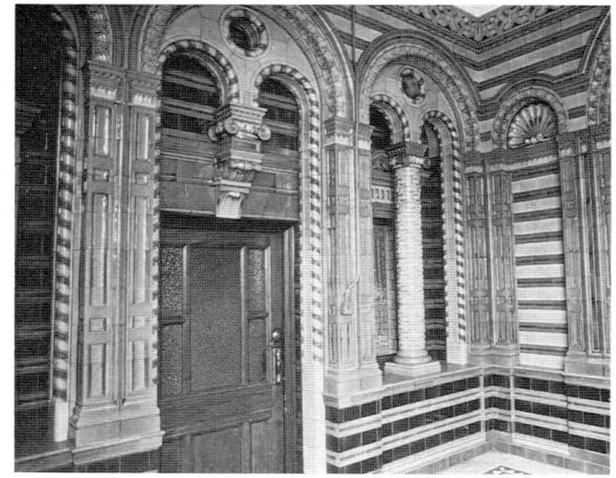

170. Manchester, Refuge Assurance Building (1891), faience decoration of the public entrance in brown, honey gold, and blue **[570]**

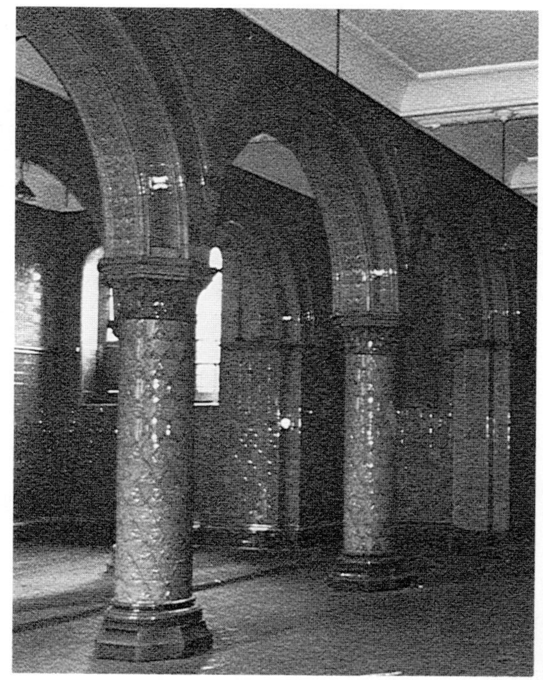

171. Liverpool, Royal Infirmary (1886), faience decoration of the chapel in green and turquoise **[532]**

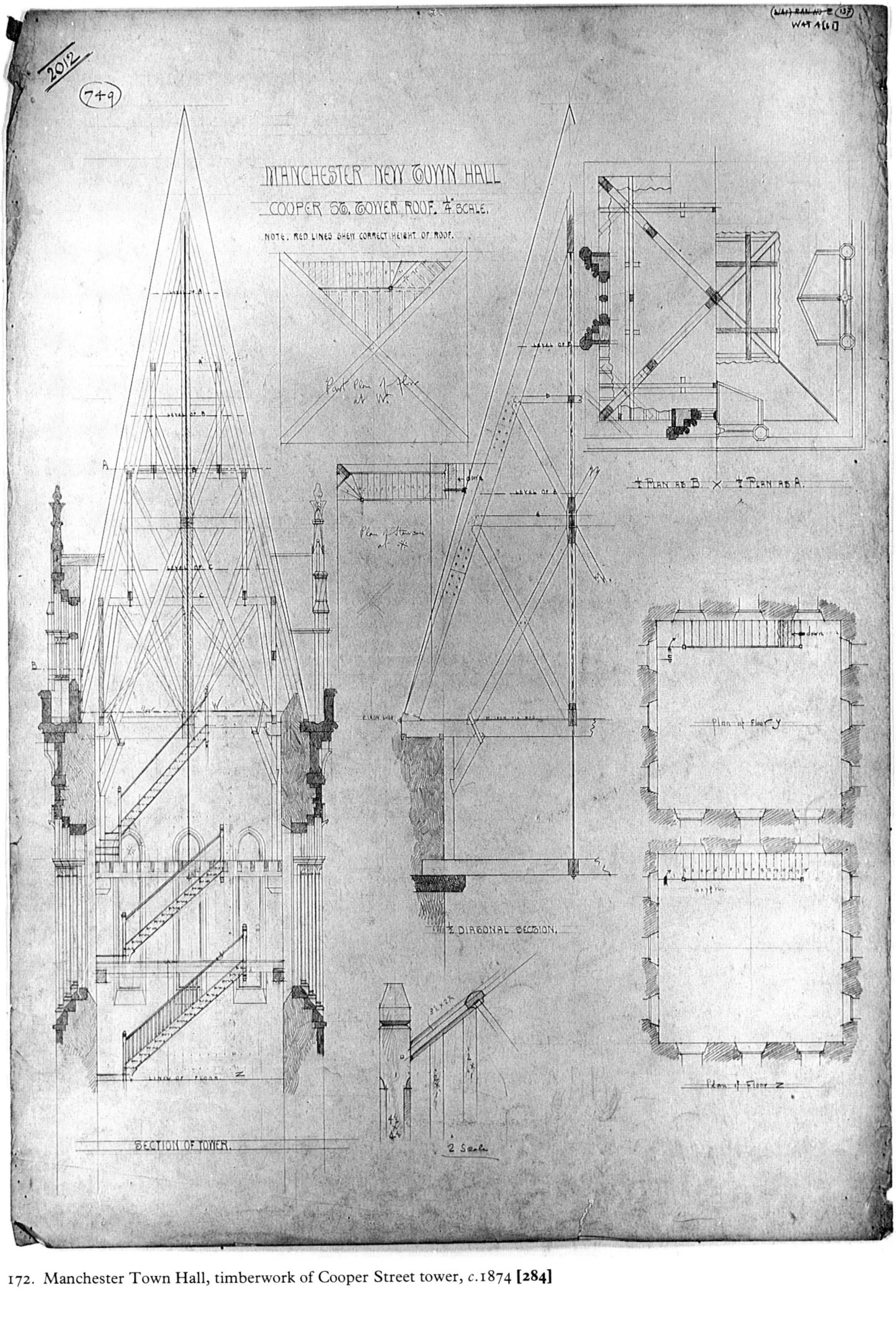

172. Manchester Town Hall, timberwork of Cooper Street tower, *c.*1874 **[284]**

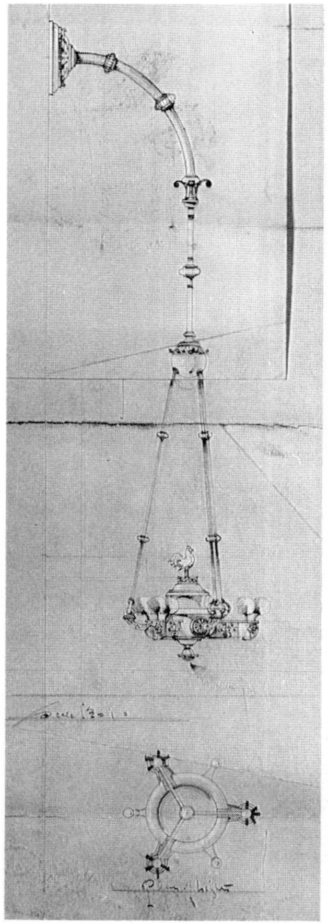

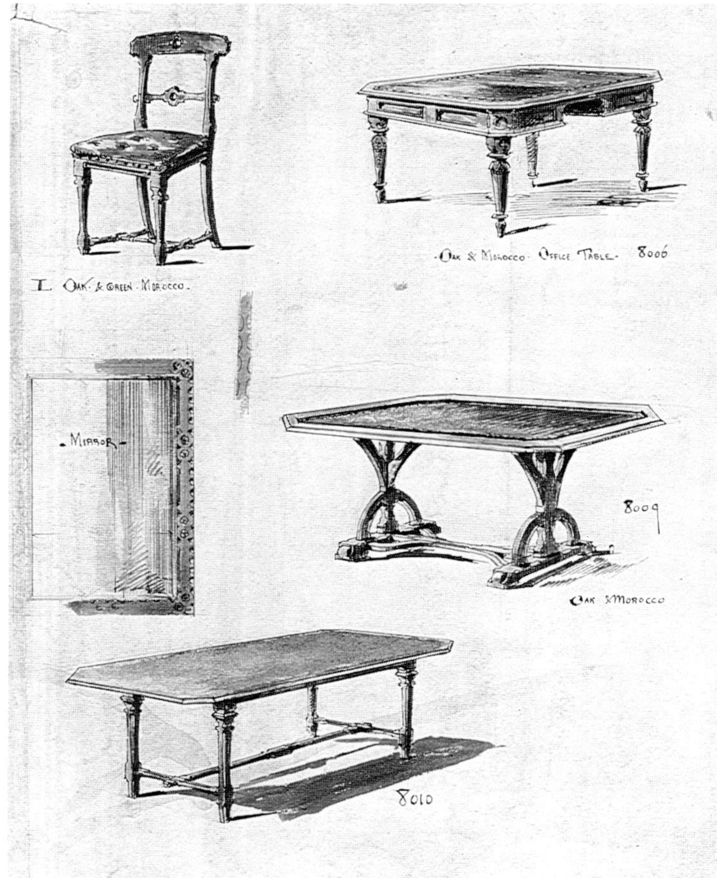

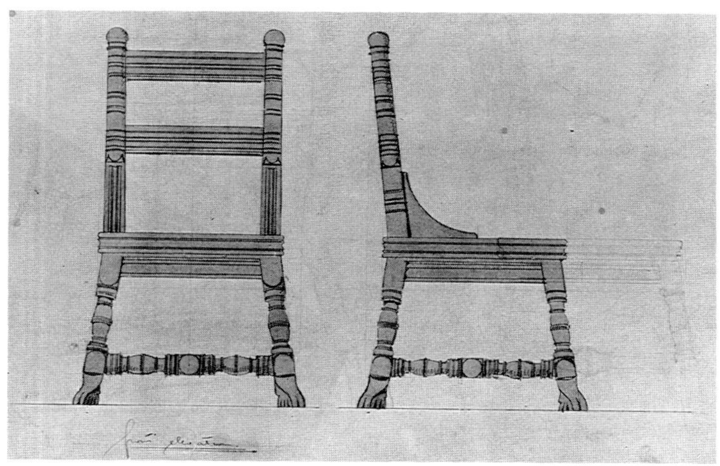

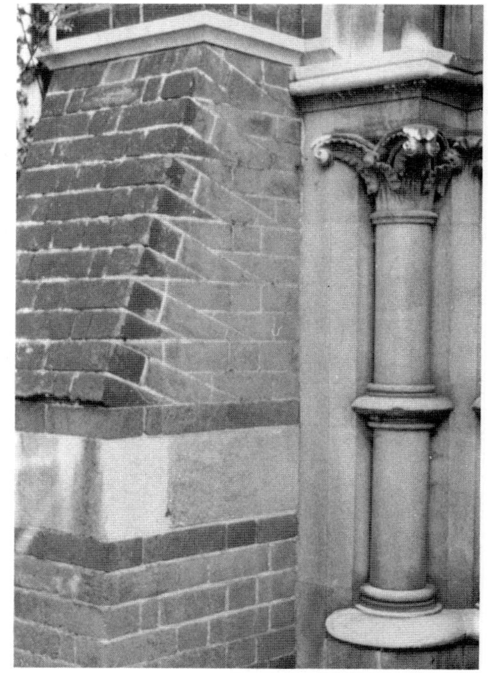

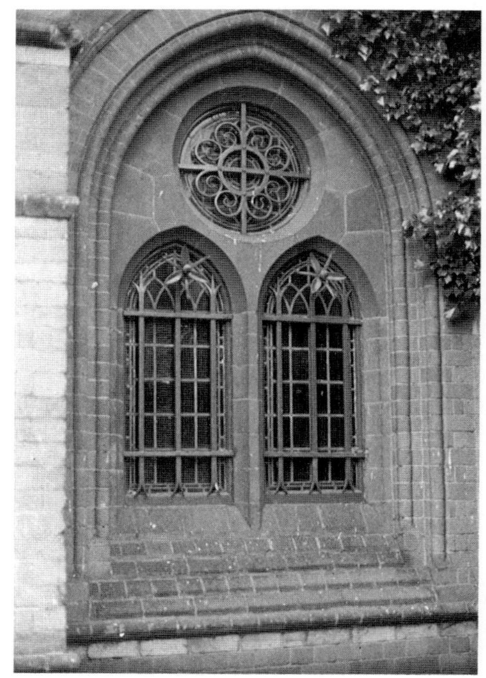

173. Above, Cambridge, Jesus College (1871), design for wall sconce for hall, showing one of three alternative stems, *c.*1876 **[347]**

174. Middle above, Manchester Assize Courts (1859), sketch designs for furniture in oak and morocco, *c.*1862 **[67]**

175. Middle below, Cambridge, Jesus College (1871), design for hall chairs, *c.*1876 **[347]**

176. Right above, Cambridge Union Society (1865), detail of brick and stone work in buttress and entrance **[197]**

177. Right below, Reading, Grammar School (1868), brick and terracotta window mouldings with wrought iron grilles **[307]**

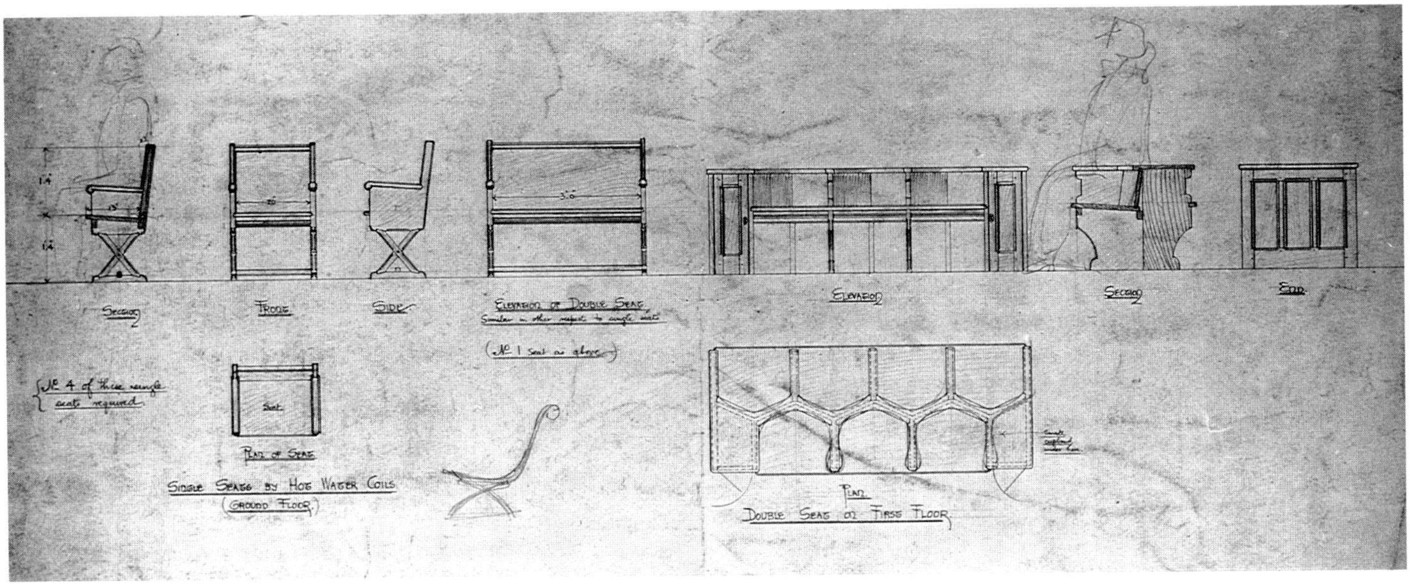

178. London, Holborn, Prudential Assurance Company (1876), designs for seats, *c.*1878. **[431]**

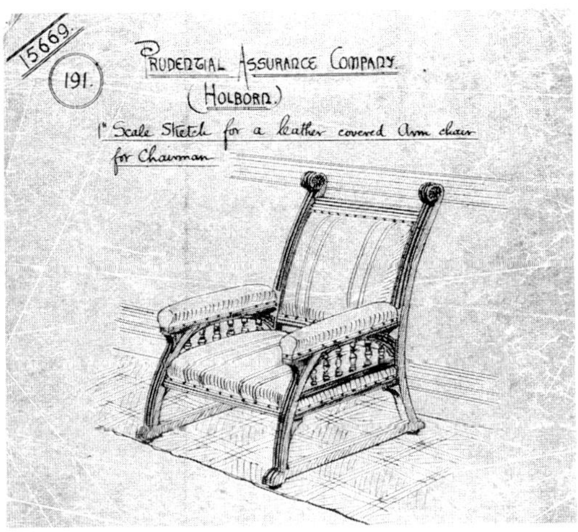

179. London, Holborn, Prudential Assurance Company (1876), design for an armchair for the Chairman, *c.*1878 **[431]**

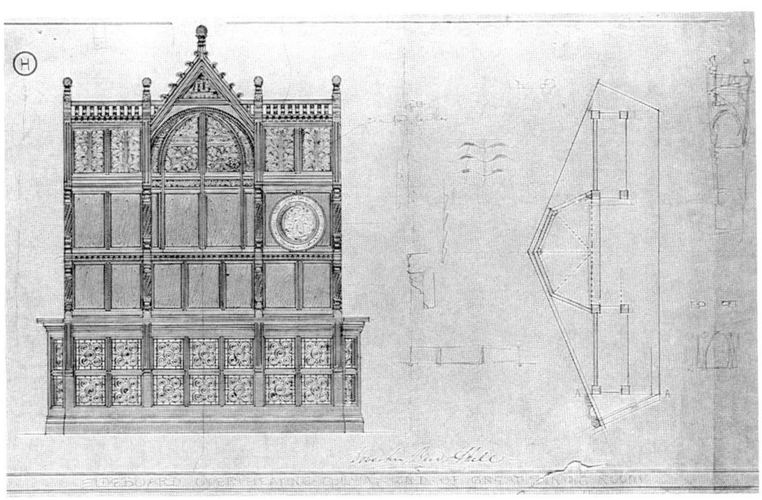

180. Manchester Town Hall (1867), design for a sideboard for the Banqueting Hall, *c.*1876 **[284]**

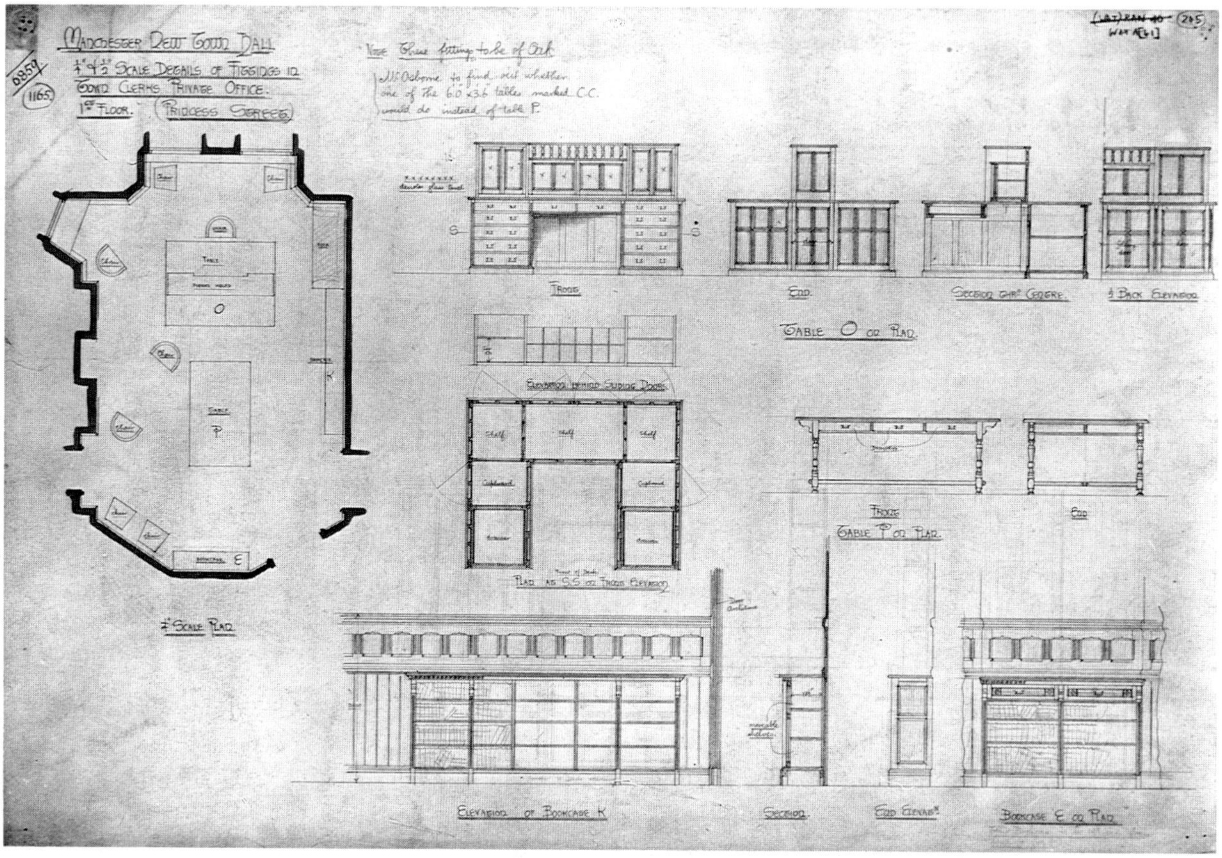

181. Manchester Town Hall (1867), designs for furniture in the Town Clerk's office, *c.*1876 **[284]**

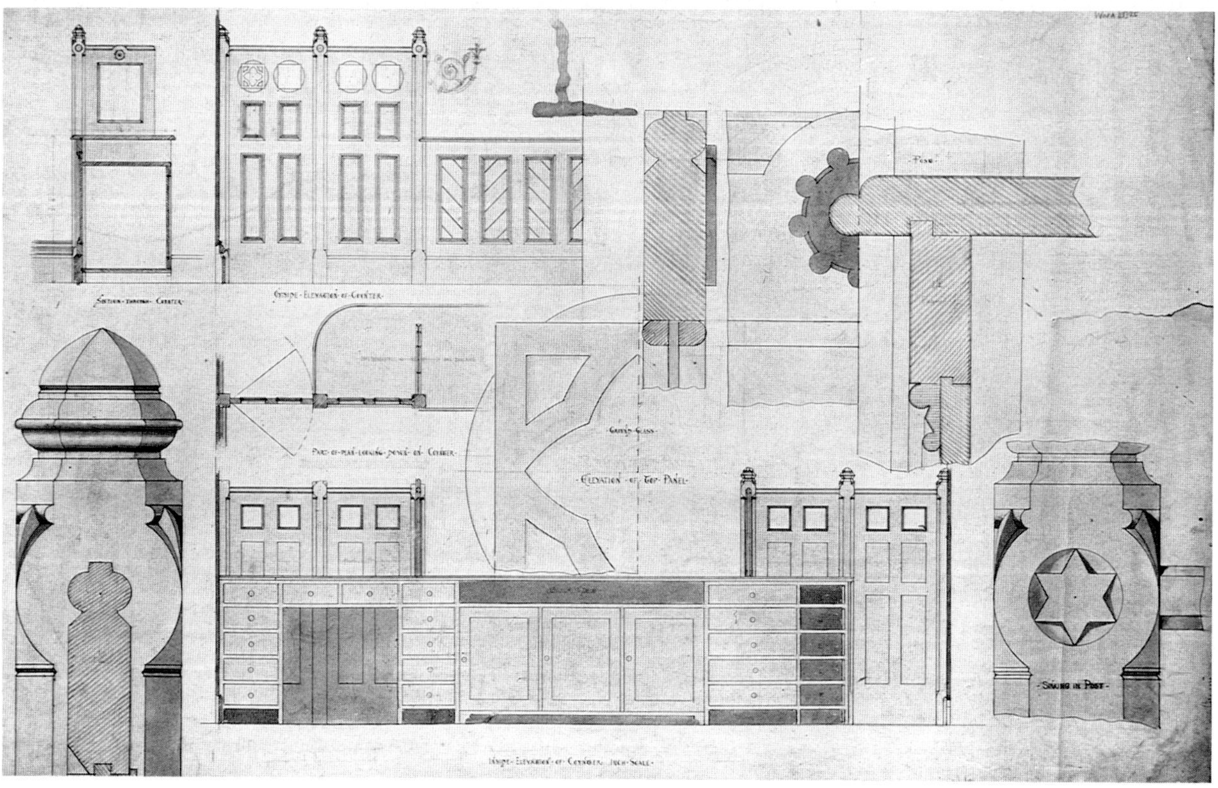

182. Cambridge, Union Society (1865), joinery details for the main office, *c.*1867 **[197]**

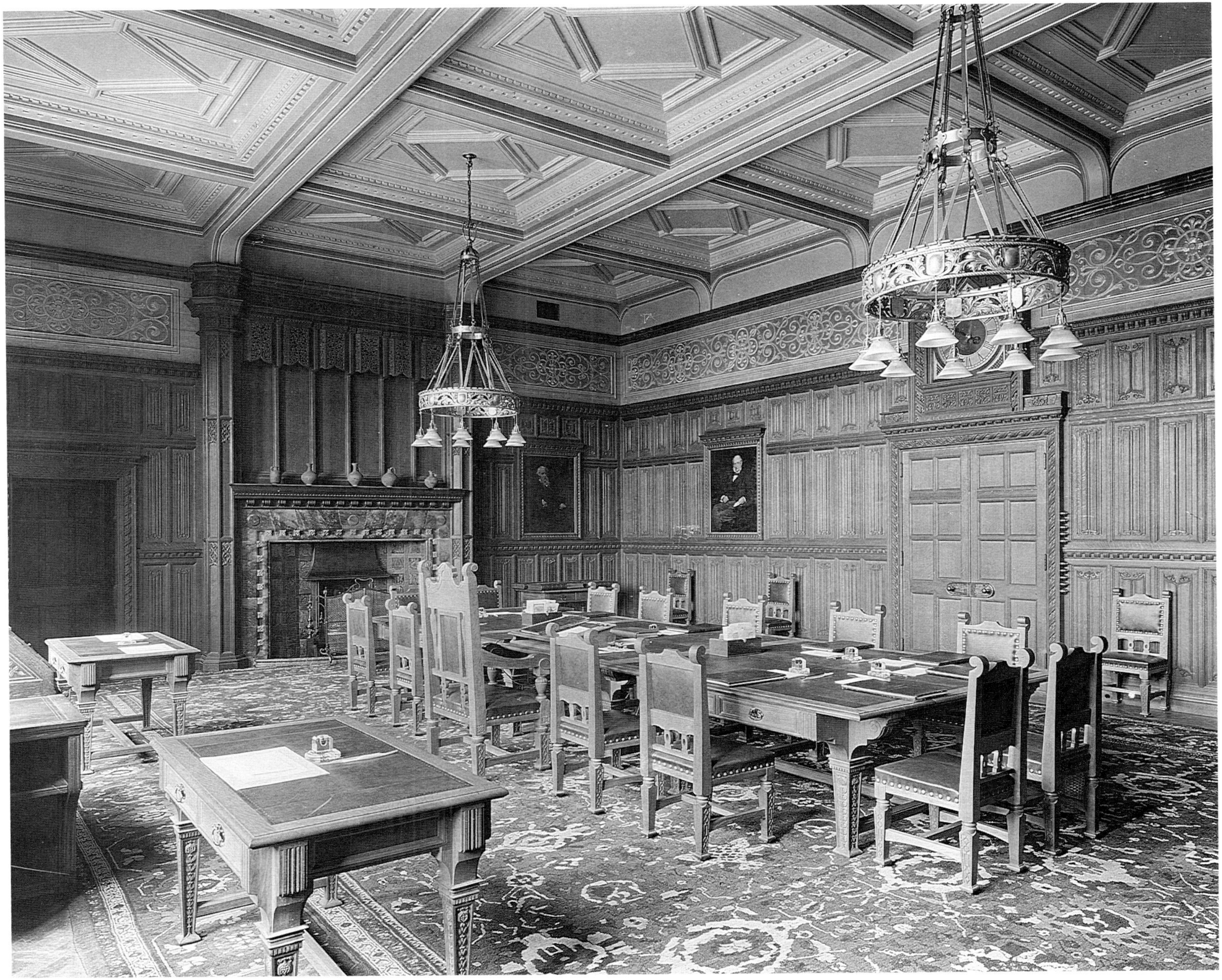

183. London, Holborn, Prudential Assurance Buildings (1895), Boardroom, 1905. The furniture and fittings were designed by Paul Waterhouse **[602]**

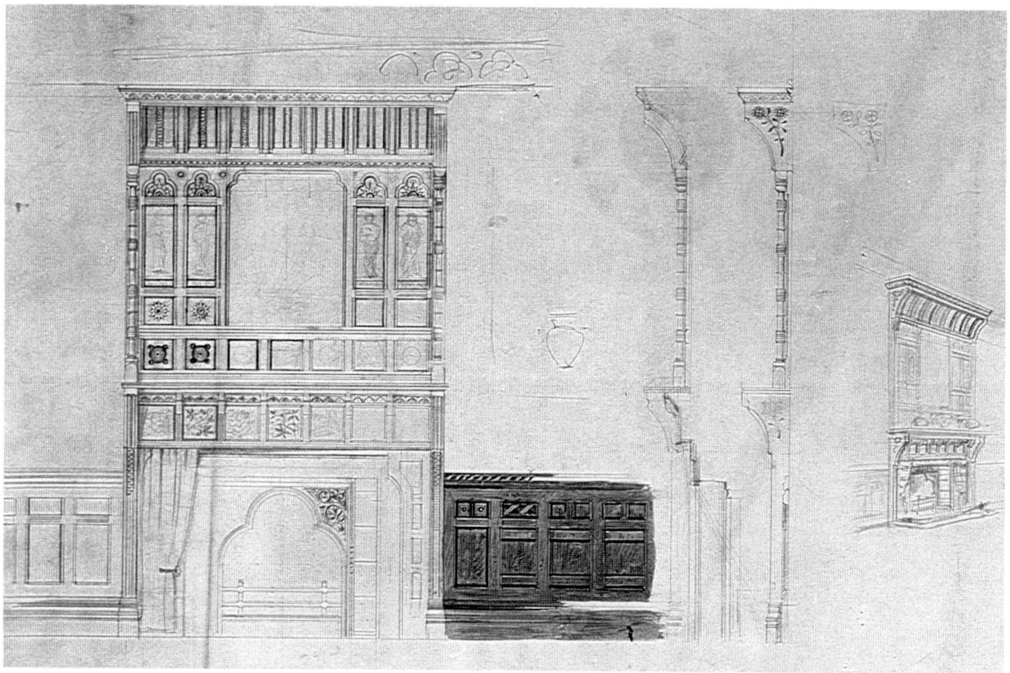

184. Eaton Hall (1869 *et seq.*), design for timber chimney-piece for Ante Drawing-Room, *c.*1877 **[315]**

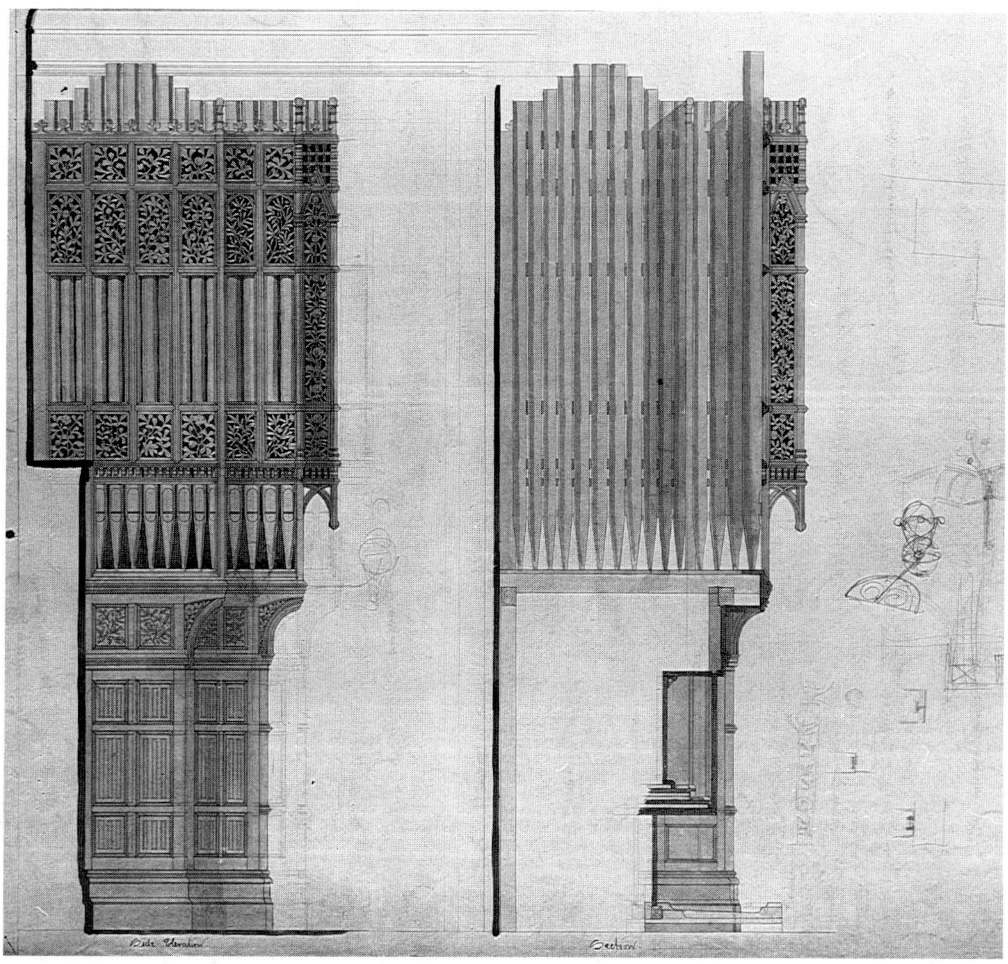

185. Eaton Hall (1869 *et seq.*), Library, design for organ case, *c.*1877 **[315]**

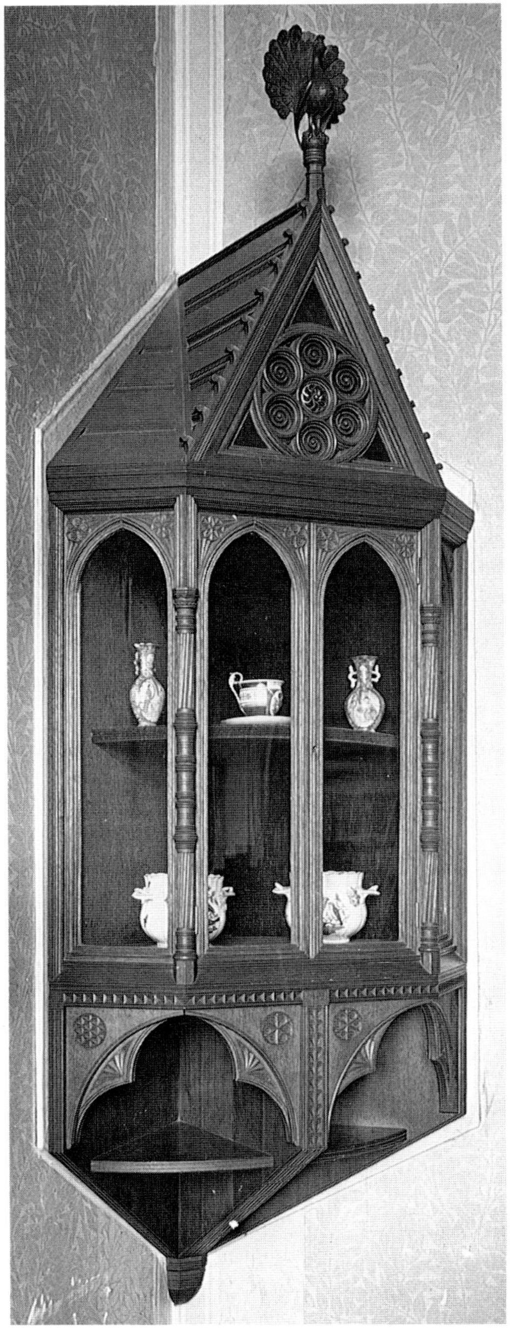

186. Blackmoor House (1866), corner cupboard, *c.*1873 **[192]**

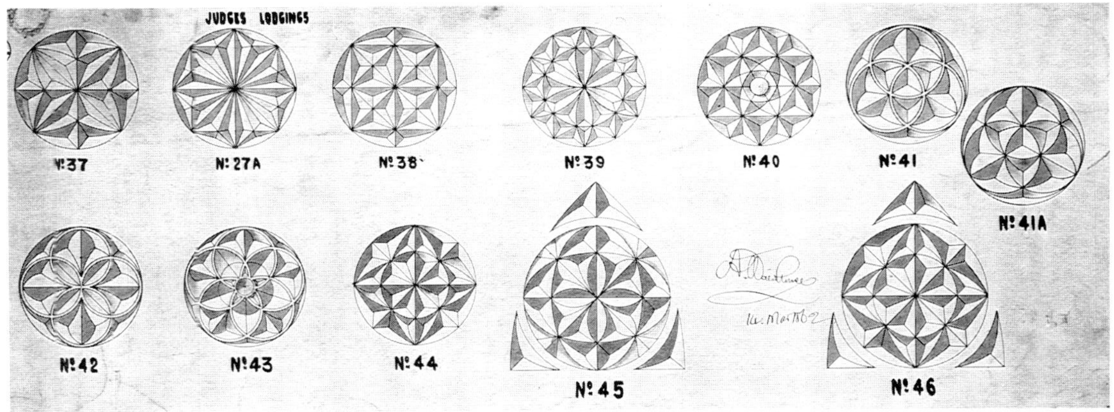

187. Manchester Assize Courts (1859), designs for newel sinkings in the Judge's Lodgings, dated 1862 **[67]**

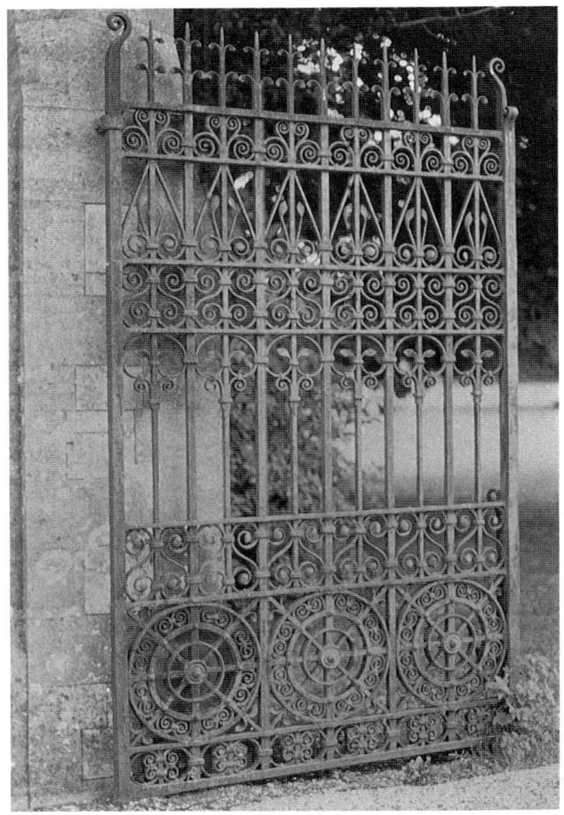

188. Iwerne Minster (1877), wrought iron gate **[445]**

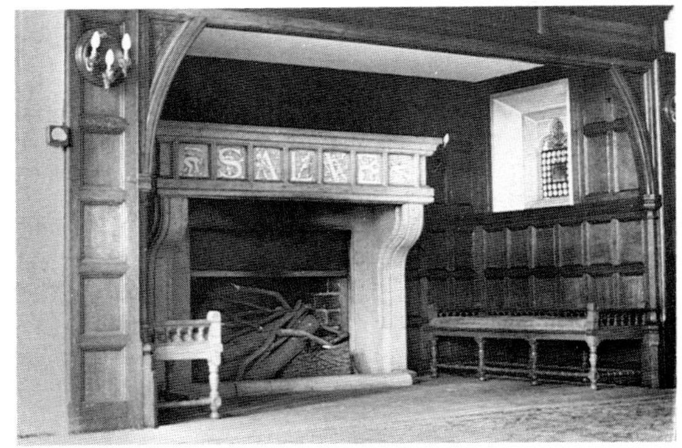

189. Blackmoor House (1866), fireplace in hall with hall seats designed by Waterhouse **[192]**

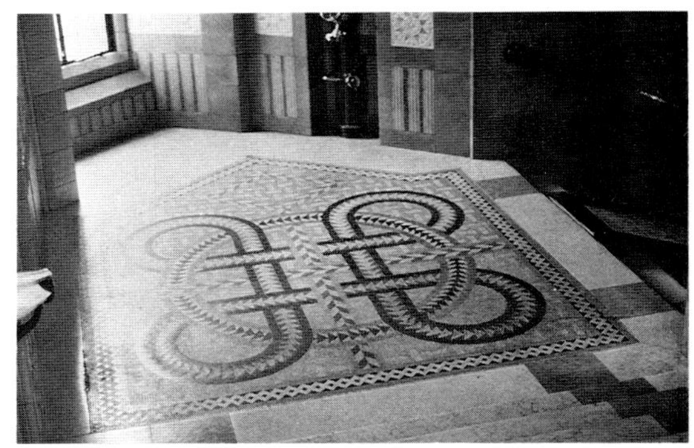

190. London, Holborn, Prudential Assurance Buildings (1895), marble landing pattern on principal staircase **[602]**

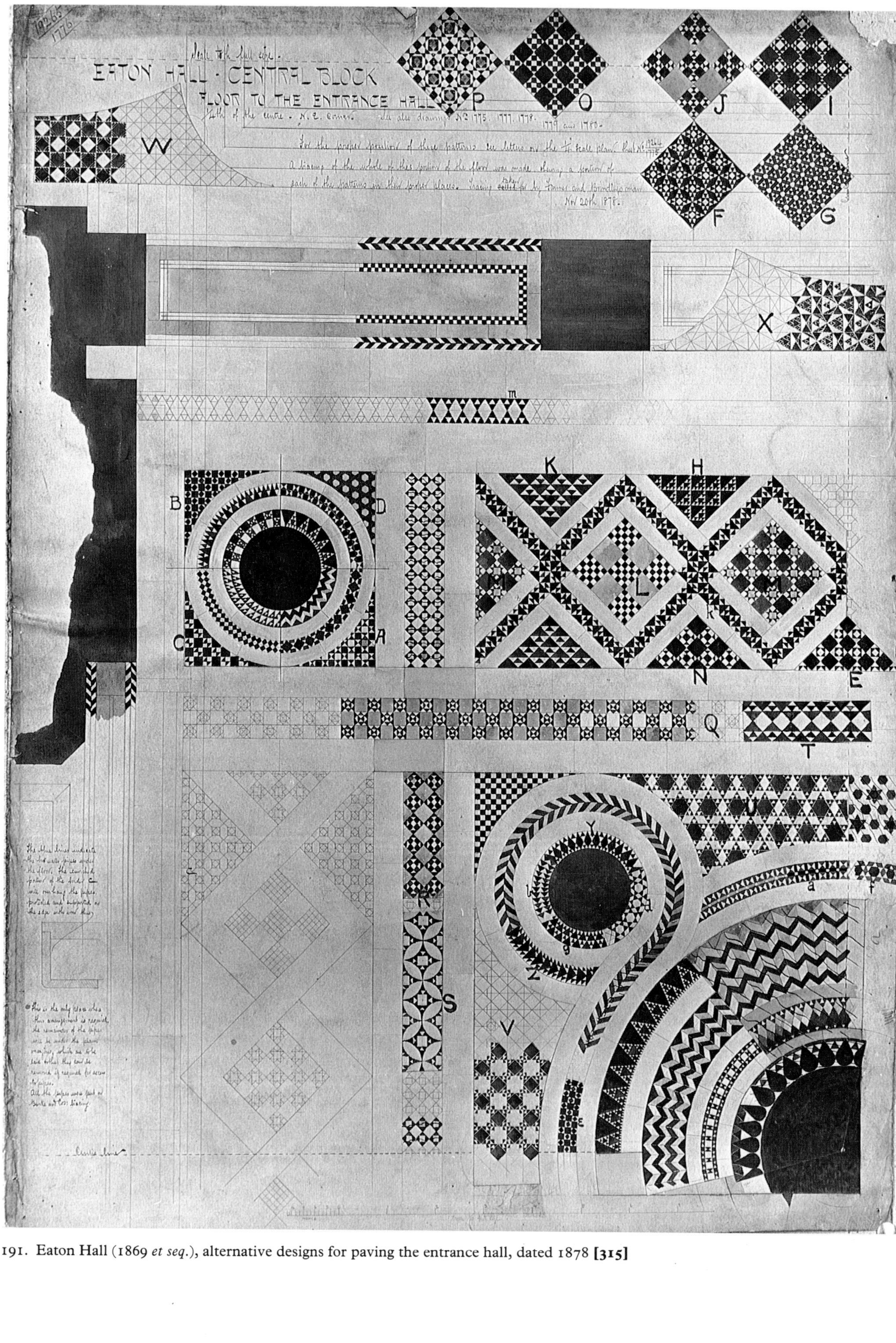

191. Eaton Hall (1869 *et seq.*), alternative designs for paving the entrance hall, dated 1878 **[315]**

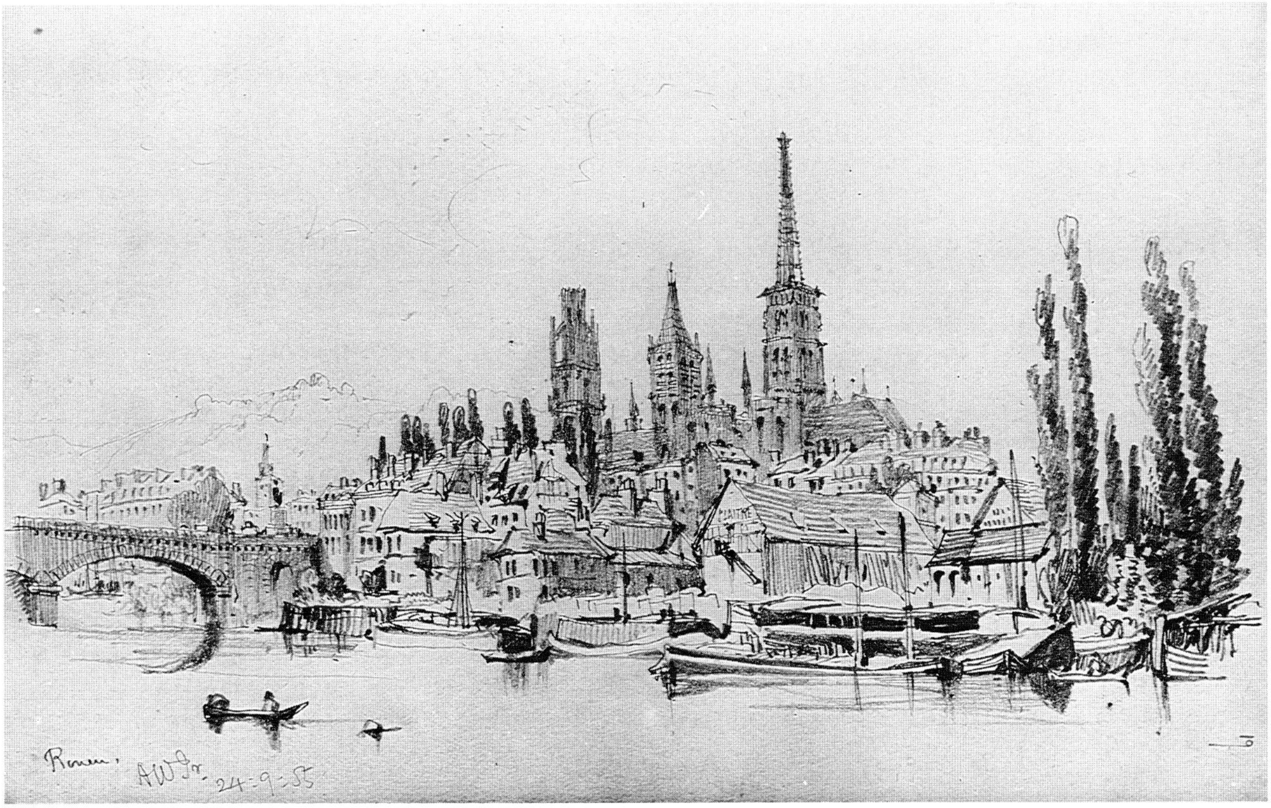

192. Sketchbook 1855, Rouen

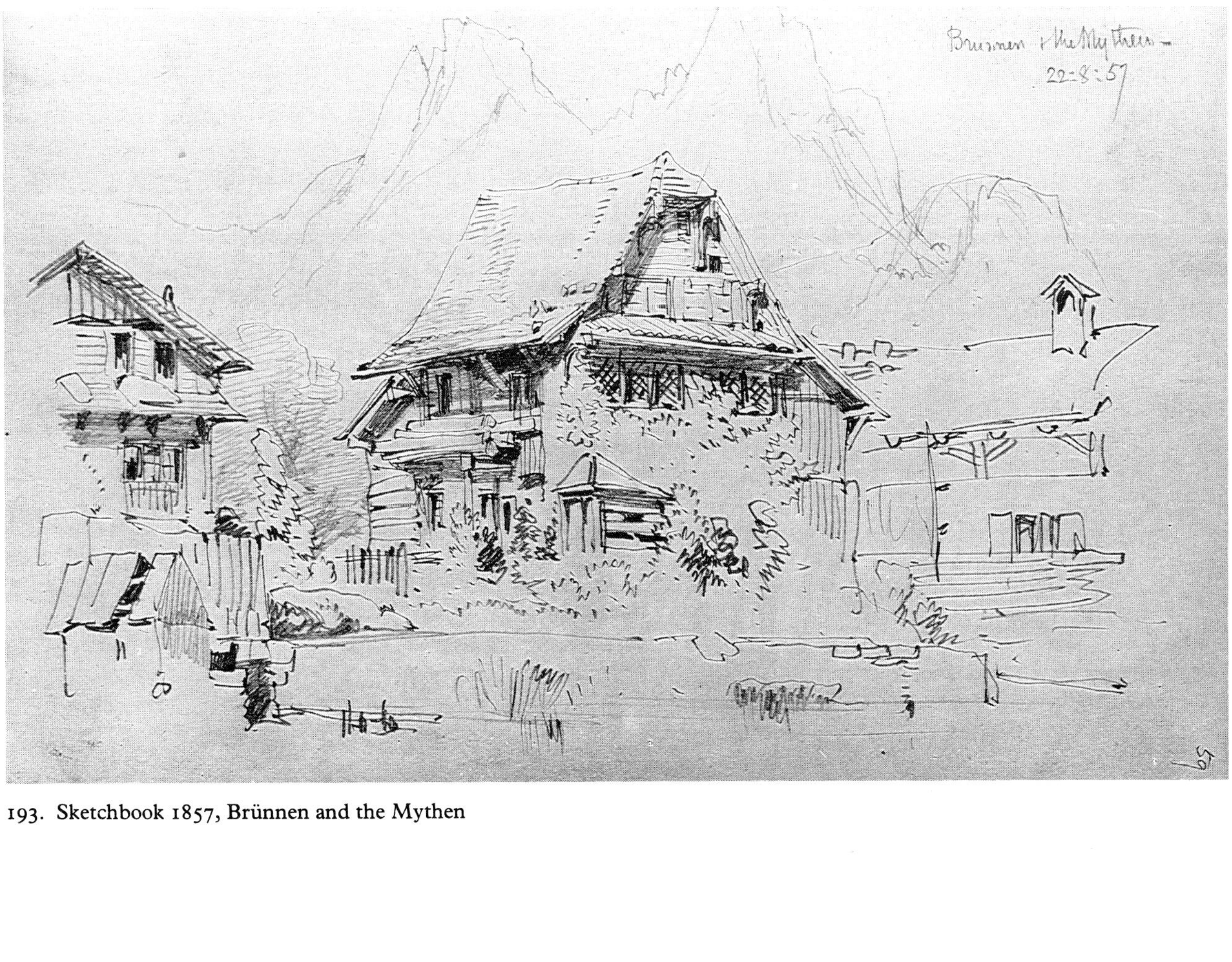

193. Sketchbook 1857, Brünnen and the Mythen

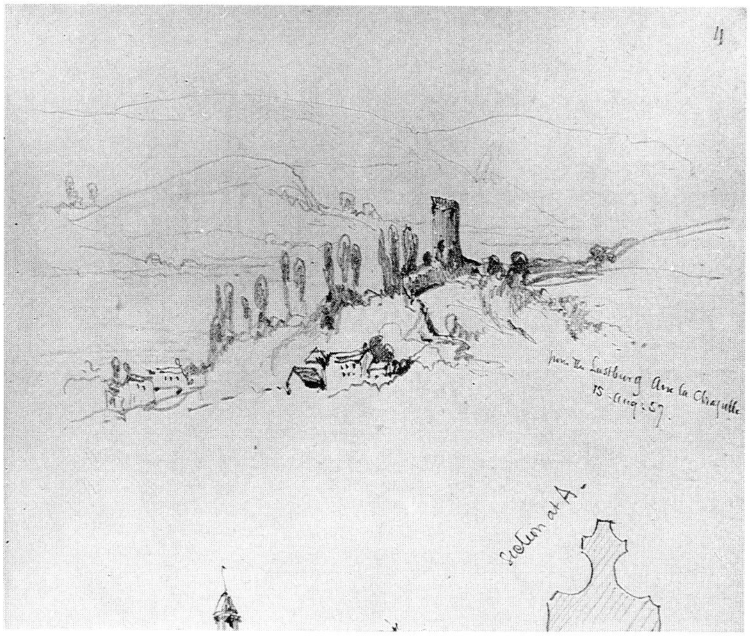

194. Sketchbook 1857, view from the Lustberg, Aix-la-Chapelle

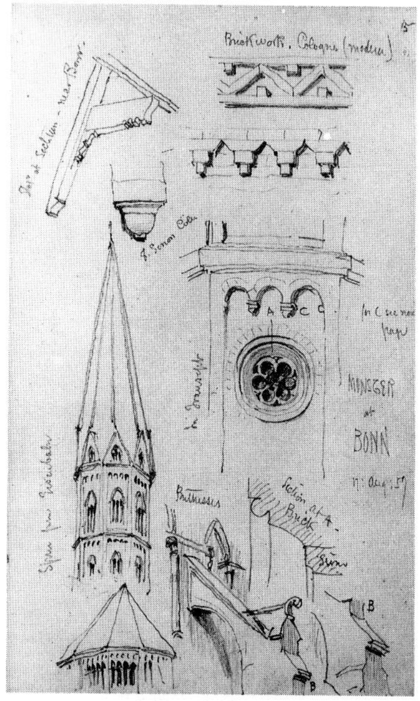

195. Sketchbook 1857, details of Bonn, Münster, and modern brickwork, Cologne

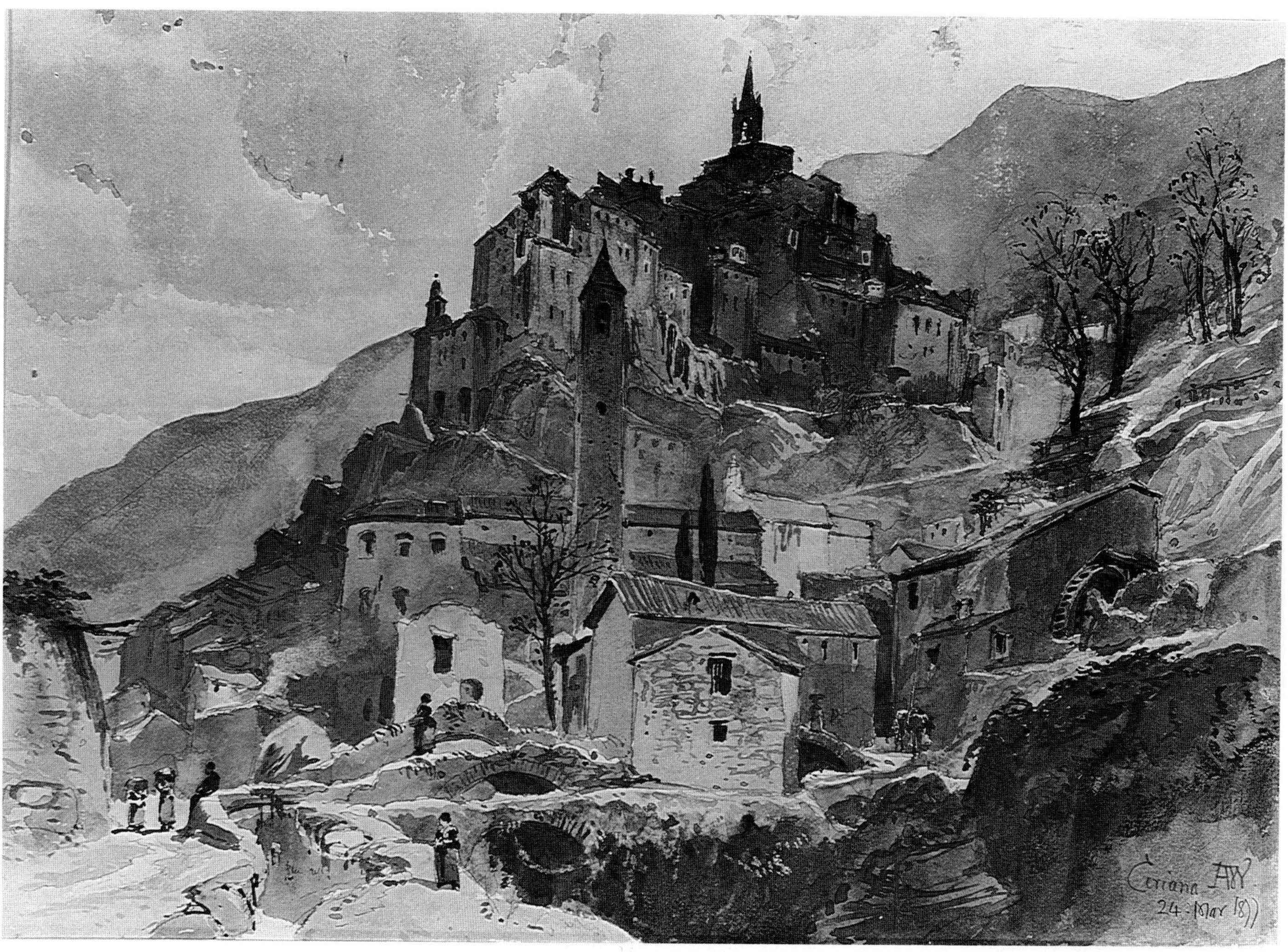

196. Ceriana, watercolour, dated 1877

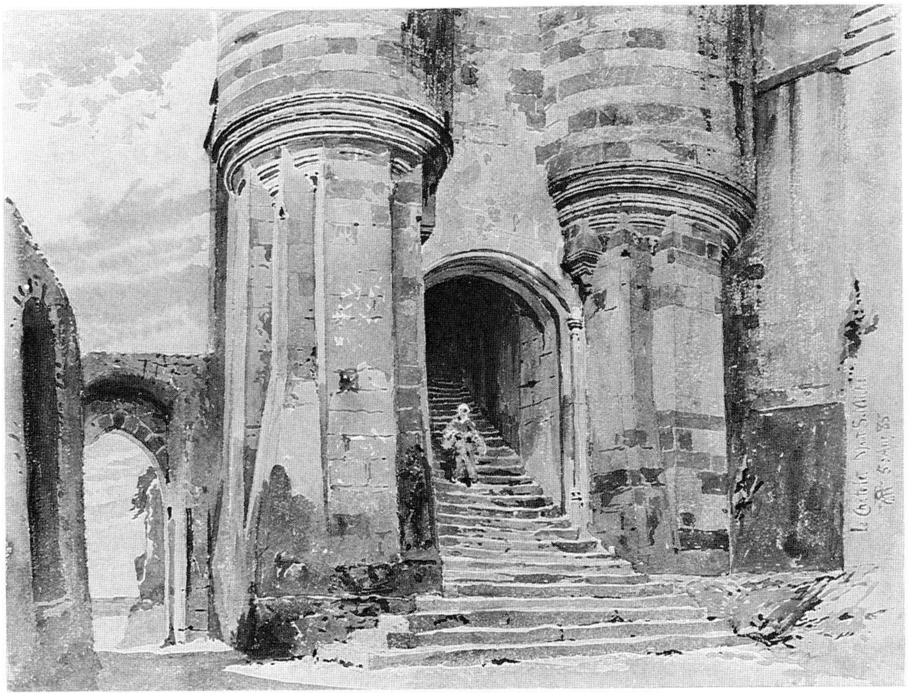

197. Mont-Saint-Michel, watercolour, dated 1885

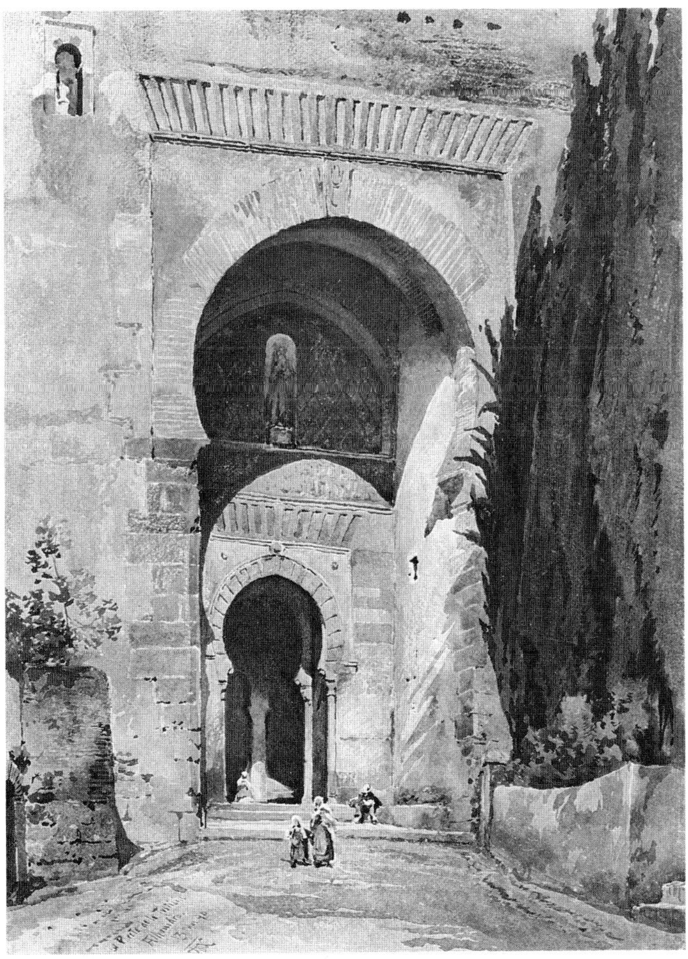

198. Granada, La Puerte de la Justicia, Alhambra, watercolour, dated 1890

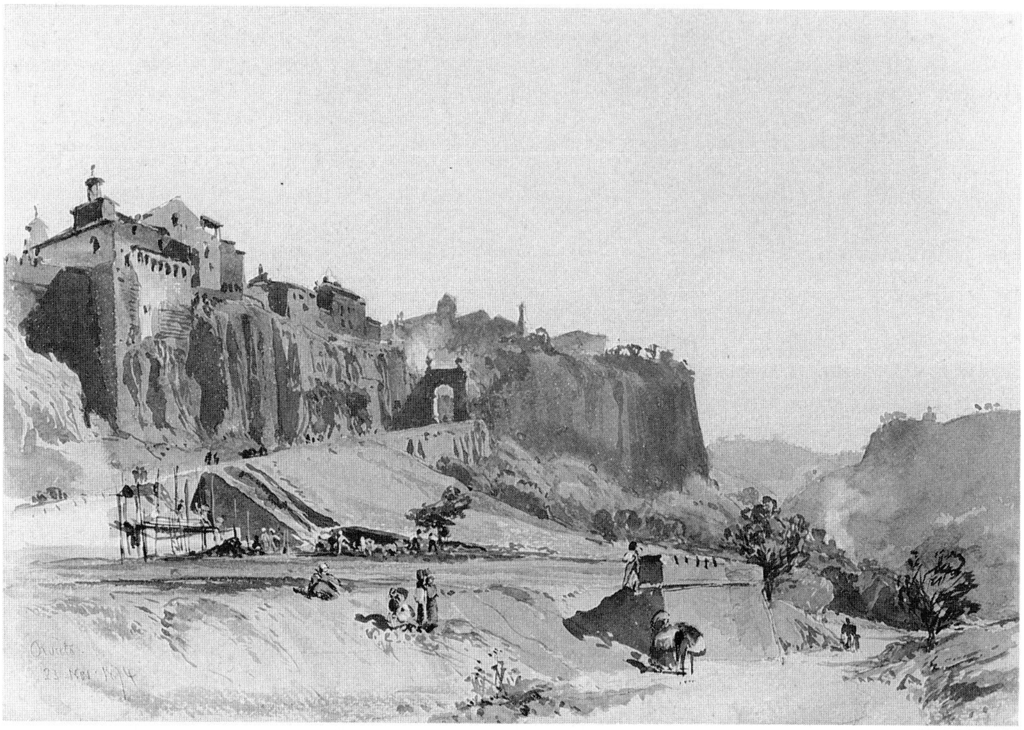

199. Orvieto, watercolour, dated 1874

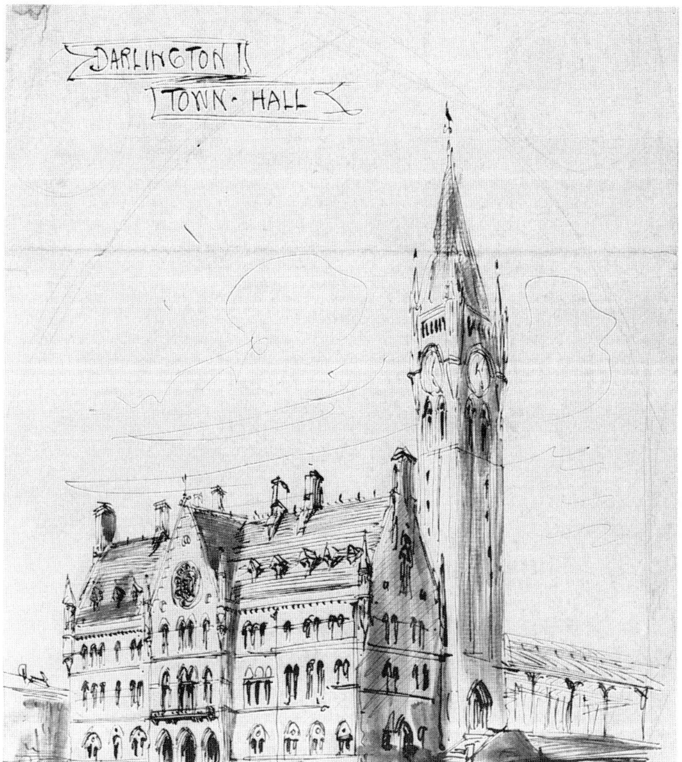

200. Darlington, Town Hall (1861), sketch of proposed design, *c.*1861 **[99]**

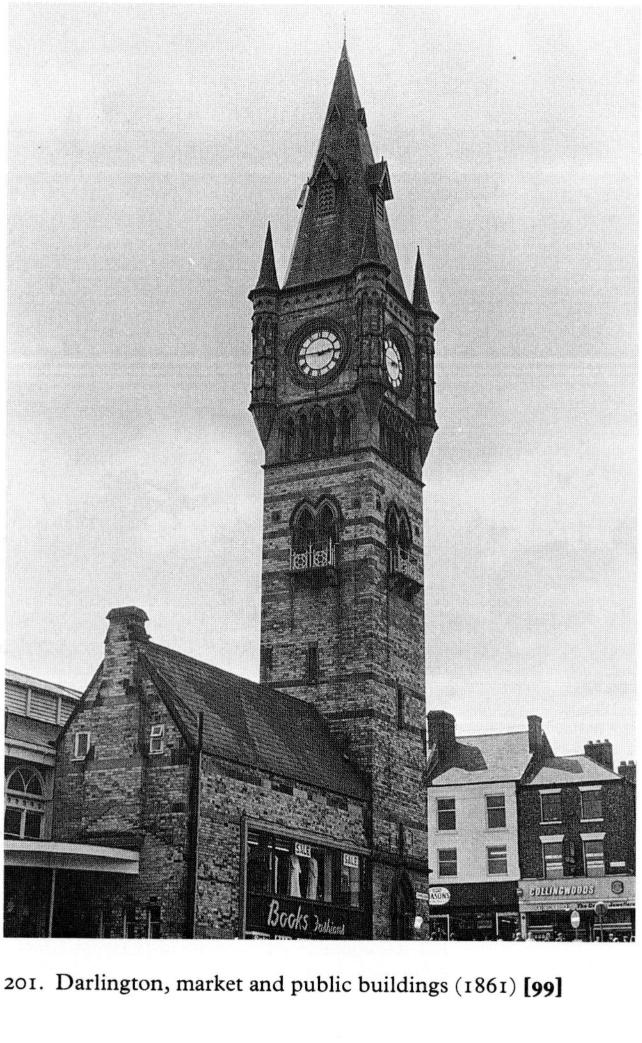

201. Darlington, market and public buildings (1861) **[99]**

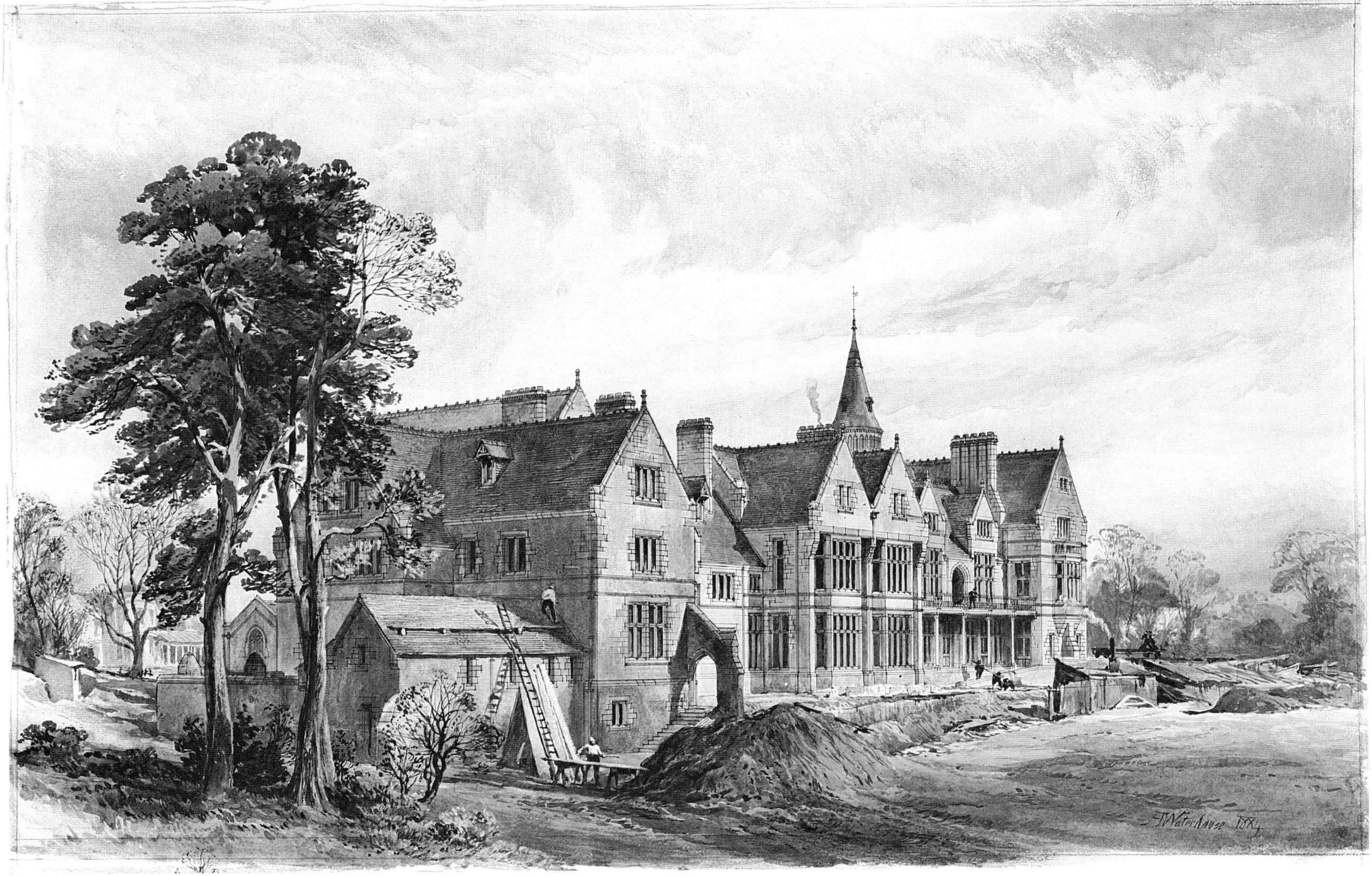

202. Liverpool, Turner Memorial Home (1882), perspective, dated 1884 **[497]**

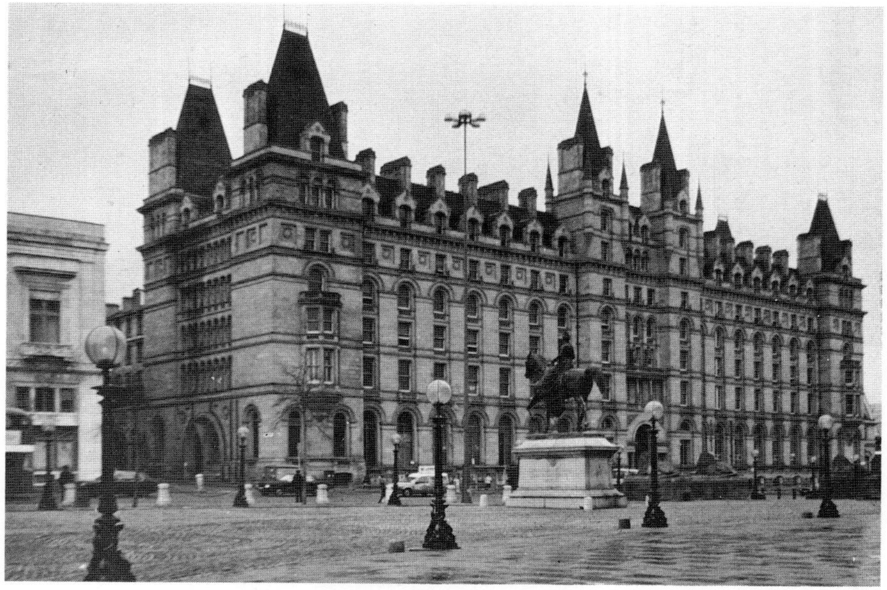

203. Liverpool, Lime Street Hotel (1867) **[274]**

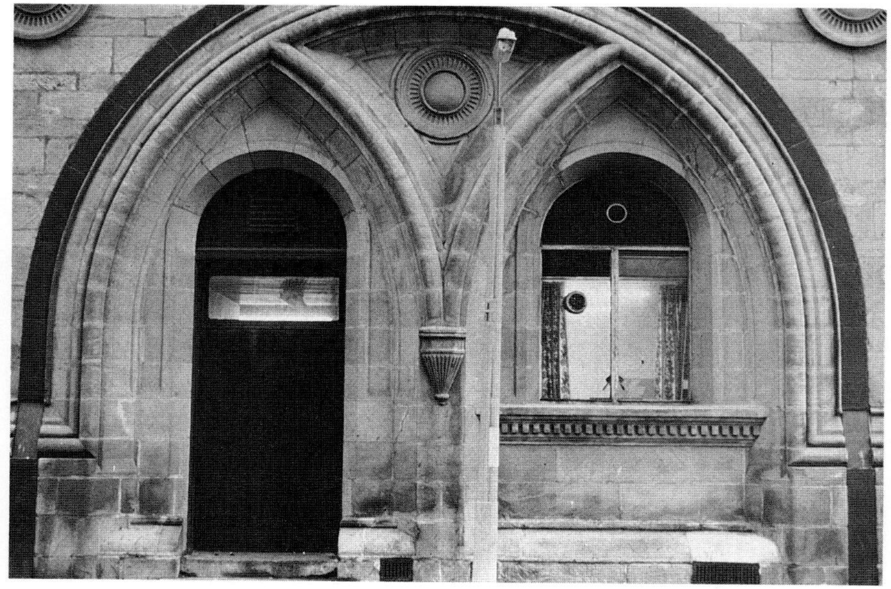

204. Liverpool, Lime Street Hotel (1867), eclectic detailing of side entrance **[274]**

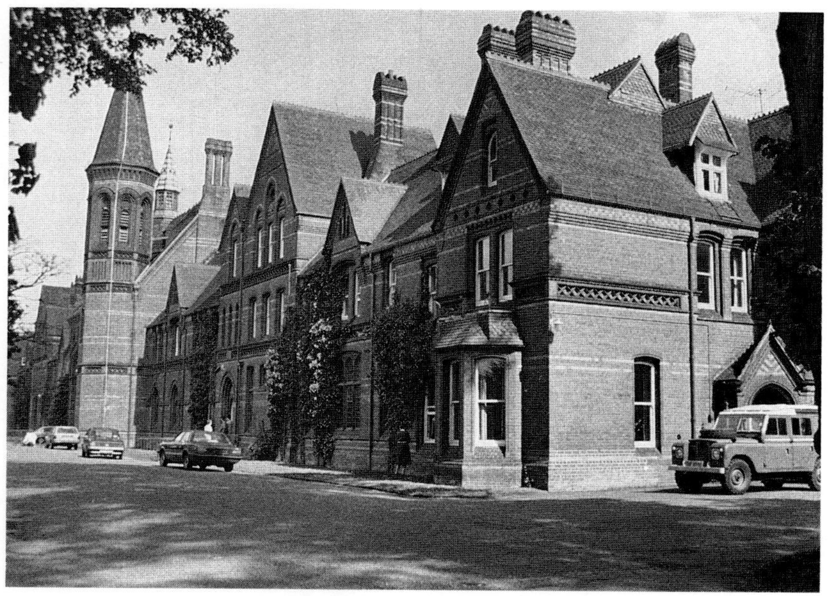

205. Reading, Grammar School (1868) **[307]**